David W. Pearson
Nigel C. Steele
Rudolf F. Albrecht (eds.)

Artificial Neural Nets and Genetic Algorithms

Proceedings of the International Conference in Roanne, France, 2003

Springer-Verlag Wien GmbH

Dr. David W. Pearson
Equipe Universitaire de Recherche en Informatique
de Saint-Etienne (Groupe de Recherche de Roanne)
Institut Universitaire de Technologie de Roanne
Université Jean Monnet, Saint-Etienne, France

Dr. Nigel C. Steele
Division of Mathematics
School of Mathematical and Information Sciences
Coventry University, Coventry, U.K.

Dr. Rudolf F. Albrecht
Institut für Informatik
Universität Innsbruck, Innsbruck, Austria

Originally published by Springer-Verlag / Wien in 2003

Camera-ready copies provided by authors and editors
Printed by Novographic Druck G.m.b.H., A-1230 Wien
Printed on acid-free and chlorine-free bleached paper
SPIN 10892027

With 226 Figures

ISBN 978-3-211-00743-3 ISBN 978-3-7091-0646-4 (eBook)
DOI 10.1007/978-3-7091-0646-4

Preface

The 2003 edition of ICANNGA marks a milestone in this conference series, because it is the tenth year of its existence. The series began in 1993 with the inaugural conference at Innsbruck in Austria. At that first conference, the organisers decided to organise a similar scientific meeting every two years. As a result, conferences were organised at Alès in France (1995), Norwich in England (1997), Portorož in Slovenia (1999) and Prague in the Czech Republic (2001). It is a great honour that the conference is taking place in France for the second time. Each edition of ICANNGA has been special and had its own character. Not only that, participants have been able to sample the life and local culture in five different European countries.

Originally limited to neural networks and genetic algorithms the conference has broadened its outlook over the past ten years and now includes papers on soft computing and artificial intelligence in general. This is one of the reasons why the reader will find papers on fuzzy logic and various other topics not directly related to neural networks or genetic algorithms included in these proceedings. We have, however, kept the same name, "International Conference on Artificial Neural Networks and Genetic Algorithms".

All of the papers were sorted into one of six principal categories: neural network theory, neural network applications, genetic algorithm and evolutionary computation theory, genetic algorithm and evolutionary computation applications, fuzzy and soft computing theory, fuzzy and soft computing applications. About half of the papers accepted for publication in the proceedings fall into the first two categories of neural network theory and applications.

Apart from the, nowadays, standard vocabulary associated with neural networks and genetic algorithms we see some new terms such as snakes and ants. There are very strong theoretical papers, looking at topological structures for example. There are interesting applications such as meteorological data modelling, control theory, pollution forecasting and modelling social phenomena. We see that interest in these methods is very much alive and research is continuing and moving forward. Nature is still inspiring researchers and will do so for a long time yet.

A lot has happened in the ten years since the first conference in 1993. Science has moved forward in leaps and bounds during this period. We salute all those who have worked for ICANNGA, participants, committee members and Springer-Verlag, in the past. Who knows what the next decade will bring for ICANNGA, we can only hope that it will be as fruitful as the previous one.

Prof. D.W. Pearson   Jean Monnet University of Saint-Etienne, France
Prof. N.C. Steele   Coventry University, England
Prof. R.F. Albrecht   University of Innsbruck, Austria

Table of Contents

Part II - Neural Network Applications

Part III - Genetic Algorithm and Evolutionary Computation Theory

Part IV - Genetic Algorithm and Evolutionary Computation Applications

Part V - Fuzzy and Soft Computing Theory

Part VI - Fuzzy and Soft Computing Applications

ICANNGA
International Conference on Artificial Neural Networks and Genetic Algorithms
Roanne, France, April 23–25, 2003

International Advisory Committee

Prof. R.F. Albrecht, University of Innsbruck, Austria
Prof. A. Dobnikar, University of Ljubljana, Slovenia
Dr. V. Kůrková, Academy of Sciences of the Czech Republic, Prague, Czech Republic
Prof. D.W. Pearson, Jean Monnet University of Saint-Etienne, France
Prof. N.C. Steele, Coventry University, United Kingdom

Programme Committee

Prof. D.W. Pearson (chairman), Jean Monnet University of Saint-Etienne, France
Dr. A.M. Alimi, University of Sfax, Tunisia
Dr. G. Andrejkova, P.J. Safarik University, Kosice, Slovakia
Prof. B. Beliczynski, Warsaw University of Technology, Poland
Dr. H. Bischof, Technical University Graz, Austria
Prof. W. Brauer, Munich University of Technology, Germany
Prof. A. Dobnikar, University of Ljubljana, Slovenia
Dr. M. Dorigo, Université Libre de Bruxelles, Belgium
Dr. G. Dray, Ecole des Mines d'Alès, France
Dr. E. Gaura, Coventry University, United Kingdom
Dr. C. Giraud-Carrier, ELCA Informatique SA, Lausanne, Switzerland
Dr. K. Hlavackova-Schindler, Vienna University of Technology, Austria
Prof. P.C. Kainen, Georgetown University, Washington, D.C., USA
Dr. H. Karatza, Aristotle University of Thessaloniki, Greece
Dr. M. Kárný, Academy of Sciences of the Czech Republic, Prague, Czech Republic
Dr. V. Kůrková, Academy of Sciences of the Czech Republic, Prague, Czech Republic
Prof. P. Larranaga, University of the Basque Country, Bilbao, Spain
Dr. A. Leonardis, University of Ljubljana, Slovenia
Prof. T. Martinez, Brigham Young University, Provo, Utah, USA
Prof. F. Masulli, University of Pisa, Italy
Dr. R. Neruda, Academy of Sciences of the Czech Republic, Prague, Czech Republic
Prof. N. Pavesic, University of Ljubljana, Slovenia
Dr. E. Pelikan, Academy of Sciences of the Czech Republic, Prague, Czech Republic
Prof. A. Pinkus, Technion, Haifa, Israel
Prof. V. Rayward-Smith, University of East Anglia, Norwich, United Kingdom
Prof. C. Reeves, Coventry University, United Kingdom
Prof. S. Ribaric, University of Zagreb, Croatia
Prof. B. Ribeiro, University of Coimbra, Portugal
Dr. M. Sanguineti, University of Genoa, Italy
Prof. H. Saxén, Åbo Akademi, Finland
Dr. J. Sima, Academy of Scienes of the Czech Republic, Prague, Czech Republic
Prof. N. Steele, Coventry University, United Kingdom
Dr. T. Tambouratzis, National Centre for Scientific Research "Demokritos", Athens, Greece
Prof. K. Warwick, University of Reading, United Kingdom

Validation of a RBFN Model by Sensitivity Analysis

Özer Ciftcioglu
Delft University of Technology, Faculty of Architecture
Berlageweg 1, 2628 CR Delft, The Netherlands
o.ciftcioglu@bk.tudelft.nl

Abstract- Radial basis functions network (RBFN) is considered as a knowledge model. The model is established from a data set by learning. For the performance assessment a novel model validation method is introduced. The method consists of sensitivity analysis integrated into a mathematical-based technique known as analytical hierarchy process (AHP). It ranks the relative importance of factors being compared where the factors are the sensitivities in this case. The relative importance of the sensitivities is computed from the model and based on this information, the consistency of this information is tested by AHP. The degree of consistency is a measure of confidence for the validity of the model.

1 Introduction

A data model generally takes a statistical form established by means of large amount of data. Such a model takes into account to the complexity and tries to "learn" from data rather than to investigate some asymptotic properties in a traditional sense. Therefore, the model generally belongs to a category called non-parametric. Many such models are largely developed for various purposes independently of inputs from statistical origin. Since such a model may represent a large amount of data possibly obtained from a complex dynamic as well as static environment, it requires more attention for the assessment of the validity of the model together with understanding of its limitations. Due to the essential "learning" character, the associated performance of the model is assessed by its generalization capability, which is the validity of the model beyond the data from which the model is obtained. Most desirably a data/information model in this context should be validated using methods independent of the method used for establishing the model itself. In this respect, the sensitivity analysis from the model is carried out and validation of this analysis, as an implicit validation of the model, is performed by a mathematical-based technique known as analytical hierarchy process (AHP). The organization of the paper is as follows. In Section 2 the development of data model and sensitivity analysis are briefly explained. In Section 3, AHP is briefly explained. In section 4, the experimental studies for validation are presented where by means of the model established, sensitivity is computed and AHP process is applied. This is followed by conclusions.

2 Data Model Development

A knowledge model aims at obtaining efficient summaries of possibly large amount of data, identifying the relevant features and relationships. By means of this process one can elicit desired information present in the data set or make predictions for the future observations. The relevant features and relationships present in the data set can be best summarized in a structural form, which can serve as a model of the data. Especially for a large amount of data with probable complex relationships, learning strategies are most suitable for the model development. Here the learning process is particularly machine learning, which is amply exercised especially in the artificial intelligence based technologies of the last decade. Among these mention may be made to feed forward neural networks [1], fuzzy logic [2] and genetic algorithms [3,4]. Among these, genetic algorithms are especially useful in combinatorial optimization, which may be a part of the learning process. Fuzzy logic methods are especially effective for processing fuzzy, that is, uncertain and/or imprecise information. However, these two distinct paradigms may have some limitations in the case of large amount of data, which results in increasing complexity thereby making the learning process difficult. Namely, in such a case fuzzy interpretations and model forming by extracting appropriate fuzzy membership functions may be extremely difficult. Alternatively the search space of combinatorial optimization involved in genetic algorithms may be prohibitively large. In contrast with these aspects, neural networks are well posed to establish a structural system for information representation as a knowledge model with well-established learning algorithms, back-propagation for instance. Therefore, unless a particular demand is imperative, to employ a neural network model as knowledge model is a straightforward approach, which is effective and also efficient. Based on this, here basically the same strategy is considered. Although such a decision is attributed to be as straightforward in a certain sense, the decision-making about the type of neural net is not proportionately straightforward. However, considering the ample modeling possibilities, the feed forward type of neural networks is most suitable for this goal. Hence, in this research, such a network is used. It is known as radial basis function network (RBFN) which is amply treated in literature due to its interesting structure and far reaching properties. Among these mention may be made to its outstanding classification and local multivariable function approximation. RBFN is briefly described below.

2.1 Radial Basis Functions Network

Consider multivariate functions f: $\nabla^p \rightarrow \nabla^q$ to be represented by means of a feed-forward neural network structure. Without any loss of generality, we assume q=1 so that the network structure has one output. For x, $c \in \nabla^p$, $y \in \nabla$ and y=f(x) , the output of the network is given by

$$f(x) = \sum_{i=1}^{M} w_i \phi(\| x - c_i \|) \quad (1)$$

where M is the number of basis functions; w_i is RBFN output weight or coefficient; $\phi(.)$ is the basis function based on the Euclidean distance metric

$$\| x - c_i \|^2 = (x - c_i)^T (x - c_i) \quad (2)$$

In case $\phi(.)$ is a radial basis function, c_i is the center location where the corresponding basis function takes the peak value. A widely used radial basis function is the Gaussian basis function

$$\phi(\| x - c_i \|) = \exp(-\| x - c_i \|^2 / \sigma_i^2) \quad (3)$$

where σ_i is the i-th width parameter that determines the effective support of the i-th basis function. In this form, unlike the multi-layer perceptron (MLP) network, which divides the input space using hyperplanes, RBFN partitions the input space using hyper ellipsoids so that it has local function approximation properties in contrast with the global approximation properties of a MLP network. Using such functions as basis, a function given by means of data points can be represented by a continuous function where such a function precisely passes through the data points given. This process is generally referred to as interpolation. Here, we consider radial basis functions due to their suitability for multivariate interpolation. To see how this is particularly implemented in an RBF network, we consider a set of N data vectors $\{x_m , m=1,...,N\}$ dimension of p in R^p and N real numbers $\{y_m , m=1,2,....,N\}$. We seek a function f(x): $R^d \rightarrow R$ that satisfies the interpolation conditions

$$f(x_m)=y_m, \; m=1,2,...N \quad (4)$$

The interpolation conditions for $y \in \nabla^q$ can be generalized as

$$f_{mk}(x) = \sum_{i=1}^{M} w_{ik} \, \phi(\| x - c_i \|) \qquad x \in R^p, \; y \in R^q; \; k = 1,...,q \quad (5)$$

Above, the mapping from input to output is

$$R^p \rightarrow R^q \quad \text{and} \quad f_{mk}(x_m) = y_{mk} \; , \; m=1,2,......,N \; ; \; k=1,2,.....,q \quad (6)$$

It has been shown that, in compact domains the function space realized by RBFN is dense in the space of all continuous functions [5], that is, it is universal approximator of continuous functions. For function approximation and interpolation there can be as many basis functions as $\{x_m, y_m\}$ pairs available. However, for the generalization capability of the learned network, the number of basis functions should be kept to a minimum for an accepted approximation error. In this respect, an orthogonal transformation method known as orthogonal least squares (OLS) [6,7] has outstanding merits for the training of the network although other feed forward neural network training algorithms are also applicable for training.

2.2 Sensitivity Analysis

The *sensitivity analysis* is used to measure the response of a model to parameter perturbations. Such information is important to understand the various relations among the parameters in a model in cooperation with the study of the model itself.

Sensitivity can be defined in various forms. The most common form of the concept is in the form of a partial derivation as adopted in this work. This is expressed by $\partial y / \partial x$, where y is the output and x is the input vector respectively. For discrete models, numerical derivation methods can be used for this purpose. The response of the network to input perturbations determines the sensitivity of the network. However, in a feed forward neural network model, the partial derivatives are dependent on the number of $\{x_m, y_m\}$ pairs used for learning. Therefore, the partial derivatives should be evaluated for each pair and the average over the total number of pairs should be computed. In order to avoid algebraic elimination in the averaging process, absolute values of the derivatives are considered so that the sensitivity S_{nm} of the output y_k with respect to input x_i is computed by

$$S_{kn} = \sum_{m=1}^{N} \left| \frac{\partial y_k}{\partial x_{nm}} \right| \quad n = 1,\ldots, p;\ k = 1,\ldots, q \qquad (7)$$

where p is the dimension of the input space and q is the dimension of the output space. After normalization, the graded sequence of the mean of the absolute derivatives indicates the global sensitivity of the network for each input variable.

3 The Analytical Hierarchy Process

The analytical hierarchy process (AHP) is developed by Saaty [8-9]. It is also called eigenvector method (EVM) or ratio-scale (RS) method, which is a technique to compute the priority vector, ranking the relative importance of factors being compared. One of the important features of the method is its ability to assess preferences over multi-attribute alternatives. The technique based on the initial determination of relative or pair wise importance of the different attributes in question. Originally these are designed as expert judgements. From these pair wise judgements, a priority ordering of the attributes of interest can be derived. In this work, the importance factors are the sensitivity labels of the variables of the model. By the AHP, the consistency of these factors will be investigated. If the priority ordering by AHP is consistent with the priority ordering established from the model by the sensitivity analysis, this will be a clear indication of the integrity of the model. The measure of consistency will be also provided by the method. Since in this work the data model is in the form of an expert system, the sensitivity outcomes from the model are referred to as expert judgements. With this terminology, the method is briefly given below.

The AHP has four axioms: (a) reciprocal comparison (b) homogenous elements in the same cluster (c) hierarchic structure (c) rank order expectations. If we denote two sensitivity inputs as s_i and s_j, we compare the ith factor with respect to the jth factor by

$$a_{ij} = s_i / s_j \qquad (8)$$

Then the relative importance of the jth factor with respect to the ith factor is represented as $1/a_{ij}$. The [n×n] dimensional A matrix obtained by arranging these pairwise comparison ratios is termed the reciprocal judgement matrix where n is the number of factors subjected to pairwise comparison:

$$A = \begin{bmatrix} \frac{w_1}{w_1} & \frac{w_1}{w_2} & \cdots & \frac{w_1}{w_n} \\ \frac{w_2}{w_1} & \frac{w_2}{w_2} & \cdots & \frac{w_2}{w_n} \\ \cdots & & & \\ \frac{w_n}{w_1} & \frac{w_n}{w_2} & \cdots & \frac{w_n}{w_n} \end{bmatrix} = \begin{bmatrix} 1 & \frac{w_1}{w_2} & \cdots & \frac{w_1}{w_n} \\ \frac{w_2}{w_1} & 1 & \cdots & \frac{w_2}{w_n} \\ \cdots & & & \\ \frac{w_n}{w_1} & \frac{w_n}{w_2} & \cdots & 1 \end{bmatrix} = \begin{bmatrix} 1 & a_{12} & \cdots & a_{1n} \\ \frac{1}{a_{12}} & 1 & \cdots & a_{2n} \\ \cdots & & & \\ \cdots & & & \\ \frac{1}{a_{1n}} & \frac{1}{a_{2n}} & \cdots & 1 \end{bmatrix} \qquad (9)$$

which satisfies the equation

$$AW = nW \qquad (10)$$

where W^T=[w1 w2 .....w_n]. The reciprocal matrix is said to be consistent if $a_{ij}a_{jk}=a_{ik}$ for all i,j,k = 1,2, ..., n, when the priority vector is simply any normalized column or row of the judgement matrix. The principal eigenvector W of A is computed from

$$AW = \lambda_{max} W \qquad (11)$$

where λ_{max} is the principal or largest real eigenvalue of A. The normalized eigenvector corresponding to λ_{max} is the priority vector P. For a consistent A matrix λ_{max}=n.

As a numerical example, consider the priority vector P given by

$P = [0.1011\ 0.3508\ 0.4033\ 0.1448]^T$

Normalizing the priority vector P as unit vector, we obtain

$P_n = 1.776 * P = [0.1796\ 0.6232\ 0.7164\ 0.2572]^T$

Then the corresponding consistent judgment matrix is

$$A = \begin{bmatrix} 1 & 0.2882 & 0.2507 & 0.6982 \\ 3.4698 & 1 & 0.8698 & 2.4227 \\ 3.9891 & 1.1496 & 1 & 2.7852 \\ 1.4322 & 0.4128 & 0.3590 & 1 \end{bmatrix}$$

The largest eigenvalue of A is found to be $\lambda_{max}=4$ and the corresponding eigenvector (i.e., priority vector) is given by

$$e_{max} = [0.1796 \;\; 0.6232 \;\; 0.7164 \;\; 0.2572]^T$$

As seen from above, EVM provides the same priority order given by the normalized priority vector. This is what one should expect, because A matrix is in this case a perfectly consistent matrix. In this case $\lambda_{max}=n$ and consequently n=4 while λ_{max} is always greater than or equal to n. For actual consistency verification, the pair-wise comparisons of n(n-1)/2 pairs should be given corresponding to the upper triangular portion of A above the main diagonal. Using this property a measure of consistency can be defined as to deviation of λ_{max} from n. In general inconsistency assessment this can be done as follows. Deviation from consistency can be determined by computing the *consistency index* (**CI**) of matrix A, given by $(\lambda_{max} - n)/(n-1)$,

The generalization of the AHP to interprets the normalized eigenvector as the priority vector even in the presence of inconsistency in the judgement ratios. This is due to the fact that for small perturbations of the pair wise comparison judgement ratios, the change in the eigenvector remains small.

In the reciprocal judgement matrix A, for each element we take

$$a_{ji} = \frac{1}{N}\sum_{k=1}^{N}[\frac{s_j}{s_i}]_k \tag{12}$$

and

$$a_{ij} = \frac{1}{N}\sum_{k=1}^{N}[\frac{s_i}{s_j}]_k \tag{13}$$

Note that, in this case the reciprocal judgement matrix is formed element-wise for all its elements. From the probability considerations viewpoint, to allow for the deviation from its true value the ratio $p'_{ij}=[s_i/s_j]$ can be modeled as a random variable in the form of

$$p_{ij}=p'_{ij} \times \varepsilon_{ij} \tag{14}$$

where p'_{ij} is the true ratio, p_{ij} is the estimated ratio and ε stands for error factor. Ideally the error factor should be unity, but it lies in the interval $[0,\infty]$. If there is no systematic error in the model, ε_{ij} can be treated as a random variable with the central value of 1, so that $\ln\varepsilon_{ij}$ is another random variable with the central value of 0. For N estimates for p_{ij}, we can write

$$(p_{ij})^N = (p_{ij})^N \prod_{k=1}^{N}(\varepsilon_{ij})^k \tag{15}$$

so that

$$\ln(p_{ij}) = \ln(p'_{ij}) + \frac{1}{N}\sum_{k=1}^{N}[\ln(\varepsilon_{ij})^k] \tag{16}$$

For large values of N, the summation of the log errors can be approximated by a normal variable ε with zero mean and certain variance σ^2. Therefore

$$\ln(p_{ij}) = \ln(p'_{ij}) + \varepsilon \tag{17}$$

or

$$\ln(p_{ij}) \rightarrow \mathrm{N}[\ln(p'_{ij}),\sigma^2] \;\; (\textit{normal distribution}) \tag{18}$$

and consequently

$$p_{ij} \rightarrow \Lambda[p'_{ij},\sigma^2] \approx \Lambda[a_{ij},\sigma^2] \;\; (\text{log-}\textit{normal distribution}) \tag{19}$$

The Lognormal probability density function is given by

$$f(x) = \frac{1}{x(2\pi\sigma_1^2)^{1/2}}\exp\left[-\frac{(\ln(x)-\mu_1)^2}{2\sigma_1^2}\right]$$

where

$$\mu_1 = \ln\left[\frac{\mu^2}{(\sigma^2+\mu^2)}\right], \quad \sigma_1 = \left[\ln\left(\frac{\sigma^2+\mu^2}{\mu^2}\right)\right]^{1/2}$$

μ and σ above are the mean and variance of the corresponding normal distribution.

From (19) it is clear that the mean of true ratios p_{ij} is estimated by a_{ij} of the reciprocal judgment matrix computed from (12). In the same way, the mean of the true ratios of p_{ji} is estimated by a_{ji} computed from (13). Hence, the reciprocal judgment matrix computed by the help of (12) and (13) is consistent since the mean values of the log-normal distribution in (19) are reciprocal, i.e. $p'_{ij}=1/p'_{ji}$. For a valid data model, the true ratio estimations given by a_{ij} and a_{ji} are approximately reciprocal providing A matrix in (9) is highly consistent, provided the model is valid. In other words, the verification of this consistency is a sound measure for the validity of the data model represented by RBFN.

From above study, it is to conclude that, for a valid model given by RBFN, the sequence of the ratios in the first row of the reciprocal judgement matrix is should approximately be the same of the sequence obtained from the largest eigenvector of the judgment matrix. This result should corroborate with the sequence obtained from the plain sensitivity analysis without recourse to the ratio-scale computation. However, since the sequence obtained from ratio-scale computation is more robust, i.e., not sensitive to the errors made during the ratio computations, the ratio-scale method provides the final outcomes to rely on.

4 Model validation

For the experimental studies a data set is used which is obtained in relation to investigations for architectural design in metro stations in the Netherlands [10]. The data set consists of 196 input-output pairs where the dimensions of the input and output vectors are 43 and 2 respectively. In particular, the outputs are variables designed as perceptional *comfort* and *safety*. For the training of the network the OLS algorithm is used. The selected number of centers for the final model is 80 for augmented generalization capability. For this model, the response of the network is shown in figure 1, together with the actual training set with the bold lines.

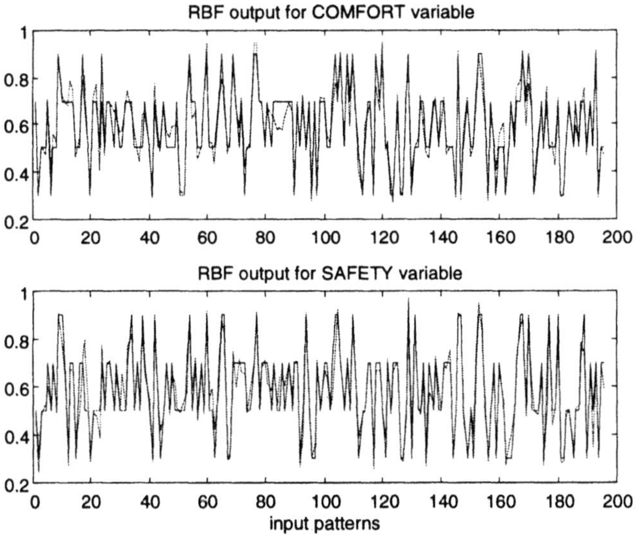

Figure 1. RBFN model of the data set whose sensitivity is subject to validation. The lines represent the data set (bold) and model response.

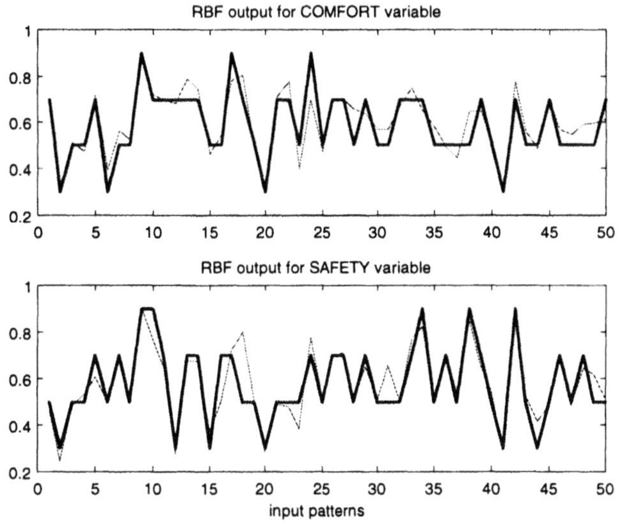

Figure 2. RBFN model of the data set whose sensitivity is subject to validation. The lines represent the data set (bold) and the model response.

The same data set with the initial 50 data samples is shown in figure 2 where the difference between the data and the estimation at the output of the RBF network is clearly seen.

The sensitivity of the output variables, namely *comfort* and *safety*, computed by the ratio-scale (RS) method with respect to 43 RBF input variables is shown in figure 3.

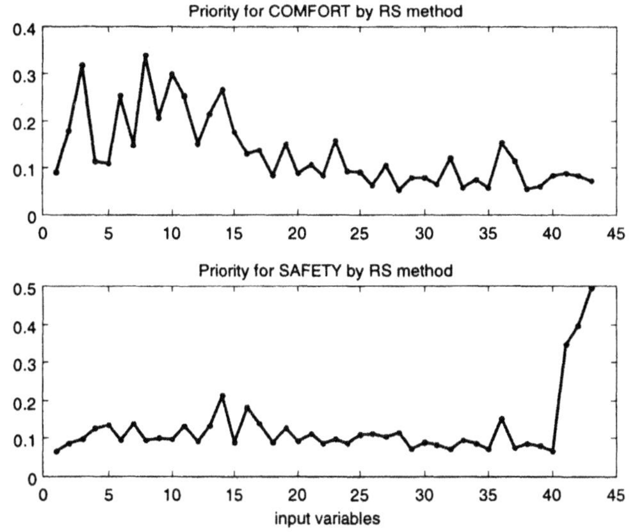

Figure 3. Sensitivity analysis results by the RS method for two outputs and 43 input variables of an RBF network model.

The sensitivity of the output variables, namely *comfort* and *safety*, by RS method and their sorted counterparts (middle plots) with respect to 43 input variables are shown in figure 4a and 4b respectively. The sequence by RS method matched the graded sensitivity sequence from (7) are also shown in this figure (lowermost ones).

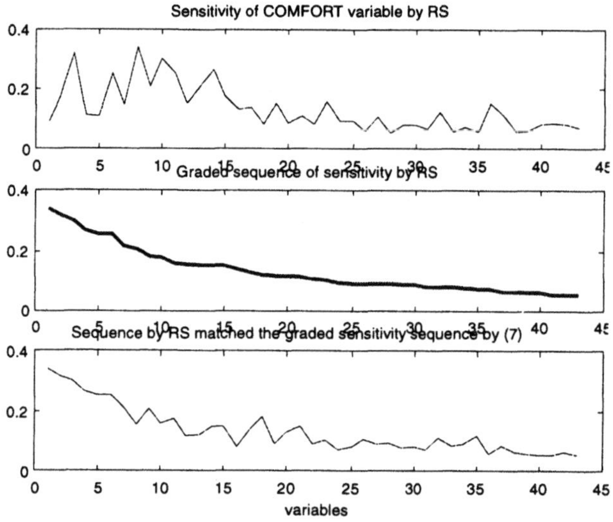

(a)

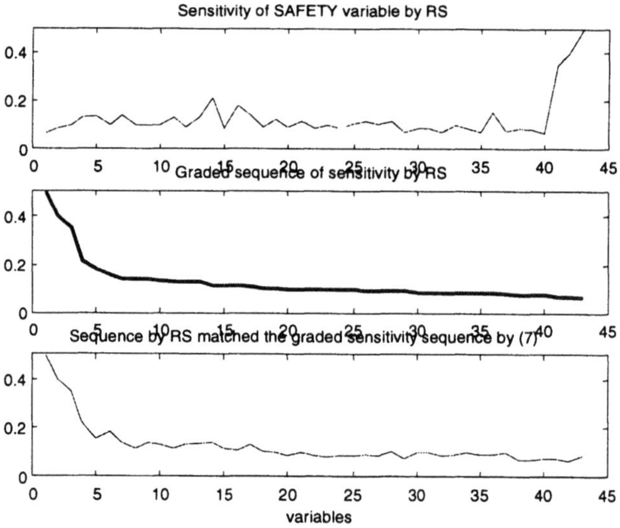

Figure 4. Sensitivity analysis results from RS for *comfort* and *safety*: The graded sequence of sensitivity (middle) and the sequence by RS matched the graded sequence obtained from (7)

In figure 4, the middle plot and the one below are shown together in one plot as presented in figure 5, for *comfort* and *safety* parameters respectively.

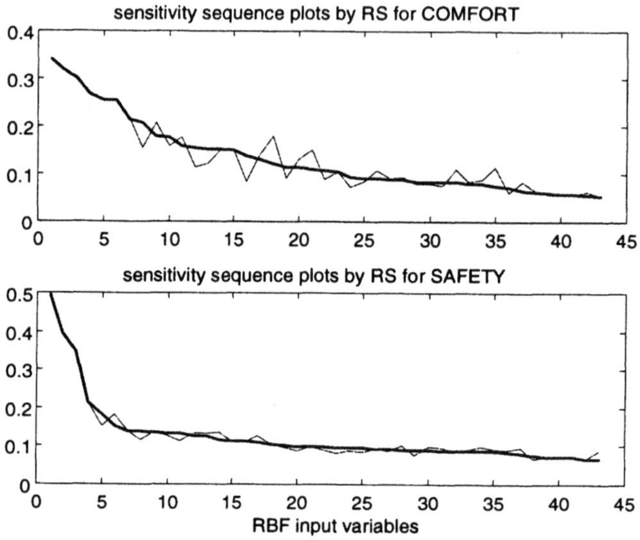

Figure 5. Sensitivity analysis results from the ratio-scale method: Sorted sensitivity results (bold) and the same sequence matched the graded sequence from the sensitivity computation by (7) for *comfort* (upper) and for *safety* (lower) respectively

In figure 5, the differences of two sets of priority results from the RS method and from (7) are easily compared. From the figure it is seen that the priority sequences from sensitivity computation by (7) and from the ratio-scale method are not the same but deviations are apparently not excessive. In particular, for variables having relatively large sensitivity, the results are the same. For variables having relatively smaller sensitivity, deviations occur indicating mainly the noise effects in identifying the sensitivities both methods. This means in such a case, the relative uncertainty is significant compared to the sensitivity itself involved. The sequences of the sensitivity amplitudes, for both outputs, are not the same. However, in each case the deviations are not significant and the trend is reasonably the same.

Similar comparisons are presented in the following figures from different viewpoints that might be of interest in different contexts, in general. Referring to this, in figure 6 the graded sequence of sensitivity from (7) and the matched sequence from RS method (bold) are shown together against the input variables for *comfort* and *safety* respectively. Conversely, in figure 7 sequence of sensitivity from (7) is matched the graded sequence from RS method (bold) and they are shown together against the input variables for *comfort* and *safety* respectively, at the output.

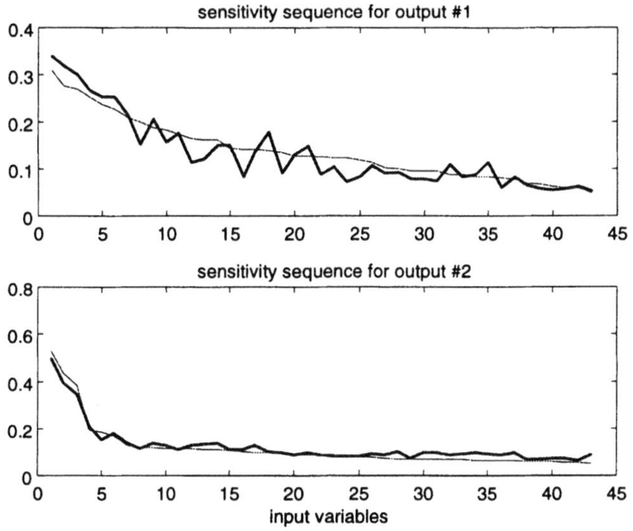

Figure 6. The graded sequence of sensitivity from (7) and the matched sequence from RS method (bold) are shown together for *comfort* and *safety* together.

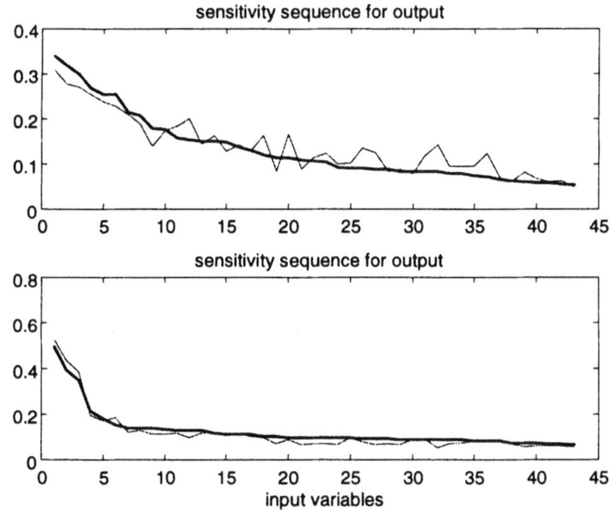

Figure 7. The sequence of sensitivity from (7) matched the graded sequence from RS method (bold) are shown together for *comfort* (upper) and *safety*(lower).

Again, although in figures 6 and 7 differences occur between the results of RS method and computations by (7), the deviations can apparently be tolerated.

As a cross check for validation of the model, the sensitivities according to the sequence the first column of the judgment matrix A and the priority vector from ratio-scale approach applied to matrix A are shown together in figure 8, for *comfort* (upper) and *safety* (lower) respectively.

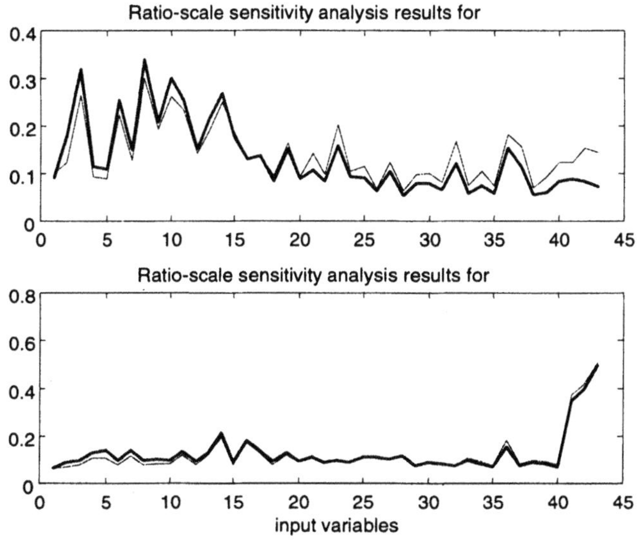

Figure 8. The sensitivities according to the sequence as seen in figure 2 and the matched priority vector from ratio-scale approach for *comfort* (upper) and *safety*(lower) respectively.

In figure 8, there are relatively small deviations, indicating the matrix A is consistent. This is also a strong indication that the probabilistic considerations given by (14) –(19) are perfectly valid.

As a final cross check, the sensitivities according to the graded sequence from RS method, that is, the priority vector from ratio-scale approach applied to matrix A in the preceding figure and the priority sequence from (7) are shown together in figure 9, for *comfort* (upper) and *safety* (lower) respectively. As it is seen from figure 9, the differences between the sensitivity/priority results by the RS method and the sensitivity computation using (7) is reasonably small so that it validates the RBF network model.

The overall results show that, the deviations in all comparisons above are to be tolerated. This indicates that the judgement matrix A is highly consistent and consequently the data model by RBFN is satisfactorily validated. It is important to note that, the validation covers the neural network structure as well as the learning process applied to the network.

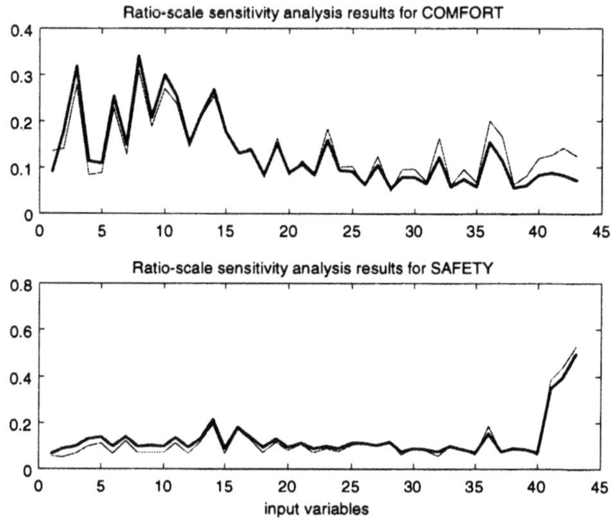

Figure 9. The sensitivities according to the graded sequence from RS method and the matched sequence from (7) are shown together for *comfort* (upper) and *safety* (lower) respectively.

5. Conclusions

Data/information model by learning is developed andvalidated. Model validation is carried out by the method of *analytical hierarchy* (AH) or *ratio-scale* (RS) method. For validation, sensitivity of the model with respect to the model variables was computed and the ratios among the sensitivities are considered for ranking the importance of the model variables with respect to the model outcomes. The same ranking is established by the ratio-scale method and compatibility between the ranking sequences is investigated for model validation. The model variables having relatively high sensitivity are found to be the same in both cases. Some of the model variables having relatively less sensitivity show some slight difference in two ranking sequences established and this difference or rather incompatibility is attributed to the relative uncertainty in the sensitivity determination. However, the incompatibility involved is identified to be insignificant and it is shown that ratio-scale method of Saaty [8] forms a novel and sound approach for data model validation for RBF network model. The RBF network implementation of AH method is particularly important since it forms a novel means of validation of neural network structure and the learning process in the context of neural networks.

References

[1] Hush D.R. and Horn B.G. (1993), Progress in supervised neural networks, *IEEE Signal Processing Magazine*, Jan., pp.8-39,

[2] Yen J. (1999), Fuzzy Logic- A modern perspective, *IEEE Trans. Knowledge and Data Engineering*, Vol.**11**, No.**1** , Jan./Feb., pp.153-165

[3] Holland J.H. (1975), *Adaptation in Natural and Artificial Systems*, University of Michigan Press, Ann Arbor, MI

[4] Goldberg D.E. (1989), *Genetic Algorithms in Search, Optimization and Machine Learning*, Addison-Wesley Publishing Co. Inc., Reading, MA

[5] Park J. and Sandberg I.W. (1991), Universal approximation using radial basis function networks, *Neural Computation*, vol.3, pp.246-257

[6 Chen] S., Billings S.A. and Luo W. (1989), Orthogonal least quares methods and their application to non-linear system identification, *Int. J. Control*, vol.50, no.5, pp. 1873-1896

[7] Chen S., Cowan C.F., and Grant P.M. (1991), Orthogonal least squares learning algorithm for radial basis functions network, *IEEE Trans. Neural Networks*, Vol.**2**, March, pp.302-309

[8] Saaty,T.L. (1980), *The Analytic Hierarchy Process*, McGraw-Hill, New York

[9] Saaty T.L (2000). *The Brain: Unraveling the mystery of how it works*, RWS Publications, Pittsburgh, PA 15213

[10] Durmisevic S. (2002), *Perception Aspects in Underground Space by Means of Intelligent Knowledge Modeling*, PhD thesis, Delft University of Technology, the Netherlands

Generalized recurrent neural networks and continuous dynamic systems

Ivan Gabrijel, Andrej Dobnikar

University of Ljubljana, Faculty of Computer and Information Science, Tržaška 25, SI-1001 Ljubljana, Slovenia
E-mail: *Ivan.Gabrijel@fri.uni-lj.si*

Abstract

The recurrent neural networks of generalized architecture (GARNN) are general continuous dynamic systems. It was shown elsewhere that they can successfully manage the problem of on-line inference of finite automata. In addition, they can successfully solve problems of a continuous nature because they are continuous systems. A frequently used problem domain to check this are the well-known trajectory tracking problems. Some new problems of this problem domain are defined in this paper. The experiments are carried out with the generalized recurrent neural networks and solutions are found for each trajectory of the problem domain.

1 Introduction

Recurrent neural networks are general continuous dynamic systems. When simulating a finite automaton, a recurrent neural network is forced either to act as discrete dynamic system or to act in a way that can be successfully reduced (discretized) to a discrete dynamic system. A finite automaton behavior is therefore only a special case of recurrent neural network processing.

The generalized architecture of recurrent neural networks is introduced to facilitate the on-line inference of finite automata [1], [2]. It consists of arbitrary connected neurons that are divided into several groups. The neurons within each group process simultaneously and the groups are processed sequentially at each time moment. A special case of this architecture is composed of two fully connected groups of neurons with complete forward and backward inter-group connections. External inputs are connected to the neurons of the first group and some neurons of the second group form network outputs. The number of neurons in the second group equals $m+r$, where m is number of neurons in the first group and r is number of network outputs ("$m+r$" heuristics). The recurrent neural networks of this specialized architecture are successfully applied to a problem of on-line inference of finite automata [3], [1], [2]. The on-line learning procedure of the GARNN is gradient based [1], [2]. It is a generalization of the real-time recurrent learning (RTRL) with teacher forcing technique, combined with decoupled extended Kalman filtering (DEKF). The on-line extraction procedure is devised to extract a structure of the underlying automaton from a successfully trained recurrent neural network [3], [1], [2]. The extraction is based on the attraction force clustering algorithm.

Recurrent neural networks are usually used for solving problems defined in continuous space because they have a continuous nature. General continuous dynamic systems that operate in discrete time are described by the equation

$$y(t) = \mathrm{f}\left(x(t),\ x(t-1),\ \ldots,\ y(t-1),\ y(t-2),\ \ldots\right) \quad , (1)$$

where x and y denote input and output signals at different time moments respectively and f describes a continuous function. It is well-known that this kind of dynamic systems can be modeled by neural networks [4], [5]. Even more general dynamic systems exist. These are spatially extended or spatiotemporal dynamic systems and are described by the equation

$$\begin{aligned} y(\mathbf{s}_i,\ t) = \mathrm{h}\big(&x(M_{i1},\ t),\ x(M_{i2},\ t-1), \ldots, \\ &y(N_{i1},\ t-1),\ y(N_{i2},\ t-2), \ldots\big) \end{aligned} \quad , \quad (2)$$

where $M_{i\bullet}$ and $N_{i\bullet}$ denote various sets of points in the neighborhood of a spatial point $\mathbf{s}_i$ at different time instants and h describes a continuous function. It is also possible to model such a complex system. Cellular automata [6] or non-parametric statistical method of local modeling [7] can be used for this purpose.

For the sake of simplicity, we have concentrated on a continuous problem domain that is frequently treated by recurrent neural networks. The GARNN, its on-line learning and the on-line extraction procedure are applied to a well-known trajectory tracking problem domain in this paper. The main purpose of this application is to show that the solution for on-line inference of finite automata is not limited to discrete systems and can also be used in continuous domains.

The paper is organized as follows. Section 2 introduces the trajectory tracking problem domain. Trajectory tracking using recurrent neural networks is described together with experiments, results and comments in Section 3. The conclusion summarizes the paper.

2 Trajectory tracking problem domain

A family of problems exists named trajectory tracking problems that are frequently used by researchers to test or evaluate their methods. These are autonomous, dynamic systems that represent a certain closed trajectory in the continuous vector space. Therefore, they have no inputs and they have to produce a closed trajectory of specific shape in their two dimensional output vector space during the course of time. A very good paper on trajectory tracking with a strong empirical and analytical basis is that by Hagner, Hassoun and Watta [8].

2.1 Description of standard problems

Two problems of this problem domain are the well-known circle and figure-eight trajectories. These trajectories are shown normalized to real vector space $[0.1, 0.9] \times [0.1, 0.9]$ in gray color in Fig. 1.

The first problem is to track a circle trajectory (Fig. 1. a.), which is defined by the equations

$$x = \sin\left(t + \frac{\pi}{2}\right) \quad ; \quad 0 \le t \le 2\pi \quad , \tag{3}$$

$$y = \sin(t) \quad ; \quad 0 \le t \le 2\pi \quad , \tag{4}$$

where t is a parameter that represents continuous time.

The second problem is tracing the figure-eight shaped trajectory (Fig. 1. b.), which is described by the expressions

$$x = \sin\left(t + \frac{\pi}{2}\right) \quad ; \quad 0 \le t \le 2\pi \quad , \tag{5}$$

$$y = \sin(2t) \quad ; \quad 0 \le t \le 2\pi \quad . \tag{6}$$

2.2 Definition of new problems

Additionally, a third problem is defined, which is to track a cardioid trajectory. The cardioid is defined by the equations

$$x = r\left(2\cos(t) - \cos(2t)\right) \quad ; \quad 0 \le t \le 2\pi \quad , \tag{7}$$

$$y = r\left(2\sin(t) - \sin(2t)\right) \quad ; \quad 0 \le t \le 2\pi \quad , \tag{8}$$

where r is a parameter that specifies starting point, $(r, 0)$, of the cardioid and can be selected as $r = 1$. The aspect ratio of the cardioid is not equal to 1 as are the aspect ratios of the other trajectories. Therefore, the transformations of the original cardioid are defined [2] in order to prevent from distortion of its shape during normalization before learning of GARNN. The transformed and normalized cardioid is drawn in Fig. 1. c. in gray color.

Another trajectory tracking problem is defined. It is depicted in Fig. 1. d. and on the basis of its shape it is named the flower trajectory. The flower trajectory is defined by the equations

$$x = \sin(t)\sin\left(4t + \frac{\pi}{2}\right) \quad ; \quad 0 \le t \le 2\pi \quad , \tag{9}$$

$$y = \cos(t)\sin\left(4t + \frac{\pi}{2}\right) \quad ; \quad 0 \le t \le 2\pi \quad . \tag{10}$$

3 Trajectory tracking with recurrent neural networks

The trajectory tracking is usually considered with recurrent neural networks that operate in discrete time. The continuous time parameter t used in equations (3) through (10) is therefore replaced by the discrete time equivalent n as prescribed in the equation

$$t = \frac{2\pi n}{N} \quad , \tag{11}$$

where N is the number of discrete time moments needed to describe the entire trajectory. N is therefore a parameter of each trajectory tracking problem. The small values of N result in fewer distinct data patterns and larger values describe a trajectory more smoothly. This parameter enables one to consider each trajectory tracking problem as a family of similar problems that differ only in timing. However, in the common interpretation the trajectory tracking domain is composed of scalable problems.

All trajectory tracking problems are naturally on-line because they are described by closed curves. The last point of each cycle along the trajectory is immediately followed by the first point of the next cycle. Therefore it is not necessary to reset the internal state of a recurrent neural network at the start of each next cycle because the old state is totally relevant.

3.1 Description of experiments

Data patterns of each trajectory are successively generated and GARNNs are applied to identify the trajectories. Two groups of neurons are used with the "$m+r$" heuristics. However, there are no inputs because trajectories are autonomous dynamic systems. The parameters of the learning procedures have the same values as used in experiments within discrete problem domains [1], [2]. 20000 data patterns are assigned for the learning, following by 2000 testing patterns. Data are normalized (transformed from real vector space $[-1, 1] \times [-1, 1]$ to real vector space $[0.1, 0.9] \times [0.1, 0.9]$) when they are applied to a generalized recurrent neural network for learning or testing. The trained recurrent neural networks are required to remain in phase and on the course of the identified trajectory as closely as possible during the testing process. Their success is determined by the mean square error measure.

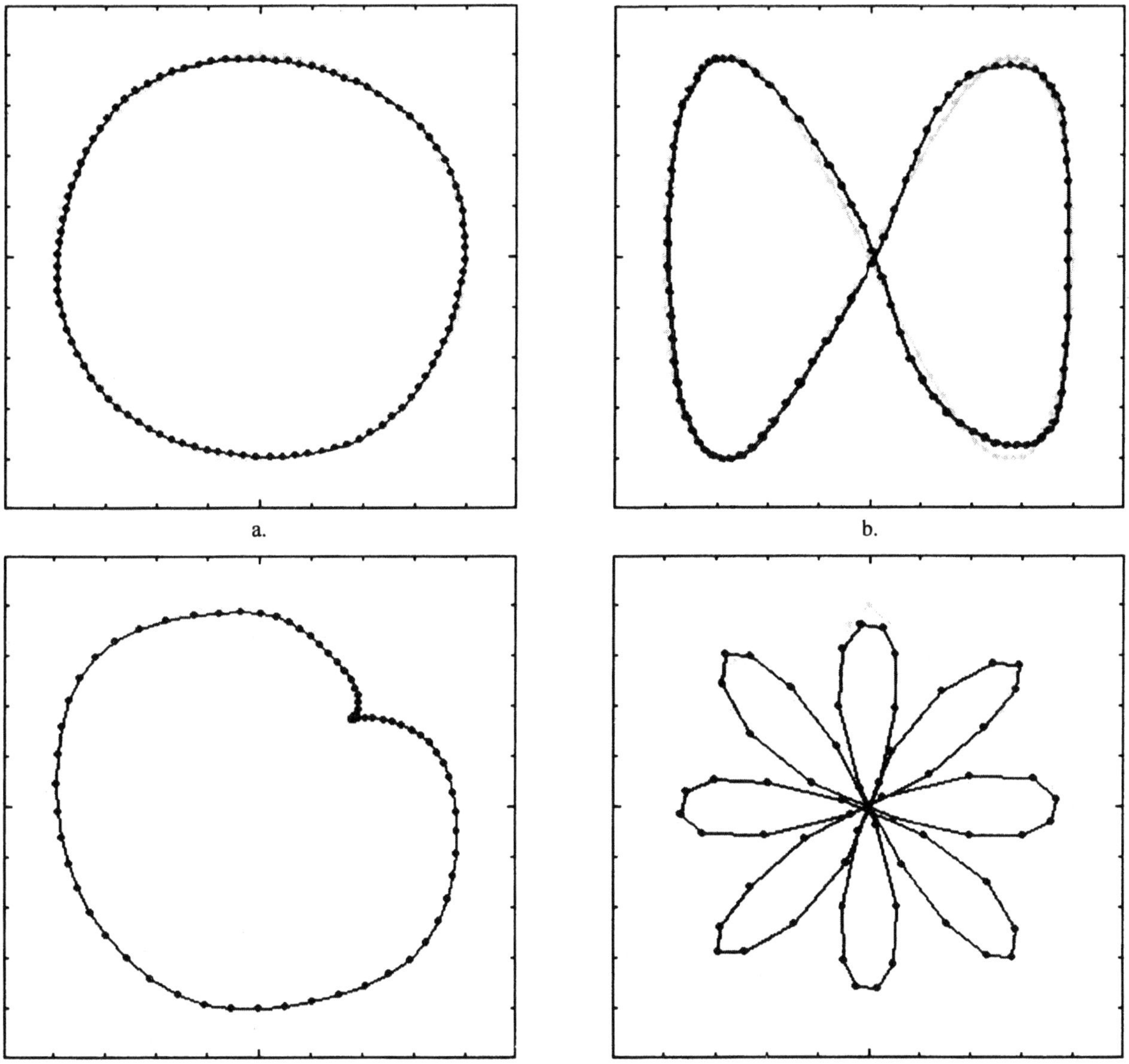

Fig. 1. The dynamics of the output neurons' activities in the real output vector space [0, 1] × [0, 1] are described for each solution from Table 1 of each problem at the beginning of the testing process at time moments t=20000+N+1, where N is a problem size. Each graph shows the last N+1 output patterns. The gray color designates and connects the desired output patterns and the black color is used for actual output patterns. The visualized problem solutions and problem sizes are: a. circle, N = 100; b. figure eight, N = 100; c. cardioid, N = 70, d. flower, N = 60.

3.2 Results of identification and comments

It was decided to explore how the GARNN size and the problem size N influence the success of the learning process. Several experiments are carried out for each problem altering the problem size N = 10, 20, ..., 100 and the GARNN size 1-3, 2-4, 3-5, ..., 20-22, where two numbers describe number of neurons in the first and in the second group, respectively. It can be seen from Table 1 and from Fig. 1 that the easiest problem is to track the circle trajectory. The figure-eight tracking problem is the second hardest, followed by the cardioid tracking problem, and the hardest problem is to trace the flower trajectory. Reliable solutions in all of the ten trials at a particular GARNN size and at a particular problem size can be found only when tracking the circle and the figure eight trajectories. Therefore, Table 1 describes only one successful trial for each problem of each problem domain. The fourth column of Table 1 represents indices of the selected successful trials among ten performed at particular combinations of problem sizes and GARNN sizes. A criterion for selecting these trials from all trials that are carried out in experiments within each particular trajectory tracking problem is a solution of largest problem size that produces a mean square error less than 1E-03 in the testing interval. The dynamics of the output neurons' activities of the selected solutions compared to desired behaviors are represented in Fig. 1 within the output vector space.

Table 1. Successful trials of each problem of the trajectory tracking problem domain

Problem	GARNN size	Problem size (N)	Index of trial	MSE
Circle	8–10	100	1	2.69E-5
Figure eight	20–22	100	5	1.74E-4
Cardioid	7–9	70	3	1.78E-4
Flower	16–18	60	9	7.49E-4

3.3 Results of extraction and comments

The extraction process is applied in the selected experiments that are used to form the results in Table 1. Parameters of the successful extraction are given in Table 2. The most general version of the extraction algorithm [2] is used. However, the basic procedure [3], [1], [2] is sufficient. Trajectories are dynamic systems that are independent of external environment. Therefore, the corresponding extracted automata contain only one special symbol in the input set: ε. This symbol represents the empty input pattern or absence of input pattern. All transitions in the automata extracted from GARNNs that identify trajectories are therefore ε-transitions. All extracted automata look like a closed chain of states connected by ε-transitions. However, the output values of states differ between trajectories.

Table 2. Extraction results and parameters of the selected successful trials of each problem of the trajectory tracking problem domain

Problem	GARNN size	Number of extracted states and output symbols	Minimal attraction force thresholds
Circle	8–10	100 100	881 1584
Figure eight	20–22	100 99	693 3168
Cardioid	7–9	70 57	6000 1.90E+07
Flower	16–18	60 34	36 1114

The last column in Table 2 describes the parameters of the attraction force clustering procedures in the extraction process. The reason for the large attraction force thresholds expressed in the cardioid problem is that successive points of the cardioid lie very close to each other at the beginning and at the end of the trajectory.

4 Conclusion

It was shown in this paper that recurrent neural networks of the generalized architecture are able to manage continuous problems. The continuous, scalable problem domain of trajectory tracking problems was considered and described. Two new problems are defined within this domain. Several generalized recurrent neural networks are tested and the results are given in numerical and in visual form. The solutions are obtained for each of the considered problems. It was found that a solution can be reliably searched for only at the easiest problem, which is to track the circle trajectory. The other problems – the figure-eight and especially the newly defined (cardioid and flower) problems – represent a real challenge for the present and for future recurrent neural systems because of scalability in problem size.

References

[1] Gabrijel, I., Dobnikar, A. On-line identification and rule extraction of finite state automata with recurrent neural networks. To be published in *Neural Networks*.

[2] Gabrijel, I. (2002) *Generalized Architecture of Recurrent Neural Networks and On-Line Identification of Finite Automata – Ph.D. Thesis*. University of Ljubljana, Faculty of Computer and Information Science, Ljubljana, Slovenia.

[3] Gabrijel, I., Dobnikar, A. (2001) On-line identification and rule extraction of finite state automata with recurrent neural networks. In: Kurkova, V., Steele, N. C., Neruda, R., Karny, M. (eds.) *Proceedings of the International Conference on Artificial Neural Nets and Genetic Algorithms*. Springer-Verlag, Vienna, Austria, pp. 78–81.

[4] Narendra, K. S. (1990) Adaptive control using neural networks. In: Miller III, W. T., Sutton, R. S., Werbos, P. J. (eds.) *Neural Networks for Control*. MIT Press, Cambridge, MA, pp. 115–142.

[5] Levin, A. U., Narendra, K. S. (1997) Identification of nonlinear dynamical systems using neural networks. In: Omidvar, O., Elliott, D. L. (eds.) *Neural Systems for Control*. Academic Press, San Diego, CA, pp. 129-160.

[6] Richards, F. C., Meyer, T. P., Norman, H. P. (1990) Extracting cellular automaton rules directly from experimental data. *Physica D* 45, pp. 189–202.

[7] Mandelj, S., Grabec, I., Govekar, E. (2001) Statistical approach to modeling of spatiotemporal dynamics. *International Journal of Bifurcation and Chaos* 11, No. 11, pp. 2731–2738.

[8] Hagner, D. G., Hassoun, M. H., Watta, P. B. (2000) Comparison of recurrent networks for trajectory generation. In: Medsker, L. R., Jain, L. C. (eds.) *Recurrent Neural Networks – Design and Applications*. CRC Press, Boca Raton, FL, pp. 243–276.

A learning probabilistic neural network with fuzzy inference

Ye. Bodyanskiy[1], Ye. Gorshkov[1], V. Kolodyazhniy[2], J. Wernstedt[2]

[1]Control Systems Research Laboratory, Kharkiv National University of Radioelectronics, 14, Lenin Av., Kharkiv, 61166, Ukraine,
E-mail: {Bodyanskiy, Yevgen.Gorshkov}@ieee.org
[2]Institute of Automatic Control and System Engineering, Technical University of Ilmenau, PF 100565, Ilmenau, 98684, Germany,
E-mail: Kolodyazhniy@ieee.org, Juergen.Wernstedt@systemtechnik.tu-ilmenau.de

Abstract

In this paper, an architecture of a fuzzy probabilistic neural network is considered. A learning algorithm for the activation function parameters is proposed. The advantages of this network lie in the possibility of classification of the data with substantially overlapping clusters, and tuning of the activation function parameters improves the accuracy of classification. Simulation results confirm the efficiency of the proposed approach in the data classification problems.

1 Introduction

Clustering and data classification are key problems of data mining, and solving these problems is important for effective knowledge accumulation on the basis of observational analysis. One of the popular techniques of data classification are the probabilistic neural networks (PNN), proposed by D.F. Specht [1, 2]. These networks are effectively applied to various problems of classification, diagnosis, pattern recognition, etc. [3, 4, 5]. However, the conventional PNN solves the classification problem according to the traditional "crisp" approach, when every observation is assigned to only one cluster. But in practice the problems with overlapping clusters are common, when the classified observation may belong to several classes simultaneously with certain degrees of membership. To solve such problems, the methods based on the fuzzy approach to clustering, classification, and pattern recognition are used [6, 7].

The neuro-fuzzy synergism [8] extends the range of the problems solved with the classifier, combining the advantages of fuzzy data analysis with the capabilities of learning from data.

In this paper, we make an attempt to modify the conventional PNN by adding the properties of fuzzy classification and learning of the activation functions parameters.

2 Network Architecture

In the general case, the probabilistic neural networks which solve the problem of Bayesian classification [9] via the recovery of unknown probability distributions by means of the Parzen kernels [9, 10], belong to the feedforward architectures and are closely related to the radial-basis function networks [11] and generalized regression neural networks [12].

The probabilistic network proposed in the paper is a three-layer structure shown in fig. 1.

The source information for the synthesis of networks is the training set of patterns, formed as a "packet" of n-dimensional vectors $x(1),\ldots,x(j),\ldots,x(N),x(N+1),\ldots,x(N+M)$, $x(j) = (x_1(j),\ldots,x_i(j),\ldots,x_n(j))^T$. The order of patterns in the packet is not significant. It is assumed that the crisp classification (the membership to one of m clusters) is known for each of the training patterns $x(j)$, and the representatives of all the possible clusters must be present in the dataset. That is, if N_l observations from the training set belong to the l-th cluster, then

$$\sum_{l=1}^{m} N_l = N. \tag{1}$$

The input layer (number 0) receives an n-dimensional vector patterns $x(k),\ k > N+M$ with unknown classification. They are transferred to the hidden layers for further processing.

The first hidden layer (the prototype layer) contains N neurons with the bell-shaped (usually Gaussian) activation functions, and their synaptic weights are determined by the components of the training patterns, i.e.

$$w_{ji} = x_i(j),\quad i = 1,2,\ldots,n,\ j = 1,2,\ldots,N, \tag{2}$$

or in the vector form

$$w_j = x(j) = (x_1(j),\ldots,x_i(j),\ldots,x_n(j))^T. \tag{3}$$

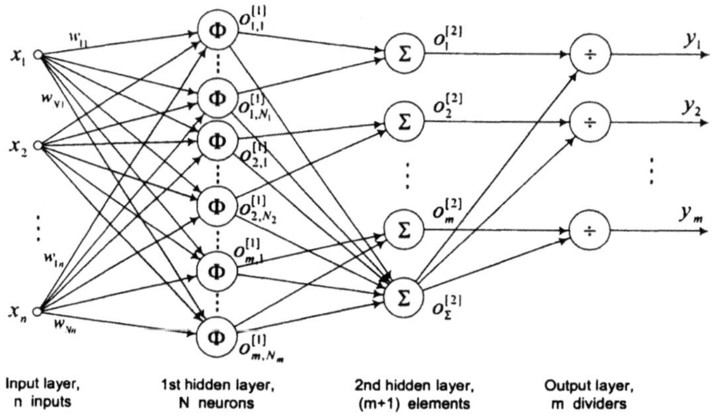

Fig. 1. Probabilistic neural network with fuzzy inference

For the convenience of notation, all the neurons in the prototype layer are divided into m groups with N_l nodes in each, corresponding to its cluster. The vector of weights of the p-th neuron in the l-th group will be denoted as $w_{l,p} = w_{l+p-1}$, $l = 1, 2, \ldots, m$, $p = 1, 2, \ldots, N_l$.

It is obvious that the training of the synaptic weights in this case is reduced to their one-time setting, that is very simple.

When the vector $x(k)$ is fed to the 0-th layer, the neurons of the 1-st layer produce the signals

$$o_{l,p}^{[1]}(k) = \Phi_{l,p}(x(k), \sigma_l) = \exp\left(-\frac{\|x(k) - w_{l,p}\|^2}{2\sigma_l^2}\right), \quad l = 1, \ldots, m;\ p = 1, \ldots, N_l;\ k > N+M, \tag{4}$$

(where σ_l^2 is the width parameter of the activation function).

If all the input vectors are pre-normalized so that $\|x\| = 1$, the expression (4) assumes the following simple form

$$o_{l,p}^{[1]}(k) = \exp\left(-\frac{1 - x^T(k)w_{l,p}}{\sigma_l^2}\right). \tag{5}$$

It is also obvious that $\|w_{l,p}\| = 1$.

Since it holds

$$-1 \le x^T(k)w_{l,p} \le 1, \tag{6}$$

the outputs of the neurons of the first layer can vary only within the interval

$$\exp(-2\sigma_l^{-2}) \le o_{l,p}^{[1]} \le 1 \tag{7}$$

as is shown in fig. 2.

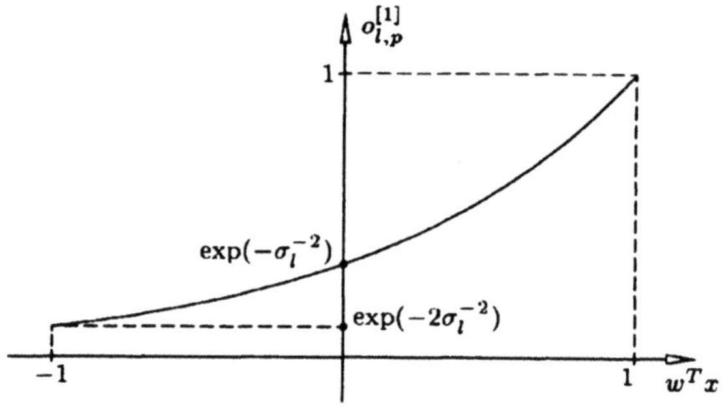

Fig. 2. Activation function of neurons in the first layer

At the level of the first layer, it is also possible to determine the diameters of the clusters formed by the training data as

$$\begin{cases} D_l^2 = \max\limits_{p,q=1,\ldots,N_l} \|w_{l,p} - w_{l,q}\|^2 \\ \quad = \max\limits_{p,q=1,\ldots,N_l} 2(1 - w_{l,p}^T w_{l,q}), \\ 0 \le D_l = \sqrt{\max 2(1 - w_{l,p}^T w_{l,q})} \le 2, \end{cases} \tag{8}$$

and to roughly estimate how much they overlap.

The second hidden layer is formed by $m+1$ elementary summing units, with first m of them receiving the outputs of the prototype layer so that

$$o_l^{[2]}(k) = \sum_{p=1}^{N_l} o_{l,p}^{[1]}(k), \tag{9}$$

and the output of the $(m+1)$-th summing unit calculates

the total sum

$$o_{\Sigma}^{[2]}(k) = \sum_{l=1}^{m} \sum_{p=1}^{N_l} o_{l,p}^{[1]}(k). \qquad (10)$$

It can be readily seen that the sums (9) are Parzen approximations [9, 10] of the unknown data distributions in the clusters.

Finally, the output normalization layer formed by m dividers calculates the vector of degrees of membership $y(k) = (y_1(k), \ldots, y_l(k), \ldots, y_m(k))^T$ of the processed observation as

$$0 \leq y_l(k) = \frac{o_l^{[2]}(k)}{o_{\Sigma}^{[2]}(k)} \leq 1, \quad \sum_{l=1}^{m} y_l(k) = 1. \qquad (11)$$

It is easy to see that the described network is a combination of the probabilistic and generalized regression neural networks, and is capable of data classification on the basis of fuzzy decision on the membership of a particular observation to a certain class.

It is also evident that the first two layers of this network are essentially a standard radial basis function architecture with fixed synaptic weights $w_{lj}^{RBF} \equiv 1$ and centers $c_j^{RBF} \equiv w_j$.

Undetermined are only the cluster width parameters σ_l^2, which significantly affect the classification accuracy and are usually chosen empirically [5].

Therefore, it is advisable to introduce the learning capabilities into the proposed network as is shown in fig. 3.

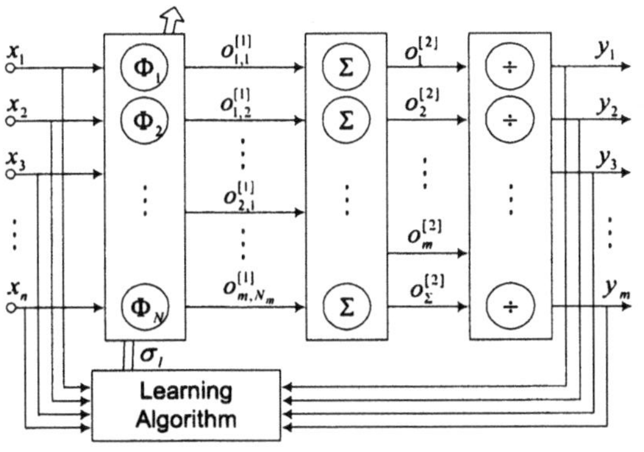

Fig. 3. Learning PNN with fuzzy reasoning

The learning algorithm adjusts the width parameters σ_l^2 to provide more accurate approximation of the data belonging to the l-th class from the observations $x(u)$, $u = N{+}1, \ldots, N{+}M$ in the training set and their crisp classification $d(u)$.

3 Learning Algorithm

Let us introduce a one-step learning criterion (error function)

$$E(u) = \sum_{h=1}^{m} E_h(u) = \frac{1}{2}\|e(u)\|^2, \qquad (12)$$

(where $e(u) = (e_1(u), \ldots, e_h(u), \ldots, e_m(u))^T$, $E_h(u) = \frac{1}{2}e_h^2(u) = \frac{1}{2}(d_h(u) - y_h(u))^2$, $d_h(u)$ is the training signal equal to 1 if the vector $x(u)$, $u = N{+}1, \ldots, N{+}M$ belongs to the h-th cluster, and 0 otherwise). The derivatives of the error function with respect to the tunable parameters will be

$$\begin{cases} \dfrac{\partial E_l(u)}{\partial \sigma_l^2} = -e_l(u)\dfrac{1 - y_l(u)}{o_{\Sigma}^{[2]}(u)} \displaystyle\sum_{p=1}^{N_l} \dfrac{\partial o_{l,p}^{[1]}(u)}{\partial \sigma_l^2}, \\ \dfrac{\partial E_h(u)}{\partial \sigma_l^2} = e_h(u)\dfrac{y_h(u)}{o_{\Sigma}^{[2]}(u)} \displaystyle\sum_{p=1}^{N_l} \dfrac{\partial o_{l,p}^{[1]}(u)}{\partial \sigma_l^2}, \quad h \neq l, \end{cases} \qquad (13)$$

or

$$\frac{\partial E_h(u)}{\partial \sigma_l^2} = e_h(u)\frac{y_h(u) - \delta_{hl}}{o_{\Sigma}^{[2]}(u)} \sum_{p=1}^{N_l} \frac{\partial o_{l,p}^{[1]}(u)}{\partial \sigma_l^2}, \quad h = 1, 2, \ldots, m, \qquad (14)$$

where $\delta_{hl} = \begin{cases} 1, & \text{if } h = l, \\ 0, & \text{otherwise.} \end{cases}$

It can be readily seen that

$$\frac{\partial E(u)}{\partial \sigma_l^2} = \frac{1}{o_{\Sigma}^{[2]}(u)} \sum_{p=1}^{N_l} \frac{\partial o_{l,p}^{[1]}(u)}{\partial \sigma_l^2} \cdot \sum_{h=1}^{m} (y_h(u) - \delta_{hl}) e_h(u), \qquad (15)$$

and since for the Gaussian membership function

$$\frac{\partial o_{l,p}^{[1]}(u)}{\partial \sigma_l^2} = \frac{\|x(u) - w_{l,p}\|^2}{2\sigma_l^4} \cdot \exp\left(-\frac{\|x(u) - w_{l,p}\|^2}{2\sigma_l^2}\right), \qquad (16)$$

the expression (15) can be re-written in the form

$$\frac{\partial E(u)}{\partial \sigma_l^2} = \frac{1}{2\sigma_l^4 o_{\Sigma}^{[2]}(u)} \sum_{h=1}^{m} (y_h(u) - \delta_{hl}) e_h(u) \cdot \sum_{p=1}^{N_l} \|x(u) - w_{l,p}\|^2 \exp\left(-\frac{\|x(u) - w_{l,p}\|^2}{2\sigma_l^2}\right). \qquad (17)$$

Minimizing (12) with the gradient-based procedure

$$\sigma_l^2(q+1) = \sigma_l^2(u) - \eta(u)\frac{\partial E(u)}{\partial \sigma_l^2}, \tag{18}$$

we can finally write the learning algorithm of the probabilistic network with fuzzy inference

$$\sigma_l^2(q+1) = \sigma_l^2(u) - \eta(u)\frac{1}{2\sigma_l^4(u)o_\Sigma^{[2]}(u)} \cdot \sum_{p=1}^{N_l} \|x(u) - w_{l,p}\|^2 \exp\left(-\frac{\|x(u) - w_{l,p}\|^2}{2\sigma_l^2(u)}\right) \cdot \sum_{h=1}^{m}(y_h(u) - \delta_{hl})e_h(u), \tag{19}$$

or taking into account (5)

$$\sigma_l^2(q+1) = \sigma_l^2(u) - \eta(u)\frac{1}{\sigma_l^4(u)o_\Sigma^{[2]}(u)} \cdot \sum_{p=1}^{N_l}(1 - x^T(u)w_{l,p}) \exp\left(-\frac{1 - x^T(u)w_{l,p}}{\sigma_l^2(u)}\right) \cdot \sum_{h=1}^{m}(y_h(u) - \delta_{hl})e_h(u), \tag{20}$$

where $\eta(u)$ is the scalar learning rate parameter.

When the learning is finished, the values $\sigma_l^2(N+M+1)$, $l = 1, 2, \ldots, m$ are used as estimates of the activation function parameters (4), (5) in the classification of objects $x(k)$, $k > N+M$ with unknown membership.

4 Experiments

Several well-known databases were chosen to test the performance of the proposed neural network and learning algorithm: the Iris data, the Wisconsin breast cancer (WBC) data, and the Thyroid disease data from the UCI repository [13].

The Iris data set consists of 150 samples with 4 attributes, which are divided into 3 classes. The WBC data set consists of 699 samples with 10 attributes of 2 classes. We used only 683 of them, because 16 had missing values. The Thyroid disease data set consists of 215 samples with 5 attributes, divided into 3 classes.

We have compared the proposed classifier with the original PNN and the PNN with fuzzy output without training. Our test is intended to examine the learning capabilities of the network.

The width parameter of the PNN network was set to $\sigma \approx 0.3$ for all datasets.

The classification results of this network do not vary significantly when the parameter σ is changed.

For the proposed network, the initial width $\sigma_l(0) = 0.5$ and the learning rate parameter $\eta \approx 0.15$ were chosen for all datasets.

It should be noted here that the optimal value of the width parameter can be achieved when the initial value is large (e.g. $\sigma_l(0) > 0.5$), while small initial values lead to numerical instability of the training procedure. During training, this parameter alternates in both directions (i.e. the delta between the initial and optimal values can be both positive and negative).

For the training of the proposed neural network M samples were randomly selected from the data set (except initial N samples). The remaining instances of the data were used as the checking data set. The original PNN was built on the same $N+M$ instances of the data (the number of hidden layer units), which were used to build and train the proposed neural network.

The sizes of the networks are given in Table I.

Dataset	Iris	WBC	Thyroid
PNN	100	100	130
PNN with fuzzy output	50	50	50
PNN with fuzzy output without training	50	50	50

Table I. Sizes of the networks

Since the one-step error function is used, the training procedure can be executed in the on-line mode. But for the small training data sets it is necessary to train the network for several epochs on the same data. In our experiments, 15–20 epochs were enough for the network to learn.

All the tests were executed 10 times each. Mean error values for all the data sets and classifiers are given in Table II (error rate on the testing data). The results for the PNN with fuzzy output without training are given to show the learning capability of the trained PNN with fuzzy output.

Dataset	Iris	WBC	Thyroid
PNN	11.4%	16.1%	20.4%
PNN with fuzzy output	8.7%	8.1%	18.3%
PNN with fuzzy output without training	19.6%	32.5%	26.5%

Table II. Test results (error rate on the testing data)

High error rate for the Iris data set can be explained by the small size of the checking data set (50 samples).

From the given results it can be seen that the proposed PNN with fuzzy output provides better classification results then the original PNN, using less computational resources since the size of the PNN at least twice bigger as the size of the PNN with fuzzy output.

5 Conclusion

Simulation results confirm high performance of the proposed learning probabilistic neural network with fuzzy output. This network provides better classification than the conventional PNN and can be further improved via the use of the finite membership function and optimization of the learning algorithm.

References

[1] D. F. Specht. Probabilistic neural networks. *Neural Networks*, 3:109–118, 1990.

[2] D. F. Specht. Probabilistic neural networks and polynomial adaline as complementary techniques for classification. *IEEE Trans. on Neural Networks*, 1(1):111–121, 1990.

[3] C. M. Bishop. *Neural Networks for Pattern Recognition*. Clarendon Press, Oxford, 1995.

[4] C. Looney. *Pattern Recognition Using Neural Networks*. Oxford University Press, N.Y., 1996.

[5] R. Callan. *The Essence of Neural Networks*. Prentice Hall Europe, London, 1999.

[6] J. C. Bezdek. *Pattern Recognition with Fuzzy Objective Function Algorithms*. Plenum Press, N.Y., 1981.

[7] F. Höppner, F. Klawonn, and R. Kruse. *Fuzzy Clusteranalyse*. Vieweg, Braunschweig, 1999.

[8] J.-S. R. Jang, C.-T. Sun, and E. Mizutani. *Neuro-Fuzzy and Soft Computing — Computational Approach to Learning and Machine Intelligence*. Prentice Hall, Upper Saddle River, 1997.

[9] K. Fukunaga. *Intruduction to Statical Pattern Recognition*. Academic Press, New York and London, 1972.

[10] E. Parzen. On the estimation of a probability density function and the mode. *Ann. Math. Statist.*, 38:1065–1076, 1962.

[11] J. Moody and C. J. Darken. Fast learning in networks of locally-tuned processing units. *Neural Computing*, 1:281–299, 1989.

[12] D. F. Specht. General regression neural networks. *IEEE Trans. on Neural Networks*, 2:568–576, 1991.

[13] P. M. Murphy and D. W. Aha. *UCI Repository of machine learning databases*. University of California, Department of Information and Computer Science, CA, 1994. http://www.ics.uci.edu/~mlearn/MLRepository.html.

Boosting Recurrent Neural Networks for Time Series Prediction

R. Boné, M. Assaad, M. Crucianu
Laboratoire d'Informatique, Université de Tours
64, avenue Jean Portalis, 37200 Tours, France
bone@univ-tours.fr, michel.crucianu@inria.fr

Abstract: We adapt a boosting algorithm to the problem of predicting future values of time series, using recurrent neural networks as base learners. The experiments we performed show that boosting actually provides improved results and that the weighted median is better for combining the learners than the weighted mean.

1 Introduction

The reliable prediction of future values of discrete-time, real-valued time series has many important applications ranging from ecological modeling to dynamical system control. To obtain a predictor, which is a model of the time series, one typically makes extensive use of the data concerning the past evolution of the series.

A prediction is commonly expressed as a function of the recent history of the time series, $\hat{x}_t = f(x_{t-1}, x_{t-2}, \ldots)$. Building a model then amounts to finding an appropriate function f. Given their universal approximation properties, multi-layer perceptrons (MLP [1]) are often successful in modeling nonlinear functions f. In this case, a fixed number p of past values is fed to the input layer of the MLP and the output is required to predict a future value of the time series according to $\hat{x}_t = f_w(x_{t-1}, x_{t-2}, \ldots x_{t-p})$. Using a time window of fixed size proves to be limiting in many applications: if the time window is too narrow, important cues may fall aside, while if the window is too wide, useless inputs may act as noise.

Ideally, for a given problem, the size of the time window should adapt to the context. This can be accomplished by employing recurrent neural networks (RNN [1]). RNNs can be trained to keep relevant information on the past in their internal state and, by consequence, are able to take into account a history of variable length when making a prediction. Not only were RNNs shown to be computationally more powerful than MLPs [2,3], but they were found to perform better and be more parsimonious in many cases.

To improve upon the performance obtained by RNNs trained with BPTT, we can attempt to develop a more appropriate learning algorithm by making use of prior information regarding the application under study. We can also adapt general procedures that were found to enhance the accuracy of various basic learners. One such procedure is known under the name of boosting and was introduced in ref. [4]. Boosting methods were further developed and applied in subsequent work by many authors. The possibly small gain a "weak" learner can bring with respect to random guessing is boosted by the sequential construction of several such learners, progressively focused on difficult examples of the original training set.

In this document we focus on the definition of a boosting algorithm for improving the prediction performance of RNNs. After a reminder of the generic boosting algorithm, we review in the next section related work on the subject. Our boosting algorithm is defined in section 3 after a discussion of the important choices involved. Section 4 presents the experimental results we obtained on two different benchmarks, showing an improvement in performance.

2 Boosting methods

The first family of boosting algorithms, introduced in ref. [4] for classification problems, exploited hierarchies of three classifiers, trained on progressively more difficult parts of the available data; their decisions are combined by a majority vote. Subsequent representatives of this family concern either classification or regression (e.g. [5], [6]).

A second family of boosting algorithms was introduced in ref. [7] and corresponds to the following generic algorithm:

1. Set the initial distribution on the training set
2. Iterate until the stopping criterion is reached
   (a) develop a weak learner given the current distribution on the training set
   (b) update the distribution on the training set
   (c) evaluate the stopping criterion
3. Combine the weak learners

Most of the recent work on boosting studied algorithms belonging to this second family and followed the introduction of the powerful AdaBoost algorithm in refs. [8,9]. Many different algorithms can be obtained from this generic algorithm by making specific choices at every stage.

The initial distribution on the training set is usually uniform, but prior knowledge concerning the problem should provide a better choice. When updating the distribution at every iteration, training examples found difficult by previous weak learners are favored. Subsequent learners can thus concentrate on the most difficult parts of the training data.

For several boosting algorithms it was shown that the expected error of the combination of learners converges to 0 as the number of learners goes to infinity. This result relies on a strong hypothesis: step 2(a) can always return a weak learner that performs (at least) slightly better than random guessing. In practice, since new learners must deal with progressively more difficult training examples, they end by performing worse than random guessing after a limited number of steps. This is the usual stopping criterion employed by the algorithm.

The generic boosting algorithm does not specify how the current distribution on the training set should be taken into account. In existing algorithms, the weak learner is trained on a subset of the training set obtained by sampling with replacement according to the current distribution. As we shall see later, this solution is not always appropriate.

Various learners were employed for step 2(a) of the algorithm: linear models, decision trees, splines, MLPs, etc. The choice of a cost function and of the method for combining the weak learners mainly depends on the type of problem: classification or regression.

Since we are concerned here with time series prediction, which is a regression problem, we now turn to existing approaches for boosting regressors. The first approach [10,11] is rather empirical and adapts AdaBoost to three cost functions that are specific to regression problems. The algorithm described in [6] belongs to the first family of boosting algorithms and employs a threshold to label an answer of a weak regressor as right or wrong. These two methods make use of the median as a way to combine the outputs of the learners.

Recently, following the view of boosting as a gradient descent procedure [12,13], a new approach to regressor boosting as residual-fitting was developed [14,15,16,17,18,19]. Instead of being trained on a different sample of the same training set, as in previous boosting algorithms, a regressor is trained on a new training set having different target values (e.g. the residual error of the sum of the previous regressors).

A different optimization approach is put forward in ref. [20]: the boosting algorithm minimizes a global exponential squared cost, which is the product of the costs of the individual regressors. The cost function is the only change with respect to algorithms like AdaBoost.

Before presenting our algorithm, we mention the few existing applications of boosting to time series modeling. In ref. [5] a boosting method belonging to the first family of boosting algorithms is successfully applied to the classification of phonemes. The learners employed are RNNs, and the authors are the first to notice the implications the internal memory of the RNNs has on the boosting algorithm. A similar type of boosting algorithm is used in ref. [6] for the prediction of a benchmark time series, but with MLPs as regressors. A residual-fitting boosting approach employing decision trees and projection pursuit as regressors is applied in ref. [21] to the estimation of volatility in financial time series. In all these cases, a significant amount of data was available for learning.

3 Boosting recurrent neural networks

While some of the existing boosting algorithms can, in principle, be used for any regression problem and with any family of weak regressors, such a black box approach does not necessarily perform well. Indeed, inappropriate choices may result in stopping after very few iterations and, as a consequence, in a limited improvement with respect to the first regressor developed. Or the algorithm may not stop so early but the progress in performance may be extremely slow.

The boosting algorithm employed should then comply with the restrictions imposed by the general context of application. In our case, it must be able to work well when a limited amount of data is available and accept RNNs as "weak" regressors.

Since only limited improvement can be expected from the first family of boosting algorithms, which combines just 3 regressors, we preferred to follow the generic algorithm of the second family. We had to decide what cost function to use for the weak regressors, how to update the distribution on the training set and how to combine the resulting regressors.

Preliminary experiments helped us finding appropriate answers to these questions. The experiments were performed on two different datasets, a natural one (sunspots [22]) and a synthetic one (Mackey-Glass [23]). The sunspots dataset contains the yearly number of dark spots on the sun from 1700 to 1979. It is common practice to use the data from 1700 to 1920 for the training set and the data from 1921 to 1979 for the test set. The Mackey-Glass benchmarks are generated by a nonlinear differential equation and are well-known for the evaluation of forecasting methods [23]. We considered here the data generated according to ref. [24] for a time constant of 17 and a sampling period of 6 (MG17). The first 500 values were employed for the learning set and the next 100 values for the test set.

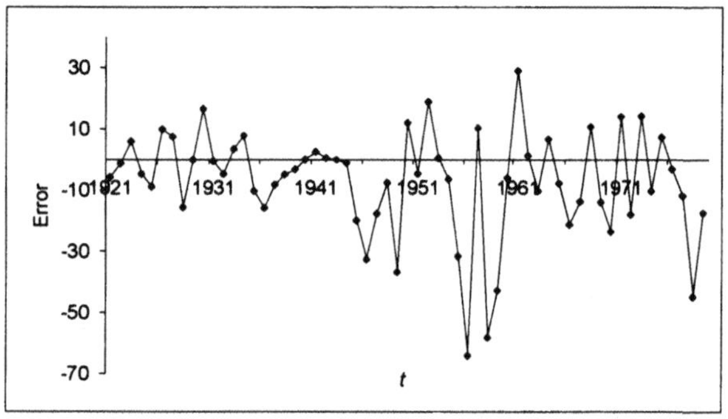

Fig. 1. Errors of a single RNN on the sunspots dataset

The preliminary experiments consisted in training on each dataset an RNN using back-propagation through time (BPTT [1]). The errors obtained at the end of training on the test sets are displayed in Fig. 1 and Fig. 2. The errors on the corresponding training sets look similarly. We notice that for the sunspots time series the big errors are few and rather isolated.

While the optimization setting in ref. [20] is attractive, we believe that the exponential squared cost favors overfitting by significantly overweighting the difficult examples. We decided to continue using the squared error for training the regressors (RNNs in our case) and to evaluate the linear, squared and saturated cost functions suggested in refs. [10,11] for updating the distribution on the training set. A saturated cost function avoids overweighting the very difficult examples and is shown in ref. [13] to be less prone to overfitting than the exponential cost of AdaBoost for classification tasks.

By employing a squared cost one can benefit from the gradient descent view of boosting applied to regression problems [17,19]. However, we believe that this residual-fitting approach cannot work well on problems similar to the sunspots dataset, for which the important residual errors are few and rather isolated.

This remark, together with the fact that RNNs have difficulties learning the time dependencies in a sequence from isolated points, also explains our choice of a method for updating and exploiting the distribution on the training set. Our updates are based on the suggestion in refs. [10,11], but we apply an affine transform to the weights before employing them (see the algorithm description below) in order to prevent the RNNs from simply ignoring the easier examples for problems similar to the sunspots dataset. Then, instead of sampling with replacement according to the updated distribution, we prefer to weight the error computed for each example at the output of the RNN with the distribution value corresponding to the example.

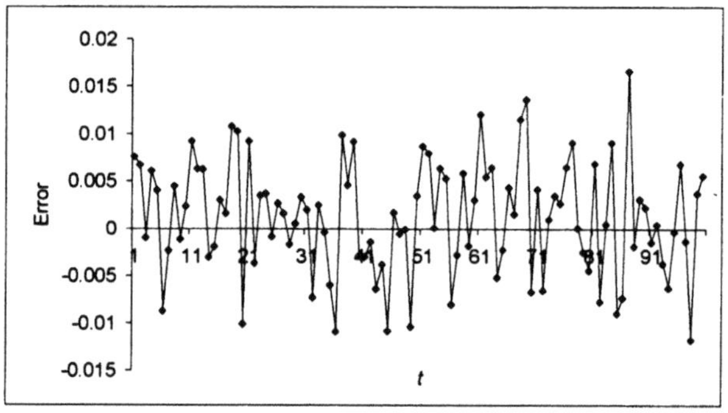

Fig. 2. Errors of a single RNN on the MG17 dataset

The choice of using all the data points from the training set for every regressor is also consistent with the requirement that the algorithm should work well with a limited amount of data.

To combine the regressors we are using the weighted median [10,11], which is less sensitive to outliers, rather than the weighted mean. The experiments presented next defend this choice. The boosting algorithm we propose can then be described as follows:

1. Initialize the weights for the examples: $D_1(i) = 1/N$, and N the number of training examples.
2. Iterate
   (a) develop an RNN h_t by using the entire training set and weighting the squared error computed for example i by $D_t(i)$, the weight of example i for the iteration t;
   (b) update the weights of the examples:
   (i) compute $L_t(i)$ for every i according to one of the three alternatives:
   - linear: $L_t(i) = \left|y_i^{(p)}(x_i) - y_i\right| / S$,
   - squared: $L_t(i) = \left|y_i^{(p)}(x_i) - y_i\right|^2 / S^2$,
   - saturated: $L_t(i) = 1 - \exp\left(-\left|y_i^{(p)}(x_i) - y_i\right| / S\right)$,
   where $S = \sup\left|y_i^{(p)}(x_i) - y_i\right|$, $i = 1, \cdots, N$;

(ii) compute $\varepsilon_t = \sum_{i=1}^{N} D_t(i) L_t(i)$ and $\alpha_t \in \Re$,

$\alpha_t = (1-\varepsilon_t)/\varepsilon_t$;

(iii) the weights of the examples become

$D_{t+1}(i) = \dfrac{1 + k \cdot p_{t+1}(i)}{N+k}$, with

$p_{t+1}(i) = \dfrac{D_t(i)\alpha_t^{(L_t(i)-1)}}{Z_t}$, Z_t being a normalizing constant;

until $\varepsilon_t < 0.5$; let T denote the last value of t;

3. Combine the weak learners: for input x, output the weighted median:

$$h_f(x) = \inf\left\{ y \in Y : \sum_{t:h_t(x)\le y} \log\alpha_t \ge (1/2)\sum_{t=1}^{T} \log\alpha_t \right\}.$$

We evaluated the linear, squared and saturated cost functions, for several values of the parameter k (Figs. 3 and 4). The results of the weighted median are displayed, together with the results of the weighted mean $h_f(x) = \sum_{t=1}^{T}[\log\alpha_t \cdot h_t(x)]$, in Tables 1 and 2. The normalized mean squared error (NMSE) is the ratio between the MSE and the variance of the time series.

These values should also be compared to the mean results (over 20 runs, with connection weights randomly initialized in $[-0.3, 0.3]$) obtained by RNNs trained without boosting. The mean NMSE is 0.24 on the sunspots test set and 8.6E-3 on the MG17 test set.

We can see that boosting with the weighted median provides better results than both boosting with the weighted mean or not using boosting at all (for appropriate values of k in the case of sunspots).

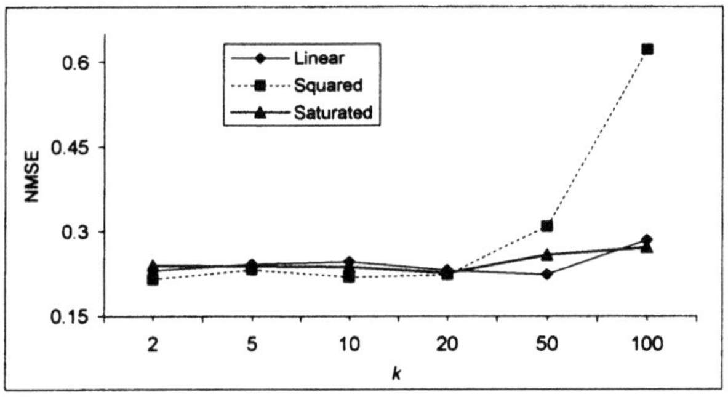

Fig. 3. NMSE of the weighted median for several values of k on the test set of sunspots

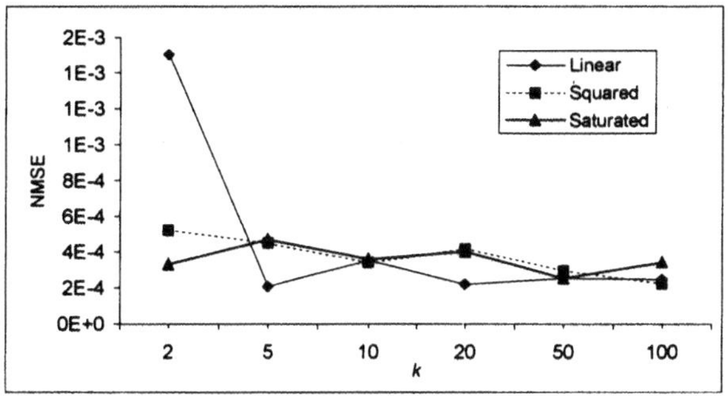

Fig. 4. NMSE of the weighted median for several values of k on the test set of MG17

Table 1. NMSE of the weighted mean and of the weighted median for several values of k on the test set of sunspots

k		2	20	50	100
Linear	Mean	0.23	0.23	0.22	0.54
	Median	0.23	0.23	0.22	0.28
Squared	Mean	0.23	0.23	0.60	0.73
	Median	0.21	0.22	0.31	0.62
Saturated	Mean	0.24	0.23	0.50	0.55
	Median	0.24	0.23	0.26	0.27

Table 2. NMSE of the weighted mean and of the weighted median for several values of k on the test set of MG17

k		2	20	50	100
Linear	Mean	6.3E-3	2.5E-4	2,7E-3	2.1E-3
	Median	1.5E-3	2.2E-4	2.6E-4	2.5E-4
Squared	Mean	9.6E-3	5.3E-3	3.6E-3	8.5E-4
	Median	5.2E-4	4.2E-4	3.0E-4	2.2E-4
Saturated	Mean	2.1E-3	5.2E-3	2.1E-3	1.9E-3
	Median	3.3E-4	4.0E-4	2.5E-4	3.4E-4

4 Experiments

The boosting algorithm described was evaluated on the sunspots and MG17 datasets. As for the preliminary experiments, we employed RNNs having a single input neuron, a single linear output neuron, a bias unit and a fully recurrent hidden layer composed of neurons with tanh activation functions. For the sunspots dataset we tested RNNs having 2 neurons in the hidden layer and for the Mackey-Glass dataset 6 neurons.

To better understand why performance depends on k and why the behavior on the two datasets is different, we must note that when $k = 0$, $D_{t+1}(i) = 1/N$, and when $k >> 0$, $D_{t+1}(i) \approx p_{t+1}(i)$. The values of $p_{t+1}(i)$ are those employed in [10,11] for updating the distribution and can be very close to 0 for the easier examples. If $k >> 0$ the RNNs are in reality trained on the examples having big errors (this fact is reinforced when the square law is used), which are few and rather isolated for the sunspots dataset. The RNNs learn then poorly on this dataset. For low k the examples have almost equal weights and boosting brings little improvement. With existing evidence, a sensible choice is to consider k in the interval $[10, 30]$, but we are currently trying to identify a simple method for adjusting k according to the distribution of the errors of the first learner.

The experiments we performed up to now do not allow us to distinguish between the three laws explored for updating the weights of the examples, as they produce rather similar results when k is in $[10, 30]$. The saturated law can nevertheless be privileged because it is expected to be less prone to overfitting.

5 Conclusion and future work

We adapted boosting to the problem of learning time-dependencies in sequential data for predicting future values, using recurrent neural networks as "weak" regressors. The experimental results we obtained show that the boosting algorithm actually improves upon the performance of the weak regressors alone and that the weighted median is a better choice for combining the regressors than the weighted mean.

Further work on this algorithm includes an evaluation on multi-step-ahead prediction and comparisons with competitive tools such as least-squares support vector machines. We are also searching for more rigorous accounts for the various choices incorporated in this boosting algorithm.

References

[1] Rumelhart D.E., Hinton G.E., Williams R.J. (1986) Learning Internal Representations by Error Propagation. In Rumelhart, D. E., McClelland, J. (eds) Parallel Distributed Processing: Explorations in the Microstructure of Cognition. MIT Press, Cambridge, pp. 318-362.

[2] Seidl D.R., Lorenz R.D. (1991) A Structure by which a Recurrent Neural Network Can Approximate a Nonlinear Dynamic System. Int. Joint Conference on Neural Networks, Seattle, USA, pp. 709-714.

[3] Santini S., Del Bimbo A. (1995) Recurrent Neural Networks Can Be Trained to Be Maximum a Posteriori Probability Classifiers. Neural Networks 8(1): 25-29.

[4] Schapire, R.E. (1990) The strength of weak learnability. Machine Learning 5: 197-227.

[5] Cook, G.D., Robinson, A.J. (1996) Boosting the Performance of Connectionist Large Vocabulary Speech Recognition. In International Conf. in Spoken Language Processing, pp. 1305-1308. Philadelphia, 1996.

[6] Avnimelech R., Intrator N. (1999) Boosting Regression Estimators. Neural Computation 11: 491-513.

[7] Freund, Y. (1990) Boosting a weak learning algorithm by majority. In third annual workshop on Computational Learning Theory. pp. 202-216.

[8] Freund Y., Schapire R.E. (1996) Experiments with a New Boosting Algorithm. In Proceedings of the Thirteenth International Conference on Machine Learning, pp. 148-156.

[9] Freund, Y., Schapire, R.E. (1997) A decision-theoretic generalization of on-line learning and an application to boosting. Journal of Computer and System Sciences 55: 119-139.

[10] Drucker H. (1997) Improving Regressors using Boosting Techniques. Fourteenth International Conference on Machine Learning, pp. 107-115.

[11] Drucker H. (1999) Boosting Using Neural Nets. In Sharkey, A. (ed.) Combining Artificial Neural Nets: Ensemble and Modular Learning. Springer, pp. 51-77.

[12] Breiman, L. (1997) Prediction games and arcing algorithms. Technical Report 504. Statistics Dept., University of California, Berkeley, 30 p.

[13] Mason, L., Baxter, J., Bartlett, P.L., Frean, M. (2000) Functional gradient techniques for combining hypotheses. In Smola, A.J., Bartlett, P.L., Schölkopf, B. and Schuurmans, D. (eds) Advances in Large Margin Classifiers. MIT Press, Cambridge, pp. 221-247.

[14] Karakoulas, G., Shawe-Taylor, J. (2000) Towards a strategy for boosting regressors. In Smola, A.J., Bartlett, P.L., Schölkopf, B. and Schuurmans, D. (eds) Advances in Large Margin Classifiers. MIT Press, Cambridge, pp. 247-258.

[15] Ridgeway G., Madigan D., Richardson T. (1999) Boosting Methodology for Regression Problems. Artificial Intelligence and Statistics, pp. 152-161.

[16] Duffy N., Helmbold D. (2002) Boosting Methods for Regression. Machine Learning 47: 153-200.

[17] Friedman, J. H. (2000) Greedy Function Approximation: a Gradient Boosting Machine. Technical Report, Dept. of Statistics, Stanford University, 36 p.

[18] Rätsch, G., Warmuth, M., Mika, S., Onoda, T., Lemm, S., Muller, K.-R. (2000) Barrier boosting. In Proceedings COLT. Morgan Kaufmann, pp. 170-179.

[19] Bühlmann, P., Yu, B. (2002) Boosting with L2-loss: regression and classification. Research Report, Seminar für Statistik, ETH Zürich, June 2002, 32 p.

[20] Zemel, R. S., Pitassi, T. (2001) A gradient-based boosting algorithm for regression problems. In Leen, T., Dietterich, T. G., Tresp, V. (eds) Advances in Neural Information Processing Systems 13. MIT Press, Cambridge, pp 696- 702.

[21] Audrino, F., Bühlmann, P. (2002) Volatility estimation with functional gradient descent for very high-dimensional financial time series, Research Report, Seminar für Statistik, ETH Zürich, June 2002, 23 p.

[22] Akaike H. (1978) On the Likelihood of Time Series Model. The Statistician 27: 217-235.

[23] Casdagli M. (1989) Nonlinear Prediction of Chaotic Time Series. Physica 35D: 335-356.

[24] Back A., Wan E.A., Lawrence S., Tsoi A.C. (1994) A Unifying View of some Training Algorithms for Multilayer Perceptrons with FIR Filter Synapses. Neural Networks for Signal Processing IV, Ermioni, Greece, pp. 146-154.

Bounds on Rates of Approximation by Neural Networks in $\mathcal{L}_p$-spaces

Terezie Šidlofová[1]

[1]Institute of Computer Science, Academy of Sciences of the Czech Republic, Pod vodárenskou věží 2, 182 07 Prague 8, Czech Republic, E-mail: terka@cs.cas.cz.

Abstract

We derive upper bounds on rates of convergence of neural network approximation in $\mathcal{L}^p$-spaces. Our bounds are based on a version of Maurey-Jones-Barron Theorem for $\mathcal{L}^p$-spaces. They are established in terms of $\mathcal{L}_1$-norm of a weight function in a neural network with continuum of hidden units representing the function to be approximated.

1 Introduction

Neural network approximation of functions from $\mathbb{R}^d$ to $\mathbb{R}^j$ has been widely studied in recent years. The existence of an arbitrarily close approximation of a continuous or $\mathcal{L}^p$ function by one-hidden-layer network with quite general units has been proven (see, e.g., Leshno et al., 1993, Park, Sandberg, 1993, [10], [13].)

These are very nice results but insufficient. We know that the approximation error goes to zero, but how fast? Linear approximation schemes exhibit the curse of dimensionality, i.e. that the number of elementary functions necessary for a given accuracy grows exponentially with the input dimension [14]. The flexibility of the nonlinear schemes in choosing the elementary approximating functions, thus improving the approximation without necessarily increasing the number of the elementary functions, hints it might be possible to overcome the curse of dimensionality.

Here we have to mention Maurey, Jones and Barron (see [2]), who coped with this problem by proving the rates of approximation of some nonlinear schemes to be independent of the input dimension (of the order of $O(1/\sqrt{n})$) for functions satisfying certain conditions on their Fourier transform. These results can be used also for neural network approximation.

In this paper we show a broad range of functions that don't exhibit the curse of dimensionality when approximated by neural networks.

2 Preliminaries

Let $(X, \|.\|)$ be a normed linear space. For Lebesgue spaces we use the notation $(\mathcal{L}_p(X), \lambda_k)$, $1 < p < \infty$, $X \subseteq \mathbb{R}^k$ where for $f \in \mathcal{L}_p(X)$ we have $\|f\|_p = \int_X |f(x)|^p \mathrm{d}\lambda_k(x)$.We denote by $B_r(\|.\|)$ the ball of radius r in norm $\|.\|$ i.e. $B_r(\|.\|) = \{x \in X; \|x\| \leq r\}$. For a function $f : X \to \mathbb{R}$ the support of f is defined by $\mathrm{supp}(f) := \mathrm{cl}_{\|.\|}\{x \in X; f(x) \neq 0\}$. If $\mathrm{supp}(f)$ is compact we call function f to be compactly supported.

For $\mathcal{G}$ a subset of $(X, \|.\|)$ and $c \in \mathbb{R}^+$ we define

$$\mathcal{G}(c) := \{wg; g \in \mathcal{G}, w \in \mathbb{R} \,\&\, |w| \leq c\}.$$

We will denote the $\|.\|$-closure of $\mathcal{G}$ by $\mathrm{cl}_{\|.\|}\,\mathcal{G}$. The linear combination of elements of $\mathcal{G}$ will be denoted by $\mathrm{span}\,\mathcal{G}$ and linear combination of at most n elements by $\mathrm{span}_n\,\mathcal{G} := \{\sum_{i=1}^n w_i g_i; w_i \in \mathbb{R}, g_i \in \mathcal{G}\}$. Convex hull of $\mathcal{G}$ shall be denoted by $\mathrm{conv}\,\mathcal{G}$ and analogously we define convex hull of at most n elements of $\mathcal{G}$: $\mathrm{conv}_n\,\mathcal{G} := \{\sum_{i=1}^n v_i g_i;\ v_i \in [0,1], \sum_{i=1}^n v_i = 1, g_i \in \mathcal{G}\}$. By $\|f - \mathcal{G}\| := inf_{g\in\mathcal{G}}\|f-g\|$ we denote the distance of f from $\mathcal{G}$. We define the notion of $\mathcal{G}$-variation as the Minkovski functional of the set $\mathrm{cl}_{\|.\|}\,\mathcal{G}(1)$:

$$\|f\|_{\mathcal{G}} := \inf\{c \in \mathbb{R}^+; f \in \mathrm{cl}_{\|.\|}\,\mathrm{conv}\,\mathcal{G}(c)\}.$$

Neural networks consist of interconnected computational units with activation functions depending on parameters and input variables: $\phi(x, a) : \mathbb{R}^d \times \mathbb{R}^k \to \mathbb{R}$, where a are parameters and x input. One hidden layer network with n ϕ-type units computes a function of d variables of the form:

$$f(x) = \sum_{i=1}^{n} w_i \phi(x, a_i),$$

where $w_i \in \mathbb{R}$, $a_i \in \mathbb{R}^k$ and $n \in \mathbb{N}^+$.

We extend this notion to continuum of hidden units obtaining

$$f(x) = \int_A w(a)\phi(x, a)\mathrm{d}a,$$

where $x \in (H \subseteq \mathbb{R}^d)$, $a \in (A \subseteq \mathbb{R}^k)$.

We would like to remind the reader of the Luzin's Theorem, that characterizes approximation of functions by continuous ones (for the proof see for example [12]).

Theorem 2.1 (Luzin) *Let (P, μ) be a locally compact space, where μ is complete Radon measure (for example the Lebesgue measure on $\mathbb{R}^n$) and f be μ-almost everywhere finite function on P. Then the following conditions are equivalent:*

(i) *f is μ-measurable;*

(ii) *for any $\varepsilon > 0$ and any compact set $K \subset P$ there is an open set E so that $\mu E < \varepsilon$ and $f\big|_{K\setminus E}$ is continuous;*

(iii) *for any $\varepsilon > 0$ and any compact set $K \subset P$ there exists a continuous function $\widetilde{f}$ on P such that*

$$\mu\{x \in K : f(x) \neq \widetilde{f}(x)\} < \varepsilon;$$

(iv) *for every compact set $K \subset P$ there exists a sequence $\{\widetilde{f}_n\}$ of continuous functions on P such that*

$$\widetilde{f}_n \to f \ \ \mu - \textit{almost everywhere on } K.$$

An easy consequence of the Luzin's Theorem for P a countable union of compact sets (any $\mathbb{R}^n$) gives a more useful form of statement (iii):

(iii') *for any $\varepsilon > 0$ there exists a continuous function $\widetilde{f}$ on P and an open set E such that $\mu E < \varepsilon$ and*

$$\widetilde{f} = f \textit{ on } P \setminus E.$$

We have one more important information about the function $\widetilde{f}$ that was derived from f: It follows from the proof of Luzin's Theorem that $\mathcal{L}^\infty$-norm of the function is conserved (see [12]).

3 Derivation of the Rates of Approximation

It is useful to consider neural network approximation in the form of its limit as an integral, i.e. instead of the finite sum $f(x) = \sum_{i=1}^n w_i\phi(x, a_i)$, to have the integral of the form $f(x) = \int_A w(a)\phi(x, a)\mathrm{d}a$, $f : H \to \mathbb{R}$, where A is a compact subset of $\mathbb{R}^k$, H a compact subset of $\mathbb{R}^d$ (compactness is reasonable to demand in real case). We use the extension to the Maurey-Jones-Barron Theorem by Darken et al. [3], reformulated in [9] in terms of $\mathcal{G}$-variation, that estimates rates of approximation by functions from $\mathrm{span}_n\, \mathcal{G}$, where $\mathcal{G}$ is a bounded subset of Lebesgue $\mathcal{L}_p$-space, $1 < p < \infty$. These estimates do not exhibit the curse of dimensionality.

Theorem 3.1 (Darken, Donahue, Gurvits, Sontag) *Let X be a finite measure space. Let $\mathcal{G}$ be a bounded subset of $\mathcal{L}_p(X)$, $1 < p < \infty$ and $s_\mathcal{G} = \sup_{g\in\mathcal{G}} \|g\|_p$. Then for every $f \in \mathcal{L}_p(X)$ and $n \in \mathbb{N}^+$*

$$\|f - \mathrm{span}_n\, \mathcal{G}\|_p \leq \frac{2^{1+1/\bar{p}} s_\mathcal{G} \|f\|_\mathcal{G}}{n^{1/\bar{q}}},$$

where $q = p/(p-1)$, $\bar{p} = \min(p, q)$ and $\bar{q} = \max(p, q)$.

Note that for $p = 2$ we get the original Maurey-Jones-Barron rates of approximation of the order of $O(1/\sqrt{n})$.

Observing the results of Theorem 3.1 we see that reasonable rates of approximation are obtainable for functions with bounded $\|f\|_\mathcal{G}$ (independently of n).

We take $\mathcal{G}$ as the set of functions computable by network units. The linear combination of n such functions can be computed by an n-hidden-unit network with one linear output unit.

Now we need to find conditions that would guarantee the desired function f to be in the convex closure $\mathrm{cl}_{\|.\|}\, \mathrm{conv}\, \mathcal{G}$ (thus the norm $\|f\|_\mathcal{G}$ be finite) to apply Theorem 3.1 for neural networks. The function f of the form $f(x) = \int_A w(a)\phi(x, a)\mathrm{d}a$ has been proven to be in the convex closure of $\mathcal{G}$ for ϕ, w continuous functions, see Kůrková et al. [6].

Theorem 3.2 (Kůrková, Kainen, Kreinowich) *Let d, k be any positive integers, H be a compact subset of $\mathbb{R}^d$ and let $f \in C(H)$ be any function that can be represented as $f(x) = \int_A w(a)\phi(x, a)\mathrm{d}a$, where $A \subseteq \mathbb{R}^k$, $w \in C(A)$ is compactly supported and $\phi \in C(\mathbb{R}^d \times A)$. Denote $\mathcal{G} = \{\phi(-, a), a \in A\}$, $M = \{a \in A; \exists x \in H :: w(a)\phi(x, a) \neq 0\}$. Then*

$$\|f\|_\mathcal{G} \leq \int_M |w(a)|\mathrm{d}a.$$

Now we present extension of this Theorem to $\mathcal{L}_p$-spaces. We make use of the Luzin's Theorem 2.1 to prove $\|f\|_\mathcal{G}$ to be finite for f computed by one-hidden-layer neural network with $\mathcal{L}^\infty$ activation functions.

Theorem 3.3 *Let k, d be positive integers, A a compact subset of $\mathbb{R}^k$ and H a compact subset of $\mathbb{R}^d$. Let $w \in \mathcal{L}_p(A, \lambda_k)$ and $\phi \in \mathcal{L}_p(A \times H, \lambda_{k+d})$ for some $1 < p < \infty$ such that there exists b so that $|w| \leq b$ λ_k-almost everywhere on A and $|\phi| \leq b$ λ_{k+d}-almost everywhere on $A \times H$. Let $f(x) := \int_A w(a)\phi(x, a)\mathrm{d}a$ and $\mathcal{G} = \{\phi(-, a), a \in A\}$. Then*

$$\|f\|_\mathcal{G} \leq \|w\|_1.$$

Proof:

The statement can be equivalently rewritten as follows:

$$f(x) \in \mathrm{cl}_p\, \mathrm{conv}\{c_i\phi(x, a_i), a_i \in A, |c_i| \leq \|w\|_1\}.$$

It suffices to show:

$$(\forall \varepsilon > 0)(\exists g \in \operatorname{conv}\{c_i\phi(x,a_i), a \in A, |c_i| \leq \|w\|_1\})$$

such that

$$\|f-g\|_p < \varepsilon,$$

where

$$f(x) := \int_A w(a)\phi(x,a)\mathrm{d}a.$$

Thus we are searching for a g of the form $g = \sum_{i=1}^m c_i\phi(x,a_i)$ that is within ε from f and $\sum_{i=1}^m |c_i| \leq \|w\|_1$.

We will make use of the Luzin's Theorem 2.1 (iii') to derive useful facts from 3.2: For all $\varepsilon_n > 0$ there exist continuous functions $\widetilde{w}_n$ and $\widetilde{\phi}_n$ on A and $A \times H$ respectively and sets E_n and F_n, such that

$$\widetilde{w}_n = w \text{ on } A \setminus E_n$$

and

$$\widetilde{\phi}_n = \phi \text{ on } (A \times H) \setminus F_n,$$

where $\lambda_k(E_n) < \varepsilon_n$ and $\lambda_{k+d}(F_n) < \varepsilon_n$. We know also $|\widetilde{w}_n| \leq b$ λ_k-almost everywhere on A and $|\widetilde{\phi}_n| \leq b$ λ_{k+d}-almost everywhere on $A \times H$.

We define an open set

$$U_n \supseteq \left\{\{a \in A; \lambda_d\{(a,-) \in F_n\} > \sqrt{\varepsilon_n}\} \cup E_n\right\}$$

and observe that $\lambda_k(U_n) < \sqrt{\varepsilon_n} + \varepsilon_n$ (use the Fubini's Theorem for the first part). Thus we have $\lambda_k(U_n) < 2\sqrt{\varepsilon_n}$ for $\varepsilon_n < 1$. Now we define

$$f_n(x) := \int_{A\setminus U_n} \widetilde{\phi}_n(x,a)\widetilde{w}_n(a)\mathrm{d}a.$$

Here we split the further proof (the formula $\|f-g\|_p$ into three parts:

$$\|f-g\|_p \leq \overbrace{\|f-f_n\|_p}^{\mathbf{A}} + \overbrace{\|f_n - \sum_{i=1}^{m_n} c_i\widetilde{\phi}_n(-,a_i)\|_p}^{\mathbf{B}} +$$

$$+ \overbrace{\|\sum_{i=1}^{m_n} c_i\widetilde{\phi}_n(-,a_i) - \sum_{i=1}^{m_n} c_i\phi(-,a_i)\|_p}^{\mathbf{C}}$$

and prove all of them to be less or equal to $\varepsilon/3$, where ε was fixed at the beginning:

(A) Without loss of generality we put $p = 1$ for better transparency.

$$\|f-f_n\|_p = \overbrace{\int_H\int_A |w(a)\phi(x,a) - \widetilde{w}_n(a)\widetilde{\phi}_n(x,a)|\mathrm{d}a\mathrm{d}x}^{(1)} +$$

$$+ \overbrace{\int_H\int_{U_n} |\widetilde{w}_n(a)\widetilde{\phi}_n(x,a)|\mathrm{d}a\mathrm{d}x}^{(2)}$$

We deal with both the parts separately (and use Fubini's Theorem):

$$(1) = \int_{\left(H\times(A\setminus E_n)\right)\setminus F_n} \underbrace{|w(a)\phi(x,a) - \widetilde{w}_n(a)\widetilde{\phi}_n(x,a)|}_{=0}\mathrm{d}a\mathrm{d}x +$$

$$+ \int_{F_n} \underbrace{|w(a)\phi(x,a) - \widetilde{w}_n(a)\widetilde{\phi}_n(x,a)|}_{\leq 2b^2 \text{ almost everywhere}}\mathrm{d}a\mathrm{d}x +$$

$$+ \int_{E_n}\int_H \underbrace{|w(a)\phi(x,a) - \widetilde{w}_n(a)\widetilde{\phi}_n(x,a)|}_{\leq 2b^2 \text{ a.e.}}\mathrm{d}a\mathrm{d}x \leq$$

$$\leq 0 + \varepsilon_n 2b^2 + \varepsilon_n\lambda_d(H)2b^2 = \varepsilon_n 2b^2(1+\lambda_d(H))$$

The second part is easier:

$$(2) = \int_{U_n}\int_H \underbrace{|\widetilde{w}_n(a)\widetilde{\phi}_n(x,a)|}_{\leq b^2 \text{ a.e.}}\mathrm{d}a\mathrm{d}x \leq 2\sqrt{\varepsilon_n}\lambda_d(H)b^2.$$

Altogether we get:

$$\|f-f_n\| \leq 2\lambda_d(H)b^2(\varepsilon_n + \sqrt{\varepsilon_n}) + 2\varepsilon_n b^2.$$

Now we can easily find n such that $\|f-f_n\| \leq \varepsilon/3$. We fix this n and continue with the second part:

(B) We have defined f_n as an integral over a compact set $A \setminus U_n$ of continuous functions $\widetilde{w}_n(a)$ and $\widetilde{\phi}_n(x,a)$.

$$f_n(x) := \int_{A\setminus U_n} \widetilde{\phi}_n(x,a)\widetilde{w}_n(a)\mathrm{d}a.$$

Thus using Theorem 3.2 we know:

$$\forall \xi > 0\, \exists m_n :: \|f_n - \sum_{i=1}^{m_n} c_i\widetilde{\phi}_n(-,a_i)\|_C < \xi$$

We find such m_n so that

$$\|f_n - \sum_{i=1}^{m_n} c_i\widetilde{\phi}_n(-,a_i)\|_C < \varepsilon/(3\lambda_d(H)).$$

Thus we get

$$\|f_n - \sum_{i=1}^{m_n} c_i\widetilde{\phi}_n(-,a_i)\|_p =$$

$$= \int_H \underbrace{|f_n(x) - \sum_{i=1}^{m_n} c_i\widetilde{\phi}_n(x,a_i)|}_{<\varepsilon/(3\lambda_d(H))}\mathrm{d}x \leq \frac{\varepsilon}{3}.$$

We proceed to the third part.

(C) We have to notice that by $F_n|_H$ we mean projection of F_n onto H.

$$\|\sum_{i=1}^{m_n} c_i\tilde{\phi}_n(-,a_i) - \sum_{i=1}^{m_n} c_i\phi(-,a_i)\|_p =$$

$$= \int_H |\sum_{i=1}^{m_n} c_i\tilde{\phi}_n(x,a_i) - \sum_{i=1}^{m_n} c_i\phi(x,a_i)|\mathrm{d}x \le$$

$$\underbrace{\int_{H\setminus(F_n|_H)} |\sum_{i=1}^{m_n} c_i\tilde{\phi}_n(x,a_i) - \sum_{i=1}^{m_n} c_i\phi(x,a_i)|\,\mathrm{d}x}_{=0}+$$

$$+\int_{F_n|_H} |\sum_{i=1}^{m_n} c_i\tilde{\phi}_n(x,a_i) - \sum_{i=1}^{m_n} c_i\phi(x,a_i)|\mathrm{d}x \le$$

$$\le 0 + \int_{F_n|_H} |\sum_{i=1}^{m_n} c_i(\tilde{\phi}_n - \phi)(x,a_i)|\mathrm{d}x \le$$

$$\le \sum_{i=1}^{m_n} |c_i| \int_{F_n|_H} \underbrace{|(\tilde{\phi}_n - \phi)(x,a_i)|}_{\le 2b \text{ a.e.}}\,\mathrm{d}x = *$$

As from $a_i \notin U_n$ we obtain the constraint on the measure of the set over which we are integrating:

$$\lambda_d(F_n|_H) \le \sqrt{\varepsilon_n}$$

for all a_i. We conclude

$$* \le 2b\sqrt{\varepsilon_n}\sum_{i=1}^{m_n}|c_i| \le 2b\sqrt{\varepsilon_n}\|w\|_1,$$

because

$$\sum_{i=1}^{m_n}|c_i| \le \int_{A\setminus U_n}|\tilde{w}_n(a)|\mathrm{d}a = \int_{A\setminus U_n}|w(a)|\mathrm{d}a \le \|w\|_1$$

It is easy to find an ε_n that secures all of the A, B, C parts to be $\le \varepsilon/3$ which is what we wanted.

■

Thus we have proven $\|f\|_{\mathcal{G}} \le \|w\|_1$ for f computed by one-hidden-layer neural network with $\mathcal{L}^\infty$ activation functions and together with Theorem 3.1 we derive input dimension independent rates of approximation for this approximation schema:

Theorem 3.4 *Let k, d be positive integers, A a compact subset of $\mathbb{R}^k$ and H a compact subset of $\mathbb{R}^d$. Let $w \in \mathcal{L}_p(A,\lambda_k)$ and $\phi \in \mathcal{L}_p(A\times H, \lambda_{k+d})$ for some $1 < p < \infty$. Additionally let $w \in \mathcal{L}_\infty(A)$ and $\phi \in \mathcal{L}_\infty(A\times H)$. Let $\mathcal{G}=\{\phi(-,a), a \in A\}$ be bounded and $s_{\mathcal{G}}=\sup_{\phi\in\mathcal{G}}\|\phi\|_p$. Let f be any function that can be represented as $f(x) = \int_A w(a)\phi(x,a)\mathrm{d}a$. Then*

$$\|f - \mathrm{span}_n\,\mathcal{G}\|_p \le \frac{2^{1+1/\bar{p}}s_{\mathcal{G}}\|w\|_1}{n^{1/\bar{q}}},$$

where $n \in \mathbb{N}^+$, $q = p/(p-1)$, $\bar{p} = \min(p,q)$ and $\bar{q} = \max(p,q)$.

Proof: An easy consequence of Theorems 3.1 and 3.3. ■

To apply the derived bounds to neural network approximation we would like to know when a function can be represented in integral form as $f(x) = \int_A w(a)\phi(x,a)\mathrm{d}a$. The problem has been dealt with and existence of such a representation has been derived for sufficiently smooth functions when approximated by perceptron type networks with Heaviside activation function (see [7]) or for functions whose Fourier transforms are compactly supported and sufficiently smooth (see [4]).

4 Conclusions

We have derived an upper bound on rates of approximation by neural networks with one hidden layer and $\mathcal{L}^\infty$-activation functions. Our results are applicable to functions that can be written as integrals of the form $f(x) = \int_A w(a)\phi(x,a)\mathrm{d}a$. The bounds are expressed in terms of $\mathcal{L}_1$-norm of a weight function w in a neural net with continuum of hidden units representing the function to be approximated.

Inspection of the proofs shows that our results for activation functions that are up to measure zero bounded, could be extended to general $\mathcal{L}_p$-functions on finite measure spaces.

5 Acknowledgments

The author would like to express her gratitude to Věra Kůrková for pointing to the subject of this paper and stimulating guidance.

This work was partially supported by GA ČR Grant 201/02/0428.

References

[1] Barron, A. R. (1992), Neural net approximation, *In proceedings of the 7th Yale Workshop on Adaptive and Learning Systems*, 69–72.

[2] Barron, A. R. (1993), Universal approximation bounds for superposition of a sigmoidal function, *IEEE Transactions on Information Theory*, **39**, 930–945.

[3] Darken C., Donahue M., Gurvits L., Sontag E. (1993), Rate of Approximation Results Motivated

by Robust Neural Network Learning, In *Proceedings of the 6th Annual ACM Conference on Computational Learning Theory*, Santa Cruz, CA, 303-309.

[4] Girosi F., Anzellotti G. (1993), Rates of Convergence for Radial Basis Functions and Neural Networks, *Artificial Neural Networks for Speech and Vision*, (Ed. R. Mammone), Chapman & Hall, 97-114.

[5] Hornik K. (1991), Approximation Capabilities of Multilayer Feedforward Networks, *Neural Networks*, **4**, 251-257.

[6] Kůrková V., Kainen P.C., Kreinowich V. (1997), Estimates of the Number of Hidden Units and Variation with Respect to Half-Spaces, *Neural Networks*, **10**, 1061-1068.

[7] Kůrková V., Kainen P.C., Vogt A. (2000), An Integral Formula for Heaviside Neural Networks, *Neural Network World*, **10**, 313-319.

[8] Kůrková V. (2002), Universality and Complexity of Approximation of Multivariable Functions by Feedforward Networks, *Softcomputing and Industry*, Springer, 13-24.

[9] Kůrková V. (to appear), High-dimensional approximation and optimisation by neural networks, In *Proceedings of Learning Theory and Practice*, (Ed. J. Stuykens), Belgium.

[10] Leshno M., Lin V., Pinkus A., Schocken S. (1993), Multilayer feedforward networks with a non-polynomial activation function can approximate and function, *Neural Networks*, **6**, 861-867.

[11] Lukeš J., *Zápisky z funkcionální analýzy* (2002) (In Czech), Karolinum, UK Praha.

[12] Lukeš J., Malý J. (1995), *Measure and Integral*, Matfyzpress, Praha.

[13] Park J., Sandberg I. W. (1993), Approximation and radial-basis-function networks *Neural Computation*, **5**, 305-316.

[14] Pinkus A. (1985), *n-Widths in Approximation Theory*, Springer-Verlag, Berlin-Heidelberg, Germany.

On input discretisation processes for tapped delay neural architecture

Bartłomiej Beliczynski[1]

[1] Warsaw University of Technology, Institute of Control and Industrial Electronics, Koszykowa 75, 00-662 Warsaw, Poland, email: b.beliczynski@isep.pw.edu.pl

Abstract

For tapped delay neural architecture impact of discretisation process i.e. sampling and amplitude measurement of the input continuous signals is analysed. By using relative matrix rank and indistinguishable signals concepts we derrive upper and lower bounds of sampling time and input signal frequencies.

1 Introduction

In many applications tapped delay neural architecture is used for input/output dynamics approximation. This architecture is a combination of line of delays called "tapped delays" with a feedforward neural network (one may mention an early publication [1]). The line of delays introduces a dynamical part to the approximation scheme, the neural net - a static, nonlinear one. We assume that an input to such neural architecture consists of several linearly independent continuous in time functions called here also "signals". Contemporary commonly used neural learning procedures require that input and output signals are discretised. The process of discretisation consists of sampling and amplitude representation. The last is influenced by measurement and quantisation. The quantisation error which in most cases could be made arbitarilly small, will be neglected in our considerations. Finally a set of input/output pairs is created and used for teaching. Both processes of discretisation i.e. sampling and amplitude measurement impose however several limitations to the performance of this tapped delay neural architecture.

Neural network parameters of the considered architecture are obtainable in a learning process based on preprepared sampled data - the input/output examples. The teaching set - a sufficiently large collection of representative data usually is formed into two matrices called then the input and output teaching matrices. Repeated data are useless in the set. Linearly dependent data are often of a little value. The input teaching matrix contains in its columns samples of the input continuous signals. Inappropriately chosen sampling time may be a reason of difficulties. Too large sampling time causes negligence of the system dynamics, too small linear or almost linear dependence in the data set what reduces dimension of the approximation space. If there exists Fourier transform of the input signal than the signal itself can be viewed as a collection of various frequency and amplitude components. It is demonstrated here that because sampling time and accuracy of measurement its effective frequency spectrum is limited from both sides.

This paper is organised as follows. In Section II the tapped delay architecture is described and commented. In Section III linear dependence of input signals and relative matrix rank are defined and their several properties demonstrated, then in Section IV interrelation between sampling time, relative matrix rank, accuracy of measurement and frequency bandwidth is shown. Finally conclusions are drawn.

2 Tapped delay neural architecture

Let consider a tapped delay neural architecture as shown in Fig.1.

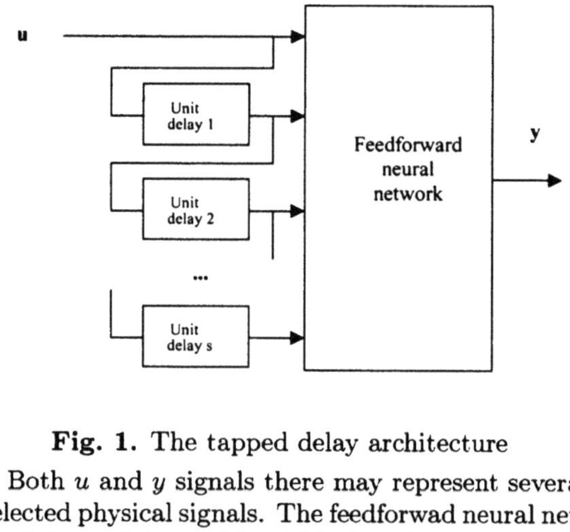

Fig. 1. The tapped delay architecture

Both u and y signals there may represent several selected physical signals. The feedforwad neural network is used for multivariable function approximation. The whole architecture could be described in the following way

$$y(k) = h(u(k), u(k-1)..., u(k-s)) \qquad (1)$$

Index k denotes kth sampling period and s is an integer number. We assume that the input and output signals are sampled with fixed sampling interval T.

The input to the system could be formed from selected measurable signals and the output, from signals to be restored. We do not exclude situation when a physical signal is included in both the input u and output y vectors.

The model represented in Fig.1 and described by (1) has number of parameters and must be tuned before being used. Delay time and number of delays s must be specified. The neural network parameters have to be selected in such way that the universal approximation network will approximate the particular function h.

Having selected u and y signals as the input and output respectively, in order to parametrise tapped delay neural model, one has to take several steps. They may be described as the following: choose sampling time for all variables, collect sufficiently large set of input/output pairs, choose number of delays, form input and output teaching matrices and teach the neural network. In this paper we will be discussing only choice of sampling time and its impact to the teaching matrix rank and useful frequency bandwidth. The output variable, the feedforward neural network architecture and learning algorithms appear to be irrelevant here.

3 Input signals and relative matrix rank

Let assume that a dynamic system to be modeled is excited by an input continuous signal denoted as $u(t)$, $t \in [0, t_{\max}]$ being collection of m measurable signals

$$u(t) = (u_1(t), ..., u_m(t)) \qquad (2)$$

$u_i(t) \in \mathcal{L}^2$, $i = 1, ..., m$.

Let construct the following matrix

$$\Gamma_c = \begin{bmatrix} <u_1, u_1> & <u_1, u_2> & \dots & <u_1, u_m> \\ <u_2, u_1> & <u_2, u_2> & \dots & <u_2, u_m> \\ \dots & \dots & \dots & \dots \\ <u_m, u_1> & <u_m, u_2> & \ddots & <u_m, u_m> \end{bmatrix} \qquad (3)$$

where $<.,.>$ denotes scalar product which for $\mathcal{L}^2$ space is usually defined as

$$<u_i, u_j> = \frac{1}{t_{\max}} \int_0^{t_{\max}} u_i(t) u_j(t) dt, \ i, j = 1, ..., m \qquad (4)$$

Matrix Γ_c property is used to define linear independence of signals. Is said that the signals $u_1(t), ..., u_m(t)$ are linearly independent if matrix (3) is nonsingular i.e.

$$rank(\Gamma_c) = m \qquad (5)$$

Specially wishful case is when Γ_c is a unit matrix. This means that all signals are of unit norms and they are orthogonal i.e.$<u_i, u_j> = 0$ when $i \neq j$ and $<u_i, u_j> = 1$ when $i = j$. Then we say that such scheme possesses the input orthonormality property.

In practice, there are two important issues associated with testing of signals linear independence. One is related to calculation of matrix Γ_c entries i.e. the scalar products of complicated signals, another to determination of rank of a matrix in a sufficiently robust way.

The scalar product (4) can be, and usually is, well approximated through samples of signals. Note that

$$<u_i, u_j> = \lim_{N \to \infty} \frac{1}{N} \sum_{k=0}^{N-1} u_i(k\frac{t_{\max}}{N}) u_j(k\frac{t_{\max}}{N}) \qquad (6)$$

Formula (6) is practically important as usually only the signal samples are available, not the signal itself.

More difficult problem is related to the input teaching matrix rank determination. When matrix (3) entries are associated with signal measurement and error prone calculations, even small error may change the calculated matrix rank. But good designer likes to be convinced that (5) is "well" fulfilled. One needs more compatible matrix rank concept and test which should take into account also finite accuracy of measurement (and calculations). Such formulation may be related to Singular Value Decomposition [2]. We start with definition of matrix rank with relation to accuracy.

Definition 1 *Rank of matrix $A \in \Re^{n,m}$ with relation to accuracy ε, denoted as $rank(A, \varepsilon)$ is*

$$rank(A, \varepsilon) = \min_{\|A-B\| \leq \varepsilon} rank(B) \qquad (7)$$

If $k < r = rank(A)$ and SVD decomposition of A is given through singular values $\sigma_1, ..., \sigma_r$ such that $\sigma_1 \geq \sigma_2, ..., \geq \sigma_r > 0$ and orthonormal vectors u_i, v_i, $i = 1, ..., r$, $A_k = \sum_{i=1}^{k} \sigma_i u_i v_i^T$ then based on [3] one can write

$$\begin{aligned} \min_{rank(B)=k} ||A - B|| &= ||A - A_k|| = \\ &= ||diag(\sigma_{k+1}, ..., \sigma_r)|| \end{aligned} \quad (8)$$

where $||.||$ denotes any unitary invariant norm, i.e. such that $||A|| = ||EA|| = ||AE||$ for any unitary E. In particular, unitary invariant norms are: σ_1 - $\mathcal{L}^2$ norm, $\sum_{i=1}^{r} \sigma_i$ - trace norm, $\sqrt{\sum_{i=1}^{r} \sigma_i^2}$– Frobenius norm.

This formulation clearly shows that if a matrix singular values beginning with $k+1$ are regarded as "small" and we require accuracy of the matrix rank determination to be $||diag(\sigma_{k+1}, ..., \sigma_r)||$ than rank of the matrix with relation to this accuracy could significantly be reduced $(r - k)$. In that sense theoretically a high rank matrix could practically be of much lower rank. It seems however that one more improvement to the problem formulation is needed. Because relative accuracy of measurement and floating point computer representation of numbers, accuracy of matrix rank should be formulated with relation to norm of the matrix. So the following definition is used.

Definition 2 *Rank of matrix $A \in \Re^{n,m}$ with relation to relative accuracy δ, $0 \leq \delta \leq 1$ is*

$$rank(A, \delta\, ||A||)$$

and will be denoted by $rank_\delta(A)$.

Proposition 3 *For any δ, $0 \leq \delta \leq 1$, $rank_\delta(A)$ is nonincreasing function of δ and $rank_0(A) = rank(A)$ and $rank_1(A) = 1$.*

If now one reformulates relative matrix rank conditions and takes unitary invariant norm simplification, then Proposition 4 could be stated as follows.

Proposition 4 *Let A be a matrix $A \in \Re^{n,m}$ with singular values $\sigma_1, ..., \sigma_r$, then $rank_\delta(A) = k$, $1 \leq k \leq r$, iff*

$$k = \max_{\frac{||diag(\sigma_i, ..., \sigma_r)||}{||diag(\sigma_1, ..., \sigma_r)||} \geq \delta} (i)$$

where $||.||$ is any unitary invariant norm.

Proposition 5 *Let Γ_c be defined as in (3) and $\mathcal{L}^2$ norm is used for relative matrix rank Definition 2, then the following statements are equivalent*

- *$rank_\delta(\Gamma_c) = m$,*
- *$cond(\Gamma_c) \leq \frac{1}{\delta}$*

The last condition in (5) may be reformulated as to be $\frac{1}{cond(\Gamma_c)} \geq \delta$, so we will be seeking of lower bound of $\frac{1}{cond(\Gamma_c)}$ starting with only two signals. The following Theorem 6 is useful.

Theorem 6 *Let $u_1, u_2 \in \mathcal{L}^2$*

$$\Gamma_c = \begin{bmatrix} <u_1, u_1> & <u_1, u_2> \\ <u_2, u_1> & <u_2, u_2> \end{bmatrix} \quad (9)$$

then condition number of matrix Γ_c is

$$cond(\Gamma_c)(s) = \frac{1 + \sqrt{1-s}}{1 - \sqrt{1-s}} \quad (10)$$

where

$$s = \alpha^2 \sin^2 \varphi, \alpha^2 = \frac{4\,||u_1||^2\,||u_2||^2}{(||u_1||^2 + ||u_2||^2)^2}, \quad (11)$$

$$\cos\varphi = \frac{<u_1, u_2>}{||u_1||\,||u_2||}, \ \varphi \in [0, \pi] \quad (12)$$

and $cond(\Gamma_c)(s)$ is a decreasing function of s, $s \in (0, \alpha^2]$, $\arg \sup_{\frac{||u_1||}{||u_2||}} \alpha^2(\frac{||u_1||}{||u_2||}) = 1$ and in such case $cond(\Gamma_c)(s) = c\tan^2\left(\frac{\varphi}{2}\right)$.

Proof: Because Γ_c is a symmetric and non-negatively defined matrix, the matrix singular values match its eigenvalues and via simple algebraic manipulations one obtains (10) where $s = \alpha^2(1 - \gamma^2)$, $\alpha^2 = \frac{4||u_1||^2||u_2||^2}{(||u_1||^2+||u_2||^2)^2}$ and $\gamma^2 = \frac{<u_1,u_2>^2}{||u_1||^2||u_2||^2}$. If now a substitution $\gamma^2 = \cos^2\varphi$ is used then (11) is obtained. Calculating $\frac{d}{ds}(cond(\Gamma_c)(s)) = -\frac{1}{2\sqrt{(1-s)}\left(1-\sqrt{(1-s)}\right)} - \frac{1}{2}\frac{1+\sqrt{(1-s)}}{\left(1-\sqrt{(1-s)}\right)^2\sqrt{(1-s)}}$ one may notice that $\frac{d}{ds}(cond(\Gamma_c)(s)) < 0$ for $s \in [0, 1)$. Also because $\alpha^2 = 1/(\frac{1}{2}(\frac{||u_1||^2}{||u_2||^2} + \frac{||u_2||^2}{||u_1||^2}))$ then its largest value is equal to 1 and is reached when $||u_1|| = ||u_2||$ and in such case $cond(\Gamma_c)(s) = c\tan^2\left(\frac{\varphi}{2}\right)$. ∎

Theorem 6 justifies again quite often used in practice input signals normalization. In such simple way maximal condition number of matrix Γ_c is achieved.

4 Sampling of input signals

Suppose that within range of time $[0, t_{max}]$ we collected $u(k)$, $k = 0, 1, \ldots$signals, sampling continuous signals u with period T and achieving the following vectors

$$U(0:N) = [U_1, \ldots, U_m] = \begin{bmatrix} u(0) \\ u(1) \\ \ldots \\ u(N) \end{bmatrix} \in \Re^{N+1,m} \quad (13)$$

where

$$U_i = \begin{bmatrix} u_i(0) \\ u_i(1) \\ \ldots \\ u_i(N) \end{bmatrix} \in \Re^{N+1,1}$$

$$u(k) = [u_1(k), \ldots, u_m(k)] \in \Re^{1,m}$$

$i = 1, \ldots, m$ and $k = 0, \ldots, N$.

Let define

$$\Gamma_d = \begin{bmatrix} <U_1, U_1> & <U_1, U_2> & \ldots & <U_1, U_m> \\ <U_2, U_1> & <U_2, U_2> & \ldots & <U_2, U_m> \\ \ldots & \ldots & \ldots & \ldots \\ <U_m, U_1> & <U_m, U_2> & \ldots & <U_m, U_m> \end{bmatrix} \quad (14)$$

where

$$<U_i, U_j> = \frac{1}{N} \sum_{k=0}^{N-1} u_i(k) u_j(k)$$

Note that

$$\Gamma_d = \frac{1}{N+1} U^T(0:N) U(0:N)$$

and $rank(\Gamma_d) = rank(U(0:N))$.

Proposition 7 $rank(\Gamma_d) \leq rank(\Gamma_c)$

Proof: *Rank of matrix Γ_d cannot be larger than m. In two cases could be smaller: either the number of samples in the observational period $[0, t_{max}]$ is less than m or sampling time has been selected in such a way that two columns become linearly dependent.* ■

For neural network teaching purpose sampled input signals are placed in U_t matrix in the following way

$$U_t = [U(s:N) \; U(s-1:N-1) \ldots U(0:N-s)] \quad (15)$$

where $s \in \{0, 1, \ldots, floor(\frac{N}{2}) - 1\}$ and $U(s:l)$, $s, l \in \{0, N\}$, $l \geq s$ is used to represent rows from s to l of the matrix $U(0:N)$.

Matrix $U_t \in \Re^{N+1-s,(s+1)m}$ containing input teaching data described in details is then the following

$$U_t = \begin{bmatrix} u(s) & u(s-1) & \ldots & u(0) \\ u(s+1) & u(s) & \ldots & u(1) \\ \ldots & \ldots & \ldots & \ldots \\ u(N) & u(N-1) & \ldots & u(N-s) \end{bmatrix} \quad (16)$$

Its relative rank characterises input data information useful for approximation. Note that every next block of columns in (16) is shifted by sampling time with relation to the previous one.

5 Bounding conditions imposed by sampling and accuracy of measurement

If an input signal $u(t)$ is such that $u(t) \in \mathcal{L}^2$ then there exists its Fourier representation $\tilde{u}(f)$ where f stands for "frequency". If a particular sampling time value T was selected for sampling signal $u(t)$ then according to Shannon's sampling theorem the signal frequency component which can be restored fulfills

$$fT \leq \frac{1}{2} \quad (17)$$

When U_t matrix is constructed to be used for teaching, then there exist also a lower frequency limitation. Let consider a signals $u(t)$ and its shifted counterpart $u(t+T)$. If T goes to zero then obviously the two items are identical. If however T is sufficiently small and the signals are measured with finite accuracy, $u(t)$ and $u(t+T)$ may be indistinguishable. We state that in Definition 8, which originally but not formally was used in [4].

Definition 8 *A signal $u(t)$, $t \in [0, t_{max}]$ is said to be indistinguishable with relation to shift in time T and relative accuracy of measurement p %> 0 iff for every $t \in [0, t_{max}]$*

$$\sup_{t \in [0, t_{max} - T]} |u(t+T) - u(t)| \leq \frac{p}{100} \sup_{t \in [0, t_{max}]} |u(t)|$$

Specially important are sinusoidal signals, because Fourier representation of any $\mathcal{L}^2$ signal could be interpreted as a collection of sinusoidal signals.

Proposition 9 *A sinusoidal signal* $\sin(2\pi ft)$ *is indistinguishable with relation to shift in time* T *and accuracy of measurement* $p\%$ *iff*

$$fT \leq \frac{1}{\pi}\arcsin(\frac{1}{2}\frac{p}{100}) \tag{18}$$

Proof: *The largest slope of the sinusoidal signal is in zero of the argument. So, let consider two sinusoidal signals defined in the range* $t \in [0, T_p]$ *shifted by time* T, *i.e.* $u(t) = (\sin(2\pi f(t+\frac{T}{2})), \sin(2\pi f(t-\frac{T}{2})))$, *where* $f = \frac{1}{T_p}$. *Directly from Definition 8 one obtains* $2\sin(\pi fT) \leq \frac{p}{100}$ *or* $fT \leq \frac{1}{\pi}\arcsin(\frac{1}{2}\frac{p}{100})$. ■

For sinusoidal signals a relation between relative accuracy, rank of Γ_c matrix and ability to distinguish the signals can easily be determined. One may expect that if a signal and its shifted counterpart are indistinguishable, matrix Γ_c formulated for those two signals is close to singularity. More precisely this is stated in the following Proposition.

Proposition 10 *If two signals* $u_1(t) = \sin(2\pi f(t+\frac{T}{2}))$ *and its shifted partner* $u_2(t) = \sin(2\pi f(t-\frac{T}{2}))$ *are indistinguishable with relation to accuracy of measurement* $p\%$ *then*

$$cond(\Gamma_c) \geq \frac{1}{\delta}, \quad where\ \delta = \frac{(\frac{1}{2}\frac{p}{100})^2}{1-(\frac{1}{2}\frac{p}{100})^2}$$

and Γ_c *is defined by (9).*

Proof: Matrix Γ_c is $\Gamma_c = \begin{bmatrix} <u_1(t),u_1(t)> & <u_1(t),u_2(t)> \\ <u_2(t),u_1(t)> & <u_2(t),u_2(t)> \end{bmatrix} = \begin{bmatrix} \frac{1}{2} & \cos^2\pi fT - \frac{1}{2} \\ \cos^2\pi fT - \frac{1}{2} & \frac{1}{2} \end{bmatrix}$ Its singular values are the following $\sin^2\pi fT, \cos^2\pi fT$. If $fT \leq 0.25$, then $||\Gamma_c|| = \cos^2\pi fT$ Finally if $fT \leq \frac{1}{\pi}arc\tan\sqrt{\delta}$ then $rank_\delta(\Gamma_c) = 1$. Combining last with (18) one obtains $\delta = \frac{(\frac{1}{2}\frac{p}{100})^2}{1-(\frac{1}{2}\frac{p}{100})^2}$. ■

Finally we may find how sampling time and accuracy of measurement limit the lower frequency of restorable signals

$$f_{\min} = \frac{1}{\pi T}\arcsin(\frac{1}{2}\frac{p}{100}) \leq \frac{1}{2\pi T}\frac{p}{100} \tag{19}$$

Taking into account (19) and (17) together we have

$$\frac{1}{2\pi f_{\min}}\frac{p}{100} < T < \frac{1}{2f_{\max}} \tag{20}$$

Formula (20) describes lower and upper bounds of the sampling time ensuring that $f_{\min}$ and $f_{\max}$ the input signal frequency componets are present in teaching data. In fact sometimes one may not find such T that conditions (20) are met. In other words the input signal frequency components bandwidth is limited as follows

$$\frac{f_{\max}}{f_{\min}} < \pi\frac{100}{p} \tag{21}$$

The formulae (20) and (21) were originally presented in [5] although in a different framework.

6 Conclusions

In this paper discretisation process of the input signals for tapped delay neural architecture was analysed. We used relative matrix rank and indistinguishable signals concepts. Assuming that input signals are linearly independent it was demonstrated that normalisation of the inputs is justified and useful. In our case in this way the relative input teaching matrix rank could have been improved. Sampling process and error prone signal amplitude measurement effected also the input teaching matrix relative rank. As consequences of that the limitations to sampling time and frequency ranges of the input signals were imposed.

Acknowledgment

This work was partially supported by Warsaw University of Technology Grant and KBN project No. 4T10A 03923.

References

[1] K S Narenda and K Parthasarathy. Identification and control of dynamical systems using neural networks. *IEEE Trans. on Neural Networks*, 1(1):4–27, 1990.

[2] G.H.Golub and C.F. Van Loan. *Matrix computation.* North Oxford Academic, 1983.

[3] Z Nehari. On bounded bilinear forms. *Annals of Mathematics*, 15(1):153–162, 1957.

[4] B Beliczynski and L Grzesiak. Induction motor speed estimation: Limitations of certain neural schemes based on time instances of periodic signals. In *Proc. IECON'01*, pages 653–657. IEEE Service Center, 2001.

[5] B Beliczynski and L Grzesiak. Induction motor speed estimation: Neural versus phenomenological model approach. *Neurocomputing*, 43:17–36, 2002.

An extended architecture of recurrent neural networks that latches input information

Branko Šter[1], Andrej Dobnikar[1]

[1]University of Ljubljana, Faculty of Computer and Information Science, Tržaška 25, Ljubljana, Slovenia, {branko.ster, andrej.dobnikar}@fri.uni-lj.si

Abstract

An attempt to facilitate learning of recurrent neural networks is made. The proposed extended recurrent neural network adaptively latches unimportant input vectors and thus effectively reduces the size of the training set. A register of latches is added as the input layer of the network. The latch is implemented with a multiplexer 2/1 whose output is differentiable with respect to all of its inputs, thus enabling the derivatives to be propagated through the network. The relevance of input vectors is learned together with the weights of the network using a gradient-based algorithm.

1 Introduction

Recurrent neural networks (RNNs) can be trained to model dynamical systems. Discrete-time RNNs were shown to be capable of behaving like finite state automata [1], thus providing a model of finite state computation by means of continuous dynamical systems. Recurrent neural networks are able to store and process context or state information, which is required in temporal processing tasks, such as sequence recognition and time-series prediction.

RNNs are harder to train than feedforward networks. A particularly difficult problem is learning long-term dependencies, since the network must be able to store relevant information over long periods of time. It was shown in [2] that in learning by gradient descent, propagation of derivatives through the network decays in time. Alternative learning paradigms which do not face this problem are non-gradient methods and evolutionary algorithms.

The problem of learning long-term dependencies has been attacked in different ways. In [3], a hierarchy of neural networks to recursively decompose sequences was proposed. When a lower-level network cannot satisfactorily predict a subsequent output, a higher-level network becomes responsible. The system detects causal dependencies in the sequence.

The algorithm called Long Short-Term Memory [4] was designed to overcome these problems by enforcing constant error flow. However, it was found in [5] that the original method was not applicable to continual input streams (on-line sequence processing), since the state may grow indefinitely. A remedy called forget gates that learn to reset the state in appropriate situations was proposed [5].

The idea in this paper is to exclude irrelevant input vectors from RNN processing and thereby to facilitate learning of the RNN. The relevance of input vectors is established through the learning process. We introduce a modification of the 2-layer recurrent neural network, where input vectors to the network are selectively latched in order to suppress irrelevant ones, and the error signals at the RNN's outputs are weighted correspondingly. When an input vector is considered unimportant, the output error of the network is considered to be unimportant, too. When this is not the case, the RNN is not expected to latch the inputs.

2 Latched recurrent neural network

First the architecture of the network is described, and then the steepest-descent learning rule is derived.

2.1 Architecture

The latched recurrent neural network (LRNN) has two feedforward processing layers and global feedback connections, as shown in Fig. 1. In addition, a register of latches in the input layer suppresses assumably irrelevant input vectors by latching previous data and context information.

The latch (Fig. 2a) is controlled through the enable input, propagating input data while active or "high", and remembering the previous data while "low". This is in fact a D-type flip-flop with clock enable. A flip-flop is an edge-triggered (synchronous) device and the RNN also operates in a synchronous manner, i.e. we can imagine a "clock" signal driving the network.

To be able to propagate derivatives through time, we have made the multiplexer's output differentiable with respect to its inputs. We applied a simple neural net with two sigmoid neurons with fixed weights and a summator,

[1]The work was partially supported by grant Z2-3402 by the Ministry of Education, Science and Sport of the Republic of Slovenia.

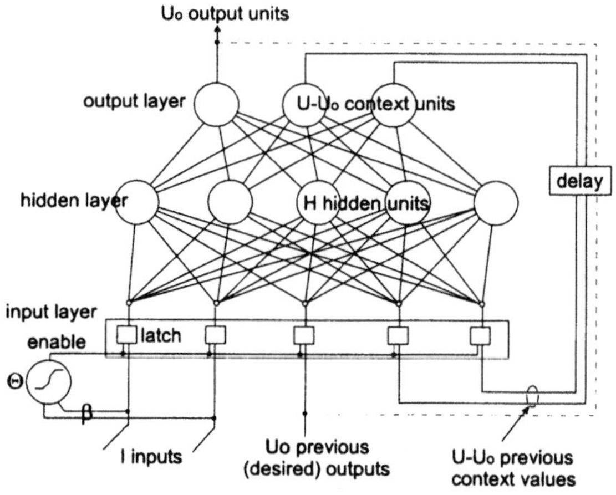

Fig. 1. Latched recurrent neural network. The dashed lines present feedback necessary only in multi-step prediction. We used the teacher-forcing technique, which applies the correct previous outputs.

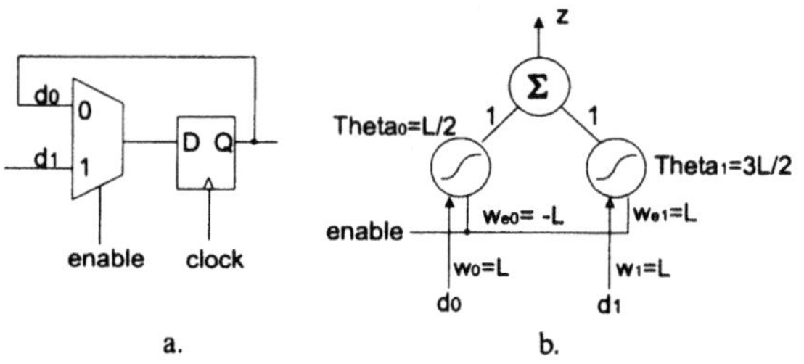

Fig. 2. (a) D flip-flop with clock enable and (b) differentiable neural-based multiplexer 2/1.

see Fig. 2b. The symbol L stands for a "large" value. The output of the multiplexer is governed by

$$\begin{aligned} z &= \mathrm{mux}(d_0, d_1, \mathrm{enable}) \\ &= \sum_{i=0}^{1} \sigma(w^i d_i + w^{ei}\mathrm{enable} - \Theta_i) = \xi^0 + \xi^1 \end{aligned}$$

with $\sigma(x) = 1/(1 + e^{-x})$ being the sigmoid activation function. When enable $= 0$, $z = \sigma(Ld_0 - L/2) + \sigma(Ld_1 - 3L/2)$ and d_0 is propagated due to a large threshold for d_1. Similarly, when enable $= 1$, $z = \sigma(Ld_0 - 3L/2) + \sigma(Ld_1 - L/2)$ and d_1 is propagated. When $0 <$ enable < 1, a mixture of both d_i is propagated.

There are Uo output units and $U - Uo$ context units in the output layer, H units in the hidden layer, $U + I$ latches in the input layer, and I external inputs (Fig. 1). In the learning stage, only the context units provide feedback. As previous outputs the desired outputs are used, i.e. the teacher-forcing technique is applied.

Outputs of the output units and of the context units are

$$y_k(t) = \sigma\Big(\sum_{l=1}^{H+1} v_{kl} \cdot u_l(t)\Big), \quad k = 1, .., U. \qquad (1)$$

The weights of the output layer are denoted as v_{kl}. Note that $v_{k,H+1}$ is the bias of the k-th neuron in the output layer, therefore $u_{H+1}(t)$ is a constant virtual input of unity. Outputs of the hidden units are

$$u_l(t) = \sigma\Big(\sum_{m=1}^{U+I+1} w_{lm} z_m(t)\Big), \quad l = 1, .., H. \qquad (2)$$

The $w_{l,U+I+1}$ is the bias of the l-th hidden neuron. Outputs of the input layer at time t are

$$z_m(t) = \mathrm{mux}(\tilde{y}_m(t-1), z_m(t-1), \mathrm{enable}) \qquad (3)$$

for $m = 1, .., Uo$ (previous desired outputs $\tilde{y}_m$),

$$z_m(t) = \mathrm{mux}(y_m(t-1), z_m(t-1), \mathrm{enable}) \qquad (4)$$

for $m = Uo + 1, .., U$ (context units), and

$$z_m(t) = \mathrm{mux}(x_m(t), z_m(t-1), \mathrm{enable}) \qquad (5)$$

for $m = U + 1, .., U + I$ (external inputs x_m). The inputs to the input (register) layer are: external inputs $x_m(t)$, delayed feedback outputs (previous desired outputs $\tilde{y}_m(t-1)$ and context values $y_m(t-1)$), and the local feedback of the latched data $z_m(t-1)$.

The derivatives of the multiplexer's output z with respect to its inputs will be further required. They are

$$\begin{aligned} \frac{\partial z}{\partial \mathrm{enable}} &= \xi^0(1-\xi^0)w^{e0} + \xi^1(1-\xi^1)w^{e1}, \\ \frac{\partial z}{\partial d_i} &= \xi^i(1-\xi^i)w^i, \quad i = 0, 1. \end{aligned}$$

The derivatives of z_m of the context units (Eq. 4) with respect to the output weights v_{ij} are

$$\begin{aligned} \frac{\partial z_m(t)}{\partial v_{ij}} &= \frac{\partial \mathrm{mux}}{\partial y_m^-}\frac{\partial y_m^-}{\partial v_{ij}} + \frac{\partial \mathrm{mux}}{\partial z_m^-}\frac{\partial z_m^-}{\partial v_{ij}} \\ &= \xi_m^0(1-\xi_m^0)w^0\frac{\partial y_m^-}{\partial v_{ij}} + \xi_m^1(1-\xi_m^1)w^1\frac{\partial z_m^-}{\partial v_{ij}}, \end{aligned}$$

where $y_m(t-1)$ and $z_m(t-1)$ are abbreviated as y_m^- and z_m^-, and similar with respect to the hidden weights. Other z_m (Eq. 3 and Eq. 5) are treated analogously.

2.2 The learning rules

In this section we derive the gradient learning rule for the LRNN. It is customary to unfold the structure of a recurrent neural network in time to be able to derive learning rules easier.

As a measure of performance, the mean-squared error (MSE) at the network's outputs is usually considered. We weighted the MSE(t) by the enable signal of the latch register at time t, which will be denoted in derivations as $f(t)$, i.e. $\text{enable}(t) \equiv f(t)$.

Let $R(\mathbf{x}) \in \Re$ be a function called relevance of the input vector $\mathbf{x} = (x_1, x_2, ..., x_I)^T$. It should be large for relevant input vectors $\mathbf{x}$ and small otherwise, such as $R(\mathbf{x}) = \sum_{i=1}^{I} \beta_i x_i$, where β_i will be called the relevance weight, denoting the relevance of the i-th input (or input line).

If the step function were applied to $R(\mathbf{x})$, then input vectors with R larger than some threshold Θ are enabled, $f(t) = 1$, and others are ignored, $f(t) = 0$. Also, the corresponding outputs either contribute or do not contribute to the error.

Instead of the step function we may utilize some similar, but differentiable step-like function such as the sigmoid function

$$f(t) = \sigma(R(\mathbf{x}(t)) - \Theta) = 1/(1 + e^{-a(R(\mathbf{x}(t))-\Theta)})$$

in order to allow the gradient descent methods to be applied. The effect is that the enable signal becomes continuous or "soft", i.e. an input vector is enabled with a degree between 0 and 1, $0 < f(t) < 1$, and the outputs contribute to the error with the same degree.

Relevant input lines x_i should have large relevance weights β_i. When input symbols are "1-of-N" encoded, this is equal to the relevance of a symbol or the corresponding input vector. There are two cases:

1. The relevances β_i of the input signals x_i are known or guessed (prior knowledge or hint) and held constant, and the threshold Θ is optimized. By adjusting Θ, the enable of input vectors and the contribution of corresponding outputs to the error is optimized. When Θ is significantly below zero and inputs are binary, $f(t)$ is always close to 1 and no latching occurs. When Θ increases, the ratio $\sigma(R_1 - \Theta)/\sigma(R_2 - \Theta)$, where $R_1 > R_2$, also increases. This is the indication that β_i are correct. By ignoring irrelevant inputs, training of the RNN is facilitated.
2. We have no clue as to what $R(\mathbf{x})$ might be. Therefore, β_i are learned and Θ is held constant.

The structure may be viewed as an additional neuron with the relevance weights β_i, see Fig. 1. The error at the output of the RNN is weighted by the "continuous enable" signal $f(t)$ and may be written as

$$\text{Err}(t) = \frac{\|y(t) - \tilde{y}(t)\|^2 f(t)}{E[f(t)]}, \tag{6}$$

where $E[.]$ denotes the expectation operator and $\tilde{y}(t)$ denotes the desired output at time t. The weights are optimized by the steepest descent learning algorithm:

$$v_{kl} \leftarrow v_{kl} - \eta \frac{\partial \text{Err}}{\partial v_{kl}}, \qquad w_{lm} \leftarrow w_{lm} - \eta \frac{\partial \text{Err}}{\partial w_{lm}} \tag{7}$$

with η being the learning step. The same holds for the threshold Θ and for the relevance weights β_i in case they are optimized. The derivative of Err(t) with respect to the output weights is

$$\begin{aligned} \frac{\partial \text{Err}(t)}{\partial v_{ij}} &= \frac{\partial \|y(t) - \tilde{y}(t)\|^2}{\partial v_{ij}} \frac{f(t)}{E[f(t)]} \\ &= 2\sum_{k=1}^{Uo} \left[(y_k(t) - \tilde{y}_k(t)) \frac{\partial y_k}{\partial v_{ij}} \right] \frac{f(t)}{E[f]}, \end{aligned} \tag{8}$$

and similar for the hidden weights. The derivatives of y_k, $k = 1, .., U$, with respect to all the weights are required.

Due to the lack of space, only the final expressions are given. For the output weights v:

$$\frac{\partial y_k}{\partial v_{ij}} = y_k' \left\{ \delta_{ki} u_j + \sum_{l=1}^{H+1} \left[v_{kl} u_l (1 - u_l) \sum_{m=Uo+1}^{U} w_{lm} \cdot \left[\xi_m^0 (1 - \xi_m^0) w^0 \frac{\partial y_m^-}{\partial v_{ij}} + \xi_m^1 (1 - \xi_m^1) w^1 \frac{\partial z_m^-}{\partial v_{ij}} \right] \right] \right\},$$

where all quantities y, u, and z are at time t, except y^- denoting $y(t-1)$, and z^- denoting $z(t-1)$. The y_k' denotes the derivative of the sigmoid activation function and is equal to $y_k(1 - y_k)$; similarly u_l'.

For the hidden weights we obtain

$$\frac{\partial y_k}{\partial w_{ij}} = y_k' \left\{ v_{ki} u_i' z_j + \sum_{l=1}^{H+1} \left[v_{kl} u_l' \cdot \sum_{m=Uo}^{U} w_{lm} \cdot \left[\xi_m^0 (1 - \xi_m^0) w^0 \frac{\partial y_m^-}{\partial w_{ij}} + \xi_m^1 (1 - \xi_m^1) w^1 \frac{\partial z_m^-}{\partial w_{ij}} \right] \right] \right\}.$$

The derivative with respect to the threshold Θ after some derivation equals

$$\begin{aligned} \frac{\partial \text{Err}(t)}{\partial \Theta} &= 2\sum_{k=1}^{Uo} ((y_k - \tilde{y}_k) \frac{\partial y_k}{\partial \Theta}) \frac{f(t)}{E[f]} + \\ & a\text{Err}(t) \cdot \left(f(t) - \frac{E[f^2]}{E[f]} \right). \end{aligned} \tag{9}$$

From Eq. 9 we require also

$$\frac{\partial y_k}{\partial \Theta} = y_k' \sum_{l=1}^{H+1} (v_{kl} u_l (1 - u_l) \sum_{m=1}^{U+I} w_{lm} \frac{\partial z_m}{\partial \Theta}) \tag{10}$$

and from Eq. 10 $\frac{\partial z_m}{\partial \Theta}$ is required, where z_m are the outputs of the multiplexers:

$$\begin{aligned} \frac{\partial z_m}{\partial \Theta} = {} & \xi_m^0 (1 - \xi_m^0) [w^0 \frac{\partial y_m^-}{\partial \Theta} - w^{e0} a f (1 - f)] + \\ & \xi_m^1 (1 - \xi_m^1) [w^1 \frac{\partial z_m^-}{\partial \Theta} - w^{e1} a f (1 - f)] . \end{aligned}$$

In a similar manner the derivatives with respect to the relevance weights β_i are obtained, but are omitted here.

To summarize, the basic procedure is calculation of the derivatives of the network's outputs $y_k(t)$ with respect to all adjustable weights v_{kl}, w_{lm}, Θ, and β_i, using also derivatives from the past. At each time step the weighted error $Err(t)$ is calculated and the weights are updated accordingly.

3 Experiments

The proposed method was tested on two tasks with long time lags.

3.1 Task 1

In the first problem [6], there are n input symbols a, b, c, etc. with the "1-of-N" or "one-hot" encoding, i.e. only one of the n input lines is 1 and all others are 0 at any moment. The task is to output a 1 immediately following the first occurrence of the symbol b after the symbol a has already appeared, no matter how long ago. All other signals have no influence and serve merely as distractors. After the occurrence of b, a is "used up", and the next time the output is 1 is following the first occurrence of b after a new a. This corresponds to a small finite automaton. The solution is simple, once the relevant signals are found. The hard part of the task is to find the distractors. The caution is necessary here: if the task were off-line and therefore with a limited number of examples, other meaningful automata might be induced.

In our experiments n was 10. RNN had 2 output units (1 output and 1 context unit), 8 hidden units, and 10 external inputs. After 10000 steps, RNN without latches fails when $n = 10$. On the reduced problem with $n = 4$, it succeeds after about 8000 steps.

In this task we chose only to test the predefined criterion $R(\mathbf{x}) = a+b-\Theta$, i.e. $\beta_a = \beta_b = 1$ and other β_i are zero. This is of course the correct criterion and it remains only to observe the course of Θ. Fig. 3a shows increase of Θ, which is understandable, since the ratio between $f(t)$ at $R = 1$ and $f(t)$ at $R = 0$ also increases. The weighted error is small, contrary to the ordinary error, which is large due to the errors when unimportant inputs appear. It is interesting to observe behavior of outputs y in Fig. 3c. After an occurrence of symbol b, e.g. at 9950 (the occurrences of a and b are indicated also in Fig. 3b), the output is 1, as desired, but y_0 continues to output a 1 until the next a occurs about 10 steps later. Since the error is weighted by $f(t)$, errors of irrelevant symbols are not considered. Behavior of the context unit is also interesting. A single a activates this unit only partially (to about 0.45), while eventual subsequent occurrences of a activate it completely.

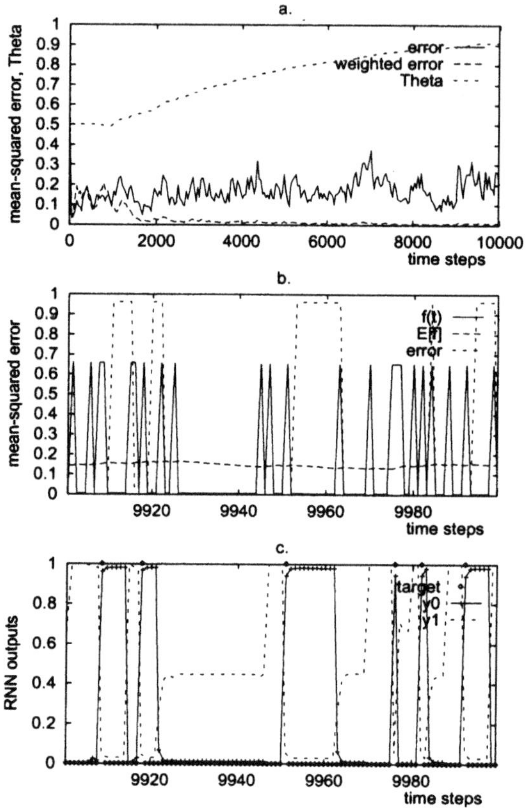

Fig. 3. LRNN on Task 1 for $n = 10$. (a) MSE, weighted MSE, and Θ, (b) enable, mean enable, and MSE during final 100 steps, (c) desired output $\bar{y}_0$, actual output y_0, and context value y_1.

3.2 Task 2

In the second task [4], there are p input symbols $a_1 = x$, $a_2 = y$, a_3, ..., a_p. Each symbol is "1-of-N" encoded with p binary input signals. The training sequence consists of random occurrences of only two similar subsequences: (x, a_3, a_4,..., a_p, x) and (y, a_3, a_4,..., a_p, y). After a subsequence is finished, the next is selected with probability 0.5 for each, and so on. The sequence is fed to the network, which always has to predict the next symbol. The prediction of the first symbol of any subsequence is meaningless, since x and y occur with the same probability. The hard part is to predict the last symbol of a subsequence, since the first symbol has to be remembered over a long time lag (depending on p, which was set to 10). We demanded the network to correctly predict the last symbol.

RNN had 12 output units (10 outputs and 2 context units), 10 hidden units, and 10 external inputs. RNN without latches failed to predict correctly the last symbol of the subsequence, the error was large, about 0.4. There

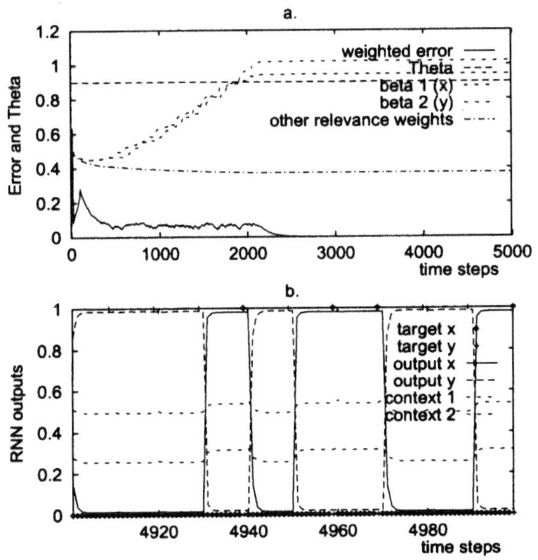

Fig. 4. LRNN on Task 2 for $p = 10$. (a) weighted MSE, Θ, and relevance weights, (b) desired outputs $\tilde{y}_x$, $\tilde{y}_y$, actual output y_0, and context value y_1 during final 100 steps.

was no improvement even after 10000 steps. When explicitly demanded, it mastered the easy part of the task, i.e. predicting a_i, $i = 3, .., p$.

Using LRNN, this time we decided to learn the relevance weights β_i, which signify the relevance of the i-th input signal. The Θ was set to 0.9 and all the β_i started at 0.5. It is clear from Fig. 4a that only β_1 (β_x) and β_2 (β_y) rise above Θ, while the others slightly decrease. This enabled the LRNN to learn the task quickly. From Fig. 4b it is obvious that the two context units remember or encode symbol x with a slightly larger activation than for symbol y. Since they encode the symbols in the same way, one of them is clearly redundant, as expected. An interesting fact is that they employ (attenuated) binary encoding, not "1-of-N", for example. Outputs are set to final target values immediately after the first symbol of a subsequence. This is the explanation why the context values are not very pronounced. The output neurons themselves remember the first symbol, which is actually the simplest way to solve the task.

4 Conclusion

An extended architecture of recurrent neural networks for dealing with temporal processing with long term dependencies was proposed. It is based on ignoring assumably irrelevant inputs using a register of latches in the input layer of the network. Latches are differentiable, such that gradient descent learning can be applied. The method yielded good results on two standard sequence processing tasks with long time lags, where irrelevant inputs were distracting symbols, rather than repeated symbols. On the latter type of problems when dealing with large finite automata with "low-frequency" properties, we intend to test the method in the near future.

References

[1] Cleeremans, A., Servan-Schreiber, D., McClelland, J. L. (1989) Finite State Automata and Simple Recurrent Networks. Neural Computation 1(3): 372-381.

[2] Bengio, Y., Simard, P., Frasconi, P. (1994) Learning Long-Term Dependencies with Gradient Descent is Difficult. IEEE Transactions on Neural Networks 5(2): 157-166.

[3] Schmidhuber, J. (1992) Learning complex, extended sequences using the principle of history compression. Neural Computation 4(2): 234-242.

[4] Hochreiter, S., Schmidhuber, J. (1997) Long short-term memory. Neural Computation 9(8): 1735-1780.

[5] Gers, F. A., Schmidhuber, J., Cummins, F. (2000) Learning to Forget: Continual Prediction with LSTM. Neural Computation 12(10): 2451-2471.

[6] Williams, R. J., Zipser, D. (1989) A Learning Algorithm for Continually Running Fully Recurrent Neural Networks. Neural Computation 1(2): 270-280.

Recurrent neural network with integrated wavelet based denoising unit

Uroš Lotrič[1], Andrej Dobnikar[1]

[1]University of Ljubljana, Faculty of Computer and Information Science, Tržaška 25, 1000 Ljubljana, Slovenia, {uros.lotric, andrej.dobnikar}@fri.uni-lj.si

Abstract

A denoising unit based on wavelet multiresolution analysis is added ahead of the multilayered perceptron with global recurrent connections. The learning algorithm is developed which uses the same cost function for setting all free parameters, those of the denoising unit and those of the neural network. It is illustrated that the proposed model outmatches the models without denoising unit and/or without recurrent connections in noisy time series prediction.

1 Introduction

Non-parametric modelling of dynamical systems from time series is an important area where neural networks can be applied. Due to their ability to establish the temporal relationship of the inputs with the help of internal states, recurrent neural networks are best suited for this problem domain.

Noise, inherently present in most real time series increases the modelling complexity. Neural networks remove noise from time series to a certain extend. However, it was recently shown that the prediction of time series with standard multilayered perceptron can be considerably improved by integrating denoising based on wavelet multiresolution analysis into the model [1, 2, 3]. The underlying idea of this integration is that denoising is not treated separately from modelling, but the same criterion is used for noise removal and for setting the neural network free parameters.

In the present work, this approach is extended to recurrent neural networks. A brief background of wavelet based smoothing is given in the next section. In the third section the integration of smoothing unit into recurrent neural network is presented and learning algorithm is derived. The capability of the model is evaluated in the fourth section. The main conclusions are drawn in the last section.

[1]This work is partially supported by the Slovenian Ministry of Education, Science and Sport under the grant Z2-3040.

2 Wavelet based denoising

Classical time series denoising approaches rooted in Fourier analysis assume noise to be manifested mainly as high frequency oscillations. The wavelet based denoising, on the other hand, assumes that analysis of time series at different resolutions might improve the separation of the true underlying signal from noise.

The wavelet multiresolution analysis is based on the scaling function $\phi(t)$ and the corresponding mother wavelet $\psi(t)$, fulfilling specific technical conditions [4]. The scaling function and the mother wavelet are localized both in time and frequency domain (Fig. 1), which

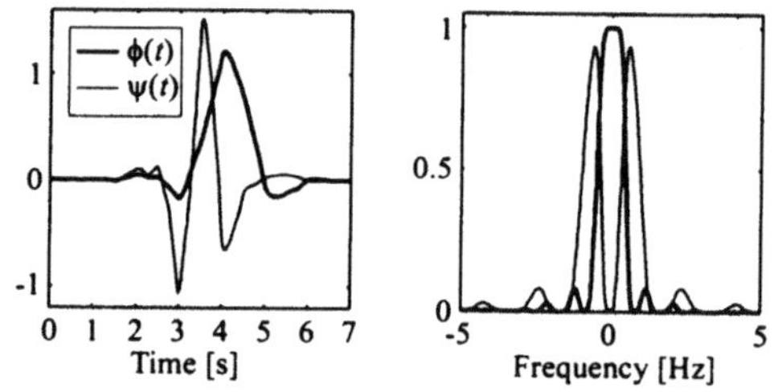

Fig. 1. Scaling function $\psi(t)$ and mother wavelet $\psi(t)$ in time domain (left) and frequency domain (right).

allows for explicit capture of the local dynamics within a time series. By dilation and translation of the scaling function and the mother wavelet the basis functions $\phi_{j,k} = 2^{-j/2}\phi(2^{-j}t - k)$ and $\psi_{j,k} = 2^{-j/2}\psi(2^{-j}t - k)$ are derived. On each scale j, the functions $\phi_{j,k}(t)$, and $\psi_{j,k}(t)$, $k \in \mathbb{Z}$, form an orthonormal basis in the spaces of the square integrable functions $L^2(\mathbb{R})$ [4]. An arbitrary time series $p(t) \in L^2(\mathbb{R})$ can be written as

$$p(t) = \sum_{k\in\mathbb{Z}} a_k^J \phi_{J,k}(t) + \sum_{j\leq J}\sum_{k\in\mathbb{Z}} d_k^j \psi_{j,k}(t) \quad , \qquad (1)$$

where the first term represents the approximation on the scale J and the second term the details on scales $1 \ldots J$, J is the number of scales the time series is decomposed to. The wavelet coefficients a_k^j of the approximation and the wavelet coefficients d_k^j of the details on adjacent scales are related by the pyramidal decomposition scheme [4].

Time series is usually sampled using a constant sampling time Δt, $p_k = p(k\Delta t)$, $k = 1, \dots, M$. Considering the linearity of wavelet transform, the pyramidal decomposition and reconstruction schemes of a discrete time series can be written in matrix notation [5], respectively,

$$c_k = \sum_{l=1}^{M} \mathbf{W}^{\mathrm{D}}_{k,l} p_l \qquad (2)$$

$$p_k = \sum_{l=1}^{C} \mathbf{W}^{\mathrm{R}}_{k,l} c_l \,. \qquad (3)$$

The vector $\mathbf{c} = (d_1^0, \dots, d_{C_0}^0, \dots, d_1^J, \dots, d_{C_J}^J, a_1^J, \dots, \dots, a_{C_J}^J)^{\mathrm{T}}$ combines all wavelet coefficients, C_j is the number of coefficients on scale j and C is the number of all wavelet coefficients $C = \sum_{j=0}^{J} C_j + C_J$. Elements of matrices $\mathbf{W}^{\mathrm{D}}$ and $\mathbf{W}^{\mathrm{R}}$ depend on the wavelet family, the wavelet order O, length of a time series M and the number of scales J.

The wavelet coefficients of the details have high absolute values only in the intervals of rapid time series change. This property led Donoho and Johnstone to propose denoising by thresholding [5], which is applied on the wavelet coefficients of the details prior to the reconstruction. To integrate denoising into neural network learning algorithm, the generalized soft thresholding function

$$T(d_k^j, \tau_j) = d_k^j + \frac{1}{2}\left(|d_k^j - \tau_j + s| - |d_k^j + \tau_j + s|\right) , \quad (4)$$

was proposed [3]. As shown in Fig. 2, it removes the wavelet coefficients of the details with small absolute values and reduces the absolute values of all other coefficients of details. The parameter s, $s \geq 0$, determines the smoothness of the generalized soft thresholding function. The denoised time series $\tilde{p}_i$, $i = 1, \dots, M$, is obtained when the thresholded wavelet coefficients of details $\tilde{d}_k^j = T(d_k^j, \tau_j)$ are used in the reconstruction formula (Eq. (3)).

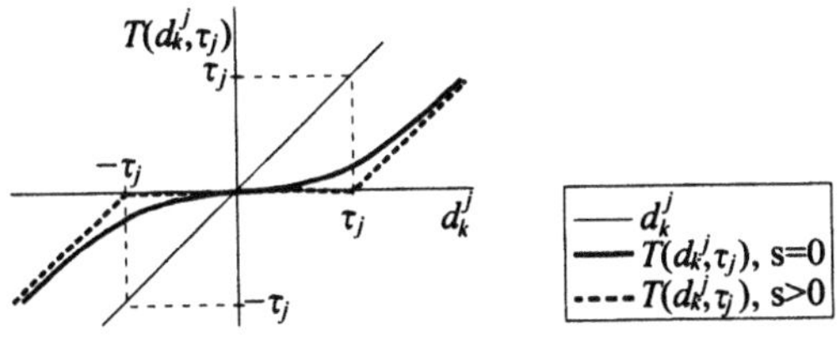

Fig. 2. Generalized soft thresholding function.

3 Model

The proposed model, shown in Fig. 3, is based on multilayered perceptron with two adaptations: (i) the outputs of neurons in output layer are connected back to all neurons in hidden layer and (ii) layers of denoising unit are added.

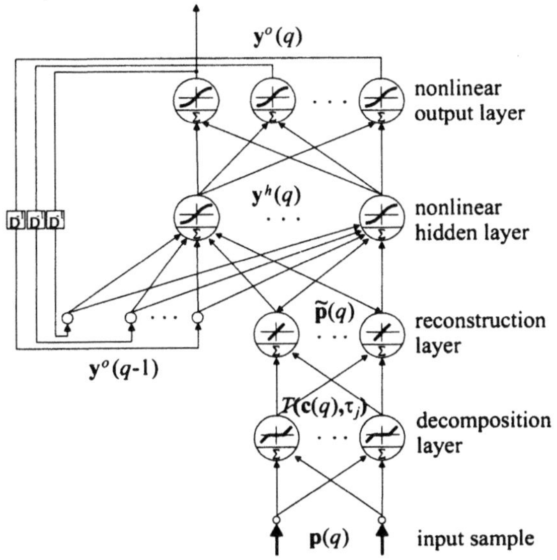

Fig. 3. Multilayered perceptron with global recurrent connections and denoising layers.

The input sample $\mathbf{p}(q) = (p_1(q), \dots, p_{N_i}(q))^{\mathrm{T}}$ is fed to the decomposition layer, where it is decomposed to the wavelet coefficients by Eq. (2) and thresholded (Eq. (4)). The denoised sample $\tilde{\mathbf{p}}(q)$ calculated in the linear reconstruction layer is merged with the last known outputs of the model, $\mathbf{z}(q) = (\mathbf{y}^o(q-1)^{\mathrm{T}}, \tilde{\mathbf{p}}(q)^{\mathrm{T}})^{\mathrm{T}}$. The output of the N_h neurons in the nonlinear hidden layer is given by

$$y_k^h(q) = \varphi^h(s_k^h(q)) \,, \; s_k^h(q) = \sum_{l=0}^{N_o+N_i} \omega_{k,l}^h z_l(q) \,, \qquad (5)$$

$k = 1, \dots, N_h$, where $\varphi^h = \tanh$ is the nonlinear activation function, $\omega_{k,l}^h$ are the weights on connections and $z_0(q) = 1$ is bias input. Similarly, the outputs of the neurons in the hidden layer, together with bias input $y_0^h(q) = 1$, are fed further to the N_o neurons in the output layer

$$y_k^o(q) = \varphi^o(s_k^o(q)) \,, \; s_k^o(q) = \sum_{l=0}^{N_h} \omega_{k,l}^o y_l^h(q) \,, \qquad (6)$$

$k = 1, \dots, N_o$, where $\varphi^o = \tanh$ and $\omega_{k,l}^o$ are the activation function and the weights, respectively.

Although the equations of the denoising layers and those of nonlinear layers are very similar, they differ in positions of free parameters. While the weights on connections between the neurons present the free parameters of the nonlinear layers, the weights on connections in the decomposition and reconstruction layer are constant and the free parameters are hidden within thresholding functions.

3.1 Learning algorithm

In the time series modelling, the model usually connects a value of a time series x_t at time t with its previous values. To establish such connection, i.e., to minimize the specified cost function, the model is trained on a set of known input–output pairs. From a time series with values $x_1, \ldots, x_N$, the input–output pairs $\{\mathbf{p}(q), \mathbf{r}(q)\}$, $q = 1, \ldots, N - N_0$, can be constructed with the elements $p_i(q) = x_{i+q-1}$, $i = 1, \ldots, N_0$, and $r_1(q) = x_{q+N_0}$.

The cost function is given as $\mathcal{E} = \sum_{q=1}^{Q} \mathbf{e}(q)^{\mathrm{T}} \mathbf{e}(q)$, where $\mathbf{e}(q) = \mathbf{r}(q) - \mathbf{y}^o(q)$. In general, to minimize the cost function, gradient algorithms need gradients of errors with respect to all free parameters [6]. Considering the relation $\nabla \mathbf{e}(q) = -\nabla \mathbf{y}^o(q)$, partial derivatives of model outputs with respect to all model free parameters must be calculated.

According to the derivation of Real time recurrent learning [7], the following dynamical systems are obtained for the partial derivatives of outputs with respect to the weights on connections to the output layer

$$\frac{\partial y_k^o(q)}{\partial \omega_{i,j}^o} = \varphi^{o\prime}(s_k^o(q)) \cdot \left\{ \delta_{k,i} y_j^h(q) + \right.$$

$$\left. + \sum_{l=1}^{N_h} \omega_{k,l}^o \varphi^{h\prime}(s_l^h(q)) \sum_{m=1}^{N_o} \omega_{l,m}^h \frac{\partial y_m^o(q-1)}{\partial \omega_{i,j}^o} \right\} \quad (7)$$

and with respect to the weights on connections to the hidden layer

$$\frac{\partial y_k^o(q)}{\partial \omega_{i,j}^h} = \varphi^{o\prime}(s_k^o(q)) \sum_{l=1}^{N_h} \omega_{k,l}^o \varphi^{h\prime}(s_l^h(q)) \cdot$$

$$\cdot \left\{ \delta_{l,i} z_j(q) + \sum_{m=1}^{N_o} \omega_{l,m}^h \frac{\partial y_m^o(q-1)}{\partial \omega_{i,j}^h} \right\} . \quad (8)$$

In Eq. (7) and Eq. (8) $\varphi^{h\prime}(s)$ and $\varphi^{o\prime}(s)$ denote derivatives of activation functions in output layer and hidden layer, respectively.

The gradient based algorithms are designed for optimization of non-constrained free parameters. For the threshold τ_j, however, only nonnegative values smaller than or equal to absolutely the largest wavelet coefficient $d_{\max}^j$ are reasonable [2]. Therefore, the unconstrained thresholds τ_j^∞, $j = 1, \ldots, J$, are used in the gradient based learning algorithm of denoising layers. They are mapped from the interval $(-\infty, +\infty)$ to the interval $(0, d_{\max}^j)$ of the thresholds τ_j, with the sigmoid function

$$\tau_j = d_{\max}^j (1 + e^{-\tau_j^\infty})^{-1} , \; j = 1, \ldots, J . \quad (9)$$

The partial derivatives of outputs with respect to thresholds are derived applying the chain rule

$$\frac{\partial y_k^o(q)}{\partial \tau_j^\infty} = \sum_{i=1}^{N_i} \frac{\partial y_k^o(q)}{\partial \tilde{p}_i(q)} \frac{\partial \tilde{p}_i(q)}{\partial \tau_j^\infty} . \quad (10)$$

The first factor in the sum is derived from Eqs. (5) and (6),

$$\frac{\partial y_k^o(q)}{\partial \tilde{p}_i(q)} = \varphi^{o\prime}(s_k^o(q)) \sum_{l=1}^{N_h} \omega_{k,l}^o \varphi^{h\prime}(s_k^h(q)) \omega_{l,N_o+i}^h , \quad (11)$$

while the second factor is obtained from Eqs. (3), (4) and (9),

$$\frac{\partial \tilde{p}_i(q)}{\partial \tau_j^\infty} = \sum_{k=1}^{C} \mathbf{W}_{i,k}^{\mathrm{R}} \frac{\partial T(c_k, \tau_j)}{\partial \tau_j} \tau_j (1 - \tau_j / d_{\max}^j) . \quad (12)$$

The parameter $s > 0$ should be used to prevent the learning algorithm from unintentional stopping [3].

4 Results

The proposed multilayered perceptron with global recurrent connections and denoising unit (DRMLP) was applied on three one step ahead prediction problems: the Feigenbaum sequence, the sunspot number and the quality control of rubber compounds. For comparison, the classical multilayered perceptron (MLP) [6], multilayered perceptron with denoising unit (DMLP) [2, 3] and multilayered perceptron with global recurrent connections and disabled denoising unit, $\tau_j = 0$, $j = 1, \ldots, J$ (RMLP), were applied to the same problems.

From each time series input–output pairs were prepared. The first 85% of the input–output pairs were included in the training set, used to set free parameters of the models, while the remaining 15% of the input–output pairs formed the testing set, used for model comparison. The models were allowed to have up to 20 inputs N_i and the number of free parameters never exceeded 30% of the number of input–output pairs. In the DRMLP and DMLP models, the wavelets from symlet family [4], S^M, $2M \leq N_i$, were used.

The root mean squared error, normalized to standard deviation σ of a time series, NRMSE $= (\sum_{q=1}^{Q} (r_1(q) - y_1^o(q))^2 / (\sigma^2 Q))^{1/2}$ was considered as a performance measure, with Q denoting the number of input–output pairs in observed set. For each model configuration, learning based on the Levenberg-Marquardt gradient algorithm [8] was repeated 20 times and only the configurations with the smallest performance measure were used in further analysis.

4.1 Feigenbaum sequence

The Feigenbaum sequence is given by the recursive relation $x_k = r\,x_{k-1}(1 - x_{k-1})$. The relation with $r = 4$ was considered, where the sequence becomes chaotic [9]. From the initial value $x_1 = 0.01$, 250 values were calculated with 15-digit precision. The sequence is graphically presented in Fig. 4.

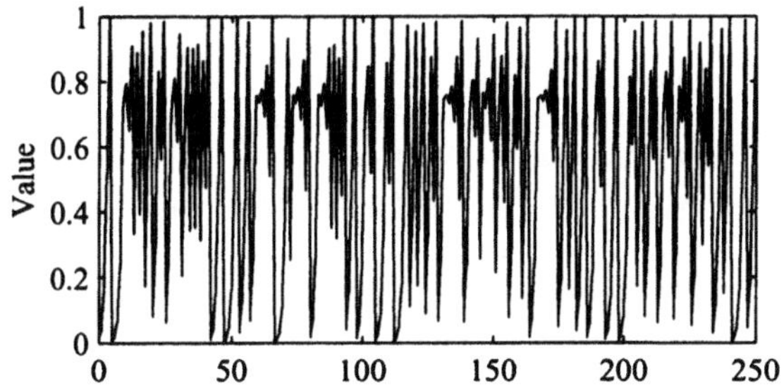

Fig. 4. Feigenbaum sequence.

Table I gives the comparison of the applied models. Models with 4 inputs, 10 neurons in hidden layer and 1 neuron in output layer managed to learn the relationship very well. The number of model free parameters is given in parenthesis.

Table I. Model comparison in prediction of Feigenbaum sequence.

Model	Structure (Free params)	NRMSE Training	NRMSE Testing
MLP	4-10-1 (61)	6.5	7.6
DMLP	4-10-1, S^2 (63)	6.6	7.8
RMLP	4-10-1 (71)	3.2	3.7
DRMLP	4-10-1, S^2 (73)	3.6	4.8

By setting thresholds to very small values, i.e., $\tau_1, \tau_2 < 0.01\, d^j_{\max}$, the models DMLP and DRMLP obviously recognized that denoising is not necessary. However, the mapping of infinite thresholds τ_j^∞, given by Eq. (9), does not allow thresholds τ_j to become zero in finite number of learning steps and leaving the input vector completely intact. Therefore, the NRMSE errors are slightly higher for the models with denoising.

4.2 Sunspot number

Sunspots are dark temporary areas of concentrated magnetic field on the sun, contracting and expanding as they move across the surface. Their number is highly correlated with the sun activity and as such observed for centuries. Yearly average of the sunspot number, based on the data collected between the years 1700 and 2001 was used in the analysis [10].

Fig. 5 shows the variation of the sunspot number together with predictions obtained with the DRMLP model in the last 100 years.

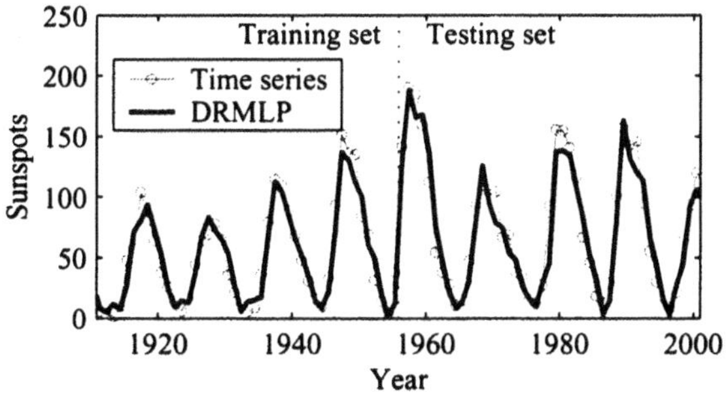

Fig. 5. Prediction of sunspot number.

Detailed comparison of the models is given in Table II. For each model the structure which resulted in the smallest error on the testing set is given. Results on testing set demonstrate the effects of specific improvements of the MLP model on the prediction performance. Both, the denoising unit and the global recurrent connections, each in its own way, helped to improve the prediction accuracy. The best result was obtained with the DRMLP model, comprising both improvements simultaneously.

Table II. Model comparison in prediction of sunspot number.

Model	Structure (Free params)	NRMSE Training	NRMSE Testing
MLP	9-1-1 (12)	0.36	0.44
DMLP	12-1-1, S^3 (18)	0.40	0.42
RMLP	6-6-1 (55)	0.23	0.39
DRMLP	4-4-2, S^2 (40)	0.27	0.37

4.3 Rubber quality

An important characteristic determining the quality of a rubber compound is its hardness. It is measured in Shore units on the scale ranging from 0 to 100. Variation of hardness for successive mixings of a compound used for bicycle and motorcycle tubes is shown in Fig. 6.

Comparison of the models in prediction of rubber quality is given in Table III. Once again, the results

Table III. Model comparison in prediction of rubber quality.

Model	Structure (Free params)	NRMSE Training	NRMSE Testing
MLP	18-1-1 (21)	0.78	0.67
DMLP	14-1-1, S^5 (20)	0.87	0.63
RMLP	4-2-1 (15)	0.82	0.69
DRMLP	18-1-4, S^8 (35)	0.74	0.61

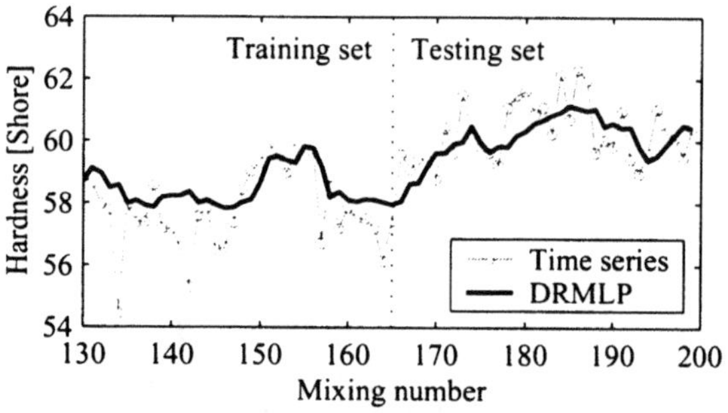

Fig. 6. Prediction of rubber quality.

on the testing set provide evidence for reasonableness of both improvements. Inferiority of the RMLP model, compared to the DMLP model and the DRMLP model shows the dominance of the smoothing unit over the global recurrent connections.

5 Conclusion

A novel approach towards modelling dynamical systems from noisy time series is proposed. It is based on multilayered perceptron to which global recurrent connections are added to enable for detecting temporal relationship of the inputs. At the same time, a denoising unit based on wavelet multiresolution analysis is added ahead of the neural network input layer. This unit is integrated into the neural network is such a way that prediction error is used as the denoising criterion. The denoising unit is treated as layers on neural network with free parameters lying within the activation function and not on connections. For these free parameters a gradient based learning algorithm is derived. Results on several time series indicate that introducing the global recurrent connection alone does not necessary improve the prediction, however, the prediction is improved when both global recurrent connection and denoising are applied simultaneously.

References

[1] Lotrič, U. (2000) Using Wavelet Analysis and Neural Networks for Time Series Prediction, PhD Thesis, University of Ljubljana, Faculty of Computer and Information Science, Ljubljana

[2] Lotrič, U., Dobnikar, A. (2001) Wavelet based smoothing in time series prediction with neural networks. In Kůrková, V., Steele, N., Nerudá, R., Kárný, M. (eds.) Artificial Neural Nets and Genetic Algorithms: Proceedings of the International Conference in Prague, Springer, Wien, pp. 43–46

[3] Lotrič, U. (2002) Wavelet based denoising integrated into multilayered perceptron. Neurocomputing, submitted for publication, http: //laspp.fri.uni-lj.si/uros/publications.htm

[4] Daubechies, I. (1992) Ten lectures on Wavelets, SIAM, Philadelphia

[5] Donoho, D. L., Johnstone, I. M. (1994) Ideal spatial adaptation by wavelet shrinkage. Biometrika 81:425–455

[6] Haykin, S. (1999) Neural Networks: A Comprehensive Foundation, 2nd edition, Prentice-Hall, New Jersey

[7] Williams, R. J., Zipser, D. (1989) A learning algorithm for continually running fully recurrent neural network. Neural Computation 1:270–280

[8] Hagan, M. T., Menhaj,M. B. (1994) Training feedforward networks with the marquardt algorithm. IEEE Transactions on Neural Networks 5(6):989–993

[9] Schuster, H. G. (1984) Deterministic Chaos. An Introduction, Physik, Weinheim

[10] Sunspot index data center, Royal observatory of Belgium (2001) Yearly definitive sunspot number, http://sidc.oma.be.

Reinforced Search in Stochastic Neural Network

Mira Trebar and Andrej Dobnikar [1]

[1]University of Ljubljana, Faculty of Computer and Information Science, Tržaška 25, 1000 Ljubljana, Slovenia
{mira.trebar, andrej.dobnikar}@fri.uni-lj.si

Abstract

A reinforced search algorithm for the stochastic feedforward neural networks is described. A stochastic neuron is used in a network as a searching unit. Reinforcement signal from environment is used for weights and variance adaptation. This is experimentally compared with more traditional techniques like gradient-based learning algorithm and evolutionary algorithm.

1 Introduction

Artificial neural networks became an efficient model of attacking 'hard' problems, where a mapping between inputs and outputs is 'weakly' defined. Two basic approaches in finding appropriate internal parameters (weights) have been put into force, namely learning (gradient based) procedures and evolutionary algorithms. As the first one is not 'biologically plausible', and the second not supposed to work in real time, significant efforts have currently been made to find a more 'natural' way of finding the weights for the problems under investigation.

There are some existing theoretical results that certainly deserve our attention because they have been proved due to some neurophysiological evidence [6]. The well known Hebbian learning paradigm suggests that the biological synaptic efficaces (weights) change locally in proportion to the correlation between the pre- and postsynaptic signals. To go on, reinforcement learning is a process of a trial end error designed to maximize the expected value of a criterion function known as a reinforcement signal. This idea has origins in psychology in connection with experimental studies of animal learning [9]. Next, there is obviously a need for a directed search for a correct mapping or at least for improving it, based on a feedback from the environment. This leads us to the stochastic nature of neurons. Several attempts were published based on similar ideas [1, 2, 5], but the problem still remains open, in particular if general network topology is considered and sequential (dynamic) problems observed. There are some attempts to prove the convergence of a stochastic search procedures [4], but in practice they are too slow to be of any practical use. Our goal would therefore be a step towards a practical reinforced search procedure in a general feedforward stochastic neural network, where the idea is to consider the above directions. Essential in the approach is avoidance of a gradient and an evolutionary based techniques.

We give first some definitions related to the stochastic neural networks and procedures for reinforced search of the parameters for correct input-output mappings. Then we introduce two problem domains and give the results of our experimental work. The results are compared with the traditional procedures like backpropagation, evolutionary algorithms and dynamic programming in terms of accuracy and speed where possible.

2 Stochastic neural network

A stochastic neural network used in this paper is defined as a feed-forward neural network composed of stochastic neurons. Stochasticity is due to the formation of the neuron's output, which is defined by a normal distribution with two parameters, the mean μ and the standard deviation σ. The neuron would first compute values of μ and σ deterministically and then draw its stochastic activation a_i from the normal distribution $N(\mu,\sigma)$ sent to the output function $f(.)$, which returns its actual output. Specifically, the output of stochastic neuron i is computed as

$$y_i(n) = f(a_i(n)) = f(N(\mu(n), \sigma(n))). \quad (1)$$

The normal distribution $N(\mu(n), \sigma(n))$ is defined by the mean value $\mu(n)$ computed as

$$\mu(n) = \Sigma_j w_{ij}(n) y_j(n), \quad (2)$$

where $w_{ij}(n)$ is the synaptic weight connecting neuron j to neuron i, and $y_j(n)$ is the input signal of neuron i. If the neuron i is in the first hidden layer of the network, then the index j refers to the input of the network. The standard deviation $\sigma(n)$ is initially set to some large value, and then updated in every step as described later.

This stochastic neuron is similar to Bernoulli semilinear unit (BSU) [11] and stochastic real-valued unit (SRV) [2] to some extent, but also differs substantially in what follows. It can have continuous output values, as opposed to BSU, variance does not follow an average response from the environment (reinforcement signal), as is the case with SRV and it uses local rule for weights modifications. Both referred units are combined with backpropagation procedure through hidden units (in case of BSU) or with hidden backpropagation units (in case of SRV) when used in multilayered networks.

The Boltzmann learning, as one of the first learning algorithms related to the stochastic neural networks, has to be referred as well. There, the activity function represents the probability for the output to be true where energy is defined as a weighted sum of products of all pairs of neuron's outputs in the net and the 'termal' parameter T is used to influence the learning similar to the simulated annealing procedure [3]. The Boltzmann equation is a result of an optimization procedure, where the goal is to minimize the relative entropy between the desired probability of the values of stochastic neurons in the net and their actual probability. The main drawback is in its global character, complex computing and therefore very slow convergence.

The basic idea in the case of the proposed algorithm is as follows. The stochastic neural network gives an answer to an input vector with output values, calculated by Eq. (1). The environment reacts by reinforcement signal $r(n)$ with possible values -1 (penalty) and +1 (reward). At the beginning the weights of the net have some random values around zero and the standard deviation in stochastic neurons is relatively large indicating wide search space. As the procedure goes on, the weights are changed according to the reinforcement signal and local inputs and outputs of the stochastic neuron. The update of weights takes place after each step of the communication with the environment:

$$w_{ij}(n+1) = w_{ij}(n) + \alpha(r(n) - 1)y_i(n)y_j(n), \quad (3)$$

where α is the rate parameter, $y_i(n)$ is the output, and y_j is the input to the stochastic neuron i, $r(n)$ is the reinforcement signal from the environment, and n means discrete time.

Though similar to Hebbian type of modification, where the weight difference depends only on pre and post synaptic values and some positive constant, we introduce the different type of modification which shows disparate dependence. The original equation is not adequate for the search algorithm. While in the learning algorithm the modification follows the local correlations, in the searching procedure it only updates the space to investigate.

The standard deviation $\sigma(n)$ used in normal distribution is also updated. It is changed according to some monotone decreasing function in every discrete time step, like:

$$\sigma(n+1) = \beta(1 - r(n)), \quad (4)$$

where β is the search rate parameter and $r(n)$ is the reinforcement signal from environment.

We are strictly using the term 'reinforced search' for the procedure of looking for suitable weights, instead of the term 'learning' which is more appropriate in the case of more smoothly graded gradient-based procedures. Searching for the parameters with the stochastic net is on the other hand similar to the way an evolutionary algorithm (a kind of oriented random search algorithm) performs the task.

3 Experimental work

We tested the proposed stochastic neural network on two problem domains. The first is a general problem of Exclusive OR (XOR) or parity problem with 2 and 3 inputs that represent non-linear separable mapping and therefore requires at least two layers of neurons In the second we simulate an autonomous robot in finding the shortest path to a goal in a two-dimensional grid with and without obstacles.

The XOR problem is simple and the reason we choose it is to compare the results (averaged over ten runs) of stochastic reinforced search (RS) with those obtained alternatively by backpropagation learning algorithm (BPG) [7] and adaptive evolutionary algorithm (AEA) [10]. Inputs and output at any particular time n are taken from the set $\{-1, +1\}$. In this experiment a threshold nonlinear function is used as a stochastic neuron's output function. The reinforcement signal is defined as $r(n) = 1 - |Error(n)|$, where $Error$ is the difference between the desired and actual output of the network, which means that $r(n)$ also takes values from the set $\{-1, +1\}$. We use a training set, an evaluation set and a testing set to accomplish the search task. During the training run where the input vectors are randomly chosen, the evaluation set composed of all different input combinations is periodically tested. When the average error within the evaluation set is zero, we stop the training run. It is followed by the testing run, composed of 20 random input vectors, without updating any parameters in the net. The initial weights taken here are random numbers between -0.05 and 0.05, and the rate parameters used are $\alpha = 0.6$ and $\beta = 2$. The results with average

Table I. Average number of steps over 10 runs.

Algorithm	Neural network	2-input XOR
BPG	MLP(2-2-1)	182
RS	SNN'(2-1-1)	180
	SNN(2-2-1)	513
	SNN'(2-2-1)	3
EAE	2-1-1	9100

number of steps over 10 runs obtained with all three algorithms, are given in Table I for 2-input XOR problem,

Table II. Average number of steps over 10 runs.

Algorithm	Neural network	3-input XOR
BPG	MLP(3-2-1)	216
RS	SNN'(3-2-1)	18839
EAE	(3-2-1)	37170

and in Table II for 3-input XOR problem. We obtain the best solution for 2-input XOR with the special feedforward stochastic neural network SNN'(2-2-1) with two hidden neurons, where the inputs are fed to both layers of neurons. The stochastic neural network SNN'(2-1-1) with one hidden neuron and inputs fed to the output neuron gives very similar results to MLP(2-2-1). All the results show that for the XOR problem back-propagation and reinforced search are faster then the adaptive evolutionary algorithm. We can not compare the number of steps in RS with those obtained with BPG and AEA. We use therefore an approximate equivalent which is the number of processings of all stochastic neurons in the net.

The second task is to simulate an autonomous robot in finding the shortest path for each starting position to a goal in a two-dimensional grid (mxm) with or without obstacles, where the space has been discretized. Primitive moving actions of the robot are: take a single step north, south, west or east. The robot is equipped with sensors that enable it to get the reinforcement signal $r(n)$ = +1 for all actions towards the goal and $r(n)$ = -1 in all other situations. In cases where the move cannot be realised (on the edge of the grid) its position remains unchanged.

The robot is supposed to have also a position-based system for localisation. This means that the stochastic neural network receives at each discrete time step a position $s(n)$ consisting of two binary vectors for x and y coordinates. A special coding is used in which the number of positive values defines the position of each direction. For example, in 3x3 grid the inputs (0.9,0.9,-0.9) for x, and inputs (0.9,-0.9,-0.9) for y give the position (x, y)=(2,1).

We use one-layered stochastic neural network with two output neurons and two-layered feed-forward stochastic neural network with two neurons in the hidden layer and two neurons in the output layer, where each output neuron represents the moving in one direction, x and y. For sigmoid output function the hyperbolic tangent function ($f(a_i)$ = tanh$(a_i/2)$) is used. According to the continuous neuron outputs the actions of the robot are coded with two outputs, e.g. north(0.9,0.9), south(-0.9,-0.9), west(-0.9,0.9), east(0.9,-0.9).

For the goal in the navigation procedure the upper right cell in the grid is taken. During one epoch the robot starts in all possible starting positions within the grid and makes for each path maximum of 20 steps towards the goal. After each move during one path, the robot gets the reinforcement signal $r(n)$ from its sensory system, and based on it updates the weights w_{ij} for all neurons. If it reaches the goal before 20 moves, the path is ended. The standard deviation is updated with the averaged reinforcement signal at the end of each path. After each epoch the evaluation is made and when the robot finds the shortest paths from all starting points to the goal state, the search run stops.

The initial weights taken in our experiments are random numbers between -0.05 and 0.05, the rate parameters are α=0.6, β=1 and the initial standard deviation is $\sigma = 0.5$. The results obtained with one-layered and two-layered stochastic neural network are given in Table III. The average number of search epochs over 10

Table III. Average number of search epochs over 10 runs.

Grid	SNN($2m-2$) without obstacle	SNN($2m-2$) with obstacle	SNN($2m-2-2$) without obstacle	SNN($2m-2-2$) with obstacle
3×3	7	38	5	32
4×4	19	39	6	39
5×5	29	96	8	59
6×6	24	164	17	81
7×7	53	162	13	81
8×8	42	108	126	50

runs is very similar for both stochastic neural networks, SNN($2m$-2) and SNN($2m$-2-2), where the first one has $2m$ inputs and 2 output neurons while the second one has also 2 hidden neurons. In most cases the stochastic search algorithm is faster for two-layered stochastic neural network where the number of adaptive weight parameters are increased, which is opposite to the situation in learning algorithms.

The results of an experiment for the 4x4 grid of cells with an obstacle of three cells are shown in Fig. 1. For every starting point (□) the robot travels to the goal (○) in the upper right corner. The resulting actions in the grid world are 'east' and 'north'. For every search run the different combination of possible actions are obtained. Finally, we define the grid 8x8 with three obstacles. The average number of epochs the RS procedure needs is 882 for SNN(16-2), and 353 epochs for SNN(16-2-2). Fig. 2 and Fig. 3 show the results after 535 epochs and the final actions that lead to the goal in one of the search runs. We compare the results of the stochastic neural network with dynamic programming method [8], where after 700 iterations the value function for the optimal sequence of actions is obtained. We must add here that only with the

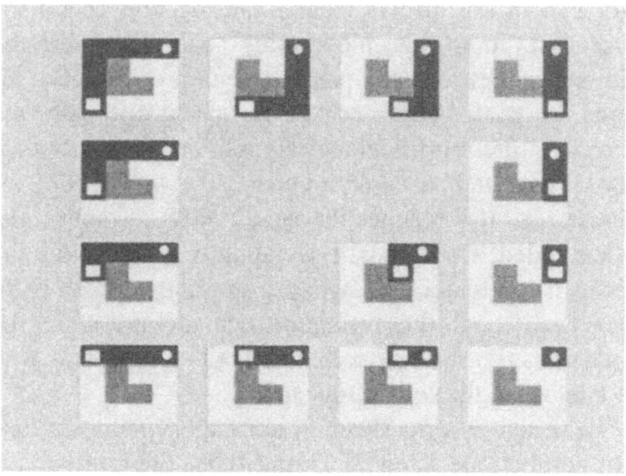

Fig. 1. The shortest paths (obstacles are in gray, shortest path to the goal is in black).

stochastic search procedure and without the given modification of the parameters we are not able to get any results, which means that the directed search is essential for the proposed procedure.

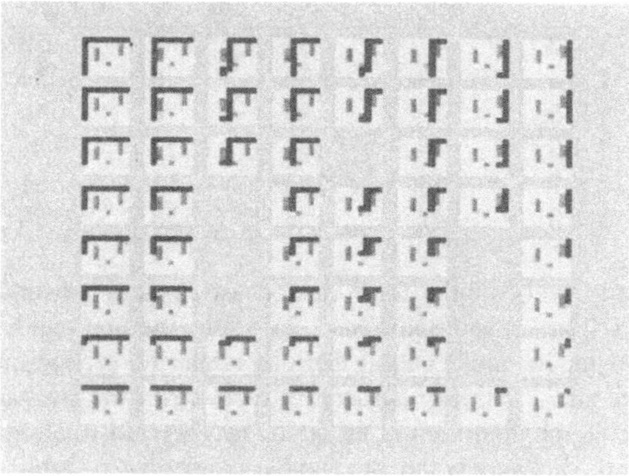

Fig. 2. The shortest paths (obstacles are in gray, shortest path to the goal is in black).

4 Conclusion

In the paper we introduced a new algorithm, which is basically a type of reinforced search algorithm with specific updating scheme, based on local modifications of the weights. We have implemented this algorithm on two tasks of stochastic feed-forward neural networks. The experiments, though obtained on rather simple tasks, show good results for all search runs and acceptable number of steps needed to find the solution of the problem (weight parameters). We believe that this kind of algorithms are 'biologically plausible' alternatives to the (gradient-based) learning algorithms and (off-line) evolutionary procedures. Our further work will be directed

→	→	→	→	→	→	→	
↑	↑	→	↑	→	↑		↑
↑	↑		↑	→	↑		↑
↑	↑		↑	→	↑		↑
↑	↑		↑	→	↑	→	↑
↑	↑	→	↑		↑	→	↑
↑	↑	→	↑	→	↑	→	↑
↑	↑	↑	↑	↑	↑	→	↑

Fig. 3. Final actions (→ - 'east', ↑ - 'north').

to more complex (recurrent) networks and to other more real life oriented tasks in order to find their universal applicability.

References

[1] Barto. A. (1997) Reinforcement learning in Neural Systems for Control, Academic Press

[2] Gullapalli, V. (1990) A Stochastic Reinforcement Learning Algorithm for Learning Real-Valued Functions. Neural Network 3:671–692

[3] Haykin, S. (1999) Neural Networks: A Comprehensive Foundation, 2nd edition, Prentice Hall, New Jersey

[4] Hassoum, M. H. (1995) Fundamentals of Artificial Neural Networks, MIT Press

[5] Hinton, G. E., Sejnovski, T. J. (1986) Learning and Relearning in Boltzmann Machines, PDP, MTI Press

[6] Markram, H., Lubke, J., Frotscher, M., Sakmann, B. (1997) Regulation of synaptic efficacy by Coincidence of postsynaptic APs and EPSPs, Science, 213-215

[7] Muller, J. J., Reinhardt, J., Strickland, M. T. (1995) Neural Network. An Introduction, Springer-Verlag, Berlin Heidelberg

[8] Šter, B., (1999) Reinforced Learning with Neural Networks in Control Problems, Ph.D Thesis, University of Ljubljana

[9] Torndike, E. L., (1911) Animal Inteligence, Hafner, Darien, Conn

[10] Vavpotič, S., Dobnikar, A., (2002) Speeding up Execution of Evolutionary Algorithms with Multiprocessor and Multicomputer Systems, Electrotechnical Review 69(3-4) : 227–233

[11] Williams, R. J. (1992) Simple Statistical Gradient-Following Algorithms for Connectionist Reinforcement Learning, Machine Learning 8: 229–256

A hybrid algorithm for weight and connectivity optimization in feedforward neural networks

F. Pettersson and H. Saxén
Faculty of Chemical Engineering
Åbo Akademi University
Biskopsgatan 8, FIN-20500 Åbo, Finland
E-mail: frank.pettersson@abo.fi, henrik.saxen@abo.fi

Abstract

In modeling with neural networks, the choice of network architecture and size is of utmost importance. The use of a too small network may result in poor performance because of lack of expressional capacity, while a too large network fits noise or apparent relations in the data sets studied. The work required to find a parsimonious network is often considerable with respect to both time and computational effort. This paper presents a method for training feedforward neural networks based on a genetic algorithm (GA), which simultaneously optimizes both weights and network connectivity structure. The proposed method has been found to yield dense and descriptive networks even from training sets of few observations.

1. Introduction

Neural networks have become popular tools for solving approximation, pattern recognition, and classification problems [1]. In networks trained by supervised learning, the weights of the connections are determined by numerical optimization, often minimizing the differences between calculated and the observed values of the output variables. Training methods range from first-order gradient techniques (such as back-propagation) to higher-order methods [2]. An important step in the modeling is the choice of network architecture and connectivity, since this affects the quality of the resulting models. Because of a large number of parameters inherent in neural networks, over-parameterization can be a severe problem; large networks often exhibit poor generalization performance due to over-fitting of the training data. Therefore, neural modeling is by and large still a manual synthesis task, and considerable efforts may be spent on finding a parsimonious network. To remedy the problem, several methods for automatic selection of network complexity have been proposed. Constructive algorithms start with a small network adding new nodes or connections during the progress of training, while destructive methods delete (prune) unnecessary nodes or connections in a large network during the training process [3,4,5,6]. Even though numerous pruning strategies have been proposed, most of the methods are based on purely heuristic criteria or on theory for linear systems. Therefore, the methods cannot be guaranteed to converge to the globally optimal network, and their performance is very problem-dependent (e.g., sensitive to the signal-to-noise ratio of the data).

By expressing the structure of a network with binary variables that indicate the existence of a node, a weight, or even a transfer function, the optimization task can be seen as a mixed integer non-linear programming (MINLP) problem. Unfortunately, existing deterministic algorithms for the solution of MINLP problems are restricted to either convex or small to medium-sized problems [7,8,9], while the network optimization problem at hand is both non-convex and large. An alternative is, therefore, to use stochastic optimization methods, such as evolutionary programming, genetic algorithms or simulated annealing. These techniques are not expected to yield the globally optimal solution, but they may constitute the only feasible alternative for problems with complex search spaces of high dimensionality and a large number of optima.

Several authors have applied stochastic techniques for training weights or for determining weights and structure of neural networks [10,11,12,13]. However, unless the problem is cast into a suitable form for the algorithms, it may require prohibitive computational efforts. To exploit the attractive features of stochastic techniques and to circumvent their drawbacks, one may combine them with methods with better local properties, such as gradient-based search [14,15]. The present paper applies a scheme where the parameters and the structure of feedforward networks are determined with GA in combination with a linear least squares procedure. The combination significantly reduces the required computational effort, and usually yields sparsely connected networks with good generalization performance. Furthermore, the actions of the resulting networks are considerably more transparent than those of fully-connected networks, so they can easily be analyzed and interpreted.

2. Genetic algorithms

Genetic algorithms are optimization techniques based on the concepts of natural selection and genetics [16,17] that work with a group of candidate solutions, called the population, evolving through time in a way resembling the evolution of a natural population of species. Each individual, or chromosome, of the population is characterized by a sequence of genes, and represents a possible solution, while the genes often correspond to the unknown variables to be determined. By selection of the fittest, reproduction and mutation, chromosomes with improved properties gradually evolve. After initialization, all individuals are evaluated to allow for comparison of their fitness, followed by a selection of individuals that are allowed to produce successive generations. In the reproduction phase, genes from (two) parent chromosomes are swapped forming offsprings, a common recombination technique being one-point crossover. Usually, the children next replace their parents in the population. Offsprings preserve fitnes of the chromosomes, but major improvements may fail to appear. Therefore, mutations are applied allowing for an evolution through random changes (to a small extent) in the chromosomes. The above-mentioned processes are continued for a number of generations or until some termination criteria, e.g., low diversity in population or lack of improvement, are satisfied. While the solution is being evolved, the fitness of the best chromosome and the coverage of the search space [18] can be monitored over the generations to illustrate the progress of the optimization.

3 Evolution of neural networks

The method applied in this paper is primarily intended for training feedforward networks of multi-layer perceptron type. Since genetic algorithms are efficient at global sampling but have poor local convergence properties, the network parameters are partitioned as indicated in Fig. 1 for a network with a single layer of hidden sigmoid nodes. The genetic algorithm works with a population, where each chromosome includes the biases and weights in the lower part of the network, while the weights to and biases of the output nodes are determined by linear least squares (LLSQ). This procedure reduces the number of variables to be manipulated by the GA, while the LLSQ-optimized part guarantees that the evolved individuals are at least sub-optimal.

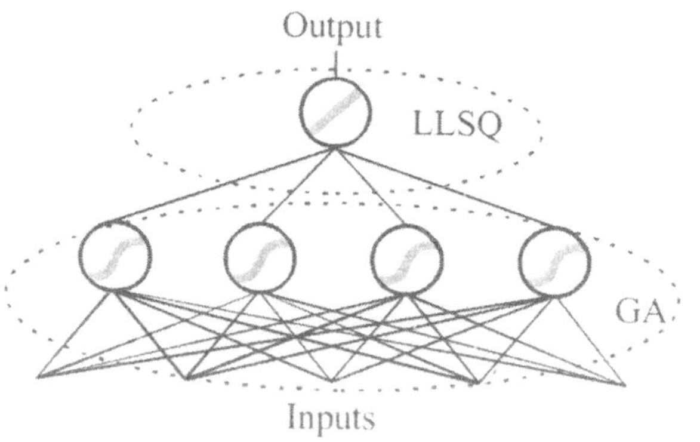

Fig. 1. Network subdivision: Parameters of the lower part are determined by a genetic algorithm and those of the upper part are solved with linear least squares.

In genetic code the weights and the biases are represented as real values in the chromosomes and the weights to each hidden node are grouped to preserve well-working nodes during the evolution. The connectivity of the network is defined by binary variables, y, for every weight of a connection to a hidden node. Biases do not have corresponding binary variables in the chromosomes; they are considered if there is at least one active incoming connection to the node in question. The length of a chromosome is fixed and does not depend on whether the connections are active or not; weights in the chromosomes are not affected by changes in the binary variables but only by crossover and mutation. An example of a network and the genetic representation of its lower part is given in Fig. 2.

The population is initialized by specifying the maximum network size, assigning random values to the genes. It is generally desirable to start with hidden nodes operating in the transition (nearly linear) regime of the sigmoid. Therefore, if the input (and also output) variables are normalized to the interval (-1,+1), the weights may be initialized as

$$w \sim N(0,\sigma^2), \quad (1)$$

i.e., as normally distributed random variables with a mean of zero and a standard deviation of σ.

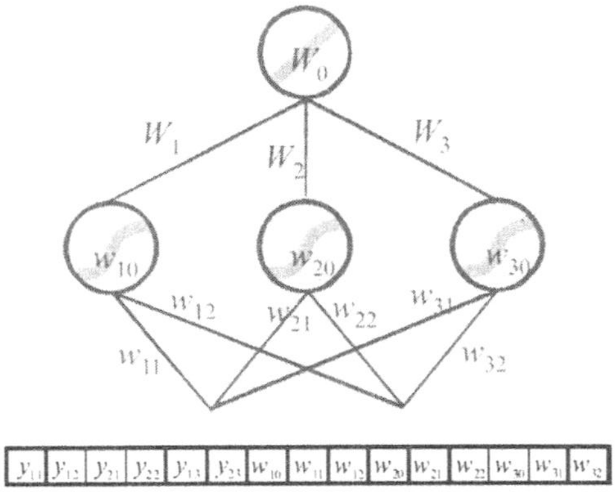

Fig. 2. Neural network and corresponding variable representation in the chromosome. Binary variables are y, lower-part weights are w, and upper-part weights (determined by LLSQ) are W.

As a measure of the fitness of a chromosome, one may use the mean square prediction error, F_1, calculated on the differences between observed and predicted outputs, f and $\hat{f}$, respectively, for the K observations in the training set. However, to limit the complexity of the network, a penalty term, F_2, was added to the objective function, yielding the minimization problem

$$\min_{\mathbf{w},\mathbf{y}}\left\{F_1 + F_2 = \sqrt{\sum_{k=1}^{K}\frac{\left(f_k - \hat{f}_k\right)^2}{K}} + q\sum_{i=1}^{I} y_i\right\} \tag{2}$$

where I is the number of possible connections and q is a user-defined penalty factor. After the genetic algorithm has evolved the lower layer weights, $\mathbf{w}$, the outputs from the sigmoids can be determined. $\mathbf{X}$ is a matrix with a first column of ones (for the bias) followed by the outputs from the hidden nodes calculated for the K input vectors. The network outputs are thus given, for a network with n nodes in the (last) hidden layer and a single output node, by

$$\hat{\mathbf{f}} = \begin{bmatrix}\hat{f}_1\\ \hat{f}_2\\ \hat{f}_3\\ \vdots\\ \hat{f}_K\end{bmatrix} = \begin{bmatrix}1 & x_{1,1} & x_{1,2} & \cdots & x_{1,n}\\ 1 & x_{2,1} & x_{2,2} & \cdots & x_{2,n}\\ 1 & x_{3,1} & x_{3,2} & \cdots & x_{3,n}\\ \vdots & \vdots & \vdots & & \vdots\\ 1 & x_{K,1} & x_{K,2} & & x_{K,n}\end{bmatrix}\begin{bmatrix}W_0\\ W_1\\ W_2\\ \vdots\\ W_n\end{bmatrix} = \mathbf{XW} \tag{3}$$

The weights, $\mathbf{W}$, can be determined with LLSQ by solving

$$\min_{\mathbf{W}} \tfrac{1}{2}\left\|\mathbf{f} - \hat{\mathbf{f}}\right\|^2 \tag{4}$$

The problem can be solved by, e.g., Householder reflections using an orthogonal-triangular factorization [19]. By varying the penalty factor, q, the size of the evolving networks can be controlled, which affects the generalization performance of the resulting models.

4. Experimental results

This section presents the performance of the method on two examples. The first test is an artificial single-input single-output relation, while the second is a problem encountered in ironmaking industry.

4.1 Staircase problem

The first test is a simple staircase-like problem, where the network has one input and one output. The problem is shown in Fig. 3 where the circles represent the data points. It is clear from the figure that three sigmoidal hidden nodes are sufficient to describe the problem exactly. The problem is difficultly trained by traditional gradient methods even when the correct network structure is used. Using the Levenberg-Marquardt gradient method, as implemented in the NNDT program package [20] and with several runs with randomly generated initial values, the best networks satisfactorily described one or maximally two of the "steps".

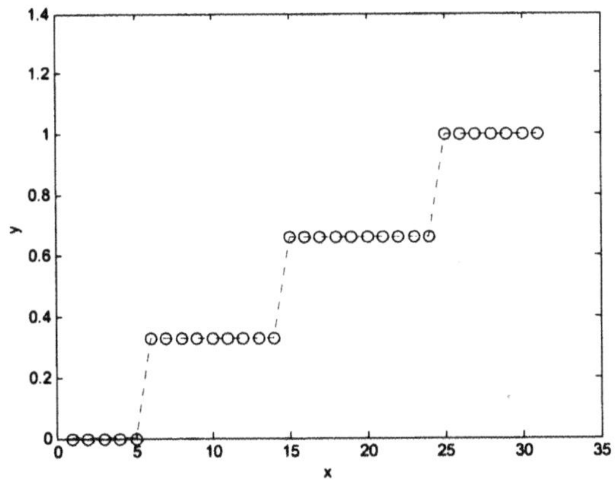

Fig. 3. Staircase problem: Observations are denoted by circles, approximation by dashed lines.

To solve the staircase problem with the proposed GA-hybrid method, a population size of 50, with 20 generations, an elitism of 40%, a mutation probability of 3%, and a maximum of ten nodes in one hidden layer were used. The selection of penalty for inclusion of a weight connection is not very critical; penalty values, q, between $2 \cdot 10^{-6}$ and $2 \cdot 10^{-3}$ have been tested and all cases result in a perfect model within the same number of generations. The computational effort is also modest (0.7 CPU min on a PC PIII 700 MHz). The evolution of the errors and the active connections selected by the model have been presented in Fig. 4.

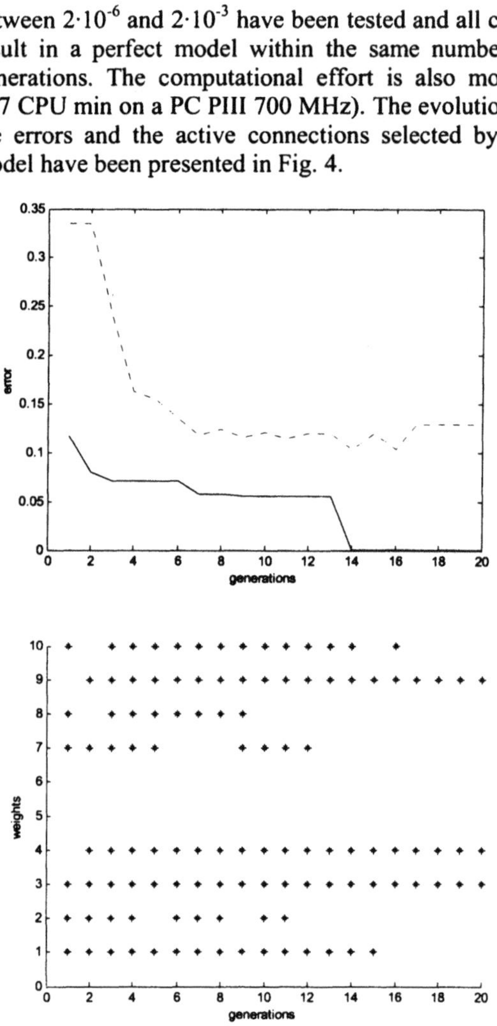

Fig. 4. Top: Evolution of the error of the best (solid line) and worst solution (dashed line) of every generation. Bottom: Nodes with active connections in each generation.

4.2 Burden-layer thickness problem

In the operation of the ironmaking blast furnace, the iron-bearing material (sinter or pellets) and coke are charged in separate layers to yield sufficient gas permeability in the dry part of the shaft. The radial distribution is important because it affects both thermal and chemical conditions in the furnace. The burden distribution can be controlled by applying different movable armor positions for the dumps in furnaces with bell-type charging equipment. Furthermore the charging sequence (i.e. the order of the dumps), the dump size and the stock level, and even the gas flow, affect the distribution.

In this example the thickness of burden layers in a blast furnace is modeled on the basis of thickness estimates calculated from stockrod signals [21]. The stockrods are sounding devices that sense the burden level after each dump and are elevated before a new dump of burden is charged into the furnace. The aim of the effort is a model that describes the interaction between the layer thickness and variables such as stock level and movable armor settings. The process data from ref. 21 is used in this study. In the data set, the movable armor (MA) positions take on the discrete values 2, 5, 6, 7, 9 and 10, while the layer thickness (z) varies between 2.3 m and 4.3 m. The data was divided into a training set and a test set, where the size of training set was varied from 50 to 200 and the test set was the same in all cases covering 800 measurements.

The problem was tackled with the GA-hybrid method, using a population size of 50 with 200 generations, and the rest of the parameters as in the example of subsection 4.1. The complexity penalty, q, was varied from 0 to $6 \cdot 10^{-5}$ in order to obtain networks of different sizes. It was found that the method was able to produce accurate solutions to the problem even with small networks. The resulting network complexities, expressed as the total number of weights and biases in the network, varied from 7 to 19 for the different values of q. Figure 5 presents the prediction accuracy of a network with two (dotted line) and six (dashed line) GA-trained weights.

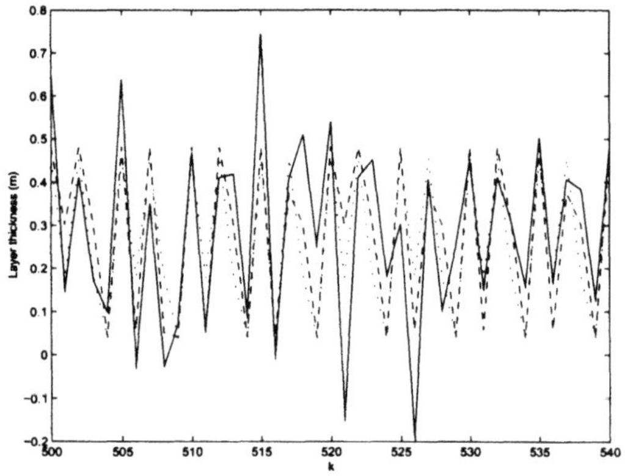

Fig. 5. Observed layer thickness (solid lines) and predictions by evolved networks with two (dotted lines) or six (dashed lines) GA-trained weights.

It was found that the method was able to solve the problem even with small-sized training sets without notable over-fitting, and most of the evolved networks were sparsely connected. An advantage of a small number of connections is, besides getting a model with good generalization capacity, that it is easier to understand the action of the model. For example, the output of the network with six GA-trained weights is studied in Fig. 6 as a function of the two inputs. The problem is seen to have been solved as a classifying task with plateaus on different levels depending on the combination of input values.

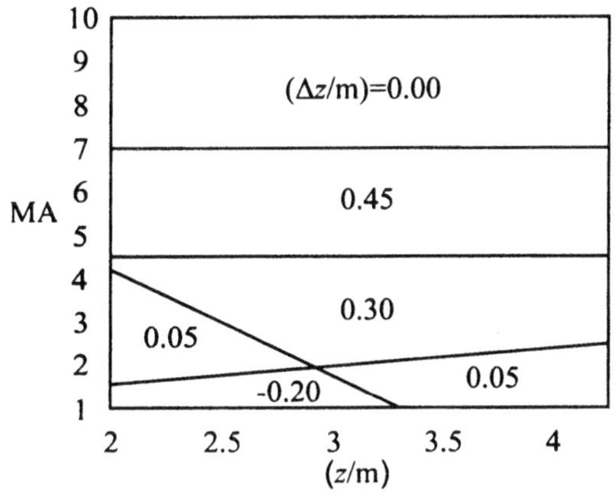

Fig. 6. Output from a network with six GA-trained weights as a function of the stock level (z) and the movable armor position (MA).

5. Conclusions

A method for simultaneous optimization of weights and connectivity in feedforward neural networks has been proposed. The method is based on a genetic algorithm that optimizes the weights of the lower part of the network, while a least-squares solution of the upper weights guarantees at least sub-optimal performance of the evolved models. Even though the training method proposed in the paper has been applied to feedforward networks, it is straightforward to apply it to other network structures. However, it is possible that the computational time grows substantially if real world problems are tackled by recurrent networks. The method has been illustrated on a test case, where an artificial data set is used, as well as on a sub-problem from the ironmaking industry. The results show that the proposed technique is efficient, and that it is worth developing it further as a tool for automatic model and input variable selection in neural network modeling.

References

[1] Haykin, S., (1994) *Neural Networks – A Comprehensive Foundation*, Macmillan Publishing Co., New York.

[2] Battiti, R., (1992) "First- and Second-Order Methods for Learning: Between Steepest Descent and Newton's Method", *Neural Computation* **4**, 141–166.

[3] Frean, M., (1989) "The Upstart Algorithm. A method for Constructing and Training Feed-forward Neural Networks", Edinburgh Physics Department, Preprint 89/469, Scotland.

[4] Fahlman, S.E., and C. Lebiere (1990) "The Cascade-Correlation Learning Architecture", in *Advances in Neural Information Processing Systems* **II**, (Ed. D.S. Touretzky), pp. 524-532.

[5] Le Chun, Y., J. S. Denker and S. A. Solla, (1990) "Optimal Brain Damage", in *Advances in Neural Information Processing Systems* **2**, ed. D.S. Touretzky, pp. 598–605, (Morgan

[6] Thimm, G., and E. Fiesler, (1995) "Evaluating pruning methods", Proc. of the *1995 International Symposium on Artificial Neural Networks (ISANN'95)*, Hsinchu, Taiwan, ROC.

[7] Duran, M. A., and I. E. Grossmann, (1986) "An Outer-Approximation Algorithm for a Class of Mixed-Integer Nonlinear Programs", *Mathematical Programming* **6**, 307–339.

[8] Westerlund, T., and F. Pettersson, (1995) "An extended cutting plane method for solving convex MINLP problems", *Computers and Chem. Eng.* **19**, S131–S136.

[9] Ryoo, H.S., and N. V. Sahinidis, (1995) "Global Optimization of Nonconvex NLPs and MILPs with Applications in Process Design", *Computers and Chem. Eng.* **19**, 551–566.

[10] Fogel, D.B., L. J. Fogel and V. W. Porto, (1990) "Evolving Neural Networks", *Biol. Cybern.* **63**, 487–493.

[11] Maniezzo, V., (1994) "Genetic Evolution of the Topology and Weight Distribution of Neural Networks", *IEEE Transactions on Neural Networks* **5**, 39–53.

[12] Reeves, C., and N. Steele, (1993) "Applications of Genetic Algorithms in Artificial Neural Networks", *Systems Science* **19**, 63–76.

[13] Angeline, P.J., G. M. Saunders and J. B. Pollack, (1993) "An Evolutionary Algorithm That Constructs Recurrent Neural Networks", *IEEE Trans. Neural Networks* **5**, 54–65.

[14] Gao, F., M. Li, F. Wang, B. Wang and P. Yue, (1999) "Genetic Algorithms and Evolutionary Programming Hybrid Strategy for Structure and Weight Learning for Multilayer Feedforward Neural Networks", *Ind. Eng. Chem. Res.* **38**, 4330-4336.

[15] Schmitz, G.P.J., (1999) *Combinatorial evolution of feedforward neural network models for chemical*

processes, Ph.D. Dissertation, University of Stellenbosch, Republic of South Africa.

[16] Holland, J.H., (1975) *Adaptation in Natural and Artificial Systems,* The University of Michigan Press, Ann Arbor.

[17] Goldberg, D.E., (1989) *Genetic Algorithms in Search, Optimization and Machine Learning*, Addison Wesley, Reading, MA, USA.

[18] Wehrens, R., E. Pretch and L.M.C Buydens, (1998) "Quality Criteria of Genetic Algorithms for Structure Optimization", *J. Chem. Inf. Comput. Sci.* **38**, 151–157

[19] Golub, G., (1965) "Numerical methods for solving linear least squares problems", *Numer. Math.* **7**, 206–216.

[20] Saxén, H., and B. Saxén, (1994) "A Tool for Modeling, Simulation and Prediction Using Feedforward and Recurrent Neural Networks", Proc. of *I Brazilian Symposium on Neural Networks*, Caxambú, MG, Brazil, pp. 55–60.

[21] Hinnelä, J., and H. Saxén, (2001) "Neural Network Model of Burden Layer Formation Dynamics in the Blast Furnace", *ISIJ International* **41**, 142–150.

Neural network learning as approximate optimization

Věra Kůrková [1] Marcello Sanguineti [2]

[1]Institute of Computer Science, Academy of Sciences of the Czech Republic, Pod Vodárenskou věží 2, 182 07 Prague, Czech Republic. E-mail: vera@cs.cas.cz [2] Department of Communications, Computer, and System Sciences (DIST), University of Genoa, Via Opera Pia 13, 16145 Genoa, Italy. E-mail: marcello@dist.unige.it

Abstract

Learning from data is studied in the framework of approximate minimization of functionals. A set of admissible solutions over which such functionals are minimized is approximated by a nested family of sets of functions computable by neural networks with n hidden units. There are derived upper bounds on the speed of convergence of infima achievable over such approximations of admissible sets to a global infimum. The bounds are expressed in terms of a certain norm tailored to the type of network units and modulus of continuity of the functional to be minimized. The results are applied to empirical error functionals regularized using stabilizers that are defined as squares of norms in reproducing kernel Hilbert spaces.

1 Introduction

Learning from a sample of empirical data, which is given by a finite set of pairs $\{(x_i, y_i) \in \mathcal{R}^d \times \mathcal{R},\ i = 1, \ldots, m\}$, can be formally modelled as minimization of a functional called the *empirical error* (or *empirical risk*) defined as $\mathcal{E}(f) = \frac{1}{m}\sum_{i=1}^{m} |f(x_i) - y_i|^2$. However for many admissible sets, the problem of minimization of the empirical error is ill-posed. A classical approach to coping with ill-posedness is regularization, which takes into account, besides of measured data, also conceptual ones (i.e., global knowledge about the solution such as smoothness) [24, 7]. Formally, minimization of the empirical error is replaced with minimization of the functional $\mathcal{E}_{\gamma,\Psi} = \mathcal{E} + \gamma\Psi$, where Ψ is a functional called *stabilizer* and γ is a *regularization parameter*. The stabilizer expresses requirements on the global behavior of a solution and the regularization parameter expresses the trade-off between fitting to empirical and conceptual data [7, 26]. The choice of a proper stabilizer (such as a strictly uniformly convex functional) guarantees existence and uniqueness of solution of the regularized problem. A common class of stabilizers is formed by functionals $\|.\|_K^2$, where $\|.\|_K$ is the norm on a Hilbert space of a particular type, namely a reproducing kernel Hilbert space (RKHS).

It has been argued in [13, p. 219] that "the regularization principles lead to approximation schemes that are equivalent to networks with one layer of hidden units". Indeed, various versions of the so-called Representer Theorem (see, e.g., [13] and [8, p. 42]) show that for suitable stabilizers (including those defined as squares of norms in RKHS) the unique function minimizing the regularized empirical error has the form of a one-hidden layer network with a linear output and hidden units computing functions corresponding to the type of stabilizer. In particular, for stabilizers defined by the Gaussian kernel, the Representer Theorem states that the minimum over the whole RKHS is achieved at a function computable by a Gaussian radial-basis function network. Moreover, the Representer Theorem gives a formula for computing network parameters as a solution of a well-posed system of linear equations. A drawback of this elegant theoretical result is that the number of hidden units in the network is equal to the number of pairs of input/output data. For large data sets, such a network might not be implementable. Moreover in typical applications of neural networks, a number of units much smaller than the number of data is chosen before learning. So minimization is performed over a smaller "approximate" admissible set formed only by functions computable by networks with a given number of hidden units.

Motivated by these complexity constrains, we investigate approximate versions of the Representer Theorem. We consider minimization over admissible sets formed by functions computable by networks with at most n hidden units corresponding to the type of the kernel. We study such approximate optimization in a more general framework of approximate optimization called the "extended Ritz method" [27, 1, 3, 21, 22, 28]. The classical Ritz method [12] approximates an admissible set by a nested family of linear subspaces of increasing dimensionality, while the extended Ritz method uses a nested family of nonlinear sets formed by variable-basis functions, a special case of which are neural networks with a n hidden units and a linear output unit.

For continuous functionals, we derive upper bounds on rates of convergence of sequences of infima over approximate admissible sets of increasing complexity to the global minimum, as well as the speed of convergence of the corresponding argminima. The bounds are expressed in terms of certain norms, tailored to the type of network units, and moduli of Tychonov well-posedness of optimimization problems. The estimates for our approximate version of the Representer Theorem depend on these special norms and moduli of continuity of functionals at argminima (for kernel stabilizers, such moduli are quadratic functions). Our estimates are not merely asymptotic: they hold for every number of hidden units and thus they can be applied to networks with a number of hidden units smaller than the number of data. Moreover, the upper bounds hold for Hilbert spaces formed by functions of any number of variables and thus they allows us to describe classes of optimization problems for which such rates do not exhibit the "curse of dimensionality".

The paper is organized as follows. Section 2 introduces notations and definitions. In section 3, estimates of rates of approximate optimization are derived for approximation of admissible sets by variable-basis functions (a special case of which are neural networks). In section 4, these estimates are applied to minimization of empirical error functionals regularized using squares of norms in RKHS.

2 Preliminaries

By a normed linear space $(X, \|.\|)$ we mean a real normed linear space; we write merely X when there is no risk of ambiguity. $\mathcal{R}$ denotes the set of real numbers, $\mathcal{R}_+$ the set of positive reals. A ball, a sphere, resp., of radius r centered at $h \in X$ is denoted by $B_r(h, \|.\|) = \{f \in X : \|f - h\| \leq r\}$, $S_r(h, \|.\|) = \{f \in X : \|f - h\| = r\}$. We write shortly $B_r(\|.\|) = B_r(0, \|.\|)$ and $B_r(h) = B_r(h, \|.\|)$, $B_r = B_r(0)$ when it is clear which norm is used; similarly for spheres. Sequences (of real numbers, sets or elements of normed linear spaces) are denoted by $\{x_n\}$ instead of $\{x_n : n \in \mathcal{N}_+\}$, where $\mathcal{N}_+$ is the set of positive integers.

A functional $\Phi : X \to \mathcal{R}$ is *continuous* at $f \in X$ if for all $\varepsilon > 0$ there exists $\eta > 0$ such that $\|f - g\| < \eta$ implies $|\Phi(f) - \Phi(g)| < \varepsilon$. A *modulus of continuity* of Φ at f is a function $\omega : \mathcal{R}_+ \to \mathcal{R}_+$ defined as $\omega(a) = \sup\{|\Phi(f) - \Phi(g)| : \|f - g\| \leq a\}$. Φ is *convex* over a convex set $M \subseteq X$ if for all $h, g \in M$ and all $\lambda \in [0,1]$, $\Phi(\lambda h + (1-\lambda)g) \leq \lambda\Phi(h) + (1-\lambda)\Phi(g)$. Φ is *strictly uniformly convex* on a convex set $M \subseteq X$ if there exists a function $\delta : \mathcal{R}_+ \to \mathcal{R}_+$, such that $\delta(0) = 0$, $\delta(t) > 0$ for all $t > 0$ and for all $h, g \in M$ and all $\lambda \in [0,1]$, $\Phi(\lambda h + (1-\lambda)g) \leq \lambda\Phi(h) + (1-\lambda)\Phi(g) - \lambda(1-\lambda)\delta(\|h - g\|)$. Any such function δ is called a *modulus of convexity* of Φ [10, Chapter 1].

Using standard notation (see, e.g., [11]), we denote by (M, Φ) the problem of infimizing a functional Φ over a subset M of X. M is called the set of *admissible solutions* or *admissible set*. By $argmin\,(M, \Phi) = \{g \in M : \Phi(g) = \inf_{g \in M} \Phi(g)\}$ is denoted the set of *argminima* of the problem (M, Φ) and for any $\varepsilon > 0$, $argmin_\varepsilon(M, \Phi) = \{g \in M : \Phi(g) < \inf_{g \in M} \Phi(g) + \varepsilon\}$ is the set of its *ε-near argminima*. A sequence $\{g_n\}$ of elements of M is called *Φ-minimizing over M* if $\lim_{n \to \infty} \Phi(g_n) = \inf_{g \in M} \Phi(g)$. By the definition of infimum, for any problem (M, Φ) with M non-empty, there always exists a minimizing sequence. The problem (M, Φ) is *Tychonov well-posed* if it has a unique global minimum in M to which every minimizing sequence converges [11, p. 1]. The *modulus of Tychonov well-posedness* of (M, Φ) at an argminimum g^o is a function $\xi : \mathcal{R}_+ \to \mathcal{R}_+$ such that for every $a \in \mathcal{R}_+$, $\xi(a) = \inf_{g \in M \cap S_a(g^o)} \Phi(g) - \Phi(g^o)$. Note that the modulus of Tychonov well-posedness is defined for any problem that has an argminimum, even when such a problem is not Tychonov well-posed.

The *linear span* of M is $span\, M = \{\sum_{i=1}^{n} w_i g_i : w_i \in \mathcal{R},\ g_i \in M,\ n \in \mathcal{N}_+\}$, the *topological interior* of M is $int\, M = \{g \in M : (\exists\, \varepsilon > 0)\, (B_\varepsilon(g) \subset M)\}$ and the *closure* of M is $cl\, M = \{f \in X : (\forall \varepsilon > 0)\, (B_\varepsilon(f) \cap M) \neq \emptyset)\}$. If $clM = Y$, then M is said to be *dense* in Y. The *Minkowski functional* of $M \subset X$ denoted by p_M, is defined for every $f \in X$ as $p_M(f) = \inf\{\lambda \in \mathcal{R}_+ : f/\lambda \in M\}$.

3 Approximate optimization by variable-basis functions

A classical approach to approximate optimization, called the *Ritz method* [12], approximates the problem (M, Φ) by a sequence of problems

$$\{(M \cap X_n, \Phi)\},$$

where X_n are n-dimensional subspaces of X. Typically, they are generated by the first n elements of a given linearly ordered subset of X (e.g., all powers of order at most $n-1$) and this corresponds to the use of *fixed-basis approximation*.

The Ritz method can be generalized by allowing $\{X_n\}$ to be any nested sequence of subsets of X. A special case, motivated by neural networks, is the *extended Ritz method* [27, 1, 3, 21, 22, 28], where a nested family of admissible sets is formed by linear combinations of at most n elements of a given set G, defined as $span_n\, G = \{\sum_{i=1}^{n} w_i g_i : w_i \in \mathcal{R},\ g_i \in G\}$. This approximation scheme is called the *variable-basis approximation* [17, 18]. It includes splines with free nodes, trigonometric polynomials with free frequencies and feedforward neural networks (see, e.g., [9] and the references therein).

To estimate the rate of convergence of a sequence of infima over variable-basis approximations of an admissible set to a global infimum, we employ a result from nonlinear approximation theory by Maurey [23], Jones [14], and Barron [5]. Here we use its reformulation from [15] in terms of a norm tailored to a given set G. For a subset G of a normed linear space $(X, \|.\|)$, G-variation, denoted by $\|.\|_G$, is defined as the Minkowski functional of the set $cl\, conv\, (G \cup -G)$. Thus,

$$\|f\|_G = \inf\left\{c > 0 \,:\, c^{-1} f \in cl\, conv\,(G \cup -G)\right\}.$$

G-variation is a norm on the subspace $\{f \in X : \|f\|_G < \infty\} \subseteq X$ and it satisfies $\|.\| \leq s_G \|.\|_G$. It is a generalization of the notion of l_1-norm as for G an orthonormal basis of a separable Hilbert space, G-variation is equal to the *l_1-norm with respect to G*, defined for every $f \in X$ as $\|f\|_{1,G} = \sum_{g \in G} |f \cdot g|$ [20, 17]. G-variation is also a generalization of the concept of total variation studied in integration theory [4]. For other properties of G-variation see [15, 17, 18]. The following theorem from [15] is a reformulation of Maurey-Jones-Barron's bound in terms of G-variation.

Theorem 3.1 *Let $(X, \|.\|)$ be a Hilbert space, G its bounded subset and $s_G = \sup_{g \in G} \|g\|$. Then for every $f \in X$ and every positive integer n, $\|f - span_n G\| \leq \sqrt{\frac{(s_G \|f\|_G)^2 - \|f\|^2}{n}}$.*

As Theorem 3.1 applied to Hilbert spaces formed by functions with any number of variables gives estimates of the order of $\mathcal{O}(n^{-1/2})$, some authors called it "dimension-independent". However, this might be misleading as both s_G and shape of balls of fixed radii in G-variation depend on the number of variables (for properties of balls in G-variation see [5, 20, 17, 18] and the references therein).

Next theorem gives estimates of rates of approximate optimization by the extended Ritz method for a problem (M, Φ) of minimization of a continuous functional over a closed convex admissible set with zero in its interior such that (M, Φ) has an argminimum. The proof of the theorem given in [19] is based on a construction of a special minimizing sequence, to which Theorem 3.1 can be applied.

Theorem 3.2 *Let $(X, \|.\|)$ be a Hilbert space, M and G be its subsets, G bounded, $s_G = \sup_{g \in G} \|g\|$, M closed, convex, and $0 \in int\, M$. Let $\Phi : X \to (-\infty, +\infty]$ be a functional, $g^o \in argmin\,(M, \Phi)$, Φ be continuous at g^o with a modulus of continuity ω, and let $\{\varepsilon_n\}$ be a sequence of positive reals such that $\lim_{n \to \infty} \varepsilon_n = 0$ and $g_n \in argmin_{\varepsilon_n}(M \cap span_n\, G, \Phi)$. Then for every integer n the following hold:*

(i) $\inf_{g \in M \cap span_n G} \Phi(g) - \Phi(g^o) \leq \omega\left((1 + c\|g^o\|)\sqrt{\frac{(s_G\|g^o\|_G)^2 - \|g^o\|^2}{n}}\right)$;

(ii) if $\|g^o\|_G < \infty$, then $\{g_n\}$ is a Φ-minimizing sequence over M and $\Phi(g_n) - \Phi(g^o) \leq \omega\left((1 + c\|g^o\|)\sqrt{\frac{(s_G\|g^o\|_G)^2 - \|g^o\|^2}{n}}\right) + \varepsilon_n$;

(iii) if ξ is the modulus of Tychonov well-posedness of (M, Φ) at g^o, then $\xi(\|g_n - g^o\|) \leq \omega\left((1 + c\|g^o\|)\sqrt{\frac{(s_G\|g^o\|_G)^2 - \|g^o\|^2}{n}}\right) + \varepsilon_n$, *where c is the Lipschitz constant of p_M on X.*

Theorem 3.2 proves that for $\|g^o\|_G$ finite, the sequence of approximate argminima $\{g_n\}$ is a Φ-minimizing sequence. Moreover, it estimates the speed of convergence of this sequence to the global argminimum as well as convergence of infima of Φ over $M \cap span_n G$ to its infimum over the whole set M. Inspection of the upper bounds from Theorem 3.2 shows that when the argminimum of (M, Φ) is in a ball in G-variation of a radius r, then rates of approximate optimization are bounded from above by $\omega((1 + c\|g^o\|)r/\sqrt{n})$. Thus when with increasing number of variables we keep a fixed bound on G-variations of global argminima (or even of all admissible functions), we can avoid the curse of dimensionality [6] in approximate optimization by the extended Ritz method. Note that for G corresponding to neural network computational units, balls in G-variation are not shrinking with increasing dimension [16].

Theorem 3.2 provides a theoretical analysis of the extended Ritz method. It supports the conjecture stated, for example, in [27, 1, 3, 21, 22, 28], that the extended Ritz method is suitable for high-dimensional optimization problems.

4 Approximate optimization in learning from data

Let $\{(x_i, y_i) \in \mathcal{R}^d \times \mathcal{R},\, i = 1, \dots, m\}$ be a finite set of input/output pairs of data. Learning from such data can be formalized as minimization of the *empirical error* functional, defined as $\mathcal{E}(f) = \frac{1}{m}\sum_{i=1}^m |f(x_i) - y_i|^2$. The minimization is usually performed over a parametrized set of functions implemented by a suitable "device", such as a neural network. Typically, such an optimization problem is ill-posed and thus it requires regularization. *Tychonov regularization* replaces the functional $\mathcal{E}$ with $\mathcal{E}_{\gamma,\Psi} = \mathcal{E} + \gamma\,\Psi$, where Ψ is a functional called *stabilizer* and γ is a *regularization parameter* (see, e.g., [24, 7, 26]).

A common class of stabilizers is formed by functionals of the form $\|.\|_K^2$, where $\|.\|_K$ is the norm on a reproducing kernel Hilbert space (RKHS). A RKHS $\mathcal{H}_K(\Omega)$

is a Hilbert space of functions defined on a set Ω such that for every $x \in \Omega$ the evaluation functional $\mathcal{F}_x$, defined for any $f \in \mathcal{H}_K(\Omega)$ as $\mathcal{F}_x(f) = f(x)$, is bounded.

Recall that a mapping $K : \Omega \times \Omega \to \mathcal{R}$ is *positive semidefinite* if for all positive integers m, all $(w_1, \ldots, w_m) \in \mathcal{R}^m$, and all $(x_1, \ldots, x_m) \in \Omega$, we have $\sum_{i,j=1}^m w_i w_j K(x_i, x_j) \geq 0$. To every RKSH one can associate a unique symmetric, positive-definite mapping $K : \Omega \times \Omega \to \mathcal{R}$ called *kernel*, such that for every $f \in \mathcal{H}_K(\Omega)$, $f(x) = \langle f, K(x,.)\rangle_K$ [2]. A kernel $K : \Omega \times \Omega \to \mathcal{R}$ is called *Mercer kernel* if Ω is compact and K is symmetric, continuous and positive semidefinite. By [8, p. 35], for Ω compact and K a Mercer kernel, the space $\mathcal{H}_K(\Omega)$ is contained in the space $\mathcal{C}(\Omega)$ of continuous functions on Ω. We denote by $\|.\|_\mathcal{C}$ the supremum norm on $\mathcal{C}(\Omega)$. A kernel K is called *convolution* or *translation invariant* if there exists a function $k : \Omega \to \mathcal{R}$ such that $K(x, y) = k(x - y)$.

Using $\|.\|_K^2$ as a stabilizer, the regularized functional obtained from $\mathcal{E}$ has the form

$$\mathcal{E}_{\gamma,K}(f) = \frac{1}{m}\sum_{i=1}^m |f(x_i) - y_i|^2 + \gamma\|f\|_K^2.$$

The following theorem is a version of the Representer Theorem from [8, p. 42]. The uniqueness of the argminimum follows from the strict uniform convexity of the functional $\mathcal{E}_{\gamma,K}$. It is easy to see that $\|.\|_K^2$ is strictly uniformly convex and the sum of a convex and a strictly uniformly convex functional is strictly uniformly convex [10, p. 10].

Theorem 4.1 *Let $\Omega \subset \mathcal{R}^d$ be compact, $K : \Omega \times \Omega \to \mathcal{R}$ be a Mercer kernel, $\mathcal{H}_K(\Omega)$ be the RKHS defined by K, $x = (x_1, \ldots, x_m) \in \Omega^m$, $y = (y_1, \ldots, y_m) \in \mathcal{R}^m$, $\mathcal{E} : \mathcal{H}_K(\Omega) \to \mathcal{R}_+$ be the empirical error $\mathcal{E}(f) = \frac{1}{m}\sum_{i=1}^m |f(x_i) - y_i|^2$, and $\gamma > 0$. Then there exists a unique argminimum of the problem $(\mathcal{H}_K(\Omega), \mathcal{E}_{\gamma,K})$ of the form $g^o(x) = \sum_1^m a_i K(x, x_i)$, where $a = (a_1, \ldots, a_m)$ is the unique solution of the well-posed linear system $(\mathcal{K}(x) + \gamma\, m\, I)a = y$, I is the $m \times m$ identity matrix, and $\mathcal{K}(x)$ is the $m \times m$ matrix defined as $\mathcal{K}(x)_{i,j} = K(x_i, x_j)$.*

Notice that the unique argminimum described in Theorem 4.1 has the form of a variable-basis function from $span_m\, G_K$, where $G_K = \{K(x, \cdot) : x \in \Omega\}$. In particular, for the convolution kernel defined by the Gaussian function, the argminimum can be computed by a Gaussian radial-basis function network with m hidden units. However, for large values of m such a network might not be implementable. Application of Theorem 3.2 allows us to estimate the quality of the approximate solution of the regularized learning problem achievable using networks with any number n of hidden units smaller than m.

Theorem 4.2 *Let $\Omega \subset \mathcal{R}^d$ be compact, $K : \Omega \times \Omega \to \mathcal{R}$ be a Mercer kernel, $s_K = \sup_{x\in\Omega} \sqrt{K(x,x)}$, $(\mathcal{H}_K(\Omega), \|.\|_K)$ be the RKHS defined by K, $\mathcal{E} : \mathcal{H}_K(\Omega) \to \mathcal{R}_+$ be the empirical error $\mathcal{E}(f) = \frac{1}{m}\sum_{i=1}^m |f(x_i) - y_i|^2$, where $(x_1, \ldots, x_m) \in \Omega^m$, $(y_1, \ldots, y_m) \in \mathcal{R}^m$, $\gamma > 0$, $g^o(x) = \sum_1^m a_i K(x, x_i)$ be the unique argminimum of $(\mathcal{H}_K(\Omega), \mathcal{E}_{\gamma,K})$ given by the Representer Theorem, $\{\varepsilon_n\}$ be a sequence of positive reals such that $\lim_{n\to\infty} \varepsilon_n = 0$ and $\{g_n\}$ be a sequence of ε_n-argminima of problems $(span_n\, G_K, \mathcal{E}_{\gamma,K})$. Then for every positive integer n the following hold:*
(i) $\inf_{g\in span_n G_K} \mathcal{E}_{\gamma,K}(g) - \mathcal{E}_{\gamma,K}(g^o) \leq \alpha\left(\sqrt{\frac{(s_K\|g^o\|_{G_K})^2 - \|g^o\|_K^2}{n}}\right)$;
(ii) if $\|g^o\|_G < \infty$, then $\{g_n\}$ is a $\mathcal{E}_{\gamma,K}$-minimizing sequence over $\mathcal{H}_K(\Omega)$ and $\mathcal{E}_{\gamma,K}(g_n) - \mathcal{E}_{\gamma,K}(g^o) \leq \alpha\left(\sqrt{\frac{(s_K\|g^o\|_{G_K})^2 - \|g^o\|_K^2}{n}}\right) + \varepsilon_n$;
(iii) $\|g_n - g^o\|_K^2 \leq \alpha\left(\sqrt{\frac{(s_K\|g^o\|_{G_K})^2 - \|g^o\|_K^2}{n}}\right) + \varepsilon_n$;
(iv) $\|g_n - g^o\|_\mathcal{C}^2 \leq \sqrt{c_K}\left(\alpha\left(\sqrt{\frac{(s_K\|g^o\|_{G_K})^2 - \|g^o\|_K^2}{n}}\right) + \varepsilon_n\right)$,
where $c_K = \sup_{x,y\in\Omega} |K(x,y)|$, $\alpha(t) = a_2 t^2 + a_1 t$, $a_1 = 2(m\|f\|_K\, c_K + mb\sqrt{c_K} + \gamma\|f\|_K)$, $a_2 = m\, c_K + \gamma$ and $b = \max\{|y_i| : i = 1, \ldots, m\}$.

Proof. (Sketch) The statement follows from Theorem 3.2 applied to the space $(X, \|.\|) = (\mathcal{H}_K(\Omega), \|.\|_K) = M$, $c = 0$ (as the Minkowski functional of $\mathcal{H}_K(\Omega)$ can be proven to be equal to zero), $\Phi(f) = \mathcal{E}_{\gamma,K}(f)$, $G = G_K = \{K(x,.) : x \in \Omega\}$, and $s_{G_K} = s_K = \sup_{x\in\Omega} \|K(x,.)\|_K = \sqrt{K(x,x)}$. Continuity of $\mathcal{E}_{\gamma,K}$ can be proven and its modulus estimated by exploiting the properties of the embedding of the space $\mathcal{H}_K(\Omega)$ into the space $\mathcal{C}(\Omega)$ described in [8]. (i) and (ii) are then obtained from Theorem 3.2 (i) and (ii), resp. with $c = 0$ and $\omega \leq \alpha$.

(iii) follows from (ii) and an estimate of the modulus of convexity of the functional $\|.\|_K^2$ (it is easy to check that for every Hilbert space $(X, \|.\|)$ the functional $\|.\|^2$ is strictly uniformly convex with modulus of convexity $\delta(t) \leq t^2$).

(iv) follows immediately from (iii) and again from the properties of the embedding of $\mathcal{H}_K(\Omega)$ into $\mathcal{C}(\Omega)$.

For the full proof see [19].

So for any Mercer kernel K (in particular for the Gaussian kernel), Theorem 4.2 can be applied to estimate rates of approximate optimization of the regularized empirical error functional $\mathcal{E}_{\gamma,K}$ over sets $span_n G_K$ corresponding to networks with n hidden units.

Acknowledgments

The authors wish to thank Paul C. Kainen and Andrew Vogt from Georgetown University for helpful discussions.

Collaboration between V. K. and M. S. was supported by grant 22 MC6 of the Scientific and Technological Cooperation Agreement between Italy and Czech Republic, project "Functional Optimization and Nonlinear Approximation by Neural Networks". V. K. was partially supported by GA ČR Grant 201/02/0428 and M. S. was partially supported by CNR - Agenzia 2000 Grant, project "New Algorithms and Methodologies for the Approximate Solution of Nonlinear Functional Optimization Problems in a Stochastic Environment."

References

[1] Alessandri, A., Parisini, T., Sanguineti, M., Zoppoli, R. (1992) Neural strategies for nonlinear optimal filtering. Proc. IEEE Int. Conf. on Systems Engineering, Kobe (Japan), pp. 44-49.

[2] Aronszajn, N. (1950) Theory of reproducing kernels. Transactions of AMS 68: 337-404.

[3] Baglietto, M., Parisini, T., Zoppoli, R. (2001) Numerical solutions to the Witsenhausen counterexample by approximating networks. IEEE Trans. on Automatic Control 46: 1471–1477.

[4] Barron, A.R. (1992) Neural net approximation. In K. Narendra (ed.) Proc. 7th Yale Workshop on Adaptive and Learning Systems, Yale University Press: 69-72.

[5] Barron, A.R. (1993) Universal approximation bounds for superpositions of a sigmoidal function. IEEE Trans. Inf. Th. 39: 930-945.

[6] Bellman, R. (1957) Dynamic Programming, Princeton University Press, Princeton, N.J..

[7] Bertero, M. (1986) Regularization methods for linear inverse problems. In: C.G. Talenti (ed.) Inverse Problems. Springer-Verlag, Berlin.

[8] Cucker, F., Smale, S. (2001) On the mathematical foundations of learning. Bulletin of AMS 39: 1-49.

[9] DeVore, R. A., Temlyakov, V. N. (1995) Nonlinear approximation by trigonometric sums. The J. of Fourier Anal. and Appl. 2: 29–48.

[10] Dontchev, A. L. (1983) Perturbations, Approximations and Sensitivity Analysis of Optimal Control Systems. Lecture Notes in Control and Information Sciences 52, Springer-Verlag, Berlin Heidelberg.

[11] Dontchev, A. L., Zolezzi, T. (1993) Well-Posed Optimization Problems. Lecture Notes in Math. 1543, Springer-Verlag, Berlin Heidelberg.

[12] Gelfand, I. M., Fomin, S. V. (1963) Calculus of Variations, Prentice Hall, Englewood Cliffs, N. J..

[13] Girosi, F., Jones, M., Poggio, T. (1995) Regularization theory and neural networks architectures. Neural Computation 7: 219-269.

[14] Jones, L.K. (1992) A simple lemma on greedy approximation in Hilbert space and convergence rates for projection pursuit regression and neural network training. Ann. of Statistics 20: 608–613.

[15] Kůrková (1997) approximation by neural networks. In K. Warwick, M. Kárný (eds.) Computer-Intensive Methods in Control and Signal Processing: Curse of Dimensionality. Birkhauser, Boston, pp. 261-270, 1997.

[16] Kůrková, V. (2003) High-dimensional approximation by neural networks. In J. Stuykens (ed.) NATO ASI on Learning Theory and Practice IOS Press, to appear.

[17] Kůrková, V., Sanguineti, M. (2001) Bounds on rates of variable-basis and neural-network approximation. IEEE Trans. on Information Theory 47:2659-2665.

[18] Kůrková, V., Sanguineti, M. (2002) Comparison of worst case errors in linear and neural network approximation. IEEE Trans. on Information Theory 48:264-275.

[19] Kůrková V, Sanguineti M. (2002) Error estimates for approximate optimization over variable-basis functions. Research Report ICS-2002-882, Institute of Computer Science, Prague.

[20] Kůrková, V., Savický, P., Hlaváčková, K. (1998) Representations and rates of approximation of real–valued Boolean functions by neural networks. Neural Networks 11:651-659.

[21] Parisini, T., Zoppoli, R. (1996) Neural networks for feedback feedforward nonlinear control systems. IEEE Trans. on Neural Networks 5:436-449.

[22] Parisini, T., Zoppoli, R. (1996) Neural approximations for multistage optimal control of nonlinear stochastic systems. IEEE Trans. on Automatic Control 41: 889–895.

[23] Pisier, G.: Remarques sur un resultat non publié de B. Maurey. Seminaire d'Analyse Fonctionelle I, no. 12. École Polytechnique, Centre de Mathématiques, Palaiseau, 1980-81.

[24] Tikhonov, A.N., Arsenin, V.Y. (1977) Solutions of Ill-posed Problems. W.H. Winston, Washington, D.C..

[25] Tjuhtin, V. B. (1982) An error estimate for approximate solutions in one-sided variational problems. Vestnik Leningrad Univ. Math. 14:247-254.

[26] Wahba, G. (1990) Splines Models for Observational Data. Series in Applied Mathematics 59. SIAM, Philadelphia, PA.

[27] Zoppoli, R., Parisini, T. (1992) Learning techniques and neural networks for the solution of N-stage nonlinear nonquadratic optimal control problems. In Isidori, A. and Tarn, T.J. (eds.) Systems, Models and Feedback: Theory and Applications, Birkhäuser, Boston, pp. 193-210.

[28] Zoppoli, R., Sanguineti, M., Parisini, T. (2002) Approximating networks and extended Ritz method for the solution of functional optimization problems. Journal of Optimization Theory and Applicatioins 112:403-440.

Binary Factorization in Hopfield-Like Neural Autoassociator: A Promising Tool for Data Compression

A. A. Frolov [1], A. M. Sirota [2], D. Husek [3], I. Muraviev [4], P. Combe [5] [1]

[1]Institute of Higher Nervous Activity and Neurophysiology of the Russian Academy of Sciences, Moscow, Russia. [2]Center for Molecular and Behavioral Neuroscience, Rutgers University, USA. [3]Institute of Computer Science Academy of Science of the Czech Republic, Prague, Czech Republic. [4]Institute of Higher Nervous Activity and Neurophysiology of the Russian Academy of Sciences, Moscow, Russia. [5]CPT, CNRS-Luminy, and Université de Provence, Marseille, France.

Abstract

Data compression of high dimentional complex patterns is one of the most challaging task of information technology today. Proposed approach is based on feature extraction procedure which maps original patterns into features (factors) space of reduced, possibily very small, dimension. In this paper, we outline that Hebbian unsupervised learning of Hopfield-like neural network is a natural procedure for factor extraction. Due to this learning, factors become the attractors of network dynamics, hence they can be revealed by the random search. The neurodynamics is modeled by Single-Step approximation which is known [5] to be rather accurate for sparsely encoded Hopfield-network. This paper is limited to the case of sparsely encoded factors, but it is not realy constraint for data compression.

1 Introduction

The feature space transformation is a widely used method for consequent data compression. Due to this transformation the original patterns are mapped into the space of features or factors of reduced dimension. One particular form of nonlinear factorization is a binary one, where a complex vector signal (pattern) has a form of a Boolean sum of weighted binary factors: $\mathbf{X} = \bigvee \beta_l \mathbf{f}^l$. Up to now there are no methods for binary factorization except the fit of the learning set by the exhaustive search. However, this method requires exponentially increasing number of trials when the dimension of the pattern space increases. In this paper we propose a method of binary factorization that naturally follows from associative learning and attractor neural networks dynamics. Particularly, Hebbian-like learning rule and properties of sparsely encoded Hopfield-like neural network dynamics will be applied to the factorization problem. Hebbian learning forms connection matrix of the network as a covariance matrix for the set of learned patterns. Neurons that tend to fire together (represent one common factor) will be more correlated and corresponding connection strengths will be larger in respect to those neurons that belong to different factors. Hence, each group of neurons that forms a factor will be tightly connected via synaptic matrix and hence might correspond to the attractor of the network dynamics. Extraction of factors from the synaptic matrix is performed during network dynamics that evolves towards and stabilizes in the attractor state, which, if a true attractor, corresponds to some factor.

2 Network description

The neural network under consideration consists of N neurons of McCulloch-Pitts type with gradually ranged synaptic connections between them. Only fully connected case is considered here.

Network is trained by a set of M patterns of the form $\mathbf{X}^m = \bigvee_{l=1}^{L} \beta_l^m \mathbf{f}^l$, where $\mathbf{f}^l \in B_n^N$ [1] are L factors (N dimensional vectors) and for every m^{th} pattern $\beta_l^m \in B_C^L$ is the corresponding factor scores vector. Thus, every factor contains exactly $n = Np$ ones and every complex pattern $\mathbf{X}^m$ contains exactly C factors, we call C the complexity of the pattern. We assumed factors and factor scores to be statistically independent. In a limit case $C = 1$ patterns become pure factors and we obtain ordinary Hopfield case.

Connection matrix $\mathbf{J}$ is formed using the correlational Hebbian rule:

$$\begin{aligned} J_{ij} &= \frac{1}{Np(1-p)} \sum_{m=1}^{M} (X_i^m - q^m)(X_j^m - q^m), i \neq j, \\ J_{ii} &= 0, \end{aligned} \qquad (1)$$

where bias $q^m = \sum_{i=1}^{N} X_i^m / N$ is the total activity of the

[1]This paper was written under grants N° 201/01/1192 and 201/00/1031 within Grant Agency of the Czech Republic and and France-Czech Republic Barrande program N° 04693QG

[1] $B_n^N = \{X | X_i \in \{0,1\}, \sum_{i=1}^{N} X_i = n\}$

m^{th} pattern.

During recall stage, on presentation of an initial pattern $\mathbf{X}^{in}$, network activity evolves until it stabilizes in some attractor. As initial patterns we used destroyed versions of factors with the same level of activity $n = Np$ as factors. The evolution of the network activity in discrete time is determined by the synchronous dynamics equation for activity vector $\mathbf{X}$:

$$X_i(t+1) = \sigma(h_i(t) - T(t)),\ i = 1, \cdots N, \quad (2)$$

where

$$h_i(t) = \sum_{j=1}^{N} J_{ij} X_j(t) \quad (3)$$

is synaptic excitation, σ – step function, and $T(t)$ – the activation threshold. The threshold $T(t)$ is chosen at each time step in such a way that the level of the network activity is kept constant and equal to n. Thus, on each step n "winners" (neurons with greatest synaptic excitation) are chosen and only they are active on the next step. To avoid uncertainty in the choice of winners when several neurons have synaptic excitations at the level of activation threshold, small random noise was added to the activation threshold of each individual neuron. The amplitude of the noise was put to be less than the smallest increment of synaptic excitation given by formula (3). This ensures that neurons with the highest excitations are kept to be winners in spite of adding random noise to neurons' thresholds. Noise to individual neurons was fixed during the whole recall process to provide its convergence. Since the number of active neurons in each factor are fixed and also equal to n, this choice of activation threshold allows for a stabilization of network activity in the vicinity of one of the factors. The stable pattern (point attractor) or first pattern of cyclic attractor was taken as a resulting pattern (final pattern $\mathbf{X}^f$) of the recall process.

Similarity between two vectors is measured in Hamming space by their overlap m:

$$m(\mathbf{X}^1, \mathbf{X}^2) = \frac{1}{Np(1-p)} \sum_{i=1}^{N} (X_i^1 - p) X_i^2$$

The level of factor's destruction in the initial state is given by the overlap $m_{in} = m(\mathbf{f}^l, \mathbf{X}(0))$ between this state and the recalled factor $\mathbf{f}^l$.

As a measure of relative informational loading we use $\alpha = LI(p)/N$, where $I(p) = -p \log_2 p - (1-p)\log_2(1-p)$ is the Shannon entropy function. The size of attraction basin around the factors is a critical initial overlap m_{ab} which separates retrieval and nonretrieval trajectories in the limit $N \to \infty$. Network factorization ability is analyzed in dependence on the following five parameters: p, C, α, N, M.

3 The size of attraction basins

The size of attraction basin is one of the most important characteristic of the neural network behavior. The larger attraction basin the larger robustness or the better resistance to pattern damage, for example.

3.1 The single step approximation

This size was estimated by single-step approximation (SS), which has been proposed by Kinzel [7] for densely encoded Hopfield network. It has been shown by other theoretical approaches [2], [1] and by Monte-Carlo simulations [6], [2] that single-step approximation is very inaccurate for dense coding. However, it becomes rather accurate when sparseness increases [5]. Thus, only the case of sparsely encoded factors $p \ll 1$ is considered.

The principal peculiarity of this approach is that at each time step one ignores the statistical dependence between network activity and connection matrix and takes into account only two macroparameters of neurodynamics: the overlap $m(t)$ between current and recalled patterns and the total network activity. Since we assumed that the network activity is constant at each time step and equal to activities of factors, the recall process in our case is described by the evolution of the only one parameter $m(t)$.

Without loss of generality, we may assume that a factor $\mathbf{f}^1$ is retrieved. Following to [3] we call neurons which are active and nonactive in $\mathbf{f}^1$ as "high" and "low" neurons, respectively. According to (1) and (3), neurons synaptic excitations at the first step of the recall process can be presented in the form $h_i = S_i^1 + S_i^0$ where

$$S_i^\mu = \sum_{m=1,M^\mu} (X_i^m - q^m) \sum_{j \neq i} (X_j^m - q^m) X_j^{in}$$

The first sum S_i^1 contains M^1 patterns of the learning set which include the recalled factor $\mathbf{f}^1$ and S_i^0 contains M^0 patterns which do not include it. Since initial pattern includes $n_1 = N(m_{in}p(1-p) + p^2)$ high and $n_0 = Np - n_1$ low neurons of $\mathbf{f}^1$, for high neurons

$$\langle S_i^1 \rangle = \frac{MC}{L}[(1-q)(1-q)n_1 + (1-q)(q'-q)n_0]$$

and for low neurons

$$\langle S_i^1 \rangle = \frac{MC}{L}[(q'-q)(1-q)n_1 + (q'-q)(q'-q)n_0]$$

where $\langle \cdots \rangle$ means averaging over all factors except $\mathbf{f}^1$ and all factor scores, $MC/L = \langle M^1 \rangle$,

$$q = \langle q^m \rangle = 1 - (1-p)^C \simeq 1 - \exp(-pC)$$

is the mean activity (the probability of the neuron to be active) of the learning set patterns and

$$q' = 1 - (1-p)^{C-1} = 1 - (1-q)/(1-p)$$

is the probability of the neuron to be active due to the choice of other factors except $\mathbf{f}^1$. Thus,

$$\langle S_i^1 \rangle = \frac{MNCp(1-q)^2}{L(1-p)}(f_i^1 - p)m_{in}.$$

Since $\mathbf{X}^{in}$ is independent of all factors except $\mathbf{f}^1$,

$$\begin{aligned}\langle S_i^0 \rangle &= M(1-C/L)N(N-1)p \\ &\quad \langle (X_i^m - q^m)(X_j^m - q^m) \rangle\end{aligned}$$

where $M(1-C/L) = \langle M^0 \rangle$. By definition of q^m, for each pattern of the learning set $\sum_{i=1,N}(X_i^m - q^m) = 0$. Hence

$$\begin{aligned}&\langle (X_i^m - q^m)(X_j^m - q^m) \rangle = \\ &-\langle (X_i^m - q^m)^2 \rangle/(N-1) = -q(1-q)/(N-1).\end{aligned}$$

For $N \to \infty$ and $L \to \infty$ $\langle S_i^0 \rangle = -MNpq(1-q)$ and, therefore, mean synaptic excitations amount to

$$\langle h_i \rangle = \frac{MNCp(1-q)^2}{L(1-p)}(f_i^1 - p)m_{in} - MNpq(1-q). \quad (4)$$

Mean synaptic excitation for high neurons ($f_i^1 = 1$) is larger than that for low neurons ($f_i^1 = 0$). Thus high neurons have higher probability to be active at the next step of network dynamics.

In the limit case $N \to \infty$ and $L \to \infty$ the dispersions of synaptic exitations are determined by the dispersion of S_i^0. This dispersion can be estimated by the same way as mean synaptic exitations. For the natural case $C^2p/L \ll 1$ it amounts to

$$\begin{aligned}D\{h_i\} &= \sigma^2 = Np(1-p)D\{J_{ij}\} = \\ &= Np(1-p)\frac{M^2C^2p^2(1-q)^4}{L(1-p)^2} \quad (5)\end{aligned}$$

In the limit case $N \to \infty$, the distributions of synaptic excitations can be approximated by normal ones. Then at the first step of the recall process

$$P\{X_i(1) = 1\} = \Phi(\theta_i)$$

where

$$\begin{aligned}\theta_i &= (T(1) - \langle h_i \rangle)/\sigma, \\ \Phi(x) &= 1/(2\pi)^{1/2}\int_x^{\infty} \exp(-u^2/2)du\end{aligned}$$

and $T(1)$ is an activation threshold. According to (4) and (5) for high and low neurons of $\mathbf{f}^1$

$$\begin{aligned}\theta^1 &= \theta - \frac{m_{in}(1-p)}{Lp(1-p)/N} = \theta - \frac{m_{in}(1-p)}{\alpha p(1-p)/I(p)}, \\ \theta^0 &= \theta + \frac{m_{in}p}{\alpha p(1-p)/I(p)}\end{aligned}$$

where $\theta = (T(1) + MNpq(1-q))/\sigma$ is a scaled activation threshold. In the model the threshold is chosen in such a way that a total level of network activity is the same as in factors, i.e. it is chosen to satisfy the condition

$$pp_1 + (1-p)p_0 = p \quad (6)$$

where $p_1 = \Phi(\theta_1)$ and $p_0 = \Phi(\theta_0)$ are probabilities for high and low neurons to be active. As a result of the first step, the overlap changes to

$$m(1) = p_1 - p_0. \quad (7)$$

In the single-step approximation these equations are assumed to be valid for all time steps (naturally, m_{in} must be replaced by $m(t)$ and $m(1)$ by $m(t+1)$ where t and $t+1$ are consequence steps of the recall process).

The obtained neurodynamic equations are completely identical to those for ordinary Hopfield network [5]. Thus, in SS approximation neurodynamics does not depend on complexity.

3.2 Computer simulation

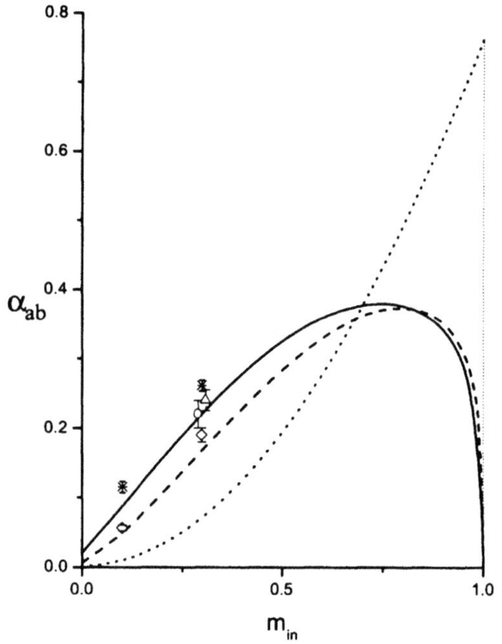

Fig. 1. Attraction basins predicted by SS (lines) and obtained by computer simulations (means ± SD); p = 0.02-solid line, p = 0.004-dashed line,$p \to \infty$-pointed line; $*$ p=0.02, C = 1, $\diamond$ p = 0.004, C = 1, $\circ$ p=0.02, C=5, Δ p=0.02, C=10.

Computer simulations were performed for N from 1000 to 10000, for $p = 0.02$ and 0.004. The program generated a set of random factors of fixed sparseness p and mixed them into the set of M patterns so that each pattern contained exactly C factors. Then network was trained with this set and tested with destroyed factors for up to 10000 simulation trials for each combination of parameters.

3.3 Discussion

The curves which characterize the behaviour of the network activity depending on α, m_{in} and p are presented in Fig. 1. Let the initial state of the network activity for a given α be characterized by the point (m_{in}, α). If this point lies under the curve, the overlap between the current pattern and the recalled pattern sways during the recall process to the right, that is to the final overlap given by the right branch of the curve. The overlap sways to the left for each point above the curve. Thus the left branch of the curve corresponds to the boundary of attraction basin.

It is well known [2], [5] that distribution of final overlaps m^f has two distinct modes: for $m^f \approx 1$ ("true") and $m^f \ll 1$ ("false"), corresponding to stabilization of the network in true and spurious attractors, respectively. For small informational loading "true" mode prevails, and as informational loading increases distribution maximum shifts into "false" mode, having sharp change in modes amplitude at some certain α_{ab} which gives the critical informational loading separated retrieval and nonretrieval trajectories for given m_{in}.

It can be seen that the size of attraction basins monotonically decreases when sparseness increases (i.e. p decreases). The curves ploted in Fig. 1 are based on SS aproximation. To approve this approach Monte Carlo method obtained estimates of α are compared with those obtained by SS model. It is shown that although SS gives underestimation of the size of attraction basins, the error is rather small and decreases when sparseness increases. Computer simulation confirms the SS prediction that the size of attraction basins around factors does not depend on complexity and it is rather large especially for small informational loading.

4 Concluding remarks

Analysis with SS approximation and computer simulations described here as well as previous work [8], show that this Hopfield-like neural networks are capable of performing binary factorization of signal, of high dimension and complexity, as a basis for data compression. The proposed method of binary factorization consists of two steps. The first is factors learning and the second is factors search. The analysis done in this contribution should help us to solve second step of the factorization procedure. We have found that for small informational loading the size of attraction basins around factors is large enough to allow for their search by starting from random initial states. Although we did not yet research the problem of factor search systematically, we can point out at least two criteria for this separation. First, as a rule, for small informational loading true attractors are point and false attractors are cyclic of the length two. Second, the time of convergence to false attractors is much larger than that for true attractors.

Let us notice that the used paradigm based on Hebbian learning has clear biological interpretation. Marr [9] was the first to formulate the problem of binary factorization in the brain memory processing. He proposed separation of functions between neocortex and hippocampus: hippocampus performs a function of intermediate processing and storage of the information collected during the day (or several days), which is further removed to the neocortex during sleep. Essential part of his model was based on autoassociative properties of the CA3 field of the hippocampus. Due to its extensive collateral excitatory system and relatively sparse activity level the field CA3 of the hippocampus is thought to be natural autoassociator [9], [10]. Hebbian associative long-term modification of synapses in CA3 forms sparsely bound cell ensembles (attractors) corresponding to factors that are encoded in incoming pattern. On the next stage during sleep synchronized activation of ensembles (factors) in CA3 (sharp waves) and consequent "ripples" in CA1 and downstream structures [4] results in replay of factors to neocortex as Marr's "classificatory units".

References

[1] Amari, S., Maginu, K. (1988). Statistical neurodynamics of associative memory. *Neural Networks*, **1**, 63-73.

[2] Amit, D. J., Gutfreund, H., Sompolinsky, H. (1987). Statistical mechanics of neural networks near saturation. *Annal of Physics*, **173**, 30-67.

[3] Bucingham, J., Willshaw, D. (1993). On setting unit thresholds in an incompletely connected associative net. *Network*, **4**, 441-459.

[4] Buzsaki, G. (1996). The hippocampo-neocortical dialog. *Cereb.Cortex* **6** (2), 81-92.

[5] Frolov, A. A., Husek, D., Muraviev, I. P. (1997). Informational capacity and recall quality in sparsely encoded Hopfield-like neural network: Analytical approaches and computer simulation. *Neural Networks*, **10**, 845-855.

[6] Hopfield, J.J. (1982). Neural network and physical systems with emergent collective computational abilities. *Proceedings of the National Academy of Science USA*, **79**, 2544-2548.

[7] Kinzel, W (1985). Learning and pattern recognition in spin glass models. *Z Physik* , **B 60**, 205-213.

[8] Sirota, A.M., Frolov, A.A., Husek D., Polyakov, P. Binary factorization in Hopfirld-like neural networks and hippocampal memory function. *Neural Network World* in print 2003.

[9] Marr, D. (1971). Simple memory: a theory for archicortex. *Phil. Trans. Roy. Soc. London* **B 262** 24-81.

[10] Rolls, R.T. (1989). The representation and storage of information in neural networks in the primate cerebral cortex and hippocampus. In R.Durbin, C. Miall and G. Mitchison, editors, *The Computing Neuron*. Addison-Weley, Reading, MA.

β_SVM a new Support Vector Machine kernel.

Tarek M. Hamdani and Adel M. Alimi

REsearch **G**roup on **I**ntelligent **M**achines (**REGIM**)

Université de Sfax, ENIS

BP. W-3038 – Sfax – Tunisia

Email: {tarek.hamdani; adel.alimi}@ieee.org.

Abstract

Support Vector Machine is a statistical learning machine developed by Vapnik in his statistical learning theory. This machine present very interesting proprieties in both classification and regression problems in the high dimensional space. In this paper, we propose β_SVM, a new kernel function for SVM, with special proprieties and high discrimination ability. We have applied this kernel in the pattern recognition, and we have compare the different performances of many other kernels, results show that the new kernel is very performant.

1 Introduction

Support Vector Machines is a statistical learning method, it's the result of evolutions in this approach.
Introduced by Vapnik in 1995 Support Vector Machines (SVM) is a learning machine for two groups classification problem.
SVM was based on the Structural Risk Minimisation principle developed in Vapnik's theory on statistical learning which goal is to find the optimum decision region.
The idea of SVM is to map the representation space to a very high dimensional space (possibly infinite), in this feature space, the linear separation of the training data is easily made.
Basically, SVM operate in the case of linear separation by mapping the original space in an augmented space using some defined kernels respecting Mercer's condition, this process was extend to the non-linear separation with the use of some particular proprieties and some kernel functions.
For regression problems SVMs was trained by solving a quadratic programming optimisation problem. The idea is to detect the optimal hyper plan maximising the distance (margin) separating the two groups of data.
In the pattern recognition case, SVM was applied in object recognition, speaker identification, face detection in images, text categorisation... Its theoretical foundation and its experimental success encourage further research on their characteristics, as well as its further use.
Defined as learning machine SVM was developed to discriminate two groups of labelled data for a supervised learning procedure, in this way we have to trait the case of multi classes problems.
We apply the SVM technique in handwriting digits images recognition. We use the NIST data presenting 20 000 digits.
Results was interesting and performance was very high in the training phase.

2 Support Vector Machine

Suppose that we have two groups of data; each group is an independent class and have a single label. We have to separate the positive from the negative data, those supposing the existence of an hyper plan separating the two groups of data.

2.1 Separable data

SVM define the optimal hyper plan H having the equation :

$$H : w \cdot x + b = 0$$

This hyper plan divide the feature space to two semi-hyper plans, each semi-hyper plan represent the mapping area of one class.
We are going to detect two subsets of data, these subsets constitute two parallel hyper plans H1 and H2, where the distance between H1 and H2 is maximum and where, there is no data points between H1 and H2.
H1 and H2 have respectively the equations :

$$H1 : w \cdot x + b = +1;$$

$$H2 : w \cdot x + b = -1;$$

H1 and H2 are parallel to H and are with equal distance to it.
The two subsets of data points, constructing H1 and H2 with the maximum distance between H1 and H2, lie on these hyper plans and are called the Support

Vectors, other data points are not considered and we can drop theme when we define our SVM.
SVM is defined by the Support Vectors data constructing the separating hyper plans and we have:

$w \cdot x_i + b \geq +1$ for the positive data $y_i = +1$;

$w \cdot x_i + b \leq -1$ for the negative data $y_i = -1$;

we can combine these two conditions as :

$$y_i(w \cdot x_i + b) \geq 1.$$

To maximise the distance between H1 and H2 we have to resolve following quadratic programming problem in (w,b) :

$$\mathrm{Min}_{(w,b)}\ 1/2\ w^T \cdot w$$

$$\text{Subject to } y_i(w \cdot x_i + b) \geq 1.$$

With introducing the positive Lagrange multipliers α_i , $i = 1,\ldots,l$, one multiplier for each constraint our quadratic programming problem became the following :

$$L(w,b,\alpha) \equiv \tfrac{1}{2}(w)^T w - \sum_{i=1}^{N} \alpha_i y_i (w x_i + b) \geq 1$$

This formulation is the primal representation of the quadratic programming problem with the respect of the primal variables w and b vanish :

$$\frac{\partial L}{\partial w} = w - \sum_i \alpha_i y_i x_i = 0,$$

$$\frac{\partial L}{\partial b} = -\sum_i \alpha_i y_i = 0,$$

and

$$\alpha \geq 0,$$

From these conditions we have :

$$w = \sum_{i=1}^{N} \alpha_i y_i x_i$$

$$\sum_{i=1}^{N} \alpha_i y_i = 0$$

When we substitute these expressions in the primal formulation we obtain the dual formulation :

$$L_D = \sum_i \alpha_i - \frac{1}{2} \sum_{i,j} \alpha_i \alpha_j y_i y_j x_i \cdot x_j$$

subject to: $\alpha_i \geq 0.$

$$\sum_{i=1}^{N} \alpha_i y_i = 0$$

In this dual formulation we have eliminate the primal variables (w and b), the problem became a maximisation problem. So we can solve α_i and compute later :

$$w = \sum_{i=1}^{N} \alpha_i y_i x_i$$

and b can be easily found by using the KKT condition.
The Karush-Kuhn-Tuker (KKT) Conditions are the optimality conditions for the primal and the dual formulations of our quadratic problem.
KKT conditions are necessary and sufficient for w, b and α to be the problem solutions.
Finding the KKT conditions solution is solving the SVM problem.
We can express the KKT conditions using the expression of the primal problem :

$$\frac{\partial L}{\partial w} = w - \sum_i \alpha_i y_i x_i = 0$$

$$\frac{\partial L}{\partial b} = -\sum_i \alpha_i y_i = 0$$

KKT conditions are :

$$\begin{cases} y_i(w \cdot x_i + b) - 1 \geq 0\ ;\ \forall i \\ \alpha_i \geq 0\ ;\ \forall i \\ \alpha_i(y_i(w \cdot x_i + b) - 1) = 0\ \forall i \end{cases}$$

Here the support vectors are detected when we have a $\alpha_i \neq 0$; and we can for any $\alpha_i \neq 0$ compute b with the KKT complementarily conditions.
For linear SVM the training process is consisting on the maximization of the dual formulation L_D subject to the constraints :

$$\sum_{i=1}^{N} \alpha_i y_i = 0$$

$$\alpha_i \geq 0$$

After training, we obtain, for each training point, a Lagrange multiplier α_i and two particular subsets of support vectors having $\alpha_i > 0$ defining H1 and H2. All other training points have $\alpha_i = 0$ and are not necessary for the construction of our learning machine.

2.2 Non separable data

In real live classification problems we have particular case and the problems are not usually linearly separable.
For these reasons we can not apply the linear SVM for the real live classifications and we have to modify the process to respond to the exigency of these cases.
Linear SVM suppose that there is no data between the two hyper plans H1 and H2, it's the perfect separation.
In the non linear case we have to tolerate some relaxations on the constraints on the two hyper plans separating the data points, for this we introduce some positive variables ξ_i, $i = 1,\ldots,l$ in the constraints

$$w \cdot x_i + b \geq +1 - \xi_i \text{ for } y_i = +1;$$

$$w \cdot x_i + b \leq -1 + \xi_i \text{ for } y_i = -1;$$

$$\xi_i \geq 0\ \forall i$$

with this relaxation of constraints we introduce the penalty C, it penalize the data points that cross the boundaries. C must be finite in the non linear case,

and when C is infinite we return in the perfect linear separation case.
The objective function is now the following :

$$\text{Min}_{(w,b)}\ 1/2\ w^T . w + C\sum_{i=1}^{N}\xi_i$$

$$\text{Subject to } y_i(w . x_i + b) + \xi_i - 1 \geq 0.$$

$$\xi_i \geq 0.$$

In this case the dual problem formulation becomes :

$$L_D \equiv \sum_i \alpha_i - \frac{1}{2}\sum_{i,j}\alpha_i\alpha_j y_i y_j x_i \cdot x_j$$

subject to : $0 \leq \alpha_i \leq C.$

$$\sum_{i=1}^{N}\alpha_i y_i = 0$$

For the KKT conditions we have to define the primal formulation of our problem :

$$L_P = \frac{1}{2}\|w\|^2 + C\sum_i \xi_i - \sum_i \alpha_i\{y_i(x_i \cdot w + b) - 1 + \xi_i\} - \sum_i \mu_i \xi_i$$

$$\frac{\partial L}{\partial w} = w - \sum_i \alpha_i y_i x_i = 0$$

$$\frac{\partial L}{\partial b} = -\sum_i \alpha_i y_i = 0$$

$$\frac{\partial L_P}{\partial \xi_i} = C - \alpha_i - \mu_i = 0$$

$$y_i(xi . w + b) - 1 + \xi_i \geq 0$$

$$\xi_i \geq 0$$

$$\alpha_i \geq 0$$

$$\mu_i \geq 0$$

$$\alpha_i\ \{y_i(x_i . w + b) - 1 + \xi_i\} = 0$$

$$\mu_i\ \xi_i = 0$$

2.3 Non linear Support Vector Machines

We have usually to trait the case of non linearly separable data and for this case we can not find an optimal hyper plan to separate the data points in the actual dimensions therefore we have to map the data points to a high dimensional space in which we can find a linear separation between the two classes this dimensional space is called Hilbert space H.
This transformation can be done with a mapping function $\Phi(.)$, and in the high dimensional space we find this dual formulation :

$$L_D \equiv \sum_{i=1}^{N}\alpha_i - \frac{1}{2}\sum_{i,j}\alpha_i\alpha_j y_i y_j \Phi(x_i)\cdot\Phi(x_j)$$

We suppose that $\Phi(x_i).\ \Phi(x_j) = k(x_i\ ,\ x_j)$. This function is called the kernel function.
The kernel function can find a dimensional space in which the data can be linearly separated.
In this dimensional space, the kernel function is the equivalent of a dot product.
To be sure that we can find the Hilbert space in which data can be separate, we must have a kernel function verifying the Mercer's condition.

2.4 Mercer's condition

We have :

$$k(x\ ,\ y) = \sum_i \Phi(x)_i \Phi(y)_i$$

If and only if, for any g(x) such that :

$$\int g(x)\, dx \text{ is finite}$$

then

$$\int k(x\ ,\ y)\, g(x)\, g(y)\, dx dy \geq 0.$$

These are some kernel functions verifying the Mercer's condition :

Kernel	*Expression*
Linear	$K(u,v) = u.v$
Sigmoid	$K(u,v) = \tanh(\alpha u.v + b)$
Polynomial	$K(u,v) = (1 + u.v)^d$
RBF	$K(u,v) = \exp(-\alpha\|u - v\|^2)$
RBF Exponential	$K(u,v) = \exp(-\alpha\|u - v\|)$

Table 1. Kernel functions

Kernel functions verifying the Mercer's condition can find, in another dimensional space the linear separation for data.
Kernels will be transformed in dot product in the high space.

3 β_SVM

β_SVM is a kernel function based on the β function used as an activation function in the β basis function neuronal network (Alimi, 1997). The use of this function is based on it's high generalisation ability and it's high flexibility for the approximation problems.
The choice of β_SVM as a kernel can be justified by the interesting propriety of approximation of a very high number of data points.
We are going to define the mono-dimensional β_SVM and the multi-dimensional β_SVM.

3.1 Mono-dimensional β_SVM

In the mono-dimensional case we define the β_SVM as following :

$$\beta(x;p,q,x_0,x_1) = \begin{cases} \left(\frac{x-x_0}{x_c-x_0}\right)^p \left(\frac{x_1-x}{x_1-x_c}\right)^q & \text{if } x \in \left]x_0,x_1\right[\\ 0 & \text{elsewhere} \end{cases}$$

with p, q, $x_0 < x_1 \in \mathfrak{R}$, and

$$x_c = \frac{px_1 + qx_0}{p+q}$$

if p > 0 and q = 0 , then :

$$\beta(x) = \begin{cases} \left(\frac{x-x_0}{x_c-x_0}\right)^p & \text{if } x \in [x_0,x_1] \\ 0 & \text{if } x < x_0 \\ 1 & \text{if } x > x_1 \end{cases}$$

if p = 0 and q > 0 , then :

$$\beta(x) = \begin{cases} \left(\frac{x_1-x}{x_1-x_c}\right)^q & \text{if } x \in [x_0,x_1] \\ 1 & \text{if } x < x_0 \\ 0 & \text{if } x > x_1 \end{cases}$$

The Beta function may be characterized by the following proprieties :

$\beta(x_0) = \beta(x_1) = 0.$

$\beta(x_c) = 1.$

$$\frac{\partial \beta(x)}{\partial x} = \left[\frac{px_1 + qx_0 - (p+q)*x}{(x-x_0)(x_1-x)}\right]\beta(x)$$

$$\frac{\partial \beta(x_c)}{\partial x} = \frac{\partial \beta(x_0)}{\partial x} = \frac{\partial \beta(x_1)}{\partial x} = 0$$

$$\frac{p}{q} = \frac{x_c - x_0}{x_1 - x_c}$$

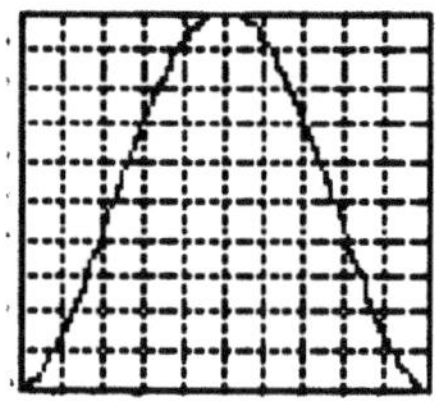

Fig. 1. β_SVM with equal p and q in the mono-dimensional case.

3.2 Multi-dimensional β_SVM

In the Multi-dimensional case, and if $X = (x^1, \ldots, x^n)$.

$X_0 = (x_0^1, \ldots, x_0^n)$; $X_1 = (x_1^1, \ldots, x_1^n)$;

$P = (p^1, \ldots, p^n)$; $Q = (q^1, \ldots, q^n)$.

In this case the Beta function is defined as the following :

$$\beta_{(X_0,X_1,P,Q)}(X) = \prod_{i=1}^{i=n} \beta_{(x_0^i,x_1^i,p_i,q_i)}(x_i)$$

4 Features extraction

In this phase, we define the features of the characters to be recognized then we carry out their extraction. These features are also called parameters or primitives. The features definition is a crucial and delicate stage in the construction of a character recognition system. The goal of this stage is to extract the index properties of the object and to express it in a numerical form or symbolic system. The representation thus obtained will be used as a basis at the later stages of training and recognition. Then, to guarantee the success of this stage, "a good" features choice thus is essential. They must be able to describe in an unambiguous way all the forms belonging to the same class of characters all while differentiating them from the other classes. The suitable features are thus those which can highlight at the same time the total and local properties of the character. They must be indifferent to the normal variations of the character and insensitive with the variations of scale, of rotation, the translation, of the curve, ...
The extraction process must have the following proprieties :

- the simplicity of the features detection process : the features selected should not require too complex calculations in order to have an execution time compatible with a "real time" implementation.

- The theoretical features discriminating capacity : the features selected must be able to describe in an unambiguous way all the forms belonging to the same class of characters all while differentiating them from the other classes.

- The features complementarily, or at least the weakest possible correlation between features: the selected features families must highlight properties different from the character and if possible from the complementary properties.

4.1 The features extractor "measurements of density"

The extractor carried out proposes to measure the densities in black pixels in various areas of the image of the character. The division of the image source in areas is the result of a successive sampling which proceeds to gather the pixels of the image according to a preset grid.
The density in each part is the report/ratio of the number of black pixels on the total number of pixels gathered in the corresponding part. The entry of this extractor is an image of the character. This image D-is successively sampled with grids of sizes 1*1, 2*2, 4*4 and 8*8. At exit, the features " measurements of density " corresponding to this image are written in a textual file which bears the same name as that of the image.

4.1.1- Determination of the vector. The features vector retained is composed of 85 elements distributed according to different D-sampling from the image of the character:

Sampling	**Nbr var**
Grid 1*1	1
Grid 2*2	4
Grid 4*4	16
Grid 8*8	64
Total	**85**

Table 2. Composition of dencity vector.

the matrix representing the image D-is successively sampled with the grids selected (1*1, 2*2, 4*4 and 8*8). Measurements of density of the consequent zones are stored in respective matrices with each D-sampling and of the same organization (dens1, dens2, dens4 and dens8). Lastly, All these matrices containing measurements are written in a textual file in the shape of a vector (of which the organization at summer defined in paragraph 1.1.2. this paper). Each characteristic holds on a line.

4.2 The Statistical features extractor

The extractor carried out proposes to detect the features corresponding to the six families restrained for the identification of a character. The entry of this extractor is an image of the character. The features corresponding to this image are written in a textual file which bears the same name as that of the image.

4.2.1- Vector determination. The features vector retained is composed of 114 elements distributed on the families in the following way :

Features Families	**Nature**	**Nbr var**
Intersection with lines *(18 variables)*	vertical sonde	6
	Horizontal sonde 1/3	6
	horizontal sonde 2/3	6
Moments Invariants	$\phi(1) \rightarrow \phi(7)$	7
Profils *(10 variables)*	Amplitudes	4
	heights (1/5,1/2,4/5)	3
	widths (1/5,1/2,4/5)	3
Projections *(20 variables)*	Vertical projection	10
	Horizontal projection	10
Extrema *(36 variables)*	Top extrema	9
	Low extrema	9
	Right extrema	9
	Left extrema	9
Strokes ends, X and Y connection *(23 variables)*	Strokes ends	9
	Jonctions X	7
	Jonctions Y	7
Total of features (Vector size)		**114**

Table 3. Statistical features vector.

5 Experimental results

SVM approach was tested on two digits images databases DB1 and DB2.

The first one is composed of vectors of 114 characteristics of the digit image, we use 5000 examples for the training phase.

The second database consist on vectors of 85 characteristics of the digit image, we use also 5000 examples for training.

In this experimentation, we use tree kernel functions presented in the following table :

Kernels	***Expression***
Linear	K(u,v) = u.v
Polynomial	$K(u,v) = ((u.v) + 1)^d$
RBF	$K(u,v) = \exp(-\alpha\|u - v\|^2)$
β_SVM	$K(u,v) = \left(\frac{u-x_0}{v-x_0}\right)^p \left(\frac{x_1-u}{x_1-v}\right)^q$

Table 4. Used kernel function.

For each database, we varied the penalty C and for each kernel we calculate for the training phase the error rate.

We define the problem for 10 classes of digits, the method used in this paper is called one class versus other. For each class we suppose that we have two classes one for the concerned digits and one for all other classes.

Linear Kernel :

Data Base	C=0.05	C=0.5	C=5	C=10
DB1	0.23	0.32	0.36	0.41
DB2	0.41	0.39	0.25	0.32

Table 5. Error rate for Linear kernel.

Polynomial Kernel :

Data Base	C=0.05	C=0.5	C=5	C=10
DB1	0.21	0.22	0.35	0.46
DB2	0.35	0.33	0.41	0.45

Table 6. Error rate for Polynomial kernel.

RBF Kernel :

Data Base	C=0.05	C=0.5	C=5	C=10
BD1	0.12	0.19	0.26	0.36
DB2	0.26	0.34	0.42	0.44

Table 7. Error rate for RBF kernel.

β_SVM Kernel :

Data Base	C=0.05	C=0.5	C=5	C=10
DB1	0.09	0.11	0.15	0.26
DB2	0.1	0.15	0.21	0.34

Table 8. Error rate for β_SVM kernel.

6 Conclusion

In this paper, we presented the β_SVM as a new Support Vector Machines (SVM) kernel for the digit recognition. We use the Sequential Minimal Optimisation (SMO) Algorithm for the training of our SVM.

The results of our experience are interesting and encourage for implementing more kernels and trying to optimise the training process to made more performance in the recognition.

SVM can be very interesting in the high dimension classification, therefore we can made progress in studying the comportment of SVM in the high dimensional classification.

7 References ;

(1) C. Cortes and V. Vapnik, (November 1995), *Support-vector networks, Machine Learning.* 20:273-297.

(2) C. Burges, (1998), *A tutorial on Support Vector Machines for Pattern Recognition.* Knowledge Discovery and Data Mining, 2(2).

(3) J. C. Platt, (1998), *Sequential minimal optimization: A fast algorithm for training support vector machines,* Technical Report Microsoft Research MSR-TR-98-14.

(4) B. E. Boser, I. M. Guyon, and V. N. Vapnik, (July 1992), *A training algorithm for optimal margin classifiers.* In D. Haussler, editor, Proceedings of the 5th Annual ACM Workshop on Computational Learning Theory, pages 144{152, Pittsburgh, PAACM Press. P. E. Gill, W. Murray, and M. H. Wright. Practical Optimization. Academic Press, 1981.

(5) T. Joachims, (1998), *Text categorization with support vector machines.* In European Conference on Machine Learning (ECML),

(6) E. Osuna, R. Freund, and F. Girosi, (1996), *Support vector machines: Training and applications. A.I.* Memo (in press), MIT A. I. Lab.

(7) E. Osuna, R. Freund, and F. Girosi, (1997a), *An improved training algorithm for support vector machines.* In J. Principe, L. Gile, N. Morgan, and E. Wilson, editors, Neural Networks for Signal Processing VII | Proceedings of the 1997 IEEE Workshop, pages 276 { 285, New York, IEEE.

(8) E. Osuna, R. Freund, and F. Girosi, (1997b), *Training support vector machines: An application to face detection.* In , editor, Proceedings CVPR'97.

(9) R. Vanderbei. *Loq, (*1994), *An interior point code for quadratic programming.* Technical Report SOR 94-15, Princeton University.

(10) V. Vapnik, (1995), *The Nature of Statistical Learning Theory.* Springer Verlag, New York.

(11) J. Werner, (1984), *Optimization - Theory and Applications.* Vieweg.

(12) G. Zoutendijk, 1970, *Methods of Feasible Directions: a Study in Linear and Non-linear Programming.* Elsevier.

Optimal neighbourhood and model quality indicators

Stefan Janaqi[1], François Hartmann[2], Meriam Chebre[3], Edith di Crescenzo[2]

[1] LGI2P Ecole des Mines d'Alès, e-mail stefan.janaqi@ema.fr, [2] Centre de Recherche TotalFinaElf Solaize, [3] Contrôle avancé - Raffinerie de Feyzin

Abstract

The construction of a good predicting model by learning algorithms does not necessarily imply a correct answer during the generalisation step. That is why one gives confidence intervals on the predicted value, often needing some hypothesis on the data's density distribution. These hypothesis can hardly be verified when a little number of samples is given, which is the most frequent case in practice. We follow a local approach on the basis of an optimal neighbourhood choice. We use this neighbourhood to predict as well as to give some simple model quality indicators for any sample.

1. Introduction

The construction of predicting models by learning algorithms, such as statistical regression or neural networks, needs a *learning set* $\mathbf{A} = \{(\mathbf{x}_k, y_k), k = 1, \dots, n\}$. The points $\mathbf{x}_k$ can have real or integer coordinates. The output of the learning algorithm is a predicting function $y = F(\mathbf{x})$ that minimizes a given error criteria such as $E(F,\mathbf{A}) = \sum_{k=1}^{n}(y_k - F(\mathbf{x}_k))^2$. Recent work on learning theory (see [1]) have shown that a "little" learning error does not necessarily imply a "little" error on a new sample $\mathbf{x}$. There are many reasons for this: the new point $\mathbf{x}$ is far from $\mathbf{A}$ or the responses y_i of its neighbours vary a lot or the model is over parametrized or under parametrized, etc. To illustrate this idea, think of $F(\mathbf{x})$ as a polynomial of a given degree p. When p is near to the size of $\mathbf{A}$, $F(\mathbf{x})$ fits the data very well, but a high degree p will add noisy oscillations even where we don't need them.

Observe that the error measure $E(F, \mathbf{A})$ is a *mean* squared error. Thus, when minimizing $E(F, \mathbf{A})$ one does not reduce the error at a given sample but just the global error. There exist methods, such as k-neighbours or case based reasoning, that evaluate locally the predicting value. These local-approach methods need the definition of an appropriate neighbourhood, most times chosen *ad hoc*, for any sample. We search an optimal neighbourhood for numerical data by minimizing an entropy criteria. This optimal neighbourhood can be used to derive a predicting function $y = N(\mathbf{x})$ but also, and here is the core of our work, one can define indicators that measure locally the quality of the predicting function.

We show how entropy depends on the distance and we give a brief description of the technical difficulties that arise and the ideas for their resolution. Then we define three simple but natural quality indicators that we have used in various industrial situations.

2. Distance and neighbourhood: search for an optimal one

When defining neighbours most of authors use classical distances, such as Euclidean or Mahalanobis. We look for a distance that is better suited in a sense to be defined latter, to the data under discussion.

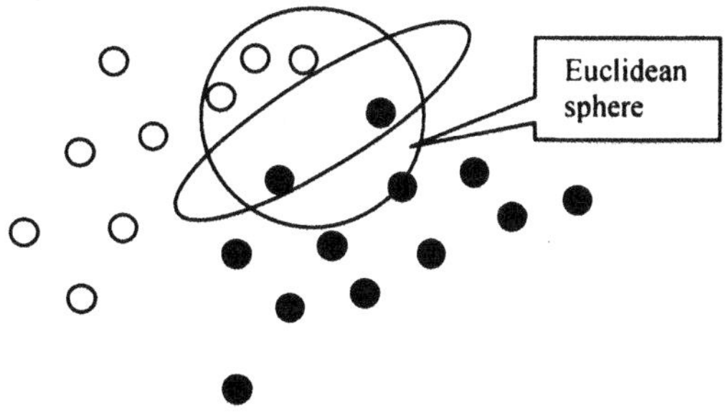

Fig. 1: The ellipsoid is better suited then the Euclidean sphere for the separation of black and white classes.

For real data we limit ourselves to the distances defined by a quadratic norm $\|x\|_M^2 = x^T \cdot M \cdot x$ where M is a positive definite matrix. The distance is then $d(\mathbf{u}, \mathbf{v}) = \|\mathbf{u} - \mathbf{v}\|_M$. For boolean and integer data we use the Hamming distance $d(\mathbf{u}, \mathbf{v}) = \sum|u_j - v_j|$. In fact, this is the only distance that takes into account the integer nature of these data. The other question to be resolved by any local approach is the search of the *radii* or *window's widths*. Given the radii h_r, h_b, $h_z \geq 0$, where r, b and z stand for real, boolean and integer, the neighbourhood V_A in **A** of a sample $\mathbf{x} = (x_r, x_b, x_z)$ is the set of samples $\mathbf{x}_i$ verifying :

$$\frac{d_r(x_r, x_{r,i})}{h_r} + \frac{d_b(x_b, x_{b,i})}{h_b} + \frac{d_z(x_z, x_{z,i})}{h_z} \leq 1$$

When there is only one data type the above definition becomes the usual one. When a radius is 0, V_A contains the points of **A** having the same coordinates as $\mathbf{x}$.

2.1. Probability density and statistical entropy

In this section we consider that **A** contains only real samples of dimension d. We are looking for a distance minimising the entropy of the density of samples of **A**. The employed method permits to find an optimal radius as well. The link between entropy and distance is given by the density function on **A**. Generally, the probability density $f(\mathbf{x})$ of data is not known and in practice we have to limit ourselves with an estimation $f^*(\mathbf{x})$ of $f(\mathbf{x})$. We use the well known *kernel* method in order to find $f^*(\mathbf{x})$. Silverman [2] gives a good deal of results and a detailed analysis of the kernel method (see also [3]). A kernel is defined as a non negative function $K(\mathbf{x})$ such that $\int_{\mathbb{R}^d} K(\mathbf{x})d\mathbf{x} = 1$. Then f^* is given by:

$$f^*(\mathbf{x}) = \frac{1}{n \cdot h^d} \sum_{i=1}^{n} K\left(\frac{\mathbf{x} - \mathbf{x}_i}{h}\right)$$

Here $\mathbf{x}_i \in$ **A** and h is the window's width. Often, $K(\mathbf{x})$ is an unimodale density probability such as the normal distribution $\alpha \exp\left(-\beta \|x\|_M^2\right)$ or the kernel of Epanechnikov $\gamma\left(1 - \|\mathbf{x}\|_M^2\right)$ for $\|\mathbf{x}\|_M \leq 1$ and 0 otherwise. The normalising coefficients α, β and γ reflect the dimension of samples and the nature of the norm induced by M. It is important to see that $f^*(\mathbf{x})$ *depends explicitly on this distance* matrix M.

The choice of the kernel K and the window's width h is realised by minimising the Mean Integrated Square Error (*MISE*):

$$MISE(f^*) = E\int (f^* - f)^2 d\mathbf{x}$$

It is shown in [2] that Epanechnikov's kernel minimise $MISE(f^*)$ and the optimal h is given by:

$$h_{opt} = \frac{A(K)}{\sqrt[d+4]{n}}$$

where the constant $A(K)$ depends only on the dimension d and the matrix M. Remark that h_{opt} is optimal for density estimation. The predicting model is *another* function on d variables and we have to adjust this initial value of h_{opt} by taking into account the observed responses y_i of samples in **A**.

2.2. Optimal distance and radius for real data

A global measure of the information contained in the learning set **A** is given by the Shannon *entropy* of the probability density $f(\mathbf{x})$:

$$\varepsilon(f(\mathbf{x})) = -\lambda \int_{\mathbb{R}^d} f(\mathbf{x}) \ln f(\mathbf{x}) d\mathbf{x}$$

As we use $f^*(\mathbf{x})$ instead of $f(\mathbf{x})$, it is clear that $\varepsilon(f^*(\mathbf{x}))$ depends on the choice of the quadratic form M. Thus, our optimal distance is defined by the matrix M that minimizes $\varepsilon(f^*(\mathbf{x}))$. The idea of using entropy measures to search an optimal distance appears also in the method of 'projection pursuit' used successfully on dimension reduction (see [3], [2]).

This is a non linear optimization problem on d^2 variables, d being the dimension of M. We have shown that the classical gradient descent methods of optimisation are not suitable for this problem. In fact, the gradient needs the calculation of at least $2(d^2 + d)$ integrals in the d-dimensional space, which is very time consuming.

In order to reduce the number of optimization variables we decompose the positive quadratic form $M = Q^T L Q$. This is a classical result of linear algebra showing that Q is an orthogonal matrix, describing a rotation and L is the diagonal matrix of eigenvalues of M.

Thus, we are looking for an orthogonal matrix Q and a diagonal positive matrix $L(\alpha)$, α is the vector of diagonal elements, that minimize $\varepsilon(f^*(Q, L))$. We can decrease the number of variables by restricting the matrix Q to be of the form $Q(\theta) = \prod_{1 \leq i < j \leq d} Q(\theta_{ij})$ where, $Q(\theta_{ij})$ denotes the rotation matrix of the plane ij. With matrices of this form, the number of variables decreases to $(d^2 + d)/2$ and the optimization problem becomes:

$$\begin{cases} \min \ \varepsilon\left(f^*\left(Q(\theta), L(\alpha)\right)\right) \\ \theta \in [0,\pi]^{\frac{d(d-1)}{2}} \\ \alpha \in R_+^d \end{cases}$$

Instead of gradient methods we use the "downhill simplex method" due to Nelder and Mead (see [4]). This method doesn't use the derivative information and needs $\frac{d^2+d}{2}+1$ integral estimations for its initialisation step and only one integral estimation at each iteration.

3. Optimal radii for boolean and integer data

The ideas of the preceding section can be extrapolated easily to the boolean and integer data. Let consider that **A** is contained in the hypercube B^d = $\{0, 1\}^d$, the set of all d-dimensional boolean vectors.

The only distance making a sense for boolean data being the Hamming distance all we need is to define an optimal radius. Following [5] we define a kernel on boolean data as

$$K(\mathbf{u},\mathbf{v},\lambda,h)=\frac{\lambda^{d-d(\mathbf{u},\mathbf{v})}\cdot(1-\lambda)^{d(\mathbf{u},\mathbf{v})}}{B(\lambda,h)}$$

when the Hamming distance $d(\mathbf{u}, \mathbf{v}) \leq h$ and 0 otherwise. Here λ is a parameter in the interval [0.5, 1] and $B(\lambda,h)=\sum_{j=1}^{h}\binom{d}{j}\cdot\lambda^{d-j}\cdot(1-\lambda)^j$.

We use this kernel in order to find a distribution of probabilities $p(\mathbf{x})$ in the space B^d. For any $\mathbf{x}$ in B^d, an estimation of the probability $p(\mathbf{x})$ is given by $p^*(\mathbf{x})=\frac{1}{n}\sum_{i=1}^{n}K(\mathbf{x},\mathbf{x}_i,\lambda,h)$. As we see, $p^*(\mathbf{x})$ depends only on h and λ. The optimal h and λ are the result of maximization of the likelihood function obtained by cross-validation:

$$CV(\lambda,h)=\frac{1}{n}\sum_{i=1}^{n}\ln(p_{-i}^*(\mathbf{x}_i))$$

where,

$$p_{-i}^*(\mathbf{x}_i)=\frac{1}{n-1}\sum_{\substack{j=1\\ j\neq i}}^{n}K(\mathbf{x}_i,\mathbf{x}_j,\lambda,h)\cdot$$

There are two ways to treat the integer samples. If the integer nature of the data is not important, we treat them as real data by applying the results of the section 2.2.

Otherwise, we transform the integer data to boolean ones by realising an isometric embedding of these data into the boolean hypercube [6]. This embedding preserves the distances and is always possible. Then we find the optimal radius by the method used for boolean data.

4. Quality indicators

At the end of the learning phase we have the *observed input* $\mathbf{x}_1, \ldots, \mathbf{x}_m$, the *observed responses* $y_1, \ldots, y_n$ and the responses, $ym_1, \ldots, ym_n$ of the *given predicting model* $y = F(\mathbf{x})$. For a new input $\mathbf{x}$, the model can predict $ym(\mathbf{x})$ but in general we don't have the observed response $y(\mathbf{x})$. We want to estimate the quality of the response $ym(\mathbf{x})$. Before giving three quality indices, we describe the procedure of adjusting the radii in order to take into account the observed responses. Remember that the radii calculated above are optimal for density estimation, but they are not necessarily good for a predicting model.

So, let $V_A(\mathbf{x})$ = $\{\mathbf{x}_1, \ldots, \mathbf{x}_k\}$ be the neighbourhood in **A** of any point $\mathbf{x}$ and $y(\mathbf{A}) = \{y_1, , y_k\}$ the corresponding observed responses. We can calculate a response $yv(\mathbf{x})$ by interpolation of values in $y(\mathbf{A})$. We can calculate in this way the responses $yv_1, \ldots, yv_n$ for all the samples in **A**. Then we vary the radii about their initial values found in the preceding sections, in order to minimise $\sum_{j=1}^{n}(y_i - yv_i)^2$.

The radii found this way are used in all ulterior calculations.

Once the optimal neighbourhood of $\mathbf{x}$ is calculated we can define the following simple but significant quality indicators on the predicted value $ym(\mathbf{x})$.

Indicator 1: Normalised density.

Generally, the learning models learn better in the zones with a high concentration of samples. Following this paradigm, our first quality indicator is:

$$C_1(\mathbf{x}) = c(n,d) \cdot \frac{f^*(\mathbf{x})}{D_{\max}}.$$

We divide by $D_{\max}$, the maximal value of the densities of the samples in **A**, in order to obtain a value between 0 and 1. The constant $c(n, d)$ takes into account the fact that the density estimation is more or less reliable and depends on the number n and the dimension d of samples.

For a reliable density calculation, a minimal size of the learning set is needed which can be found in [2].

Indicator 2: Agreement of predictions.

This indicator compare the values of the predictions $ym(\mathbf{x})$ and $yv(\mathbf{x})$. If these values are "very" different, then at least one of the predictions is wrong and the indicator will have a low value. When $ym(\mathbf{x})$ is near to $yv(\mathbf{x})$ we can suppose that the predictions are good and the chances that both models go wrong are slim.

As for the fist indicator, in order to normalize the value of the indicator between 0 and 1, we find an interval $[y_{\min}, y_{\max}]$ containing $ym(\mathbf{x})$ et $yv(\mathbf{x})$. The agreement between these two predictions is then:

$$C_2(\mathbf{x}) = 1 - \frac{|ym(\mathbf{x}) - yv(\mathbf{x})|}{|y_{\max} - y_{\min}|}.$$

Indicator 3 : Local variability.

Let $\mathbf{x}_1, \ldots, \mathbf{x}_k$ be the neighbours of $\mathbf{x}$. The local variability is simply the standard deviation of the observed responses $y_1, \ldots, y_k$. The standard deviation describe the local variability of the (unknown) function $y = F(\mathbf{x})$ that we try to learn. It is clear, at least intuitively, that the prediction is less precise in the zones of high variability. The quality indicator $C_3(\mathbf{x})$ is in inverse proportion to the local variability.

These three indicators are complementary. So, one has to consider these indicators simultaneously in order to have a global idea about the quality of predictions.

5. Implementation

We have implemented the presented ideas in a software named "CONFIDENT". All the numerical libraries and interfaces were realised with Matlab. The figure 2 is one screen capture of the software containing about 12 screens. A simplified run-independent version was written in C++ and Fortran 90 with the IMSL Library. This version is used as an early-drift detector in an oil refinery.

The indicator C_3 permits to adapt the window's width (see the dot line in fig. 2). So, we can choose the neighbourhood minimizing the predicting error for the present sample. It correspond to the narrow part of the bottleneck which typically has the form appearing in fig. 2.

The values of the indicators C_1 and C_2 are shown in the low-left corner of the window (see fig. 2). The question of giving a single indicator calculated on the basis of C_1, C_2 and C_3 remains opened.

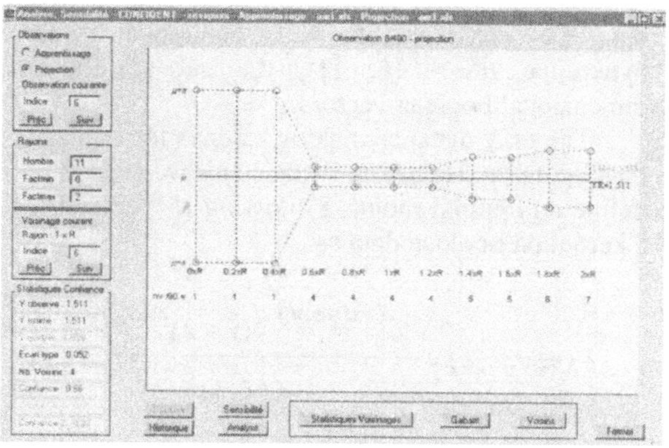

Fig. 2 : The typical bottleneck shows that the quality of the predictions is better for an intermediary window's width.

6. Conclusion

The ideas presented in the preceding sections give a satisfying answer to the problems we have met with a number of industrial models.

The indicators have been used already to the early-drift diagnostic. Also, they have been used to measure the quality of ozone air pollution predictors. In fact, a statistical model and a neural net model forecast the ozone level for the next 24 hours. This two models use historical data (collected during 5 years) as well as meteorological forecast. These models perform globally well but, 8 errors persisted. We have shown with "CONFIDENT" that this 8 samples were simply outliers.

The software is used actually for the prediction of missing values in incomplete data.

After these first comforting results, we plan to deepen the theoretical as well as practical aspects of our approach. A better comprehension of the indicators is necessary. Other significant indicators can be formulated. It is important and useful to melt these indicators into a single quality indicator.

Moreover, in a short-term we plan to exploit the indicators in the real-time monitoring of the neural models used to estimate on-line the properties of the products in an oil refinery.

7. References

[1] Guermeur, Y., Paugam-Moisy, H. (1999) Théorie de l'apprentissage de Vapnik et SVM, Support Vector Machines, in Sebban, M. & Venturini, G. (Eds) Apprentissage automatique, 109-138. HERMÈS.

[2] Silverman B. W., (1986) Density Estimation for Statistics and Data Analysis, Chapman & Hall.

[3] Scott D. W., (1992) Multivariate density estimation. Theory, Practice and Visualisation, Rice University, Huston, Texas, John Wiley.

[4] Flannery B. P., Press W. H., Teukolsky S. A., Vetterling W. T., (1986) Numerical Recipes, The art of scientific computing, Cambridge University Press.

[5] Aitchison J., Aitken C. G., (1976) Multivariate binary discrimination by the kernel method, Biometrika, 63, 3, 413-420.

[6] Janaqi S., (1995) Quelques éléments de la géométrie des graphes, Thèse de l'Université Joseph Fourier, Médiathèque IMAG, Grenoble.

[7] Aamodt A., (1995) Case-Based Reasoning, Methods and Applications, Tutorial Notes, Al Petro, Lillehammer.

[8] Bagatelj V., Bren M., (1996) *Determining the Distance Index*, In Ordinal and Symbolic Data Analysis, Diday E., Lechevallier Y. and Opitz O. Eds., Springer.

[9] Bath P., Morris C., Willett P., (1993) Effect of standartization on fragment-based measures of structural similarity, Journal of Chemometrics, Vol. 7, 543-550.

[10] Blumenthal L. M., (1970) Theory and applications of distance geometry, Chelsea, New York.

[11] Boussaid O., Rabaseda S., Sebban M., (1996) Contribution of Related Geometrical Graphs In Pattern Recognition, In Ordinal and Symbolic Data Analysis, Diday E., Lechevallier Y. and Opitz O. Eds., Springer.

[12] Dasarathy B. V., Nearest Neighbour Norms : NN Pattern Classification Techniques, IEEE Computer Society Press, Los Alamitos, California.

[13] Diu B., Guthmann C., Lederer D., Roulet B., (1989) Eléments de physique statistique, Hermann Ed.

[14] Heskes T., (1997) Practical confidence and prediction intervals, Advances in Neural Information Processing Systems, 9, Cambridge, MIT Press.

[15] Kohonen T., (1990) The Self-Organizing Map, IEEE, Vol. 78, No. 9.

[16] Kramer M. A., Leonard J. A., Ungar L. H., (1992) Using Radial Basis Functions to Approximate a Function and Its Errors Bounds, IEEE Transactions on Neural Networks, Vol. 3, No. 4.

[17] Martens H., Naes T., (1993) Multivariate Calibration, John Wiley & Sons.

[18] Martinez T. M., Ritter H. J., Schulten K. J., (1989) Topology-Conserving Maps for Learning Visuo-Motor-Coordination, Neural Networks, Vol. 2, 159-168.

[19] Neural Computing, (1993) A Technology Handbook for Professional II/PLUS and NeuralWorks Explorer.

[20] Saporta P., (1990) Probabilités, Analyse de données et Statistique, Editions TECHNIP.

[21] Zagoruiko N. G., (1996) Knowledge Analysis Using the Metrics in the Knowledge Space, In Ordinal and Symbolic Data Analysis, Diday E., Lechevallier Y. and Opitz O. Eds., Springer,.

Manufacturing process quality control by means of a Fuzzy ART neural network algorithm

Massimo Pacella[1*], Quirico Semeraro[2], Alfredo Anglani[1]
[1]Dip. di Ingegneria Innovazione – Università degli Studi di Lecce, Via per Monteroni, Lecce, 73100 Italy.
[2]Dip. di Meccanica – Politecnico di Milano, Via Bonardi 9, Milano, 20132 Italy.
[*]corresponding author, e-mail: massimo.pacella@unile.it

Abstract. Neural networks are potential tools that can be used to improve process quality control. In fact, various neural algorithms have been applied successfully for detecting groups of well-defined unnatural patterns in the output measurements of manufacturing processes. This paper discusses the use of a neural network as a means for recognising changes in the state of the monitored process, rather than for identifying a restricted set of unnatural patterns on the output data. In particular, a control algorithm, which is based on the Fuzzy ART neural network, is first presented, and then studied in a specific reference case by means of Monte Carlo simulation. Comparisons between the performances of the proposed neural approach, and those of the CUSUM control chart, are also presented in the paper. The results indicate that the proposed neural network is a practical alternative to the existing control schemes.

1 Introduction

The main goal of quality control in manufacturing is to maintain a constant and acceptable level of some process characteristics. Usually, a certain amount of variability affects measurements of the quality parameters of interest. Two sources of variability may influence the outcomes of a process; commonly they are referred to as *unassignable* and *assignable* causes [1]. The variations due to unassignable causes are the result of numerous unremarkable changes that may occur in a process. Often, this kind of variation is inevitable without a profound revision of the whole production procedure. When only unassignable causes are in effect, a process is considered to be in a natural state (i.e. in control). On the other hand, the variations due to assignable causes are generated by factors that lie outside the process. New methods and different machines, or changes in the measurement instruments, are common examples of assignable causes. In such cases, the process is said to be in an unnatural state (i.e. out of control), and quality improvement is possible by detection and removal of the assignable causes.

Among the Statistical Process Control (SPC) methods, the control charts are the most common tools to reveal unnatural variations in the monitored measurements [1]. However, with the growing exploitation of automatic on-line data-collection methods, nowadays a demand exists to automate the analysis of process data.

In the last decade, the artificial neural networks have been widely used for data analysis in quality control applications [2]. The neural networks appear suitable for quality control because of their ability to elaborate large amounts of data in real-time, and their capacity for handling noisy, uncertain or fuzzy data. Many different neural networks and learning algorithms have been proposed in the literature [2]. Hwarng and Hubele [3] proposed a multilayer perceptron (MLP) trained with back-propagation algorithm (BP) to detect six unnatural patterns. Smith [4] described a similar algorithm in order to analyse both mean and variance shifts. Guh and Tannock [5] developed a MLP BP neural network for concurrent unnatural pattern recognition. Cook *et al.* [6] discussed the development of a MLP BP neural network to identify changes in the variance of serially correlated process parameters.

The neural network for quality control, which has been proposed by researchers in almost all the published works, is the MLP BP [2]. The MLP BP has been studied thoroughly, and has been exploited successfully in various applications. However, the use of a supervised neural network means that both a set of well-defined patterns and an adequate number of examples are available for neural network training. Frequently, in various industrial cases, training patterns are not available because unnatural process behaviours cannot be manifested by the appearance of predictable patterns and thus, the mathematical models are not readily available or they cannot be formulated.

With the exception of two published works [7,8], little attention has been devoted by researchers to the development of quality control systems based on the Adaptive Resonance Theory (ART). ART neural networks are competitive learning pattern classifiers. Competitive learning is an unsupervised training strategy that accomplishes a clustering task, which is based on a function optimisation (e.g. a distance between vectors of an *n*-dimensional space). The ART neural network can be used to monitor process under the assumption that no knowledge on the unnatural state is available in advance for network training.

Recently, we have investigated on the use of ART for quality control applications [9]. A neural network approach, based on a simplified Fuzzy ART that is capable of fast and cumulative learning, has been proposed for quality control. In this paper, the Fuzzy ART control system is firstly presented, and then it is applied to identify a special pattern of process data: the upward trend. The purpose of this paper is to analyse the performance of the proposed system in recognising trend of process data when network training is limited on the natural target and on a specific unnatural off-target value.

The Fuzzy ART algorithm is based on the fuzzy set theory operations, thus the values of the input nodes, as well as of the weights of the network, can range between zero and one. The reader is referenced to the papers 10 and 11 for further details on Fuzzy ART.

The rest of the paper is organised as follows. Section 2 gives an overview of the neural network algorithm for process monitoring. Section 3 explains the training phase of the proposed neural network while in section 4 the testing phase is discussed. Section 5 discusses the performance of the artificial neural network for the reference test case. Finally, conclusions are given.

2 Outline of the neural algorithm

The use of a control chart as well as of a neural network algorithm for process monitoring resembles hypothesis testing. Usually, the process is analysed to verify a constant mean with some natural inherent variation. The null hypothesis H_0 and the alternative hypothesis H_1 of the test can be formulated as follows.

$$\begin{aligned} &H_0 : \text{the process is under control.} \\ &H_1 : \text{the process is out of control.} \end{aligned} \quad (1$$

In order to investigate the Fuzzy ART performances for quality control applications, in the present work the output of a generic manufacturing process has been synthetically reproduced by means of Monte Carlo simulation. In the reference test case, the measurements of the quality parameter are collected at regular interval of time. Let $\{Y_t\}$ be the random time series of the mono-dimensional process output. Generally, the outcomes of a manufacturing process in a natural state may be realistically modelled by a random time series, which values are distributed normally, independently, and identically (NID). Without loss of generality, it is assumed that the mean and the variance of such a distribution are equal to zero and one respectively, i.e. NID(0,1). Moreover, it is assumed that when the process starts drifting from the natural state, a form of a special disturbance signal overlaps the series of output measurements.

Let $\{Z_t\}$ be the time series of the natural process data and let $\{S_t\}$ be the time series of the special disturbance signal. At each instant of index t, the statistical test can be re-formulated as reported by the following equation 2.

$$\begin{aligned} &H_0 : Y_t = Z_t \\ &H_1 : Y_t = Z_t + S_t \end{aligned} \quad (2$$

The proposed neural system for quality control and the simulation model of the reference manufacturing process are both depicted by figure 1. At the time of index t, the control system accepts as input the process output Y_t, and produces the binary signal $b_{nn,t}$ that is the result of the test performed by Fuzzy ART on the state of the process. In particular, the algorithm produces $b_{nn,t} = 1$ if the process is considered in a natural state, $b_{nn,t} = 0$ otherwise.

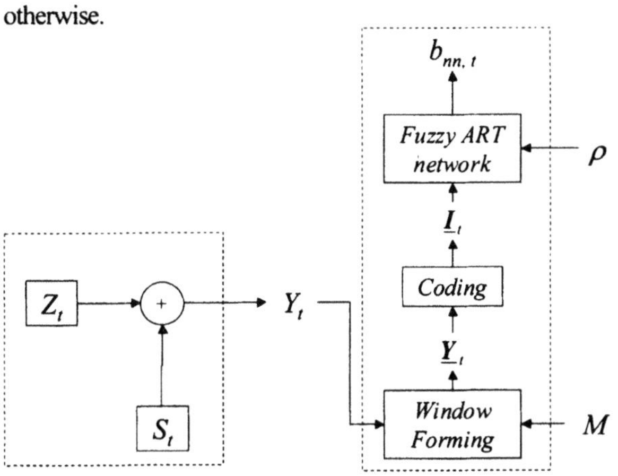

Fig. 1. The proposed neural system for quality control

As depicted by figure 1, some pre-processing of the input data takes place before they are presented to the Fuzzy ART neural network. The first stage (*Window Forming*) depends on the integer parameter $M \geq 1$ (called the *Window Size*). It transforms the time series $\{Y_t\}$ of process output data into *M*-dimensional vectors. In particular, the most recent *M* observations are collected to form the vector $\underline{Y}_t$ that is equal to:

$$\underline{Y}_t = [Y_{t-M+1}, Y_{t-M+2}, \ldots, Y_{t-1}, Y_t] \qquad t \geq M \quad (3$$

The second pre-processing stage (called *Coding*) takes as input an *M*-dimensional input pattern $\underline{Y}_t$ and transforms it into the corresponding *M*-dimensional output vector (say $\underline{I}_t$) whose components fall into the interval $[0,1]$. The implemented *Coding* stage is a linear re-scaling of the process data. Specifically, let $\underline{I}_t$ be:

$$\underline{I}_t = [I_{t-M+1}, I_{t-M+2}, \ldots, I_{t-1}, I_t] \qquad t \geq M \quad (4$$

then we have for $t - M + 1 \leq \tau \leq t$ that

$$\begin{cases} I_\tau = 0 & Y_\tau < -l; \\ I_\tau = \frac{1}{2}\left(1 + \frac{Y_\tau}{l}\right) & -l \leq Y_\tau \leq l; \\ I_\tau = 1 & l < Y_\tau; \end{cases} \quad (5$$

In the reference test case, the parameter l has been fixed to $l = 3$. This is motivated by the assumption that, when the process is a natural state, the output time series values are modelled as NID(0,1). We expect that about 99.74% of the natural observations fall into the interval $[-3,3]$.

The Fuzzy ART neural network accepts as input the vector $\underline{I}_t$. The neural network consists of two major subsystems, the attentional and the orienting subsystem. Three fields of nodes denoted as *F0*, *F1* and *F2* compose the attentional subsystem. On the other hand, the orienting subsystem consists of a single node called the reset node.

In the most simplified terms, the layer *F1* acts as a feature detector that receives external input patterns. The layer *F2* acts as a category classifier that receives internal patterns. The application of an input vector leads to a neural activity that results in the formation of a pattern in both the layers *F1* and *F2*. The orienting subsystem is responsible for generating a reset signal to *F2* when the bottom-up input pattern and the top-down template mismatch according to a vigilance criterion. The vigilance criterion depends on the vigilance parameter ($\rho \in [0,1]$). The choice of high values for the vigilance parameter implies that only a slight mismatch will be tolerated before a reset signal is emitted. On the other hand, a small value implies that large mismatches will be tolerated.

In the *F0* field, an additional pre-processing stage on the incoming input vectors $\underline{I}_t$ is implemented. This pre-processing stage accepts an *M*-dimensional vector, and it produces the following *2M*-dimensional output vector.

$$\underline{I}_t^c = (\underline{I}_t, \underline{1} - \underline{I}_t)$$
$$\underline{I}_t^c = [I_{t-M+1}, \dots I_t, 1 - I_{t-M+1}, \dots, 1 - I_t] \quad (6$$

The above transformation is called the *Complement Coding*.

3 Training phase

In the present work, it is assumed that a predetermined list of natural and/or unnatural input patterns is not available for network training. Instead, it is assumed to know the *target* of the process, and a specific *off-target* value that we want to detect quickly. The target is the nominal mean of the process, i.e. the output that the process should have if both assignable and unassignable causes of variation are not present. The off-target is a specific deviation from the natural target that we want to reveal promptly. If both the natural target and the shifted target can be considered constant over time (steady-state response), then the Fuzzy ART training list consists of two *M*-dimensional vectors only: the steady-state natural process mean (the target), and the unnatural shifted mean (the off-target).

During training, we want that Fuzzy ART stores both the vectors and thus the vigilance parameter is set to its maximum value ($\rho = 1$). In such way, the network learns two different categories that reproduce the specific training patterns: the first one represents the natural target and the second one the shift (Perfectly Learned Patterns – PLP training approach). The number of list presentations for the Fuzzy ART training can be reduced to one because once a cluster has been formed the weights of this category cannot change during the subsequent list presentations if the vigilance parameter is set to $\rho = 1$ [10,11].

4 Testing phase

Let us assume that at time of index $t \geq M$ an *M*-dimensional input pattern $\underline{I}_t$ is presented at the *F0* field of the Fuzzy ART. The appearance of the *2M*-dimensional pattern $\underline{I}_t^c$ across the *F1* field produces bottom-up inputs that affect the nodes in the *F2* layer. The bottom-up inputs activate a competition process among the *F2* nodes, which eventually leads to the activation of a single node in the *F2*, namely the node that receives the maximum bottom-up input from *F1*. In particular, let $\underline{w}_\mu^c$ ($\underline{w}_s^c$) be the top-down weight vector of the committed node in the *F2* layer that stores the natural (unnatural) cluster. The natural cluster wins the competition on the unnatural one if the following condition is satisfied.

$$\frac{|\underline{I}_t^c \wedge \underline{w}_\mu^c|}{\alpha + |\underline{w}_\mu^c|} \geq \frac{|\underline{I}_t^c \wedge \underline{w}_s^c|}{\alpha + |\underline{w}_s^c|} \quad (7$$

Where α is a constant called the *choice parameter*; $|\underline{x}|$ is the size of a vector $\underline{x}$, (i.e. the sum of the absolute value of its components $|\underline{x}| = \sum_i |x_i|$); $\underline{x} \wedge \underline{y}$ is the vector whose i^{th} component is the minimum between the ith component of the vector $\underline{x}$ and the ith component of the vector $\underline{y}$, thus: $\underline{x} \wedge \underline{y} = [\dots, \min(x_i, y_i), \dots]$. The operation $\wedge$ is called the fuzzy min operator.

Then, the neural network classifies the input pattern $\underline{I}_t^c$ natural (i.e. as a member of the natural cluster) if the following check is passed [10,11].

$$\frac{|\underline{I}_t^c \wedge \underline{w}_\mu^c|}{|\underline{I}_t^c|} \geq \rho \quad (8$$

The vigilance parameter ρ, which is used in this phase, can be different from that used in the training phase.

The size of each input vector $\underline{I}_t^c$ is equal to *M* since it results that:

$$|\underline{I}_t^c| = |(\underline{I}_t, \underline{1} - \underline{I}_t)| = |\underline{I}_t| + M - |\underline{I}_t| = M \quad (9$$

Moreover, since the PLP training approach has been used, it results also that $|\underline{w}_\mu^c| = |\underline{w}_s^c| = M$. Therefore, both the equations 6 and 7 can be rewritten as follows.

$$\begin{cases} |\underline{I}_t^c \wedge \underline{w}_\mu^c| \geq |\underline{I}_t^c \wedge \underline{w}_s^c| \\ |\underline{I}_t^c \wedge \underline{w}_\mu^c| \geq M\rho \end{cases} \quad (10$$

In other words, the input pattern $\underline{I}_t^c$ is recognised by the neural network as a natural pattern (i.e. the output is set to $b_{nn,t} = 1$) if the following check is passed:

$$\left|\underline{I}_t^c \wedge \underline{w}_\mu^c\right| \geq \max\left\{M\rho, \left|\underline{I}_t^c \wedge \underline{w}_s^c\right|\right\} \qquad (11$$

Otherwise, the input patter is classified as unnatural and the output result is set to $b_{nn,t} = 0$.

As underlined by the above equation 11 the performance of the neural network for quality control depends on two parameters, namely the window size M and the vigilance parameter ρ.

5 Testing results

In order to evaluate the performances of the proposed neural algorithm for quality control, two characteristics are calculated by means of computer simulation. The first is the ability to model unassignable causes of variation without creating Type I errors (i.e. false alarms), which indicate that the process is out of control when it is in fact not. This property is measured by reporting the mean of the false alarm occurring in process data having only unassignable sources of variation (say $\hat{\alpha}$). The second performance measures the control system ability to detect unnatural patterns in the process output data. This property is calculated experimentally by reporting the mean of the Type II errors (say $\hat{\beta}$, i.e. the non-alarm signals) occurring in the process data when a special disturbance, with a controlled magnitude, is introduced. In particular in this work, an upward linear trend has been used to simulate a special disturbance of the process mean that we want to detect by using the Fuzzy ART neural network. In figure 2 the effect of such a pattern on a control chart is depicted.

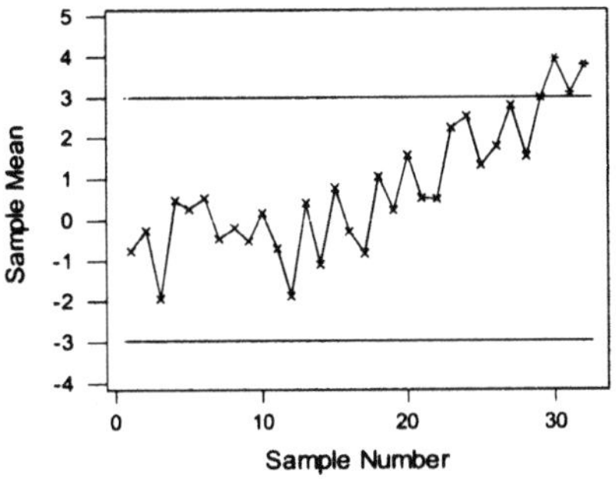

Fig. 2. The linear trend pattern depicted on a control chart

In order to evaluate the performance of the neural network, in table 1 the simulation results are compared to those of a SPC benchmark: the CUSUM control chart [1]. The CUSUM parameters have been chosen in order to both detect changes in the process mean of one unit of standard deviation, and to obtain a Type I error of about 0.27%. For the comparison to be unbiased, the CUSUM alarms, which occur during the first $M-1$ observations, have been neglected, and the performances have been measured for time indexes $t \geq M$. Furthermore, the Fuzzy ART neural network has been firstly tuned to give comparable performance in terms of Type I error to that of the reference CUSUM chart. In particular, the window size has been set to $M = 75$ and the vigilance parameter ρ has been adjusted to give a Type I error level comparable to that of the CUSUM control chart (that results in $\rho = 0.8375$). The comparison is based on the Type II error estimators for the linear upward trend pattern. Twenty levels of magnitude (ranging in $[0.001, 0.020]$ with steps of 0.001 unit of standard deviation) have been considered in the simulations.

Table 1 presents both the Type I and Type II error estimator points of the neural network ($\hat{\alpha}_{nn}$, $\hat{\beta}_{nn}$ respectively) and of the CUSUM chart ($\hat{\alpha}_{cc}$, $\hat{\beta}_{cc}$ respectively), as well as the confidence intervals (coverage 95%) of the difference between them. Specifically, the column labelled as $\hat{\alpha}_{nn} - \hat{\alpha}_{cc}$ reports the difference between the Type I error point estimator of the neural network ($\hat{\alpha}_{nn}$) and that of the control chart ($\hat{\alpha}_{cc}$). The column marked as $\hat{\beta}_{nn} - \hat{\beta}_{cc}$ reports the difference between the Type II error estimator points.

	CUSUM k=0.5 h=4.7749	Fuzzy ART M=75 ρ=0.8375	Fuzzy ART – CUSUM		
	$\hat{\alpha}_{cc}$%	$\hat{\alpha}_{nn}$%	$(\hat{\alpha}_{nn} - \hat{\alpha}_{cc})_-$	$\hat{\alpha}_{nn} - \hat{\alpha}_{cc}$	$(\hat{\alpha}_{nn} - \hat{\alpha}_{cc})_+$
	0.269	**0.268**	-0.047%	-0.001%	0.045%
	$\hat{\beta}_{cc}$	$\hat{\beta}_{nn}$	$(\hat{\beta}_{nn} - \hat{\beta}_{cc})_-$	$\hat{\beta}_{nn} - \hat{\beta}_{cc}$	$(\hat{\beta}_{nn} - \hat{\beta}_{cc})_+$
0.001	**99.358**	**98.991**	-0.425%	-0.367%	-0.309%
0.002	**98.860**	**98.196**	-0.735%	-0.664%	-0.593%
0.003	**98.255**	**97.054**	-1.305%	-1.201%	-1.097%
0.004	**97.459**	**95.560**	-2.041%	-1.899%	-1.757%
0.005	**96.269**	**93.261**	-3.207%	-3.008%	-2.809%
0.006	**94.635**	**90.030**	-4.877%	-4.605%	-4.333%
0.007	**92.002**	**84.931**	-7.452%	-7.071%	-6.690%
0.008	**87.831**	**77.471**	-10.835%	-10.360%	-9.885%
0.009	**81.436**	**66.932**	-15.231%	-14.504%	-13.777%
0.010	**71.347**	**53.706**	-18.457%	-17.641%	-16.825%
0.011	**57.216**	**38.954**	-19.207%	-18.262%	-17.317%
0.012	**39.172**	**24.756**	-15.453%	-14.416%	-13.379%
0.013	**21.739**	**13.946**	-8.690%	-7.793%	-6.896%
0.014	**10.106**	**7.153**	-3.529%	-2.953%	-2.377%
0.015	**3.835**	**3.419**	-0.823%	-0.416%	-0.009%
0.016	**1.294**	**1.416**	-0.080%	0.122%	0.324%
0.017	**0.336**	**0.623**	0.162%	0.287%	0.412%
0.018	**0.095**	**0.219**	0.058%	0.124%	0.190%
0.019	**0.023**	**0.068**	0.011%	0.045%	0.079%
0.020	**0.000**	**0.030**	0.009%	0.030%	0.051%

Table 1. Comparison between Fuzzy ART (M=75, ρ=0.8375) and CUSUM (k=0.5, h=4.7749) chart (simulation results).

The results of table 1 show that the neural network performance is comparable to the CUSUM chart in terms of Type I errors since the confidence interval includes the zero value. This implies that there is no statistical evidence to reject the hypothesis $\alpha_{nn} = \alpha_{cc}$. The neural network has better performances than those of the CUSUM to recognise upward trend of small and medium slope, i.e. 0.001 - 0.015 units of standard deviation. In fact, the confidence intervals on the difference between the point estimators include only negative values and thus we can statistically conclude that $\beta_{nn} < \beta_{cc}$. On the other hand, the performances of the neural network are either approximately similar to those of the control chart for higher magnitude (0.016-0.020 units of standard deviation) or slightly worse.

In figure 3, the simulation results are graphically depicted. In addition, figures 4 and 5 show in more detail the simulation results in the case of high errors of Type II (i.e. higher than 95%) and low errors of Type II (i.e. lower than 5%) respectively. In particular, in figure 4 the case of low slopes of the trend pattern (less than 0.005 units of standard deviation) is considered, while in figure 5 only slopes of higher magnitudes (i.e. higher than 0.01 units of standard deviation) are taken into account.

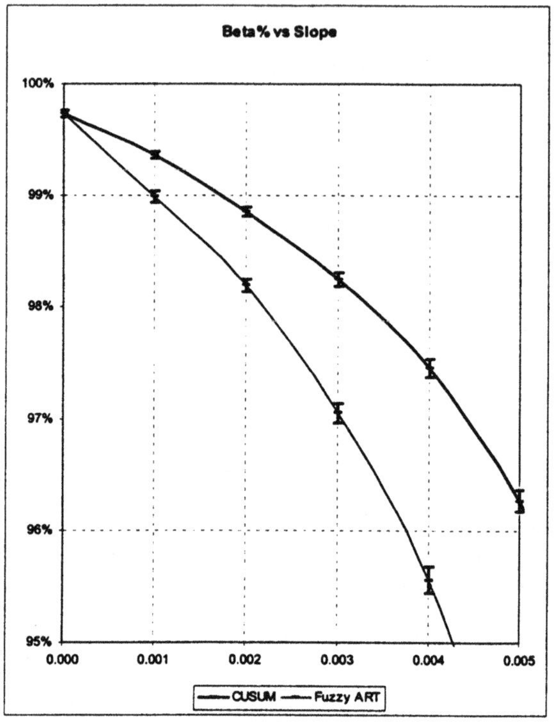

Fig. 4. Detail. Neural network (thin line) and CUSUM (bold line) type II point estimators (>95%, ordinate) vs. trend slope (abscissa) and interval estimators at coverage 95% (simulation results).

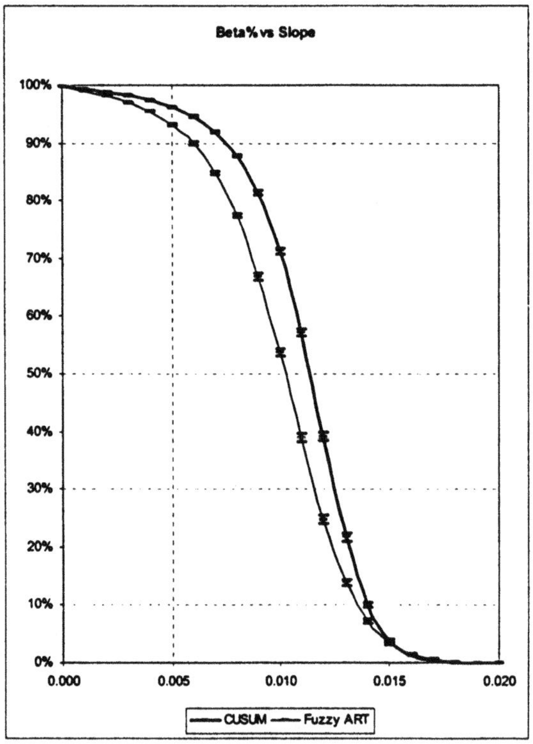

Fig. 3. Neural network (thin line) and CUSUM (bold line) type II point estimators (ordinate) vs. trend slope (abscissa) and interval estimators at coverage 95% (simulation results).

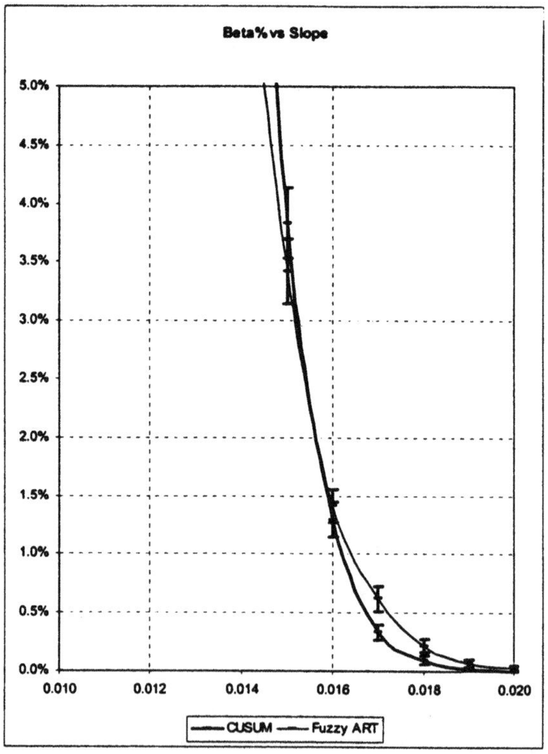

Fig. 5. Detail. Neural network (thin line) and CUSUM (bold line) type II point estimators (<5%, ordinate) vs. trend slope (abscissa) and interval estimators at coverage 95% (simulation results).

6 Concluding remarks

In this paper, the application of Adaptive Resonance Theory for quality control tasks has been briefly analysed. Several proprieties of ART-based neural network make it a practical tool for quality control applications over supervised ones. Since ART networks are self-organising, the number of training iterations needed to mach the performances of supervised neural networks is lower. Thus, training times in the development of a neural-based control system are significantly reduced.

The main advantage of this approach is that it requires no previous information about unnatural pattern appearances, related mathematical models, or probability distribution functions. This neural network can be potentially adopted to signal any types of unnatural pattern, so it provides a powerful diagnostic tool for detecting assignable causes in real processes.

We recommend the proposed Fuzzy ART neural algorithm when probabilistic/mathematical models of either the natural or unnatural process output are not available. Especially when a new process is starting up for which earlier data are not sufficient to obtain an adequate number of training examples for a supervised neural network control system.

Acknowledgement

This work has been partially funded by the Ministry of Education, University and Research of Italy (MIUR) and by the National Research Council of Italy (CNR).

References

[1] Montgomery, D. C. (2000) Introduction to Statistical Quality Control, 4th ed. J. Wiley, New York.

[2] Zorriassantine, F., Tannock, J. D. T. (1998) A review of neural networks for statistical process control. J. of Intelligent Manufacturing, 9: 209-224.

[3] Hwarng, H. B. and Hubele, N. F. (1993) Back-propagation pattern recognizers for X-bar control charts: methodology and performance. Comp. & Ind. Eng., 24: 219-235.

[4] Smith, A. E. (1994) X-bar and R control chart interpretation using neural computing. Int. J. of Production Research, 32: 309-320.

[5] Guh, R. S., Tannock, J. D. T. (1999) Recognition of control chart concurrent patterns using a neural network approach. Int. J. of Production Research, 37(8): 1743-1765.

[6] Cook, D. F., Zobel, C. W., Nottingham, Q. J. (2001) Utilization of neural networks for the recognition of variance shifts in correlated manufacturing process parameters, Int. J. of Production Research, 39(17): 3881-3887.

[7] Hwarng, H. B., Chong, C. W. (1995) Detecting process non-randomness through a fast and cumulative learning ART-based pattern recognizer. Int. J. of Production Research, 33, 1817-1833.

[8] Al-Ghanim, A. (1997) An unsupervised learning neural algorithm for identifying process behavior on control charts and a comparison with supervised learning approaches. Comp. & Ind. Eng., 32(3): 627-639.

[9] Pacella, M., Semeraro, Q., Anglani, A., (2002) On the use of adaptive resonance theory based neural algorithms for manufacturing process quality control. Submitted to Int. J. of Production Research.

[10] Huang, J., Georgiopoulos M., Heileman, J. L. (1995): Fuzzy ART Proprieties. Neural Networks, 8(2): 203-213.

[11] Georgiopoulos, M., Fernlund, H., Bebis, G., Heileman, G. L. (1996) Order of search in Fuzzy ART and Fuzzy ARTMAP: effect of the choice parameter. Neural Networks, 9(9), 1541-1559.

Hybrid Models for Forecasting Air Pollution Episodes

Harri Niska, Teri Hiltunen, Mikko Kolehmainen, Juhani Ruuskanen

Department of Environmental Sciences
University of Kuopio, P.O. Box 1627, FIN-70211 Kuopio, Finland
E-mail: mikko.kolehmainen@uku.fi

Abstract

Urban air pollutants have emerged as a severe problem which causes health effects and even premature deaths among sensitive groups. Therefore a warning system for air pollution episodes is widely needed to minimize negative health effects. However the forecasting of air pollution episodes has been observed to be problematic partly due their rareness and short-term nature. The research presented here aims to evaluate different neural network based models for forecasting urban air pollution (NO_2) hourly time series and particularly the episode peaks. The performances of three multi-layer perceptron (MLP) models namely basic MLP and two hybrid models were compared by calculating several statistical indicators. In the hybrid models evaluated here, training data set was clustered to several air quality episodes using the k-means (KM) and fuzzy c-means (FCM) algorithms and then several MLP models were applied to the clustered data, each one representing one cluster. The results showed that the hybrid models have some advantages over a basic MLP model in the forecasting air quality episodes, but the performance achieved also show that architectural issues cannot solely solve the model performance problems.

1 Introduction

It has been observed that urban air pollution and particularly air pollution episodes where some air pollutants such as ozone and nitrogen dioxide have high concentrations cause severe health effects especially among sensitive groups such as asthmatics and elderly people [1]. Therefore, efforts have been focused on the developing suitable methods for modelling air pollution time series. However, more important than to develop methods which are capable of forecasting air pollution in low concentration levels, is to find methods which will be efficient and robust in the forecasting rare and short-term high concentrations. This is important in order to start preventive actions before and during the episodes and in this way to minimize possible health effects.

In recent years, statistical models such as neural networks based on the long-term measurement data from different sources such as pollution, meteorology and traffic have been used in the forecasting air quality parameters [2]. The results in general have shown that neural network based models perform better than or as well as conventional linear methods [3], [4], [5]. However, there have been difficulties in forecasting episodes using neural networks. This is at least partly due to the restricted episode data in the training of neural network. In this study, the focus was on the evaluating of three different multi-layer perceptron (MLP) models for the episode forecasting task. Main interest was in the comparing the performances of two hybrid models [6] to the one of basic MLP [2]. However, in the analysis of model performances, also the ability of models to predict lower air pollutant concentrations was considered.

2 Hybrid modelling

Hybrid modelling techniques contain a wide range of applications where several methods or properties of methods are combined to a more sophisticated model in order to achieve better performance or some advantages in specific situations. A promising and fairly new area in hybrid modelling is computational intelligence which includes a number of forms of computation techniques such as neural networks, genetic algorithm and fuzzy logic. By combining intelligent and traditional methods together in specific way, new adaptive and efficient methods can be found. One example of using hybrid modelling to simulate real world processes is the model used here to forecast urban air quality for the next day using airborne pollutant, meteorological and timing variables.

The basic idea of the hybrid model (see Fig. 1) applied to the air quality forecasting used here was developed and tested by Kolehmainen *et al.* [6]. In this scheme pre-processed air quality data was first analysed by using Self-Organizing Map (SOM) [7], finding the kernels of *n* pollution episodes where the pollutant averages reach their highest values locally and calculating fuzzy membership values for each data row and episode. Then several possible overlapping MLP models, each representing one cluster (episode), were trained with data selected based on membership values. Finally at the modelling stage, the air quality prediction was calculated by finding the best matching unit (BMU) in the SOM map and calculating membership values for MLP input vector consisted of timing and meteorological variables. The final estimate was performed as membership-weighted sum of several submodels above a limit membership.

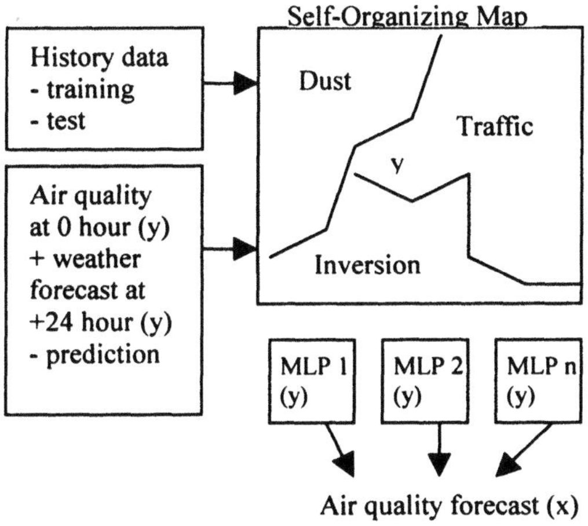

Fig. 1. General principle of the hybrid neural network [6] for +24 hours forecasting where *y* is an input vector (air quality at +0 hour; weather at +24 hour) for the MLP(s) and *x* the predicted air pollutant concentration at +24 hour.

3. Material and methods

3.1 Data

In this study we were able to use more applicable data for the episode modelling namely the data set used contained more meteorological variables and air pollution episodes with higher pollution levels compared to previous study [6]. The test data set was extracted from the APPETISE (Air pollution Episodes: Modelling Tools for Improved Smog Management, http://www.uea.ac.uk/env/appetise/) database. The pollutant data used was monitored in Helsinki, Finland (air quality monitoring station Töölö) during the years 1996-1999. The meteorological data contains routine weather and sounding observations from the stations of Finnish Meteorological Institute at Helsinki-Vantaa airport (about 15 km north of Helsinki downtown), Helsinki-Isosaari (an island about 20 km south of Helsinki) and at Jokioinen observatory (90 km northwest of Helsinki). The atmospheric turbulence parameters and mixing height was evaluated using a meteorological pre-processing model [8].

The data set consisted of 30 timing, meteorological and airborne pollutant variables. The meteorological variables were sea level pressure, temperature at 2 meters, relative humidity, total cloudiness, dew point temperature, wet bulb temperature, amount of rain, visibility, amount of low clouds, height of low clouds, wind speed, wind direction (sine of wind direction, cosine of wind direction), hourly amount of sunshine, albedo, solar elevation, solar radiation, moisture parameter, inverse of Monin-Obukhov length, temperature scale, friction velocity, turbulent heat flux, net radiation, latent heat flux, mixing height, mixing height2 (lower altitude), convective velocity scale and gradient of potential temperature. The pollutant variables were NO_2 and ozone. In addition there were timing variables: month and hour in the form of sine and cosine components. The predicted pollutant variable was nitrogen dioxide (NO_2).

3.2 Basic MLP model

The parameters of basic MLP model were defined based on the experience of earlier studies by Kolehmainen *et al.* [2], [6] and previous experience. Thus, the number of hidden neurons was 16, the learning algorithm used was Levenberg-Marquardt and the performance function of mean squared error (MSE) was applied. The transfer functions selected were hyperbolic sigmoid tangent for the hidden layer and linear for the output layer. The weights of MLP were scaled to a floating range of numbers [-0.7, 0.7]. The basic MLP was trained using the states of timing, pollution at +0 hour and meteorological variables at +24 hour (simulating a weather forecast) as network inputs *y* and the actual level of airborne pollutant NO_2 at +24 hour as output *x*. Thus the network was "forced" to learn a possible time dependency between input measures and the predicted air pollutant. The settings presented here were used also in the MLP models of the hybrid models.

3.3 Hybrid models

The hybrid models used here follow the main

principle of the original hybrid model tested by Kolehmainen *et al.* [6]. The architectural exception in the model was that the clustering was done algorithmically by applying k-means and fuzzy c-means clustering algorithms to the air quality data self-organized by the SOM with 512 neurons.

The k-means clustering (KM) [10] is a classical algorithm which is used to partition the input space into k partitions by minimizing the distance of each point from the centre of the cluster to which the point belongs by using the error function J (Eq. 1).

$$J = \sum_{i=1}^{m} \sum_{j=1}^{K} u_{ij} d^2(x_j, c_i), \tag{1}$$

where m is the number of data points, K is the number of clusters, u_{ij} is element of partition matrix which is equal to one if object j belongs to cluster i (zero otherwise), x_j is the jth n-dimensional feature vector, c_i is the centre of the ith cluster and $d^2(x_j, c_i)$ is any inner product metric (distance between x_j and c_i).

The fuzzy c-means (FCM) was introduced by Bezdek (1981) as an improvement on earlier clustering methods like k-means algorithm. The basic assumption is that an object can be classified to more than one cluster which causes that several overlapping MLP models wil! be trained in the hybrid model. This property reflects the uncertainty of real world data where class boundaries are not clear-cut. In the FCM the following cost-function (Eq. 2) is minimized, with respect to U, a fuzzy K-partition of the data set, and to C, a set of K prototypes (cluster centres):

$$J_q(U, C) = \sum_{j=1}^{m} \sum_{i=1}^{K} (u_{ij})^q d^2(x_j, c_i), \tag{2}$$

where u_{ij} is the degree of membership of x_j in the ith cluster, q is any real number greater than 1 and weighting exponent for u_{ij} which controls the "fuzziness" of the resulting clusters.

3.4 Optimising the number of clusters

An essential point in the using of the clustering techniques concerns selecting appropriate number of clusters. In the hybrid model the number of clusters was decided in specific way. Firstly, the SOM map was partitioned several times using the number of cluster in the range of [2, 12] (based on experience). Secondly, the evaluation of clustering performance was done for each cluster number and the most appropriate number of clusters was then selected. In the evaluation phase the validity indexes were calculated by using the Davis-Bould index [12], [13] for the k-means clustering (Eq. 3) and the PC index (Eq. 4) for the fuzzy c-means clustering [14]. The index of Davis-Bould was calculated according to Eq. (3).

$$\frac{1}{K} = \sum_{k=1}^{K} \max_{l \neq k} \left\{ \frac{S_c(Q_k) + S_c(Q_l)}{d_{ce}(Q_k, Q_l)} \right\}, \tag{3}$$

where S_c within-cluster distance, d_{ce} between clusters distance and Q is a set of clusters and the PC index according to Eq. (4).

$$PC = \frac{1}{N} \sum_{i=1}^{n} \sum_{j=1}^{K} u_{ij}^2, \tag{4}$$

where PC values range in [1/K, 1]. The PC values were scaled to the range of [0, 1] due to the variation in minimum values.

3.5 Evaluating the performances of the models

In the model evaluating phase several statistical indicators were calculated to provide a numerical description of the goodness of the estimates particularly in the episode cases. The number of predicted episodes, correctly predicted episodes and false alarms were used to describe the ability of method to predict episodes from different point of view. On the ground of these parameters the success index was determined simply by dividing the number of correctly predicted episodes by the number of episodes. Furthermore the index of agreement (d) was calculated using Eq. (5).

$$d = 1 - \left[\frac{\sum_{i=1}^{N} (P_i - O_i)^2}{\sum_{i=1}^{N} \left(|P_i - \overline{O}| + |O_i - \overline{O}| \right)^2} \right], \tag{5}$$

where N is the number of data points, O_i the observed data point, P_i the predicted data point and $\bar{O}$ is the average of observed data.
In addition, commonly used statistical indicators were calculated [15]. The root mean squared error (RMSE) was used to provide the average error of model according to Eq. (6). Also systematic and unsystematic components of RMSE were calculated by fitting a least-squares regression line and the decomposing the RMSE using a least-squares estimate.

$$RMSE = \left(\frac{1}{N} \sum_{i=1}^{N} [P_i - O_i]^2 \right)^{1/2} \quad (6)$$

The index of agreement (Eq. 5) was used to make cross-comparisons between models [15] and the coefficient of determination (R^2) to describe how much of the observed variability is accounted for by the estimate model. R^2 was calculated according to Eq. (7).

$$R^2 = \frac{\sum_{i=1}^{N} (P_i - \bar{O})^2}{\sum_{i=1}^{N} (O_i - \bar{O})^2} \quad (7)$$

4 Results and discussion

4.1 General performance

The results (Table 1.) showed that proposed hybrid models are not significantly better in forecasting air pollutant than the basic MLP when overall performance is considered. However, it can be seen that the hybrid model with fuzzy c-means clustering yields the best results.

Table 1. The average performances of the models where MLP+FCM is the hybrid model with fuzzy c-means clustering and MLP+KM the hybrid model with k-means clustering.

Estimator	MLP	MLP+FCM	MLP+KM
RMSE	14,63	14,20	15,63
$RMSE_s$	7,83	7,79	7,89
$RMSE_u$	12,35	11,87	13,49
d	0,85	0,86	0,84
R^2	0,74	0,76	0,71

4.2 Episodes

The results in Table 2 showed that slight improvement in the episode performance was achieved compared to the basic MLP. The hybrid model with k-means clustering seemed to give best results with almost 30 percent success index. In contrast, the success index of fuzzy hybrid model was the worst of all, although the index of agreement was at the same level as for the other models.

Table 2. The performances of the models for the episodes where MLP+FCM is the hybrid model with fuzzy c-means clustering and MLP+KM the hybrid model with k-means clustering.

Estimator	MLP	MLP+FCM	MLP+KM
predicted	3,13	0,75	2,75
correct	1,00	0,63	1,50
false	2,13	0,13	1,25
success	19,98	12,50	29,99
d	0,40	0,42	0,42

4.3 Scatter plots

The performance was also visualised using the plots of predicted versus observed signals of NO_2 (see Fig. 2.). It can be seen that the scatter for the hybrid model utilizing the K-means clustering (c) has the best overall shape namely the narrowest one. However, all the models showed bias and in general, it could be seen that there is no significant difference in them.

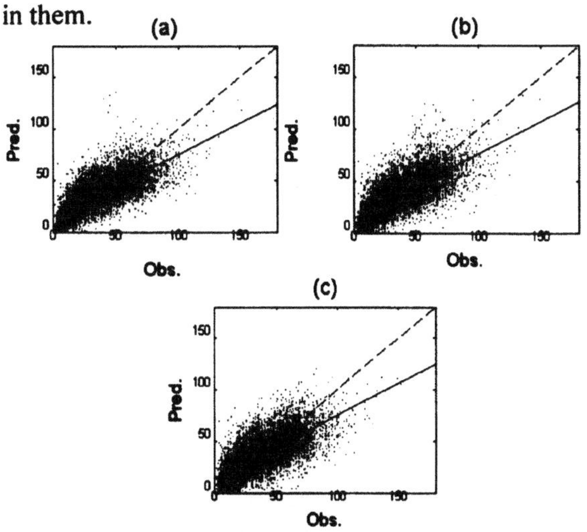

Fig. 2. The performance of the models: (a) MLP, (b) MLP+FCM and (c) MLP+KM presented by the plots of predicted versus observed values. The dashed line gives the perfect match and the solid line the fitting using least-squares.

5 Conclusions

Three models for forecasting especially the episodes of air quality data were compared in the study. The first model was the MLP neural network applied in a straightforward manner. The other two models consisted of several MLP models, which were constructed using partial data of the original dataset. The first hybrid model used data that was partitioned with fuzzy c-means clustering and the second one using K-means clustering.

The results showed that the hybrid model with K-means clustering has some advantages for forecasting the episode concentrations while the hybrid model with fuzzy c-means partitioning of the data performs best concerning overall performance. However, it must be concluded that architectural issues presented in the study do not enhance the models adequately by themselves. Therefore, further work is needed and the most suitable component of the model for that is the MLP network itself. Thus, by enhancing the error term of the network using, for example, regularization factors in it to better consider the nature of the episodes, the issues studied here could have a more substantial effect.

Acknowledgements

This research as a part of the FORECAST project was funded by the Academy of Finland, and also as a part of the APPETISE project funded by the EU. More information at the web addresses: http://www.uea.ac.uk/env/appetise/

References

[1] Tiittanen, P., Timonen, K. L., Ruuskanen, J., Mirme, A., Pekkanen, J. (1999) Fine particulate air pollution, resuspended road dust and respiratory health among symptomatic children. *European Respiratory Journal*, **12**, pp. 266-273.

[2] Kolehmainen, M., Martikainen, H., Ruuskanen, J. (2001) Neural networks and periodic components used in air quality forecasting, *Atmospheric Environment*, **35**, pp. 815-825.

[3] Gardner, M. W., Dorling, S.R. (1999) Artificial neural networks (the multi-layer perceptron) – a review of applications in the atmospheric sciences. *Atmospheric Environment*, **33**, pp. 709-719.

[4] Sharda, R., Patil, R. (1990) Neural networks as forecasting experts: An empirical test. *Proceedings of the 27th Symposium on the Interface of Computer Science and Statistics*, pp. 352-360.

[5] Tang, Z., Fishwick, F. (1993) Feed-forward neural nets as models for time series forecasting. *ORSA Journal of Computing*, **5(4)**, pp. 374-386.

[6] Kolehmainen, M., Martikainen, H., Hiltunen, T., Ruuskanen, J. (2000) Forecasting air quality parameters using hybrid neural network modelling. *Environmental Monitoring and Assessment*, **65**, pp. 227-286.

[7] Kohonen, T. (1997) Self-Organizing Maps, 2nd Edition, Springer-Verlag, Berlin Heidelberg, Germany.

[8] Karppinen, A., Joffre, S. and Vaajama, P. (1997): Boundary layer parametrization for Finnish regulatory dispersion models. *International Journal of Environment and Pollution*, **8**, pp. 557-564.

[9] Haykin, S., (1994) Neural networks: a comprehensive foundation. Prentice-Hall, Englewood Cliffs, NJ.

[10] MacQueen, J. (1967) Some methods for classification and analysis of multivariate observations. *Proceedings of the 5th Berkeley Symposium on Mathematical Statistics and Probability, Volume I, Statistics,* University of California Press.

[11] Bezdek, J.C., (1981): Pattern recognition with fuzzy objective function algorithms. Plenum Press, New York.

[12] Davies, D. L., Bouldin, D. W. (1979) A Cluster Separation Measure. *IEEE Transaction on Pattern Analysis and Machine Intelligence*, PAMI-1, no. 2, pp. 159-179.

[13] Vesanto, J., Alhoniemi, E. (2000) Clustering of the Self-Organized Map. *IEEE Transaction on Neural Networks,* **11**, pp. 586-600.

[14] Bezdeck, J.C, Ehrlich, R., Full, W. (1984): FCM:Fuzzy C-Means Algorithm, *Computers and Geoscience*.

[15] Willmott, C. (1982) Some comments on the evaluation of the model performance. *Bulleting Ameriacan Meteorological Society*, **63**, pp. 1309-1313.

Influence of Language Parameters Selection on the Coarticulation of the Phonemes for Prosody Training in TTS by Neural Networks*

Jana Tučková[1] and Václav Šebesta[2],

[1] Faculty of Electrical Engineering, Czech Technical University,
tuckova@feld.cvut.cz

[2] Institute of Computer Science, Academy of Sciences of the Czech Republic,
and Faculty of Transportation, Czech Technical University
vasek@uivt.cas.cz

Abstract. This contribution describes the influence of the Czech language parameters selection on the coarticulation of the phonemes for the modelling of prosody features by the artificial neural network (ANN) in a text-to-speech (TTS) synthesis. The GUHA method and neural network pruning can be used for this reason. In our work we analyzed the errors between the target and calculated values of F_0 and D from the point of view of the different context of speech units. The context of three phonemes combinations CCC, VVC, VCV, CVV, VCC, CCV, and CVC (C = consonant, V = vowel) were analyzed for the determination of a next improvement of prosody. The qualitative criteria have been found in this contribution.

1. Introduction

The automatic speech synthesis is an interdisciplinary part of artificial intelligence. One must take into account the knowledge from acoustics, phonetics, phonologics, linguistics, physiology, psychology, signal processing and informatics for a successful solution.

The Text-to speech synthesis (TTS) enables to do a synthesis of an arbitrary text irrespective of the length of the text and the diversity of the themes. In the past, synthesizers based on the LPC (linear predictive coding, part of the "source coding") were used very often. The LPC is a parametric method; the speech units are described by the parameter sequence based on the speech analysis. It is based on the simplified prototype of the speech production. Speech patterns are mathematically expressed by a linear combination of the preliminary patterns and excitations. During the several transformations a great part of the speech signal information is loosed.

The synthetic speech is very well intelligible, but it is less natural. Better results from the point of view of intelligibility of synthetic speech (e.g. a melody and a rhythm) can be obtained by non-parametric methods based on the concatenation of the saved sampled speech segments. All versions of the PSOLA method (Pitch Synchronous Overlap Add), see [1], belong to this category. The quality of the synthetic speech obtained by PSOLA is usually better than by the LPC, but the consumption of the computer memory is many times higher because of the speech unit database storage. An advantage of the synthesizers which operate with a small database and with a low memory claim are decreasing as a consequence of the computer technology development.

After the concatenation of the speech units the resulting speech is usually monotonous and the special block for the prosody control is incorporated to the synthesizer (see Fig.1).

Grammatical rules were used for the prosody control in the past but recently more sophistic methods based on artificial neural networks (ANN) or Markov's chains (HMM – hidden Markov's model) are investigated and integrated to the syntheses process. The basic prosodic parameters (the fundamental frequency of voice F_0 – called pitch, segment duration D and signal intensity I) can be extracted from the natural speech signal [2]. The size of the speech unit database can be also partially decreased by the autonomous prosody control.

As in the human brain also in the synthesizer one must extract suitable parameters for the training of the neural network which will calculate the prosody parameters. The GUHA method and neural network pruning can be used for this reason.

* This research was supported by grant GA CR No 102/02/0124.

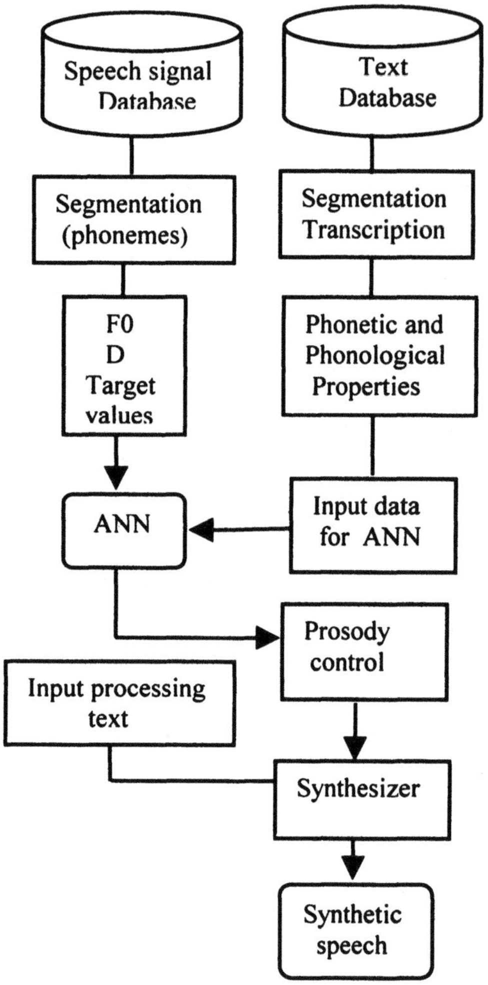

Fig.1. Simplified scheme of the prosody control in TTS synthesis.

2. Basic description of GUHA Method

The GUHA method can be used for the determination of relations in experimental data. The processed data form a rectangular binary matrix of attributes, where the rows correspond to the different speech units (phonemes, diphones, triphones). Each speech unit is represented by a set of language parameters. The columns correspond to different investigated parameters (e.g. the type of the phoneme, the type of the sentence, the number of phonemes in the word, etc.). Parameters may be either dichotomous (either 1 or 0) or real numbers. In the case of dichotomous parameters, the attributes are equal to the parameters. In the case of real parameters the entire interval of possible values of the parameter must be divided into several subintervals according to the selected cut-points and the value of corresponding attributes is equal to one in only one relevant subinterval of a parameter. The values of attributes are equal to zero in all other subintervals of a parameter. It means that real parameters are crumbled into several attributes, one of them is equal to one and the others are equal to zero.

Attributes are split into antecedents (corresponding to input language parameters) and succedents (corresponding to output prosodic parameters).

The program generates and evaluates the hypotheses about association in the form $A \rightarrow S$ (A is associated with S), where A is an elementary conjunction of antecedents, S is an elementary conjunction of succedents and $\rightarrow$ is a quantifier of implication. The implicational quantifier estimates, in some sense, the conditional probability $P(S \mid A)$. Each quantifier is a decision function having the value "true" if the corresponding numerical characteristic supersedes the a priori stated limits. The definition of several quantifiers is based on statistical tests but the interpretation is not statistical.

The user must specify the minimal and the maximal number of elements in the antecedent's conjunctions and succedent's conjunctions and the program generates for all tuples of antecedents and succedents all possible hypotheses in Tab.1 in the form of a "four-fold table":

Table 1.

Succedent / Antecedent	S (1)	non S (0)	total
A (1)	**a**	**b**	**a + b**
non A (0)	**c**	**d**	**c + d**
total	**a + c**	**b + d**	**n**

where a, b, c and d are observed frequencies (i.e. the frequencies of the "1"-value occurrence), calculated from the data and $n = a + b + c + d$ is the total number of the objects. For more details see [3].

3. Results analyses

An ANN with one hidden layer (trained with adaptive learning rate and momentum) was applied for the determination of prosody parameters (F_0 and D). A sigmoid activation function in hidden neurons and linear activation function in output neurons are used.

The optimal topology of ANN was obtained by pruning, where the input layer was pruned by the GUHA method and the hidden layer was pruned in dependency on the minimal values of synaptic weights. The optimal number of input parameters is somewhere among 17 and 19 (see Fig.2). The optimised scale of final ANN is 18-22-2.

The language parameters of Czech selected by the GUHA method are connected to the input layer. The ANN outputs are the fundamental frequency and the segment duration.

The target values of prosodic parameters were extracted from the natural speech signal automatically, as opposed to the past experiments, where the hand extraction was used [3], [4] and [5].

We analyzed the achieved results (i.e. the conformity between the target and the calculated values of F_0 and D) from the point of view of the trained phonemes quality in dependency on the coarticulation. By the coarticulation we will mean the influence of the previous and the following phonemes on the current phoneme. We have analyzed the errors for different context of speech units. The context of three phonemes combinations CCC, VVC, VCV, CVV, VCC, CCV, and CVC (C = consonant, V = vowel) were analyzed for the determination of a next improvement of prosody. The qualitative criteria are the following:

- the sum of absolute values of errors, in Fig. 3, 4 marked as "average",
- the sum of square roots of errors "MSE",
- the maximal error "MAX".

The results of the original network with 30-25-2 neurons before pruning and canonical network (with optimal topology after pruning) with 18-22-2 neurons were compared.

The following facts can be extracted:

- The pruning of the input layer of ANN has the great influence to the final prosody. The best improvement can be seen in that of vowels (a, e, i, o , u) and diphthong (ou) in comparison with consonants. Especially the improvement of vowel "i" can be heard, because even a small error of F_0 for vowel "i" is very disagreeable for the listeners (see Fig.2).
- The amelioration can be recognized also for consonants, especially for explosives"p, d". This result is particularly significant, because the target values for "p, d" cannot be extracted accurately from the speech signal. The duration of the explosives is very short from the point of view of the acoustic form. It is difficult to label their beginning by hand and even more automatically.
- The values of the error function (difference between the target and the output values) for the coarticulation C-C-V and V-C-V are increasing according to the following schema:

	C-C-V	V-C-V
max. error	nasals	explosives
↓	explosives	nasals
	semi-expl.	semi-expl.
min. error	fricatives	fricative

The most difficult modeling was found for nasals and explosives also after using the data mining methods.

- The error is decreased by application of the data mining methods in the majority of phonemes combinations. The most dramatic improvements were recognized in the case of the maximal errors, because the errors in the phoneme extraction from the natural speech signal were contrived. The only exceptions are some V-C-C (all types of errors) and C-C-V (max errors) combinations. The main reason can be the complicated pronunciation of two or more consonants consequently.
- Errors of fundamental frequency of vocals follow the vocalic triangle, known by the experts in phonetics.

	vowels	vocalic triangle			
max. error	i				
↓	o	i			u
	e		e	o	
	a			a	
	u				
min. error	ou				

Generally, the greatest errors were recognized in the case of C-i-C coarticulation.

- The results confirmed the opinion of the phonetic experts [6] that the

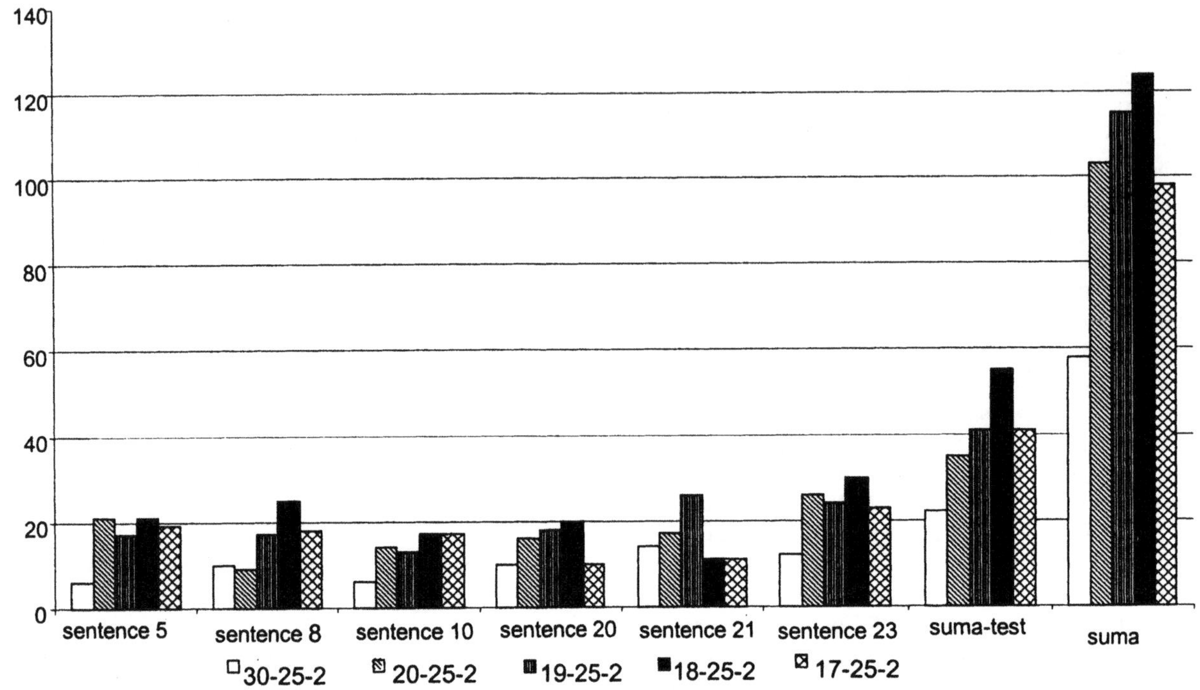

Fig.2. The histogram of the numbers of the best attempts (the nearest to the target value) according to the selected sentences and ANN topology. Each phoneme is taken as one attempt on the F_0 scale and 1 attempt on the duration scale.

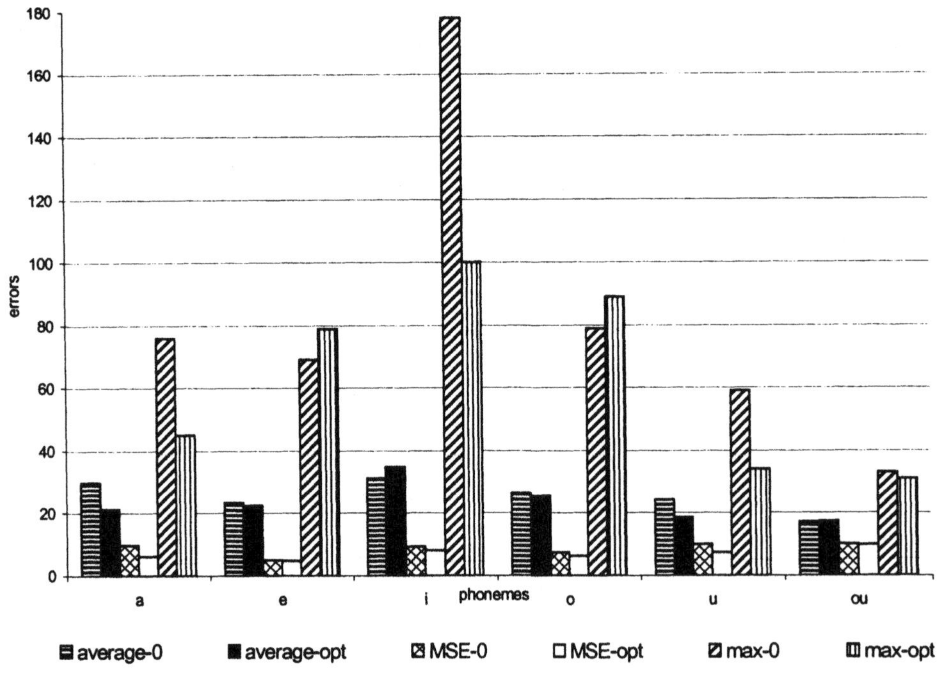

Fig.3. The improvement of the final prosody for vowels

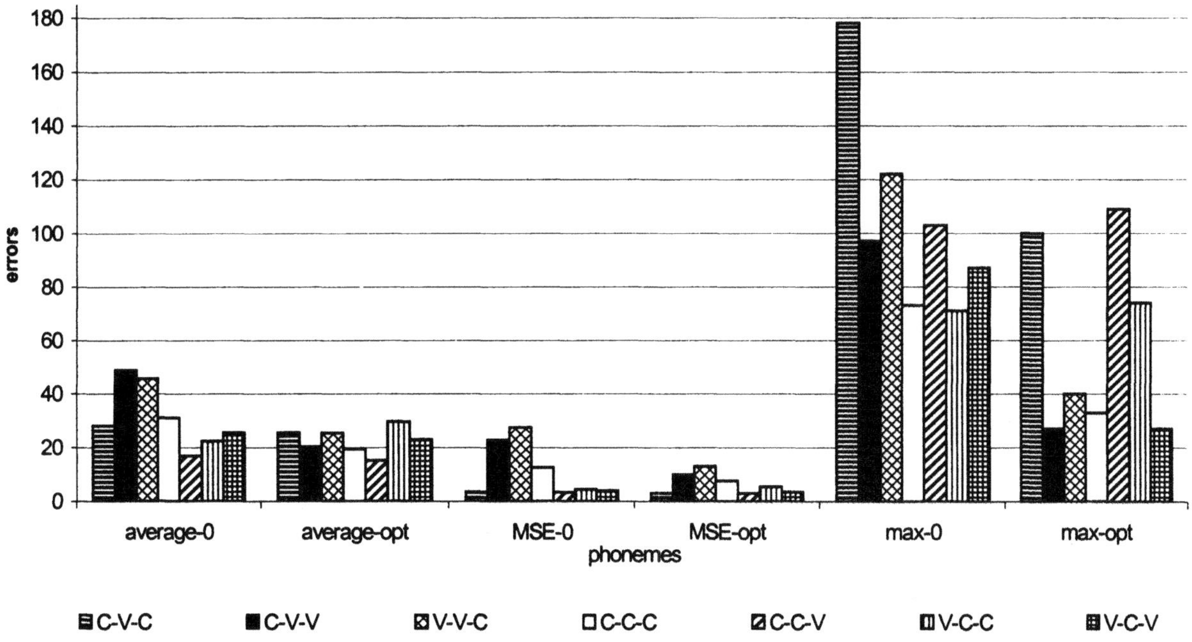

Fig.4. The influence of the triphon coarticulation diagrammed by the sum of absolute values of errors, the sum of square roots of errors and the maximal error

influence of the neighbouring consonants on vowels in coarticulation C-V-C is not very important.

After the described analysis we can summarize that the most suitable criterion is the maximal error (see Fig.3. and Fig.4). The determination of the quality of the phonemes after ANN training and optimization was approved by listening tests. Vice versa, the sum of square errors was recognized as more or less inappropriate. It is caused by the listener's better ability to perceive the great differences between the target and the real pronunciation of the phonemes. The listener's impress is better in the case of a great amount of small differences than in the case of a small amount of great differences.

4. Conclusion

The only mathematical evaluation of the synthetic speech quality is not fully adequate. As we have shown above, also the evaluation of human listeners is not objective either. The both approaches must be combined for the best evaluation of the synthetic speech quality.

The unsystematic errors during the hand labeling process caused by individual human abilities, by the physical and mental state of the approbator, can be eliminated by the automatic approach. It is possible to note that a full automatic determination of the beginning and ending of speech units (labeling) is controversial. During the training of ANN the automatically labeled signal needed for the determination of the target values have to be checked and corrected by a person, but in the real synthesizer utilization it is not possible. Therefore, the resulting signal can have some audible anomalies.

The perception of speech is seriously dependent on the individual features of listeners. Therefore the determination of target values of the fundamental frequency F_0 and phonemes duration are also individual and the automation of this process is not simple.

The Neural Network Toolbox of Matlab, ver.5.2 for LINUX and Windows 98 (2000) was used for the neural network training. The fundamental frequency contour F_0 and the duration contour were tested by comparison of the labeled values extracted from the natural speech and the output values from the ANN training. The Czech synthesizer for prosody modeling ARTIC [7] has been used.

References

1. Moulines, E. (1990) Algorithmes de codage et de modification des parametres prosodiques pour la synthèse de parole a partir du texte. In: Thèse de Docteur, l'Ecole National Superieure des Telecommunications, TELECOM Paris 90 E 004

2. Tuckova,J., Sebesta,V. (2001) Data Mining Approach for Prosody Modelling by ANN in Text-to-Speech Synthesis. In: Proc. of the Int. Conf. IAESTED AIA2001, Marbella, Spain, September 2001, pp. 161-166, ISBN:0-88986-301-6

3. Sebesta,V., Tuckova,J. (2001) Optimisation of Artificial Neural Network Topology Applied in the Prosody Control in Text-to-Speech Synthesis. In: ICANNGA'2001, Prague, Avril 2001, pp. 420-430, ISBN:3-540-41348-0

4. Tuckova,J., Sebesta,V. (2000) Prosody Modeling for a Text-to-Speech System by Artificial Neural Networks. In: Proc. of the Int. Conf. IAESTED SIP'2000, Las Vegas, USA, November 2000, pp.307-312, ISBN: 0-88986-308-3

5. Sebesta,V., Tuckova,J. (1999) Selection of Important Input Parameters for a Text-to-Speech Synthesis by Neural Networks. In:IJCNN'99(CD-ROM), Washington, D.C., USA, July 1999, IEEE Catalog Number: 99CH36339C,ISBN:0-7803-5532-6

6. Palkova, Z.(1994) Phonetics and phonologics of the Czech language (in Czech: Fonetika a fonologie češtiny). Univerzita Karlova-Praha, 1994, ISBN: 80-7066-843-1.

7. Matoušek, J. (2000) Text-to-Speech Systém Using Statistical Approach to Speech Segment Database Construction. PhD dissertation, UWB in Plzeň, Czech Republic, (in Czech).

Vertical Vector Fields and Neural Networks: An Application in Atmospheric Pollution Forecasting

David William Pearson[1] Mireille Batton-Hubert [2] and Gérard Dray[3]

[1]EURISE (Roanne Research Group) Jean Monnet University of Saint-Etienne, I.U.T. de Roanne, 20, avenue de Paris, 42334 Roanne, France. email - david.pearson@univ-st-etienne.fr [2]SITE, Ecole des Mines de Saint-Etienne, cours Fauriel, Saint-Etienne, France. email - batton@ensmse.fr [3]LGI2P, EMA Site-EERIE, Parc Scientifique G. Besse, 30319 Nimes, France. email - gerard.dray@site-eerie.ema.fr

Abstract

In this paper we look at the role that vertical fields can play in enhancing the performance of a feedforward neural network. Vertical fields help us to determine zones in the input space that are mapped onto the same output, they act in a similar way to kernels of linear mappings but in a nonlinear setting. In the paper we illustrate our ideas using data from a real application, namely forecasting atmospheric pollution for the town of Saint-Etienne in France.

1 Introduction

We have been investigating the problem of forecasting atmospheric pollution for some time [1, 2, 3]. We have worked on data from two towns in France: Montpellier and Saint-Etienne. Although the local conditions are not the same for the two towns and pollution levels are different, the problems facing potential forecasters are the same for these and indeed most towns.

In general we have a large database available, data are collected for many different variables about every 10 minutes or so 24 hours a day throughout the year. We need to plunge into the databases in order to find relationships between variables. These relationships are usually expressed in the form of linear or nonlinear regression formulae, neural networks, fuzzy inference systems, or indeed many other types of mathematical tools. One then uses the relationship in order to forecast pollution levels. We are particularly interested in ozone levels.

One thing which intrigues us is that there are instances where "similar" days produce the same levels of ozone and "dissimilar days" produce the same levels of ozone. Similar and dissimilar being defined by Euclidian distances between data vectors containing meteorological data, actual levels of pollution etc.

For this reason, we looked toward vertical vector fields to see whether they could enlighten us. This article illustrates some of our preliminary trials.

2 Vertical vector fields

Given a mapping $\pi : X \to Y$, where the topological spaces X and Y have dimensions n and m respectively with $n > m$, and where we assume that the mapping is at least C^1. We denote the differential of this mapping by $\pi_\star$ (the matrix of first partial derivatives) then a vector field $v : X \to X$ is said to be a vertical vector field if

$$\pi_\star v(x) = 0 \qquad (1)$$

where the 0 on the right hand side of (1) represents a vector of zeros of dimension m. In this article we will be taking the input space to an open subset of $\mathbb{R}^n$ and we will assume that $\pi_\star$ has full rank (i.e. m) on this subset. That being the case, there will be $n - m$ independent vertical fields satisfying (1).

The interesting thing about vertical vector fields is that if $\pi(x_0) = y_0$ then

$$\pi(\exp(tv)x_0) = y_0 \text{ for } t > 0 \qquad (2)$$

where $\exp(tv)x_0$ denotes the operation of exponentiating the vector field, i.e. solving the differential equation

$$\dot{x} = v(x) \text{ with } x(0) = x_0$$

where $\dot{x} := \frac{dx}{dt}$.

What is more, due to the fact that any trajectory defined by the $n - m$ vertical vector fields is mapped by π onto the same value, we can make use of the following construction. Let the mapping $\phi : S \to X$, where S is an open neighbourhood of the origin in $\mathbb{R}^{n-m}$, be defined as follows

$$\phi(s) = \exp(s^{n-m}v_{n-m}) \cdots \exp(s^1 v_1)x_0 \qquad (3)$$

where the s^i are the elements of a vector $s \in S$ and x_0 is such that $\pi(x_0) = y_0$. The concatenation in (3) means that we start at the point x_0 and go in the direction of the vector field v_1 for s^1 time units, then we go in the direction of vector field v_2 for s^2 time units etc. The result is that $\pi(\phi(s)) = y_0$. As s goes through all of S then the trajectories form a hyper-surface in X and this hyper-surface is usually called the leaf above y_0 and denoted L_{y_0}.

In our case, the mapping π is a standard feedforward neural network with analytic activation functions. We assume that the neural network has been trained on some subset of the available training data, taken to be a collection of data pairs (x_i, y_i), $i = 1, \ldots, N$. In particular, we assume that the subset used for training is of very low cardinality when compared to all the data available and that the network has been trained as a sort of classifier. So, for example, if we initially train the network to identify two classes then $\pi(x) = y_1$ if $x \in \mathcal{X}_\infty$ and $\pi(x) = y_2$ if $x \in \mathcal{X}_\in$ where $\mathcal{X}_\infty$ and $\mathcal{X}_\in$ are two disjoint subsets of X.

Once the vertical vector fields have been determined for $\pi(x_0) = y_0$ we can determine whether or not any of the other data points $x_i \in L_{y_0}$, if such is the case then $\pi(x_i) = y_0$ of course and so x_i need not be included in the training data.

We believe that determining whether or not data points lie in L_{y_0} (for various y_0) will give us a better picture of the structure of the data. For example, we could retrain the neural network using data selected on the basis of whether or not they are related by vertical vector fields. If two points belonged theoretically to the same class but not to the same L_{y_0}, then we could consider including them in the training data set but with them mapping to very slightly different output values. In this way, the two data points should belong to two leaves in vertical spaces very close together but not coincident. The separation or nearness of leaves could be used to increase robustness properties of the neural network.

3 Estimating the vertical vector fields

We need to find a set of vector fields satisfying (1). This is quite a task because these vector fields could be nonlinear, this is indeed the case with feedforward networks. Not only that, it would be extremely difficult to find a set of vector fields which satisfy the condition everywhere, i.e. $\forall x \in X$. We are therefore confined to making approximations and assumptions about the vector fields.

- The first, and probably the strongest, assumption that we make is that the vertical vector fields are linear so that in (1) $v(x) = Ax$ for some $n \times n$ matrix A.

- The principal approximation that we use is to satisfy (1) at a single point x_0, by continuity this will mean that the condition is satisfied in a neighbourhood of x_0.

- The second assumption that is made about the vector fields is that they form an involutive Lie algebra at the point x_0 and thus in a neighbourhood of x_0.

To begin the calculation we select a point x_0 and evaluate $\pi_\star(x_0)$ by using a second order central difference approximation for the partial derivatives. We then determine a basis for the kernel of $\pi_\star(x_0)$ by using the singular value decomposition [4], that we denote by Ker($\pi_\star(x_0)$)=$\{v_1, \ldots v_{n-m}\}$. Making use of the above listed approximation and to satisfy the first of the assumptions we need to find matrices $A_1, \ldots A_{n-m}$ such that

$$A_k x_0 = v_k \text{ for } k = 1, \ldots, n-m \tag{4}$$

Then, for the second of the above assumptions the following conditions have to be satisfied [5]

$$[A_i, A_j]\,(x_0) = \sum_{k=1}^{n-m} c_{ij}^k A_k x_0 \quad \text{for } i, j = 1, \ldots n-m \text{ and } i \neq j \tag{5}$$

where $[A_i, A_j] = A_i A_j - A_j A_i$ is the Lie bracket and the c_{ij}^k are scalars.

In fact, we employ a numerical optimisation method to calculate the elements of the matrices A_i and the scalars c_{ij}^k in order to ensure that the following condition is satisfied.

$$\frac{d\pi(x(t))^\nu}{dt^\nu}_{|x=x_0} = 0 \quad \text{for } \nu = 0, 1, 2, \ldots$$

The complete method is explained in [6].

Having calculated the vertical vector fields, we can then evaluate ϕ as in (3) for a fixed s.

4 Application

Space does not allow us to fully explain the application, the interested reader is invited to consult [1] for some idea of the objectives and available data. Basically, today at about 16h we want to forecast the maximum level of ozone for tomorrow based on data collected today and forecast meteorological data for the town of Saint-Etienne which is just to the west of Lyon in France. By experience we have found that some of the best results are obtained with the space X of dimension 7 with components

1. maximum ozone today
2. temperature at 6h today
3. temperature at 12h today
4. average wind speed today
5. forecast temperature at 6h tomorrow
6. forecast temperature at 12h tomorrow
7. forecast average wind speed tomorrow

The output is the class of maximum values of ozone forecast for tomorrow, hence a one-dimensional space Y. The classes are

1. maximum ozone < 100
2. $100 \leq$ maximum ozone < 120
3. $120 \leq$ maximum ozone < 140
4. $140 \leq$ maximum ozone < 160
5. $160 \leq$ maximum ozone < 180
6. $180 \leq$ maximum ozone < 200
7. maximum ozone ≥ 200

We chose a data point in each class from the second and fifth classes, with required outputs of 0.2 and 0.5 for each class respectively. We then trained a neural network of the following form on the data

$$\pi(x) = W_3 \tanh(W_2 \tanh(W_1 x + b_1) + b_2) + b_3$$

where $W_1 \in \mathbb{R}^{7\times 7}$, $b_1 \in \mathbb{R}^7$, $W_2 \in \mathbb{R}^{3\times 7}$, $b_2 \in \mathbb{R}^3$, $W_3 \in \mathbb{R}^{1\times 3}$ and $b_3 \in \mathbb{R}$. The function tanh is assumed to operate individually on each vector component.

We then estimated the vertical vector fields at one of the points in the fifth class by solving (4) and (5). To determine the members of $L_{0.5}$ we applied the following algorithm

- $\forall x_i \neq x_0$
- if $\|x_i - x_0\| < \epsilon$ then
- if $\exists s \in \mathbb{R}^6 : \|\phi(s) - x_i\| < \delta$ then
- $x_i \in L_{0.5}$
- end end

Using this algorithm with $\epsilon = 20$ and $\delta = 2$ of the 41 data vectors in the fifth class we were able to determine 11 which were not equal to the one used for training but which were members of $L_{0.5}$.

The two vectors used for training the neural network were the following, the one on the left comes from the second class and the one on the right from the fifth class

$$\text{input} = \begin{bmatrix} 88 & 136 \\ -0.4 & 11.3 \\ 12.1 & 20.2 \\ 1.083 & 1.25 \\ 0.8 & 14.4 \\ 10.3 & 26.7 \\ 0.9166 & 1.166 \end{bmatrix}$$

To illustrate how the method enables us to determine part of the structure of the data points, other than simple Euclidian distance, we present two vectors belonging to $L_{0.5}$

$$\begin{bmatrix} 152 & 120 \\ 16.2 & 11.3 \\ 29.2 & 25.9 \\ 0.875 & 0.8333 \\ 16.2 & 9.4 \\ 29 & 25.5 \\ 0.875 & 0.6666 \end{bmatrix}$$

One can easily calculate the distances between x_0 (the vector on the right of input above) and these two vectors, they are respectively 19.23 and 17.76. They are, however, both mapped to 0.5 by π. The following vector is only 9.29 distant from x_0, is mapped to 0.5 by π, but doesn't belong to $L_{0.5}$

$$\begin{bmatrix} 136 \\ 16.9 \\ 26.8 \\ 0.9583 \\ 17.3 \\ 28.4 \\ 1 \end{bmatrix}$$

5 Conclusion

We have shown how vertical fields can be used to analyse data structure via neural network training (or indeed other types of classifiers or regression methods). We have also shown how vertical fields can be estimated in a numerical sense.

We are at present investigating how we can best put to use the extra knowledge (about the data structure) supplied by vertical fields. We believe that this is one of the points that will lead us to a better understanding of atmospheric pollution data and thus lead to better forecasting of pollution.

6 Acknowledgements

The authors would like to acknowledge the cooperation with AMPASEL, the agency responsible for monitoring the air quality in the Loire department of France, and the Conseil Regional of the Rhône-Alpes region for its financial support of this research.

References

[1] D.W. Pearson, G. Dray, M. Mesbah and B. Vuillot (2000), Ozone, Systmes Dynamiques et "Shadowing", Journées thématiques "Automatique et Environnement", Nancy, France.

[2] D.W. Pearson, M. Batton-Hubert and G.H. Garcia (2002), Predicting Ozone Peaks: A combined CBR and cell mapping approach, IEMSS02, Lugano, Switzerland.

[3] N. Peton, G. Dray, D.W. Pearson, M. Mesbah and B. Vuillot (1998), Modelling and Analysis of Ozone Episodes , International Conference on Air Pollution Modelling and Simulation, Paris, France.

[4] G.H. Golub and C.F. Van Loan (1983), Matrix Computations, North Oxford Academic.

[5] P.J Olver (1986), Applications of Lie Groups to the solution of differential equations, Springer Verlag.

[6] D.W. Pearson (1996), Approximating Vertical Vector Fields for Feedforward Neural Networks , Applied Mathematics Letters, Vol 9, No. 2, pp 61-64.

A systematic method to neural network modeling with application to sintermaking

Petteri Laitinen, Henrik Saxén

Heat Engineering Laboratory
Åbo Akademi University
Biskopsgatan 8, FIN-20500 Åbo, Finland
E-mail: plaitine@abo.fi, hsaxen@abo.fi

Abstract

In developing data-driven models of complex real-world systems, a common problem is how to select relevant inputs from a large set of measurements. If the observations of the outputs to be predicted by the model are scarce, which may be the case if the outputs are indices determined in toilsome laboratory tests, strict constraints may be imposed on the number of model parameters. In neural network modeling, such a limitation in practice also restricts the number of input variables, since the dimension of the weight vector strongly depends on it. This paper presents a systematic method for the selection of input variables for feedforward layered neural networks. The method is illustrated on a problem from ironmaking industry, where sinter quality indices are predicted on the basis of raw material properties. Furthermore, an inversion technique of the resulting network models is proposed, where an optimization problem is solved to maximize the performance of the sintering operation by manipulating the inputs.

1 Introduction

An efficient and economical operation of industrial processes is today intimately connected with a good quality control which, in turn, requires that the cause-effect relationships in the system are known. However, many industrial processes are extremely complex, and it is therefore impossible, or at least economically infeasible, to develop mechanistic or first-principle models of them. If measurements from the process are available, it is possible to use a data-driven approach where black-box models are developed. In cases where linear techniques prove insufficient, an alternative is to use neural networks: Feedforward networks with one hidden layer have been shown to be able to approximate any continuous differentiable function to arbitrary accuracy, if the number of hidden nodes is large enough. [1] However, a common problem faced in neural network modeling of real-world processes is that there are plenty of potential input variables to choose between, but only a limited number of observations on the inputs and outputs. A method for selection of the most relevant input quantities to a neural network is proposed and illustrated on data from a Finnish sinter plant. To address the problem of lack of observations, the networks have multiple outputs, yielding more degrees of freedom in the parameter estimation (weight training). The input variables that are detected as being most important are used to develop a multitarget network, and this network is applied to study the optimal static operating conditions of the sinter plant.

2 Methods

Let the number of potential input variables be $m_{\max}$. From these inputs the m most relevant quantities are to be chosen to predict the n outputs (targets). If the number of observations, k, is relatively small, the dimension of the input vector (m) must not be very large. A general rule of thumb is that the number of parameters, p, in the network should not exceed a tenth of the number of observations. [2] This problem is partly relaxed if it is possible to include multiple outputs in the network, yielding the condition $p < 0.1\,k\,n$. In a network with h neurons in a single hidden layer, the number of parameters equals $h(m+n+1)+n$, so the number of neurons should satisfy

$$h < \frac{n(0.1k-1)}{m+n+1}. \qquad (1)$$

The advantage of using a multitarget network instead of n single-output networks becomes clear from the following simple reasoning: For m inputs and n outputs, we can have h hidden nodes in the multitarget network with the same number of parameters as in n networks with m inputs, H hidden nodes and one output, where

$$h = H\frac{n(m+2)}{(m+n+1)}. \qquad (2)$$

If for example $n = m = 3$, this equation gives that it is possible to have twice as many nodes in the multitarget network as in the single-output networks. Naturally, a prerequisite for an efficient use of a multitarget network is that the outputs are (at least slightly) correlated with each others. However, this is often the case if variables (e.g., quality indices) from the same process are predicted, since there are usually common underlying mechanisms and phenomena that govern their values.

If we use a fixed architecture of the neural network with m inputs, there are $\binom{m_{max}}{m}$ different choices for the m inputs among m_{max}. If a network for each of these m-tiples is trained, the training errors can be compared, and the inputs that occur most frequently in the elite may be considered relevant variables. The performance of the worst m-tiple of the elite, in turn, provides a measure of how difficult the modeling problem is. The most important input variables thus detected can be used, possibly together with additional input quantities selected on the basis of other criteria, such as controllability, process know-how, etc., in developing the final multitarget network. The final modeling step can involve more detailed optimization of the network structure (e.g., the number of hidden nodes) by cross-validation on independent test sets. If the generalization ability is considered sufficient, the network can be accepted as a model of the system studied.

The resulting model can be used for simulation and sensitivity analysis of the system. Since the model is compact (with a small number of inputs and hidden nodes), it is also possible to use it in more advanced investigations. Assuming that the output variables, $y_i, i = 1, ..., n$, be normalized quantities, the model may be used for maximization of the performance of the system by solving the optimization problem

$$\max_{\mathbf{x}} \left\{ f(\mathbf{x}) = \sum_{i=1}^{n} \alpha_i y_i(\mathbf{x}) \right\}, \tag{3}$$

where $\mathbf{x} = [x_1, ..., x_m]$ is the vector of (normalized) input variables and α_i are factors that facilitate a weighing of the contribution to the objective function from each output. Note that negative values of α may be used to minimize certain quantities (e.g., energy consumption). It should be noted that eq. (3) is a simple formulation of the objective function, and that more sophisticated alternatives, such as multi-criteria optimization [3], could be used as well. To yield a meaningful outcome of the optimization, it is often motivated to impose constraints on the feasible input space. Depending on the system studied, it may also be motivated to fix one or several inputs, optimizing the values of the remaining variables, or to impose additional constraints (e.g. dependencies between the input variables).

3 Application

3.1 Sintering process and sinter quality

The methods described above have been applied on data from a Finnish sinter plant. The raw material fines are taken from large bedding piles, which hold mainly ore and recycled materials. In the sintering process the fines and some additives, e.g., coke, lime and water, are micro-pelletized and fed on the sinter strand. The mix is ignited from above in the beginning of the strand, which slowly moves forward. Combustion gases are drawn through the strand by applying a low pressure in the wind boxes below the strand by means of a suction fan. In the strand a combustion zone moves downward, binding the small particles into a clinker-like aggregate. After the strand the sinter cake is crushed, cooled and screened and sinter particles with a diameter exceeding a certain limit go to the blast furnace, while the rest is recycled. The sinter is used as a main iron source in the blast furnace, which produces molten iron that is futher processed to steel. A detailed description of the sintering process is given in [4].

On its way from the sinter plant to the blast furnace, the sinter must withstand collision forces and in the blast furnace the conditions are extremely harsh, with enormous mechanical load and wear, high temperatures, and pressure. Moreover, the iron oxides in the sinter should be readily reduced in the blast furnace, and the sinter should melt within a well-defined temperature range. In order to characterize the sinter, a number of laboratory tests have been developed, each mimicking the conditions that the sinter will face after it has been produced. The degradation is tested in a drum test, called ISO +6.3mm, referred to as ISO in what follows. For the ability to withstand the conditions before the furnace and in its upper part, a test for low-temperature breakdown (LTB) is used, while the softening-melting properties are captured in a softening index (TK50) [5]. Other, or alternative tests exist, but results from three tests mentioned above were used in this study.

3.2 Data preprocessing

Process data from Rautaruukki Steel's sinter plant at the Raahe Steel Works in Finland was used in the study. Even though the data was available from a quite long period of time (1996-2002), the data set consisted of only 160 bedding piles, since a pile lasts for approximately two weeks. The fractions of 19 different raw materials in the mix were the potential model inputs while the sinter quality indices, (i.e. ISO, LTB and TK50) formed the outputs.

After removing observations with missing variables, every original input and output variable, $\tilde{z}$ (=$\tilde{x}$ or $\tilde{y}$), was normalized by subtracting its mean value, $\overline{z}$ and dividing the result by the estimated standard deviation, σ_z, of the variable, i.e.,

$$z = \frac{\tilde{z} - \overline{z}}{\sigma_z}. \tag{4}$$

yielding normalized variables, z, with zero mean and unit standard deviation. After this measure, observations where any of the variables (inputs or outputs) deviated more than three times the standard deviation from the mean value (i.e., $|z| > 3$), were considered as outliers and therefore removed. This further reduced the data set to 150 observations left for a further analysis.

3.3 The quantitative search

The preprocessed data set was used with the method outlined in section 2 to detect the most relevant inputs. For each of the outputs, $\binom{19}{3} = 969$ triples were studied by training networks with three sigmoidal nodes (that were considered sufficient) in the hidden layer with the Levenberg-Marquardt (LM) method using Matlab's 'trainlm' function [6]. In order to explore the search space more globally, the training algorithm was hybridized by a simple genetic algorithm with a (constant) population size of 16 using binary encoding of all the weights of the network. All the 16 individuals were trained with the LM method, followed by a tournament selection of eight survivors. Offsprings were produced by a cross-over and the whole procedure was repeated for 10 generations. Every weight was crossed at one point, the weights of the first layer shared one cross-over point, the weights of the second layer another point, and bias weights had a third cross-over point. These three cross-over points were created using uniformly distributed random numbers. An accuracy of 27 bits (i.e. 8 decimals) was used.

Table 1 Possible inputs and their frequencies in the elite models of the three quality indices.

Quantity	ISO	TK50	LTB
Error limit	0.6	0.32	0.48
XF	14	15	10
MAF	10	17	25
POA	11	18	22
MPF	13	15	19
SP	13	12	13
SYO	17	17	12
GKA	*115*	13	12
POKA	19	15	16
OLVI	10	16	18
MN	11	25	13
KUPA	13	13	14
KYPA	24	17	20
MAPA	10	11	12
VAHI	23	*106*	18
SOTK	16	14	6
PERO	17	16	8
FEOK	15	0	18
MVP	20	27	*112*
KOST	10	14	13

The 127 best-performing triples (of totally 969) were taken to belong to the elite. The results of the quantitative search are presented in Table 1. The error limits (i.e., the performance of the worst elite triple) reported on row two indicate that TK50 was easiest and ISO was most difficult to predict. Because all of the 127 elite models used three inputs, the average of the frequencies was $127 \cdot 3/19 \approx 20$.

Two or three input variables in each column appear with a frequency of 20 or more and one variable occurs more than 100 times in each column. The variables of the former category, namely magnetite fines (MAF), crushed coke (POA), manganese ore (MN) and cold return fines (KYPA) can be taken as possible inputs, while those of the latter group, lime stone (GKA), mill scale (VAHI) and pellet fines (MVP), should definitely be included.

3.4 Developing the final model

After an analysis of the physical and chemical meaning of the most relevant variables, they were traced to three other quantities: the basicity of the sinter, a fraction of the sinter feed size, and the hematite/magnetite ratio of the sinter feed. Since these variables are more easily comprehended and physically interpreted, they were used as inputs in the final model. It should be noted that sinter feed size and the hematite/magnetite ratio are determined by the composition of the bedding pile, while the basicity is still affected at the dosing plant. According to process know-how and because it is a control variable, the moisture of the sinter mix was included as a fourth input variable. In addition to the three quality indices, it was decided to predict two additional quantities, namely the productivity of the sinter plant and the specific fuel consumption, since these are key variables in sinter making. Thus, a neural network with four inputs, sigmoidal nodes in one hidden layer and five linear output nodes was trained.

Approximately one fourth of the 150 observations

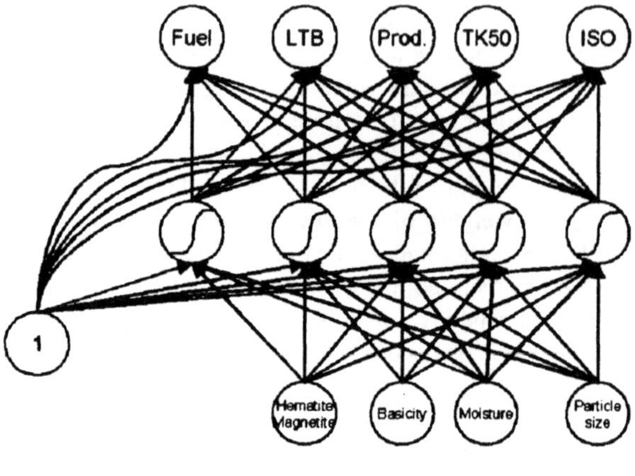

Fig. 1. The architecture of the network in the second stage.

available were left for the test set to evaluate the network's ability to generalize. For the training phase, eq. (1) with $k = 110$ observations, $n = 5$ outputs and $m = 4$ inputs indicates that the number of hidden neurons should not exceed five, so $h = 5$ hidden nodes were selected. The network, shown in Figure 1, was trained by gradient descent with momentum, i.e., without genetic search. The momentum method was chosen because this method was found to avoid getting stuck in local minima, which often happens when higher-order methods, such as the Levenberg-Marquardt algorithm (without GA) are used. The desired and predicted outputs of the training and test sets are illustrated in Figures 2 and 3. It can be seen that the fuel consumption has the best fit in the training data, and the tumble index seems to have the poorest fit in the training data. In the test data tumble index has the best fit, a fact that refers to a good generalization. With respect to the other outputs the performance of the model is poor. If we take a closer look at softening index, we see that the test set contains many extreme spikes; such values are almost absent from the training data. This is a consequence of unfortunate choice of test points and explains the poor test result obtained.

Table 2 Input and target variables in the second stage.

i	Input variables x_i	Output variables y_i
1	Hematite/magnetite	Fuel consumption
2	Moisture	LTB
3	Basicity	Productivity
4	Particle size	TK50
5		ISO+6.3mm

3.5 Study of optimal operating conditions

The resulting network was next used in different problem formulations. With $n = 5$ and $m = 4$ in eq. (3), and with the weighing set by $\alpha_1 = -1$ and $\alpha_2 = ... = \alpha_5 = 1$, the sinter plant productivity and sinter quality is maximized, while minimizing the specific fuel rate (cf. Fig. 1). The optimization problem was solved by constrained nonlinear programming using sequential quadratic programming with BFGS update of the Hessian, under the constraints $-3 \leq x_i \leq 3$. There was no guarantee of convexity of the goal function so the search was started from different initial points to get as good local minimum as possible. Results of the search are given in Table 3. Two locally optimal operating points, $\mathbf{x}^I$ and $\mathbf{x}^{II}$ were found. They have in common a quite low hematite/magnetite ratio and a low basicity. The corresponding values of the objective function are $f(\mathbf{x}^I) = 8.44$ and $f(\mathbf{x}^{II}) = 5.07$ at these points. Even though the moisture and the particle size are very different in these points, the targets are quite similar. One reason could be the micropelletization, where adding water into the sinter feed causes small particles to stick to each other thus forming larger particles, micropellets. The first point, $\mathbf{x}^I$, stresses low fuel consumption and sinter quality, whereas the second point, $\mathbf{x}^{II}$, emphasizes productivity.

A typical task at the sinter plant would be to control the moisture and basicity of the sinter mix to get as high productivity, low fuel consumption and good quality indices as possible, given (fixed values of) the hematite/magnetite ratio and the particle size distribution. This problem can be formulated as

$$\begin{aligned} \max \quad & f(\mathbf{x}) \\ \text{s.t.} \quad & x_1 = k_1, x_4 = k_4. \end{aligned} \tag{5}$$

Table 3 Target values y_i and input values at the local optima $\mathbf{x}^I$ and $\mathbf{x}^{II}$.

i	$\mathbf{x}_i^I$	$\mathbf{x}_i^{II}$		$y_i(\mathbf{x}^I)$	$y_i(\mathbf{x}^{II})$
1	-0.23	-0.38		-2.99	-1.79
2	-3.00	0.43		1.74	0.74
3	-1.17	-0.78		0.89	1.12
4	2.67	-1.89		1.13	0.60
5				1.69	0.82

Table 4 The local optima and ouput values y_i when hematite/magnetite ratio and particle size index are held at their mean values $x_1 = x_4 = 0$.

i	$\mathbf{x}_i^a$	$\mathbf{x}^b$	$\mathbf{x}_i^c$		$y_i(\mathbf{x}^a)$	$y_i(\mathbf{x}^b)$	$y_i(\mathbf{x}^c)$
1	0	0	0		-1.39	-0.62	-1.28
2	-0.73	2.78	-1.36		2.26	0.01	1.25
3	-1.39	3.00	-1.07		1.23	1.44	0.78
4	0	0	0		0.31	0.33	0.49
5					1.09	1.35	1.23

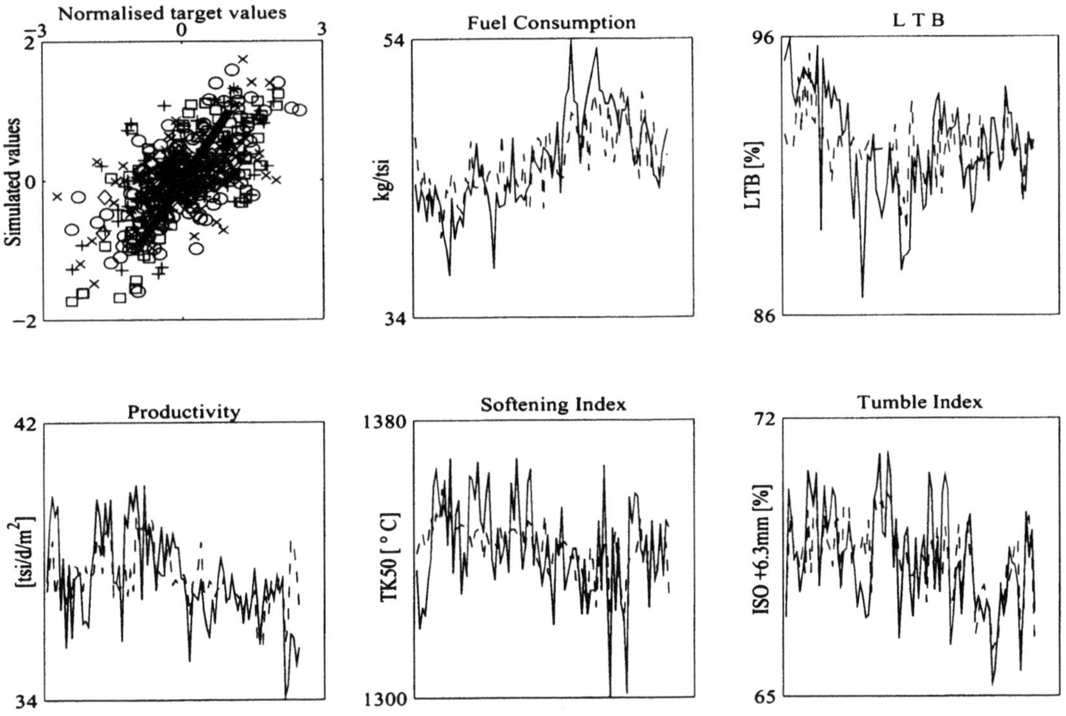

Fig. 2. Training set results for multitarget network. Solid lines give actual values, dashed the predictions.

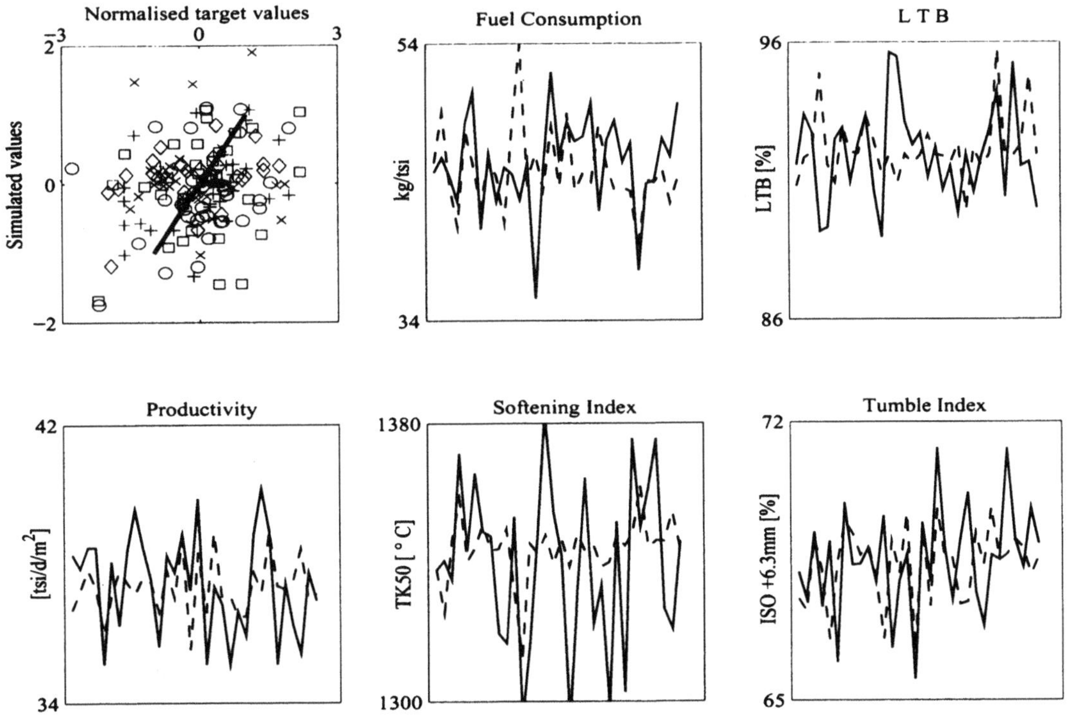

Fig. 3. Test set results for multitarget network. Solid lines give actual values, dashed the predictions.

Table 4 reports the results for such an optimization where the particle size and hematite/magnetite ratio were fixed at their mean values, i.e., $k_1 = k_4 = 0$.

In the constrained case the search method found three local optima at the points $\mathbf{x}^a, \mathbf{x}^b, \mathbf{x}^c$. The target values and the components of these three points are given in Table 4. The corresponding values of the objective function are $f(\mathbf{x}^a) = 6.28$, $f(\mathbf{x}^b) = 3.75$ and $f(\mathbf{x}^c) = 5.03$, respectively. None of the absolute values of the (normalized) outputs in Tables 3 and 4 exceeds 3. The optimal points were far away from each other, probably because there were too few observations in the training phase to cover the input space properly. This can be seen in the upper left diagrams in Figures 2 and 3, where the predicted values have been depicted against the target values. The point $\mathbf{x}^a$ stresses LTB most, but TK50 is quite low. $\mathbf{x}^c$ is quite close to $\mathbf{x}^a$, but emphasizes more TK50 and ISO at the cost of LTB and productivity. $\mathbf{x}^b$ suggests a high moisture and basicity of sinter mix and weights productivity most at the expense of fuel consumption and LTB. Also the tumble strength is good in the operating point $\mathbf{x}^b$, probably due to the high basicity together with the poor permeability of the sinter mix on the strand, which in turn could depend on high moisture. This is in agreement with the results of Ref. 5.

4 Discussion and Conclusions

A method for choosing the most relevant input variables for neural networks has been proposed and applied successfully on data from the field of sinter making. In the application, the three most important input variables of 19 possible were detected and used together with an additional variable in modeling the state of the sintering process. The resulting model was used to find an optimal sinter mix. The methods gave promising results and are to be applied on combination of bedwise and daily, operational data to avoid the problem of scarce data [7].

Acknowledgements

We are grateful to Rautaruukki Oy for providing the data that was used in this study.

References

[1] Hornik K.,Stinchcombe M., White H.(1989) MLPs are universal approximators, Neural Networks, 2:359-366

[2] Principe J. C.,Euliano N.R., Lefebvre W.C.(1999) Neural and adaptive systems: Fundamentals through simulations, John Wiley & Sons, New York.

[3] Sawaragi Y., Nakayama H.,Tanino T.(1985) Theory of multiobjective optimization, Academic Press

[4] Ball D.F., Dartnell J., Davison J., Grieve A., Wild R.(1973) Agglomeration of iron ores, Heinemann Educational Books, London

[5] Heinänen K.(1993) Mineralogy and Metallurgical Properties of Iron Ore Sinter Based on Magnetite Fines, Raahen kirjatyö

[6] Demuth H., Beale M.(2000) Neural Network Toolbox User's Guide, Version 4, The MathWorks, Inc.

[7] Laitinen P.(2003) History-based Prediction Model for Sinter Plant, Proceedings of ISSTECH 2003, Indianapolis

A framework for neural quality control systems design

N. Costa, B. Ribeiro
CISUC – Centro de Informática e Sistemas da Universidade de Coimbra
Dep. Informatics Engineering, University of Coimbra, Polo II, P-3030 Coimbra, Portugal
Tel: (+351) 239 790000 Fax: (+351) 239 701266
E-mail: nfrc@netc.pt, bribeiro@dei.uc.pt

Abstract: Stimulated by the growing demand for improving system performance and reliability, fault-tolerant system design has been receiving significant attention. This paper proposes a new framework for fault-tolerant and quality control design based on the learning capabilities of neural networks. In highly nonlinear systems, with slow process variation, and a high degree of reactivity, quick changes (in the environment, for example) cause fast responses. This class of systems is broad enough so that it is not only of theoretical interest but also of practical applicability. Moreover, the fault-tolerance ability of the adaptive controller will be further improved by exploiting information estimated from a fault-detection and diagnosis unit designed by a pattern-matching strategy in multiple faults models interfaced with the overall system. Results of the approach are presented in a practical case of a plastic injection moulding process in industry.

1 Introduction

The development of a control system for the plastics injection moulding industry has been a case of study and focus of interest over the past years. An increase in production is no longer the goal to achieve. Instead, competition is now based on the quality of the final products. Although much effort has been put in the field of quality control, the basis of any intelligent diagnosis method is built upon a representative and reliable model of the overall system behaviour in normal operation conditions [1]. The proposed quality control system comes as a sequence of previous work [2][3] and is expected to be a step forward in the direction of the desired global stable and robust quality control system.

Section II describes the proposed quality control system. In section III system identification is performed using neural networks. Once the model is derived, it is applied to the detection and diagnosis of faults, as explained in Section IV. Finally, in section V, a neural adaptive controller is designed to act on the process in order to keep the setup parameters stable and within normal conditions.

2 Overview and Proposed scheme

Until recently, Statistical Process Control (SPC) was the main tool used in the monitoring of industrial processes and, in particular, the injection moulding process. Some produced parts were taken into a lab, from time to time, and their characteristics measured. If their specifications didn't meet the predefined requisites then some corrective actions would be taken in order to prevent further defects on the moulded parts. This technique was clearly inefficient as it was quite time consuming since the corrective actions, when taken, were too late to prevent the production of several defective parts.

Alternative diagnostic systems capable of earlier fault detection and parameter auto adjustment have been under the eye of many researchers over the past two decades. Such systems are commonly called Fault Detection and Diagnosis (FDD) systems and usually have the capability to predict the process's future state, which would allow a control device to act in the whole process behaviour and to reduce the anomalies.

In general, the FDD methods can be classified into pattern recognition, information flow graphs, analytical redundancy categories, that make use of available quantitative mathematical models, and more recently, the knowledge-based system and neural networks (NNs) approaches, using qualitative models. There are several survey papers in this area and many methods of FDD have been presented [4]. The dynamic behaviour of a system can be described either by a quantitative or qualitative model. Faults in processes are detected and diagnosed using analytical redundancy, by comparing the measured and estimated signal of the system. A problem with these methods, based on residual generation, parameter estimation, state estimation and parity equations, and used to generate test quantities [5], has to do with the heavy computational power required. Some of these methods, like the state space approach, assume that the process model is known quite well which, in practice, is quite difficult as modelling errors may be interpreted as faults. Others like rule-based expert systems require a set of rules about the faults, generated by process experts, operators and engineers. The accuracy of such approach depends on how rich and accurate the rule base is, which may become quite time consuming [6][7]. Recently, analytical redundancy and knowledge-based methods have been combined in order to increase the robustness of the whole FDD scheme. The idea is to use NNs to represent the knowledge and to realize a pattern recognition matching task.

In the case of the plastic injection moulding process, its specific characteristics and the extreme necessity of a real-time implementation have, however, conditioned many of the studies and schemes proposed so far. In spite of being a quite stable process, with slow process variation, it has a high degree of reactivity, since quick changes (in the environment, for example) cause fast responses.

The framework described in the paper proposes a fully neural system to deal with the nonlinearities and the request of real-time implementation. NNs, are good modelling tools for strongly nonlinear input output relationships due to the fastness of response and their capability to deal with nonlinearities and easy learning from examples, without the need of the process analytical description, constituting an important alternative to the classic fault diagnosis and quality control design methodologies.

3 Process Modelling

Chemical processes behaviour is normally nonlinear and difficult to model. Even if all their dynamics are well understood, the final system would, most probably, not be suitable for a real-time implementation and, therefore, not so useful in practice, which is our main concern. Traditional approaches, like mathematical modelling, although desirable, have high equation development costs, due to the complex dynamics of the injection moulding process. These costs involve not only the time spent in deriving the differential equations, but also the computational overload such a model consumes despite all the simplifications normally involved. The strong interdependencies among the process parameters, the material properties and the moulded part quality, together with the environmental fluctuations and the possibility of having more than one parameter setup, contribute to the complexity of the task in hands. Therefore, no ideal state of constant quality can be reached [8].

In such a process, where efficiency and fastness are crucial requirements, the capability to predict in advance the process parameters is the key to success. Knowledge of what will happen in the future allows us to take preventive measures that might be sufficient to avoid performance degradation which would, with no doubt, lead to poor quality produced parts.

Table 1 – Process Parameters Description.

Symbols	Description
DOS	Dosage
In2P	Beginning of second pressure
V1	Injection velocity 1
V2	Injection velocity 2
PI	Injection pressure
2P	Second pressure
CP	Counter pressure
CL	Cycle time
TPDOS	Dosage time
TPI	Injection time
AL	Cushion

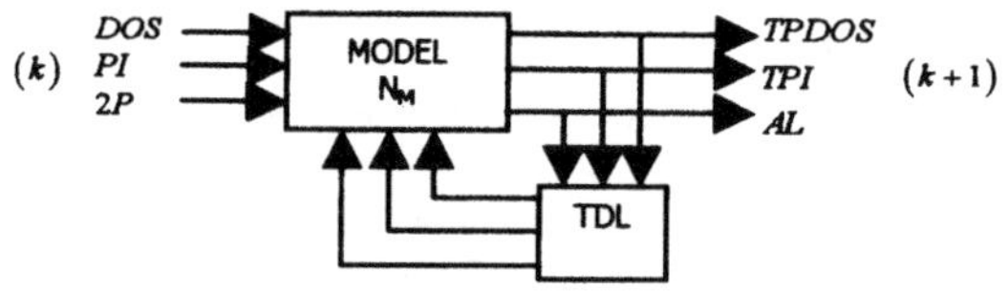

Figure 1 – Neural model (TDL - tapped delay line).

In the first stage, the *process window* was determined and the process parameters like dosage time, injection time and cushion (see table 1 for a comprehensive description of process parameters) were identified as being related with part defects. A large number of experiments were performed with several neural models, having different architectures and learning algorithms, trained according to figure 1, using as inputs the injection pressure, second pressure and dosage, as well as the delayed output vector. Several delay units have been used and tested in the TDL, but experience has shown that no benefits came from using more than one delay unit, since the process reacts rapidly to changes and, therefore, there is no need for a medium or long-term memory.

The results shown in figure 2 were obtained with a two hidden layer $\aleph_{6,15,9,3}$ feedforward neural network, trained with the Levenberg-Marquardt algorithm. The network had, respectively, 15 and 9 nodes in the first and second hidden layers, both having hyperbolic tangent sigmoid transfer functions, and 3 nodes in the linear output layer. All data used in training and testing was normalized to the interval [-1; 1]. As it can be seen, the network's outputs follow quite well the original signal.

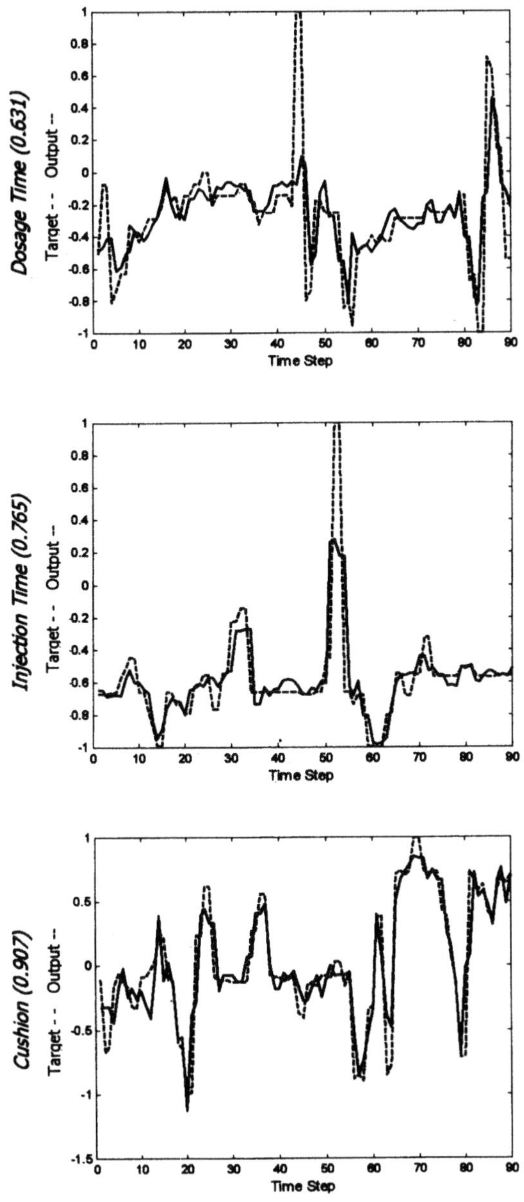

Figure 2 – Neural model's response.

4 Model-Based Fault Diagnosis

Model-based fault diagnosis consist on the use of a model for comparison with the actual system's response. Faults are detected if a so called *residual signal* [9], generated from the difference between real measurements and their estimates using the model, exceeds a predefined threshold. If this happens, then the residual may be analysed and the proper measures taken, in order to stabilise the parameter values, using some kind of fuzzy system not yet developed, but under study (see figure 3).

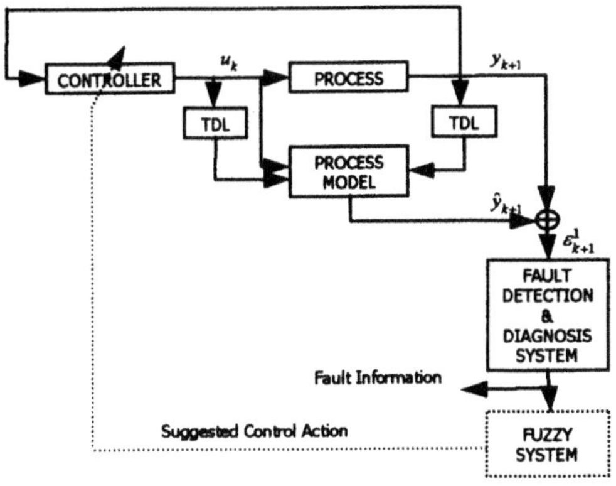

Figure 3 – Intelligent quality control proposed scheme.

The second step is the creation of a Fault Diagnosis and Detection system that, once trained to recognize the cause-effect relationships, would be able to identify faults like part dimensions, spots, burn marks, flash, unfilled parts and warpage (see figure 4). This system should be capable of responding in useful time and to deal with incomplete information or noisy data.

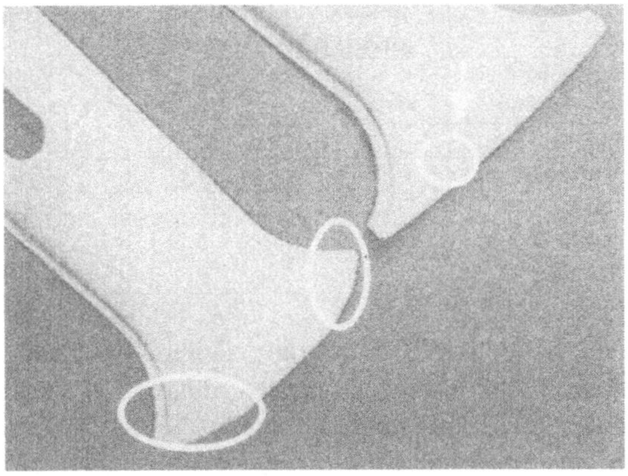

Figure 4 – Common defects: spots and unfilled parts.

As stated before, a deviation in the residual signals may indicate an initial manifestation of an anomalous situation [10] and its analysis allows us to detect abnormal process behaviour, which may possibly indicate a fault. However, it is no secret that modelling errors, δ, may influence the residual signal ε. Instead of an optimal situation where $\varepsilon = 0$, we should expect to get $\varepsilon \leq \delta$ since $|y - \hat{y}| \leq \delta$, where $\hat{y}$ is the model's output. This means the residual threshold should be greater than δ.

If we consider that $\hat{y}$ is a two-dimension vector formed by the predicted values of cushion and injection time (the two parameters most related with the defects we are interested in), we easily see that our problem will be $\Re^2 \rightarrow \Re^6$. Needless to say that such an approach has a high degree of complexity. Besides, many network architectures, such as the Linear Vector Quantization or the Probabilistic Neural Network only allow one class per input vector. Since the same produced part may contain several defects, such an approach is not appropriate to our case. To overcome the difficulties mentioned a modular scheme was adopted and several neural networks were trained using different topologies and learning algorithms. Six neural networks were then selected, one for each possible fault mentioned above. This way, we get six maps of the kind $\Re^2 \rightarrow \Re^1$ and are able to reduce the complexity of the system. Table 2 shows the results obtained when predicting the faults mentioned above, using the residual signals, for a $\aleph_{2,15,1}$ feedforward neural network trained with the Levenberg-Marquardt algorithm. The network had 15 nodes and a hyperbolic tangent sigmoid transfer function in the hidden layer, and 1 linear output.

Table 2 – Neural network classifier's performance.

Defects	**Number of defects (errors)**	**Undetected defects (errors)**	**Error Percentage**
Part dimensions	33	9	27,3%
Spots	4	2	50%
Burn marks	0	0	0%
Flash	37	4	10,8%
Unfilled parts	34	4	11,8%
Warpage	3	2	66,7%

Notice that in 198 samples, faults like flash and unfilled parts show a greater degree of correct classifications. This might be due to the fact that these faults are more related with the cushion and injection time parameters than with the others. The classifier's performance may be explained by the sample distributions. Some faults are harder to classify which could be due to the sample dispersion. There are no common properties and, therefore, no clusters were identified, which explains the results shown above. Notice also that warpage has a high error percentage, due to the lack of representative samples collected.

5 Process Control

No quality control system would be complete without a device capable of correcting, in a short period of time, the causes of malfunctioning. We attempt to derive an intelligent, adaptive, neural-based controller, which should also supply the process parameters set-points.

There are many approaches to process control in the literature. One of them, Model Reference Adaptive Control (MRAC) has been extensively used and seems to be adequate to our pretensions. This methodology attempts to make the output of the designed system behave in an identical way to the output of a model used as a reference. In order to build such a reference model, knowledge about the process window is essential. Once identified the process parameters related to quality defects, a reference model or simulator is derived using as a reference input signal the process setup vector (*vide* figure 5).

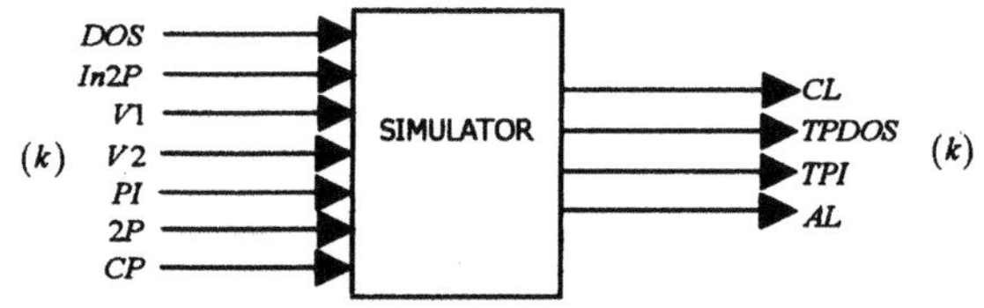

Figure 5 – Process simulator.

Notice that cycle time (CL) is also predicted. However, since it is also dependent on the operator's efficiency, it can not be used to train the model, as it would introduce spurious behaviour.

If only fault detection was required, the system would be complete. However, for a quality control system to be effective, it must have some regulation device capable of reconfiguring the system. To do this a controller is trained to perform the map $\Re^3 \rightarrow \Re^3$, using dosage time, injection time and cushion as inputs to estimate the dosage, injection pressure and second pressure parameters. For logistical and safety reasons, the controller cannot be tuned using the real process. Instead, the process model derived in section 3 is used, but only the weights of the controller are adjusted during the learning stage. This way, since the model reacts only to the control signal, the controller has to learn the proper actions. Once the train is completed, the "connection" between the two networks is broken and the controller can then be used with the real process. Figure 6 shows the process model's outputs when subjected to the control actions. Notice that although the correlation coefficients are lower than the ones in section 3, when the model was used in open-loop, the outputs follow the reference signals showing, not only good generalization capabilities from the model, but also a reasonable reaction from the controller.

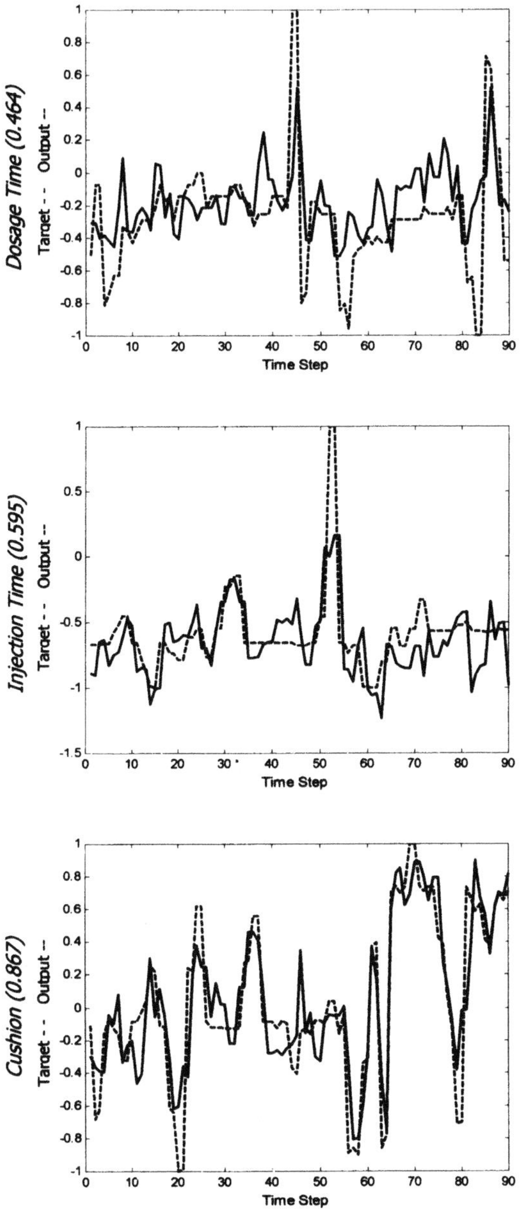

Figure 6 – Model's response when using the controller.

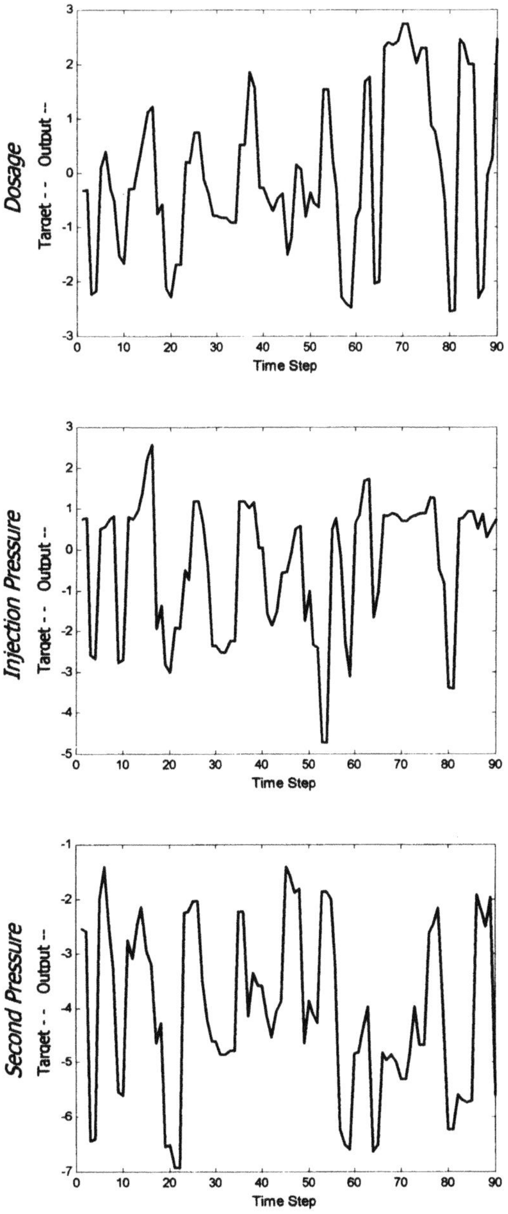

Figure 7 – Control parameters.

A look into the controller's outputs shows, however, a lot of activity in spite of the model's response. It is believed this may be caused by the use of, what might be called, second class parameters, like cushion and times, to predict pressures and dosage.

6 Conclusions

Quality control systems have been a focus of interest during the past years. We have seen that for such systems to be used in practice, they should react to process deviations in real-time, so that faults may be quickly corrected before quality degradation happens.

A framework for developing an intelligent neural quality control system has been proposed in this paper. Neural networks learning capabilities have been exploited with success, in several research areas like system identification, fault detection and diagnosis and process control. Many of the results shown herein are, however, conditioned by a lack of representative data samples. The system could be further improved for application in the thermoplastic moulding injection industrial process provided that more precise and representative parameters were used, like pressures and temperatures.

Future research will comprehend the development of a fuzzy decision mechanism, capable of using the information estimated from the fault-detection and diagnosis unit to control the process. This way it would be possible to incorporate the operator's know-how and develop a more user-friendly and perceptible system.

Acknowledgements

The authors would like to thank PLASFIL - Plásticos da Figueira, Lda. for the availability of the mould and for making the moulding data available. The authors would also like to express their gratitude to PRAXIS XX1, Sub-Programa Ciência e Tecnologia do 2° Quadro Comunitário de Apoio, which financially supports this project.

References

[1] Fuminori Yakuwa, Shingo Satoh, Muhammad Shafique Shaikh, and Yasuhiko Dote (2002) *Fault Diagnosis for Dynamical Systems Using Soft Computing*.

[2] Nuno Costa and Bernardete Ribeiro (1999) *Monitoring an Industrial Plastic Injection Moulding Machine Using Neural Networks*. In *Artificial Neural Nets and Genetic Algorithms*, Proceedings of the International Conference ICANNGA'99, Portorož, Slovenia.

[3] Nuno Costa and Bernardete Ribeiro (1999) *A Neural Prediction Model for Monitoring and Fault Diagnosis of a Plastic Injection Moulding Process*. European Control Conference ECC'99, Karlsruhe, Germany.

[4] Qiugui Yang (1997) *A Review Of Methods For Fault Diagnosis*, Presented In Mechanical Engineering Research Colloquium, Mechanical Engineering, Engineering Faculty, University of Glasgow, July 1997.

[5] Mattias Nyberg (1999) *Model Based Fault Diagnosis: Methods, Theory, and Automotive Engine Applications*. Linköping Studies in Science and Technology. Department of Electrical Engineering. Linköping University, Sweden. ISBN 91-7219-521-5 ISSN 0345-7524. Dissert. No. 591.

[6] J. C. Hoskins, K. M. Kaliyur, and David M. Himmelblau (1991) *Fault Diagnosis in Complex Chemical Plants Using Artificial Neural Networks*. AIChE Journal, vol. 37, no. 1, pp. 137-141.

[7] Kajiro Watanabe, Ichiro Matsuura, Masahiro Abe, Makoto Kubota, and David M. Himmelblau (1989) *Incipient Fault Diagnosis of Chemical Processes via Artificial Neural Networks*. AIChE Journal, vol. 35, no. 11, pp. 1803-1812.

[8] Tatiana Petrova and David Kazmer (1997) *Development of a Hybrid Network for Quality Control of Injection Molding*. University of Massachusetts Amherst.

[9] Ali Alawi and R. Doraiswami (1997) *Real time fault diagnosis*. IFAC Conference SAFEPROCESS'97, vol. 1, pp 432-437, Hull, UK.

[10] O. Vadam, J. Montmain and J. C. Cassar (1997) *Fault Detection Using Parallel Simulations*. IFAC Conference SAFEPROCESS'97, vol. 1, pp 121-126.

Prediction of natural gas consumption with feed-forward and fuzzy neural networks

Nguyen Hoang Viet[1], Jacek Mańdziuk[2] *

[1]Institute of Fundamental Technological Research, Polish Academy of Sciences. ul. Świętokrzyska 21, 00-049 Warsaw, Poland. [2]Faculty of Mathematics and Information Science, Warsaw University of Technology. Plac Politechniki 1, 00-661 Warsaw, Poland.

Abstract

In this work several approaches to prediction of natural gas consumption with neural and fuzzy neural systems are analyzed and tested. The data covers daily natural gas load in a certain region of Poland. Prediction strategies tested in the paper include: single neural net module approach, combination of three neural modules, temperature clusterization based method, and application of fuzzy neural networks. The results indicate the superiority of temperature clusterization based method over modular and fuzzy neural approaches. One of the interesting issues observed in the paper is relatively good performance of tested methods in the case of a long-term (four week) prediction compared to mid-term (one week) prediction. Generally, the results are significantly better than those obtained by statistical methods currently used for this task in the gas company under consideration.

1 Introduction

The paper discusses several neural and fuzzy neural approaches to the problem of prediction of natural gas consumption in a certain region of Poland. The area is mostly rural with several small cities and therefore the consumers are mostly individual or they belong to small industry (bakeries, restaurants, laundries, etc.).

Predicting the consumption of natural gas is an interesting, non-trivial and highly economically-motivated prediction task [1, 2]. The conventional approach to solving it is based on applying statistical methods which are usually efficient enough provided the large amount of past (historical) data is available. An alternative approach is to exploit soft-computing methods - and especially neural and fuzzy neural network models.

Neural networks are well known to be universal non-linear approximators [3, 4] and therefore are in general capable of close approximation of the prediction model without the need of its explicit (mathematical) formulation as oppose to statistical approaches. The other advantage of using neural networks over statistical methods in the application domain considered in this paper is the problem of gas market violatility in Poland in recent years. On one hand several new consumers are attracted by the relatively low costs of this type of energy but on the other hand some gas consumers (especially in the rural regions) switch to alternative energy sources, likewise wood or coal. This type of "structural instability" of natural gas consumers is very harmful for statistical approaches, which do not implement any adaptation mechanisms.

The paper is organized as follows: in the next section network architectures and training methods used in this work are briefly introduced. The data files and preprocessing methods are described in section 3. Experimental results are presented in section 4. Conclusions and directions for future development of this work are placed in the last section.

2 Network architectures and training methods

Generally, two types of neural architectures are tested in this work: feed-forward, perceptron-type networks and fuzzy neural networks.

2.1 Feed-forward networks

Feed-forward networks applied here are two hidden layer architectures with sigmoidal units. Each prediction network takes the amount of daily gas consumption together with the daily average temperature measured in the given region as the inputs. These values are scaled to some predefined ranges before being passed to the networks. Moreover the time factor, coded in a special way, is also included as part of the input. For details on data preprocess-

*The work was supported by the grant no. 503G 1120 0009 002 from the Warsaw University of Technology. The support from the Mazovian Regional Gas Distribution Company - Poland is also gratefully acknowledged.

ing please refer to section 3.

The size of the input layer depends on the horizon of prediction. Three types of prediction are considered in this work: one-day prediction (denoted by **D-type**), one-week prediction (denoted by **W-type**) and four-week prediction (denoted by **4W-type**).

Let's denote by $G(t)$ the actual gas consumption on day t, by $\hat{G}(t)$ the prediction of gas consumption made on day t and by $T(t)$ the average temperature on day t. It should be noted that all references to temperature values in this work concern the observable average temperatures in a given period in the past. No explicit temperature prediction was made since this kind of prediction was virtually embedded in the gas load prediction network, due to its specific architecture (see description bellow).

2.1.1 Short-term prediction. The network architecture used for short-term prediction, i.e. $\hat{G}(t+1)$ - the next day's load, is schematically presented in Figure 1. The first k neurons in the input layer represent the daily gas load for k previous days: $G(t), G(t-1), \ldots, G(t-k+1)$. The second group of k neurons refers to the average temperatures: $T(t), T(t-1), \ldots, T(t-k+1)$. The next input item d denotes the type of the next day (i.e. day $t+1$) where $d = -1$ for working days and $d = +1$ for non-working days[1]. The last two inputs represent the time factor or the "season of the year" for the day $t+1$ coded in a special way described in section 3. The value of $k = 3$ was chosen for one-day prediction in this work[2]. This means that the daily gas load and the average temperature from three previous days were taken into account in order to make the next day's load prediction.

The first hidden layer also consists of three groups of neurons. Neurons belonging to the first group are fully connected to neurons in the first group in the input layer (i.e. the *gas units*). Similarly, neurons in the second group are connected to neurons in the second input group (the *temperature inputs*). The last group is connected to the rest of the inputs (representing day type and season coding). In this paper the numbers of neurons in the first and the second groups are set to be three. The third group in the first hidden layer contains two neurons.

[1]Actually, for the sake of simplicity, only weekend days (Saturdays and Sundays) are treated as non-working days. All holidays that occurred in weekdays are treated as working days.

[2]In each type of prediction, the sizes of input and hidden layers were selected experimentally based on preliminary tests.

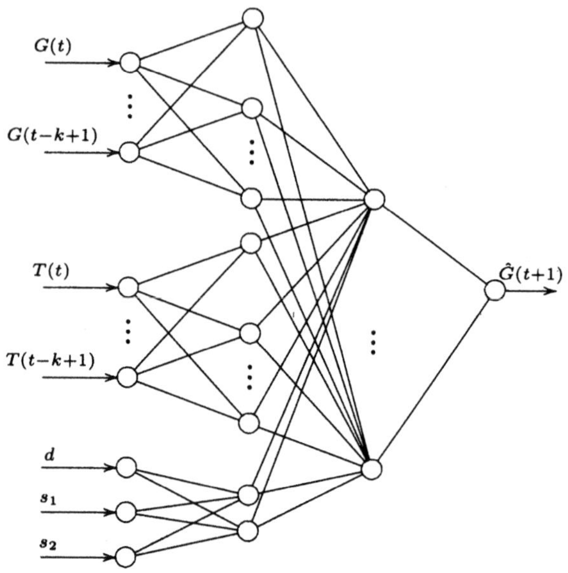

Fig. 1. The architecture of feed-forward neural network used for one-day prediction. The networks used for one-week and four-week prediction are of similar structure except that they do not have the *day type* input (denoted by d in the figure). See description within the text for details.

All first hidden layer neurons are fully connected to the neurons in the second hidden layer (there are four of them). The second hidden layer neurons are in turn connected to the output neuron, which produces $\hat{G}(t+1)$.

The goal of the first hidden layer neurons is to aggregate the input data of each type independently from the inputs of the other types. This information is then aggregated in the second hidden layer, where no restrictions are imposed on connections to the previous layer.

2.1.2 Mid-term prediction. The network used for the mid-term prediction (the next week's load) differs from the short-term prediction network described above by the sizes of the input and hidden layers. The first two input sections are composed of data concerning 5 previous days, i.e. $G(t-i)$ and $T(t-i)$, $i = 0, \ldots 4$, resp. The neuron representing the type of the day is not used in this case. The season of the year is coded in a similar way as in D-type prediction.

The first hidden layer is composed of 10 neurons (4 neurons in each of the first two groups, 2 neurons in the third group). The second hidden layer is composed of 4 units. The output of the network is equal to the predicted average daily consumption in the following week (i.e. of the days $t+1, t+2, \ldots, t+7$)[3].

2.1.3 Long-term prediction. For the long-term prediction (i.e. in a four week horizon), the input layer is composed of 16 neurons: past load values of the last 7 days $G(t-i), i = 0, \ldots, 6$, the respective average daily temperatures $T(t-i), i = 0, \ldots, 6$ and the season coding - the last two inputs. The structures of the hidden layers are the same as in the W-type prediction case. Similarly to the mid-term prediction, the output of the network represents the predicted average daily natural gas consumption in the period of the next four weeks.

All feed-forward neural networks described above are trained by the *Scaled Conjugate Gradient* method [5], which proved to be a fast and effective training method.

2.2 Fuzzy neural networks (FNNs)

Several FNN models have been developed and successfully used in various applications [6, 7]. The advantages of fuzzy over conventional neural networks are their high generalization abilities and the capability of dealing with imprecise data. In the gas load prediction problem the temperature input plays a significant role. The use of fuzzy neural network for this task is, among other things, motivated by the fact that the average daily temperature is only an approximation of the exact temperature which varies throughout the day. In this section a fuzzy neural network model is firstly described together with its training algorithm. The rest of this section discusses the use of this model to our prediction task.

2.2.1 The FNN model. The fuzzy neural network implemented in this work is represented by a single hidden layer feed-forward architecture with N input units, K hidden units and M output ones (Figure 2).

Every connection between input and hidden units has a weight equal to 1. Each unit in the hidden layer represents a fuzzy set over R^N. The output value of the i-th hidden neuron for a given input vector $\mathbf{X} = [x_1, x_2, \ldots, x_N]^T$ can be interpreted as the degree of membership of $\mathbf{X}$ to the fuzzy set represented by this neuron. The neurons in the hidden layer are fully connected with the neurons in the output layer.

[3]Calculation of the week's load is then straightforward and equal to network's output multiplied by 7.

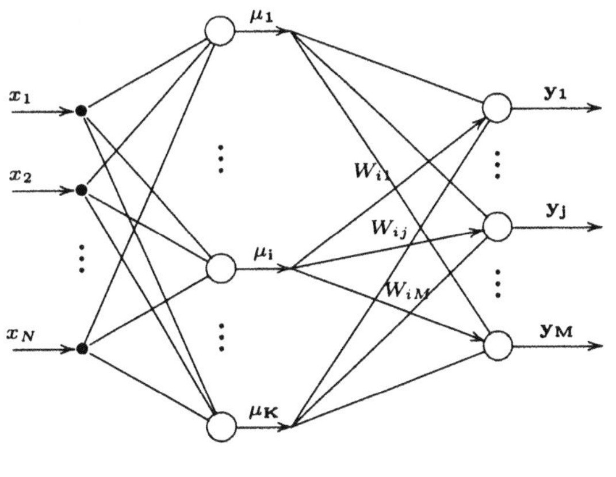

Fig. 2. The architecture of fuzzy-neural network used in simulations.

Assuming the Gaussian form of the membership function of fuzzy sets represented by hidden neurons, the output of the i-th hidden neuron given $\mathbf{X}$ as the input is equal to:

$$\mu_i = \exp - \left(\frac{\| \mathbf{X} - \mathbf{C}_i \|}{\sigma_i} \right)^2, \tag{1}$$

where $\| \cdot \|$ is the Euclidian norm, $\sigma_i > 0$, $\mathbf{C}_i = [c_{i1}, c_{i2}, \ldots, c_{iN}]^T$ are parameters associated with the given neuron.

Denoting by $\mathbf{W} = [W_{km}]^{K \times M}$ the weight matrix of connections between hidden layer and output layer, the input value for the j-th ouput neuron is:

$$u_j = \sum_{k=1}^{K} W_{kj} \mu_k. \tag{2}$$

Let:

$$s = \sum_{k=1}^{K} \mu_k, \tag{3}$$

the output value of that neuron is then defined as:

$$y_j = \frac{u_j}{s} = \frac{\sum_{k=1}^{K} W_{kj} \mu_k}{\sum_{k=1}^{K} \mu_k}. \tag{4}$$

2.2.2 FNN training. The fuzzy neural network described above can be trained in supervised mode. For a single training sample $(\mathbf{X}, \mathbf{d})$, $\mathbf{X} = [x_1, x_2, \ldots, x_N]^T$, $\mathbf{d} = [d_1, d_2, \ldots, d_M]^T$, the error function defined as:

$$E = \frac{1}{2} \sum_{j=1}^{M} (d_j - y_j)^2 \tag{5}$$

is to be minimized w.r.t. σ_i, $\mathbf{C}_i$ $\left(i=\overline{1,K}\right)$ and $\mathbf{W}$. This minimization task can be solved by any gradient based optimization algorithm. The partial derivatives of E can be computed in the following:

$$\begin{aligned}
\frac{\partial E}{\partial W_{ij}} &= -\left(d_j - y_j\right)\frac{\partial y_j}{\partial W_{ij}}, \\
\frac{\partial E}{\partial c_{il}} &= -\sum_{j=1}^{M}\left(d_j - y_j\right)\frac{\partial y_j}{\partial c_{il}}, \\
\frac{\partial E}{\partial \sigma_i} &= -\sum_{j=1}^{M}\left(d_j - y_j\right)\frac{\partial y_j}{\partial \sigma_i}
\end{aligned}$$

for $i=\overline{1,K}$, $j=\overline{1,M}$ and $l=\overline{1,N}$. Using (4) one can write:

$$\begin{aligned}
\frac{\partial y_j}{\partial W_{ij}} &= \frac{\mu_i}{\sum_{k=1}^{K}\mu_k} = \frac{\mu_i}{s}, \\
\frac{\partial y_j}{\partial c_{il}} &= \frac{\partial y_j}{\partial \mu_i}\frac{\partial \mu_i}{\partial c_{il}} = \frac{W_{ij}s - u_j}{s^2}\frac{\partial \mu_i}{\partial c_{il}}, \\
\frac{\partial y_j}{\partial \sigma_i} &= \frac{\partial y_j}{\partial \mu_i}\frac{\partial \mu_i}{\partial \sigma_i} = \frac{W_{ij}s - u_j}{s^2}\frac{\partial \mu_i}{\partial \sigma_i}.
\end{aligned}$$

From (1) we have:

$$\begin{aligned}
\frac{\partial \mu_i}{\partial c_{il}} &= 2\frac{\mu_i}{\sigma_i^2}\left(x_l - c_{il}\right), \\
\frac{\partial \mu_i}{\partial \sigma_i} &= 2\frac{\mu_i}{\sigma_i^3}\sum_{n=1}^{N}\left(x_n - c_{in}\right)^2.
\end{aligned}$$

Finally:

$$\frac{\partial E}{\partial W_{ij}} = \delta_j^{(1)}\mu_i, \tag{6}$$

$$\frac{\partial E}{\partial c_{il}} = \delta_i^{(2)}\left(x_l - c_{il}\right), \tag{7}$$

$$\frac{\partial E}{\partial \sigma_i} = \delta_i^{(2)}\frac{1}{\sigma_i}\sum_{n=1}^{N}\left(x_n - c_{in}\right)^2, \tag{8}$$

where

$$\begin{aligned}
\delta_j^{(1)} &= -\frac{d_j - y_j}{s}, \\
\delta_i^{(2)} &= 2\frac{\mu_i}{\sigma_i^2}\sum_{j=1}^{M}\left(\delta_j^{(1)}\left(W_{ij} - \frac{u_j}{s}\right)\right)
\end{aligned}$$

are the error signals in output and hidden neurons, resp. which can be computed effectively in a back-propagation manner. The network parameters can then be adapted, e.g. by the standard steepest descent algorithm. In this work the Möller's variant of conjugate gradient method [5] was applied.

2.2.3 FNNs for gas load prediction. Analogously to the case of feed-forward networks, the fuzzy neural networks used here receive the daily gas load and the average daily temperature from the last few days as the input: three days for short-term, five days for mid-term and seven days for long term prediction.

In all FNNs used here, the number of hidden units was arbitrarily set to four[4]. Each network contains one output neuron that produces the predicted average daily gas load (c.f. section 2.1).

One important issue worth mentioning here is the method of initializing FNN's parameters before training. Since each hidden neuron in such network represents a fuzzy set of Gaussian membership function, the parameter vectors $\mathbf{C}_i$, $i=\overline{1,K}$ should be set *close* to the medians of some fuzzy sets that *cover* the input space. The training process *moves* these sets to an optimal configuration. A self-organizing map of the training dataset was first built using the neural gas technique [8]. Then the vectors $\mathbf{C}_i$ were set to be the centers of clusters in that map.

3 Data files

3.1 Data collection

The main problem in this particular prediction task is the lack of historical data, which has been collected by a natural gas distributing company for less than three years only. The available data ranged from Jan. 01, 2000 to Jul. 31, 2002 was initially divided into two sets: the first set covering the years 2000 – 2001 to compose the training set and the second one with year 2002 data to form the test set. Furthermore 10% of the training data was randomly chosen for validation.

Each data record represents the daily cumulative load of natural gas provided by the telemetric system as well as the average daily temperature. The daily load distribution registered during the years 2000 and 2001 is presented in Figure 3. It can be easily observed that the gas consumption is clearly seasonal: the highest loads occur in winter periods and the lowest ones in summer time. Figure 4 shows the strong dependency of daily gas consumption on the average daily temperature.

Due to the relatively small amount of past data, the sliding window mechanism was used in order to artificially enlarge the datasets. Namely, for W and 4W-type predictions the target periods are overlapping and subsequent target periods are de-

[4]This parameter was selected experimentally based on preliminary tests.

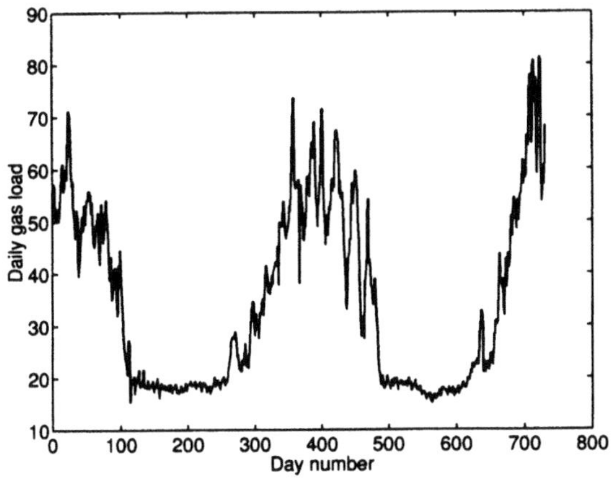

Fig. 3. Daily gas load data from Jan. 01, 2000 to Dec. 31, 2001 used as the training and validating set.

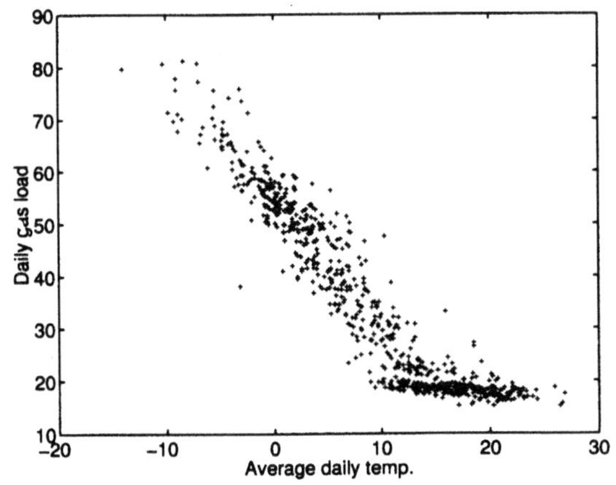

Fig. 4. The dependency of the daily gas load on the average daily temperature. The data covers the period from Jan. 01, 2000 to Dec. 31, 2001.

fined as $[t+1, t+n]$ - for prediction made on day t, $[t+2, t+n+1]$ - for prediction made on day $t+1$, $[t+3, t+n+2]$ - for prediction made on day $t+2$, etc., where $n=7$ or $n=28$ for W-type and 4W-type predictions, resp.

3.2 Data scaling

The data, before being input to the network, is scaled to some predefined range. The maximum and minimum temperatures in the training set, called t_{max} and t_{min}, resp. were determined. Afterwards the temperature range $[t_{min}, t_{max}]$ was uniformly extended by 20% from each side into the range $[T_{min}, T_{max}]$. Temperature values were finally scaled to the interval $(-1, 1)$ using T_{min} and T_{max}. The daily load values were scaled to $(0, 1)$ by dividing each value $G(t)$ by G_{max}, where G_{max} is the extended (as in the case of temperature) maximum daily load over the training data set. The output (target) values in the training set, which are the average daily loads for a given one-day, one-week or four-week period - depending on the type of prediction - were scaled similarly, i.e. were divided by G_{max}.

Since the data is highly seasonal, it is important to properly code the time factor of each prediction period. For a n day period: $[t+1, t+n]$ the following two variables are applied:

$$s_1 = \sin\frac{2\pi\, D_c}{366} \quad \text{and} \quad s_2 = \cos\frac{2\pi\, D_c}{366}, \tag{9}$$

where D_c is a real value indicating the *center day* of the period under consideration, i.e.:

$$D_c = t + \frac{n+1}{2}. \tag{10}$$

The use of (9) allows smooth coding of the season of the year which is especially important in the case of mid-term and long-term prediction.

4 Experimental results

For each prediction horizon the following experiments were performed: naive prediction, prediction using a single neural network module, prediction using a combination of three neural modules, prediction using three neural modules - each of which is devoted to a predefined temperature range, single neural network prediction performed for the working days only (concerns D-type prediction only) and prediction based on a single fuzzy neural network module.

The test data set for W and 4W-type predictions was also artificially enlarged by using the sliding window mechanism - analogously to the case of the training data. Several experiments with various neural architectures were performed. In all tests the same error measure - the Mean Absolute Percentage Error (MAPE) - commonly used in prediction tasks [2, 9] - was applied.

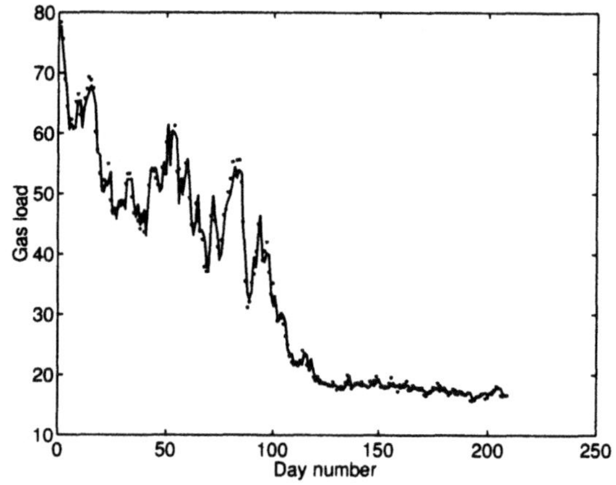

Fig. 5. Test results for D-type prediction in the period Jan. 01, 2002- Jul. 31, 2002.

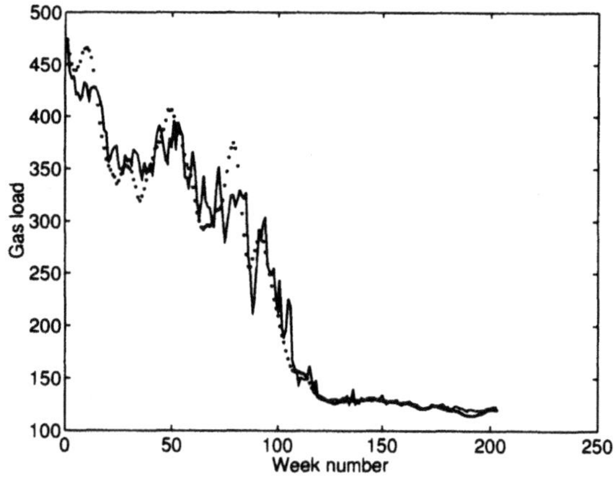

Fig. 6. Test results for W-type prediction in the period Jan. 01, 2002- Jul. 31, 2002.

Naive prediction (*Naive*). In this experiment no neural network is involved. The predicted average daily load in the period $[t+1, t+n]$ is simply assumed to be equal to the average daily load in the preceding period, i.e. $[t-n+1, t]$, where $n = 1, 7, 28$, resp. for D, W and 4W-type prediction.

Prediction using a single neural network (*SingleN*). In this experiment, for each prediction

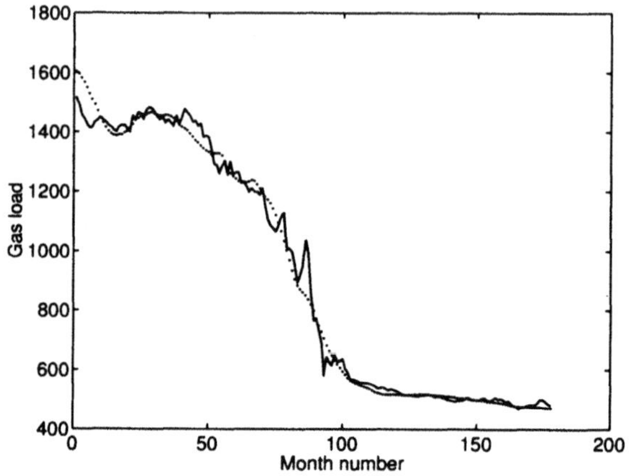

Fig. 7. Test results for 4W-type prediction in the period Jan. 01, 2002 - Jul. 31, 2002.

horizon, 50 neural networks were trained and tested as single predictors.

Prediction using a combination of three neural modules (*3AvgN*). Here all combinations of three different networks among the above 50 were tested. The output value was set to be the average value of the three modules' outputs.

Prediction using three neural modules - each of which is devoted to a specific temperature range (*3TempN*). In this experiment the training data was divided into three subsets. Depending on the average temperature in the last 3, 5 or 7 days (resp. for D, W and 4W-type prediction) the input sample belonged to either "low" (**L**), "medium" (**M**) or "high" (**H**) temperature data subset[5]. The subsets **L**, **M** and **H** had an equal number of samples and had some pairwise overlapping areas (except for the pair (**L**, **H**)), i.e.

$$\mathbf{L} \cap \mathbf{M} \neq \emptyset, \qquad \mathbf{M} \cap \mathbf{H} \neq \emptyset, \qquad \mathbf{L} \cap \mathbf{H} = \emptyset. \quad (11)$$

In effect several samples belonged to two data subsets. With each subset a set of 20 neural networks was independently trained. In the test phase all possible combinations of three modules (one of each type) were tested (8000 combinations in total). *Please note, that due to some overlapping areas in the training sets the effect of gluing the modules was achieved in a straightforward way* (cf. [10, 11]). For a given test sample the corresponding average input temperature was calculated and then depending on this value one or (usually) two appropriate networks were activated. The ultimate prediction was the av-

[5]Certainly, the notion of "low", "medium" and "high" temperature is subjective, hence the use of quotation marks.

D-type	*AVG*	*MIN*	*MAX*	*STDEV*
Naive	4.12	−−	−−	−−
SingleN	3.81	3.53	4.29	0.0015
3AvgN	3.65	3.43	3.94	0.0006
3TempN	3.45	3.27	3.63	0.0005
WorkD	3.60	3.32	4.37	0.0021
FuzzyN	3.74	3.54	4.31	0.0015

Table I. The average, minimum and maximum MAPE (in percent) and its standard deviation for D-type predictions.

W-type	*AVG*	*MIN*	*MAX*	*STDEV*
Naive	11.94	−−	−−	−−
SingleN	5.98	4.80	7.27	0.0058
3AvgN	5.49	4.45	6.70	0.0034
3TempN	5.18	4.61	5.77	0.0020
FuzzyN	5.85	5.18	7.05	0.0037

Table II. The average, minimum and maximum MAPE (in percent) and its standard deviation for W-type predictions.

erage value of the outputs of the modules involved in prediction.

Prediction for the working days only (*WorkD*). This experiment was analogous to the single network prediction except that it was made for the working days only. The experiment was performed for D-type prediction only.

Prediction based on a fuzzy neural network module (*FuzzyN*). An ensemble of 50 FNNs was trained and tested analogously to the experiment with single neural modules.

The results. Figures 5-7 show examples of D-type, W-type and 4W-type predictions, resp. using a combination of three networks, each of which was trained for some temperature range. The solid lines indicate the exact values taken from the test set, the dot lines indicate the predicted values produced by neural modules. For each experiment mentioned above the *average, min, max* (in percent) and *standard deviation* of MAPE over all tested networks (or all their combinations, where applicable) were calculated. The results are presented in Tables I-III.

5 Conclusions

Several conclusions can be drawn from Tables I-III. The main of them are the following:

- The best performance is observed for the combination of three temperature-based modules. The second one is the combination of three non-temperature-based networks. The next one is the neuro-fuzzy module. These three outperform the single neural network approach.

- In the case of D-type prediction an interesting issue is to compare the working days prediction versus all days prediction. The results for working days are only slightly better than the respective results for all days. The main reason for such a little improvement is the rural character of the area with very few industrial consumers. Therefore the consumption profile over the week is more stable than in highly industrialized regions.

- Surprisingly enough 4W-type prediction occurred to be much "easier" than W-type one. In all cases (except the naive approach) 4W-type results outperformed W-type ones. This phenomenon could be explained by the fact that due to the averaging effects happening in the 28 day period, the four-week gas load curve is much *smoother* than the one-week one, what can be observed by comparing Figures 6 and 7.

- In the case of W and 4W-type predictions significant improvement was achieved by all neural methods in comparison with the case of naive prediction.

4W-type	*AVG*	*MIN*	*MAX*	*STDEV*
Naive	31.82	−−	−−	−−
SingleN	4.76	3.30	7.69	0.0089
3AvgN	4.02	2.60	6.37	0.0052
3TempN	3.99	2.77	4.93	0.0031
FuzzyN	4.20	3.42	4.84	0.0046

Table III. The average, minimum and maximum MAPE (in percent) and its standard deviation for 4W-type predictions.

In summary, the results achieved by neural networks and fuzzy neural networks are very encouraging and acceptable from the natural gas company's viewpoint. Statistical methods used so far by the company yield the average MAPE error for monthly prediction[6] in the period 01.2000 – 12.2001 equal to 12.86% [7].

Another comparison can be fairly made with the naive prediction approach, which was clearly outperformed by neural methods in the case of mid-term and long-term predictions.

[6] This prediction horizon is actually used in practice.

[7] Unfortunately, the more recent data is not available.

The comparison of our results with the literature is ambiguous since the datasets used in other works are different from ours and also the consumer profiles may be different. One example is [2] where the MAPE error of 3.78% for D-type prediction involving temperature data is reported.

There are several directions in which this work can be continued. In the future we plan to test the efficacy of partially recurrent and RBF networks as well as other models of fuzzy neural networks for this task. Another issue is to test the prediction methods on the data gathered in a more industrialized area.

Acknowledgments

The authors would like to thank R. Galbarczyk and M. Łowczykowski from the Mazovian Regional Gas Distribution Company - Poland for their help with collecting and preparing the data.

References

[1] Brown, R.H., Martin, L., Kharouf, P., Piessens, L.P. (1996) Development of artificial neural-network models to predict daily gas consumption. A.G.A. Forecasting Review 5: 1-22,

[2] Khotanzad, A., Elragal, H., Lu, T-L. (2000) Combination of artificial neural-network forecasters for prediction of natural gas consumption. IEEE Transactions on Neural Networks 11(2): 464-473,

[3] Kolmogorov, A.N. (1957) On the representation of continuous functions of many variables by superposition of continuous function sof one variable and addition. Dokl. Akad. Nauk ZSRR 114: 953-956,

[4] Kurkova, V. (2000) Rates of approximation by neural networks. In: Sincak, P., Vascak, J. (eds.) Quo Vadis Computational Intelligence. Springer, Berlin, pp. 23-36,

[5] Möller, M.F. (1993) A scaled conjugate gradient algorithm for fast supervised learning. Neural Networks 6: 525-533,

[6] Buckley, J.J., Hayashi, Y. (1994) Fuzzy neural networks: A survey. Fuzzy Sets and Systems 66: 1-13,

[7] Gupta, M.M., Rao, D.H. (1994) On the principles of fuzzy neural networks. Fuzzy Sets and Systems 61: 1-18,

[8] Martinetz, M., Berkovich, S., Schulten, K. (1993) "Neural-gas" network for vector quantization and its application to time series prediction. IEEE Transactions on Neural Networks 4: 558-569,

[9] Zhang, G., Patuwo, B.E., Hu, M.Y. (1998) Forecasting with artificial neural networks: The state of the art. International Journal of Forecasting 14: 35-62,

[10] Sharkey, A.J.C. (1997) Modularity, combining and artificial neural nets. Connection Science 9(1): 3-10,

[11] Waibel, A. (1989) Consonant recognition by modular construction of large phonemic time-delay neural networks. In: Touretzky, D. (eds.) Advances in NIPS 1. Morgan Kaufmann, pp. 215-223.

Systems identification with GMDH neural networks: a multi-dimensional case.[1]

Marcin Mrugalski, Eugen Arinton, Józef Korbicz[1]

[1]Institute of Control and Computation Engineering, University of Zielona Góra, ul. Podgórna 50, 65-246 Zielona Góra, Poland, e-mail: {*M.Mrugalski,E.Arinton,J.Korbicz*}*@issi.uz.zgora.pl*

Abstract

This paper presents a relatively new identification method based on Artificial Neural Networks, which can be used for multi-input multi-output systems. In particular, a Group Method of Data Handling neural network with dynamic neurons is considered. The final part of this work contains an illustrative example regarding the application of the proposed approach to a fault detection system.

1 Introduction

The complexity and reliability demands of contemporary industrial systems and technological processes require the development of new fault diagnosis approaches. The early detection of faults may help to avoid system breakdowns and material damages. A fault can generally be defined as an unexpected change in a system of interest, e.g a sensor malfunction. All the unexpected variations that tend to degrade the overall performance of a system can also be interpreted as faults. During the last few decades many investigations have been made using analytical approaches, based on mathematical models [9]. In a fault diagnosis task, the model of the real system of interest is utilized to provide estimates of signals. Then, in the most usual case, the estimates of the measured signals are compared with their originals, i.e. the difference between the original signal and its estimate is used to form a residual signal. This residual signal can then be employed for *Fault Detection and Isolation* (FDI). Thus, the better the model used to represent system behaviour, the better the chance of improving the reliability and performance in diagnosing faults. If the residues are properly generated, then fault detection and isolation becomes a relatively easy tasks. This means that the problems of system identification and fault diagnosis are closely related to each other.

One way out of this problem is to apply artificial neural networks or fuzzy neural networks. Many works confirm their effectiveness and recommend their use [7, 8]. On the other hand, there is no efficient approach to selecting structures of such networks. Thus, many experiments have to be carried out to obtain an appropriate configuration. An alternative approach, which seems to avoid such a difficulty, is to employing the *Group Method of Data Handling* (GMDH) [2, 3]. The concept of this approach is based on the iterative processing of an operation defined sequence leading to the evolution of the resulting network structure. The GMDH network is a combination of the black box nonlinear system identification concept, and the inductive and probabilistic approaches. The assumptions of GMDH neural networks give a lot of freedom in defining the particular elements of the algorithm of synthesis. The mentioned possibilities relate to, for example, the definition of the elementary model, in particularly transition function, evaluation criteria of processing accuracy or selection methods. The concept of the group method of data handling also allows to develop the formula of the GMDH network, through the application of additional procedures, which can be used to extend the scope of the application. The GMDH network can be used to identify of static and dynamic systems. Moreover, the GMDH approach is especially attractive in the multi dimensional systems identification task. It follows from the fact, that the group method of data handling was developed to design models with a large number of inputs.

In this work, the structure of the GMDH network is presented. In particular the concept of the synthesis of the *multi-input, multi-output* (MIMO) GMDH network is considered. For such a network type, the classical least square method is employed for parameter estimation. As an example, which confirms the effectiveness of the proposed approach in the system identification and fault detection tasks, the DAMADICSC benchmark problem is considered.

[1]This work was supported by the EU FP 5 RTN project DAMADICS and in part by the State Committee for Scientific Research (KBN) in Poland.

2 Synthesis of the multi-input multi-output GMDH neural network

The concept of the synthesis of the GMDH network is based on the iterative processing of a defined sequence of operations leading to the evolution of the resulting structure, which generates the output signal and is the best approximation of the real system output. The process is completed when the optimal degree of complexity is achieved. The GMDH neural network is constructed through the connection of a given number of elementary models (neurons) with the application of the appropriate selection methods. The main characteristic of the GMDH method is the separate parameter estimation of each neuron. The parameters of the neurons are estimated in such a way that their output signals are the best approximation of the real system output. The training proceeds with introducing an input pattern to the input layer and adapting the weights of each neuron according to a suitable learning algorithm. It is assumed that at least two input signals $u_1^{(l)}, ..., u_m^{(l)}$ constitute the stimulation which results in the formation of the neuron output signal $y_n^{(l)}$:

$$y_n^{(l)} = f(\boldsymbol{u}) = f(u_1^{(l)}, \ldots, u_m^{(l)}), \tag{1}$$

where $n = 1, \ldots, N$, and N is the number of neurons in the layer, $l = 1, \ldots, L$, and L denote the number of layers. The number of neurons N in the first layer of the network depends on the number of external inputs m. In a general case, a network of m inputs is built from neurons that have p inputs $(m > p)$. In this case, N new elements are formed in the following way:

$$N = \begin{pmatrix} m \\ p \end{pmatrix} = \frac{m!}{p!(m-p)!}. \tag{2}$$

The assumptions of the GMDH approach presented earlier lead to the formation of the neural network of a multi input $u_1, u_2, \ldots, u_m$ and a single output y. However, systems of the multi input $u_1, u_2, \ldots, u_m$ and the multi output $y_1, y_2, \ldots, y_R$ are found in practical applications most often. The synthesis of this model can be performed in a similar way like in the case of multi-input single-output models. In the first step of network synthesis, based on all combinations of inputs, the system output $y_1^{(1)}$ is obtained. Next, based on the same combinations of input signals, the remaining outputs $y_2^{(1)}, \ldots, y_R^{(1)}$ are obtained.

$$\begin{cases} y_{1,1}^{(1)} = f(u_1^{(1)}, u_2^{(1)}) \\ \quad \ldots \\ y_{1,m-1}^{(1)} = f(u_1^{(1)}, u_m^{(1)}) \\ y_{1,m}^{(1)} = f(u_2^{(1)}, u_3^{(1)}) \\ \quad \ldots \\ y_{1,2m-3}^{(1)} = f(u_2^{(1)}, u_m^{(1)}) \\ \quad \ldots \\ y_{1,N}^{(1)} = f(u_{m-1}^{(1)}, u_m^{(1)}) \\ \quad \vdots \\ y_{R,1}^{(1)} = f(u_1^{(1)}, u_2^{(1)}) \\ \quad \ldots \\ y_{R,m-1}^{(1)} = f(u_1^{(1)}, u_m^{(1)}) \\ y_{R,m}^{(1)} = f(u_2^{(1)}, u_3^{(1)}) \\ \quad \ldots \\ y_{R,2m-3}^{(1)} = f(u_2^{(1)}, u_m^{(1)}) \\ \quad \ldots \\ y_{R,N}^{(1)} = f(u_{m-1}^{(1)}, u_m^{(1)}) \end{cases} \tag{3}$$

The first layer of the GMDH network, for $p = 2$, has the structure shown in Fig. 1,

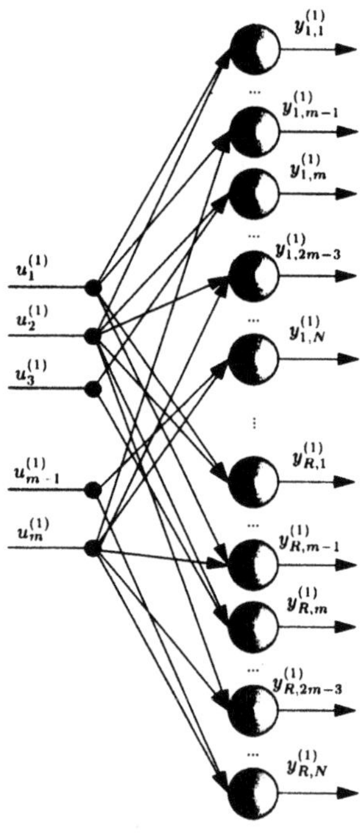

Fig. 1. First layer of the MIMO GMDH network

The definition of the evaluation criterion $Q_r(y_{r,n}^{(l)})$ of neurons is a preliminary task in designing a GMDH network [6]. It allows any neuron to define the quantity of a processing error of each output.

The independent evaluation of any of the processing errors $Q_1, Q_2, \ldots, Q_R$ is performed after the generation of each layer of neurons. Moreover, based on the defined evaluation criterion it is possible to make the selection of neurons in the layer:

$$\left\{\begin{array}{ll} Q_1(y_{1,1}^{(l)}), & \ldots, Q_R(y_{1,1}^{(l)}) \\ & \ldots \\ Q_1(y_{1,m-1}^{(1)}), & \ldots, Q_R(y_{1,m-1}^{(1)}) \\ Q_1(y_{1,m}^{(1)}), & \ldots, Q_R(y_{1,m}^{(1)}) \\ & \ldots \\ Q_1(y_{1,2m-3}^{(1)}), & \ldots, Q_R(y_{1,2m-3}^{(1)}) \\ & \ldots \\ Q_1(y_{1,N}^{(1)}), & \ldots, Q_R(y_{1,N}^{(1)}) \\ & \vdots \\ Q_1(y_{R,1}^{(l)}), & \ldots, Q_R(y_{R,1}^{(l)}) \\ & \ldots \\ Q_1(y_{R,m-1}^{(1)}), & \ldots, Q_R(y_{R,m-1}^{(1)}) \\ Q_1(y_{R,m}^{(1)}), & \ldots, Q_R(y_{R,m}^{(1)}) \\ & \ldots \\ Q_1(y_{R,2m-3}^{(1)}), & \ldots, Q_R(y_{R,2m-3}^{(1)}) \\ & \ldots \\ Q_1(y_{R,N}^{(1)}), & \vdots \ldots, Q_R(y_{R,N}^{(1)}) \end{array}\right. \tag{4}$$

The selection of best performing neurons in terms of their processing accuracy in layer is realized with the application of selection methods [4], before the formed layer is added to the network. According to the chosen selection method, elements that introduce too big a processing error of each output $y_1, \ldots, y_R$ are removed (Fig. 2).

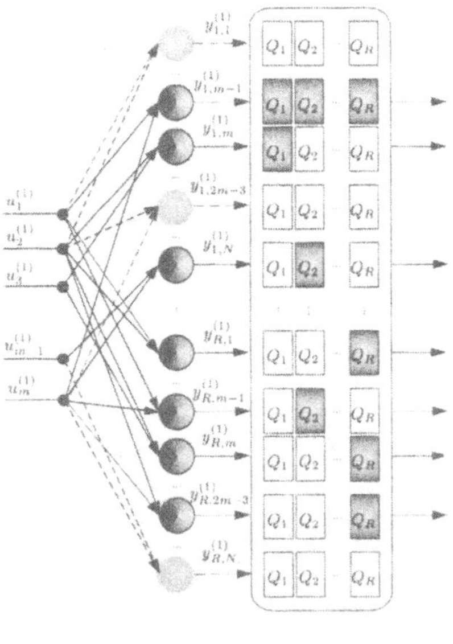

Fig. 2. Selection in the MIMO GMDH network

The effectiveness of a neuron in processing at least one output signal is sufficient to leave the neuron in the network. Based on all selected neurons, a new layer is created. The parameters of the neurons in the newly created layer are „frozen" during further network synthesis. The outputs of the selected neurons become the inputs to other neurons in the next layer:

$$\left\{\begin{array}{l} u_1^{(l+1)} = y_{r,1}^{(l)} \\ u_2^{(l+1)} = y_{r,2}^{(l)} \\ \quad \ldots \\ u_m^{(l+1)} = y_{r,N}^{(l)} \end{array}\right. \tag{5}$$

During the synthesis of the next layers, all selected outputs from the previous layer must be used to generate each output $y_1, \ldots, y_R$. It follows from the fact that in real industrial systems outputs are usually correlated, so the output $y_{r,n}^{(l)}$ should be obtained based on all potential outputs $y_{1,1}^{(l-1)}, \ldots, y_{R,N}^{(l-1)}$. In analogous way, new layers of the network are introduced. During the synthesis of the GMDH network, the number of layers suitably increases. Each time when a new layer is added, new neurons are introduced. The synthesis of the GMDH network is completed when the *optimum criterion* is obtained (Fig. 3).

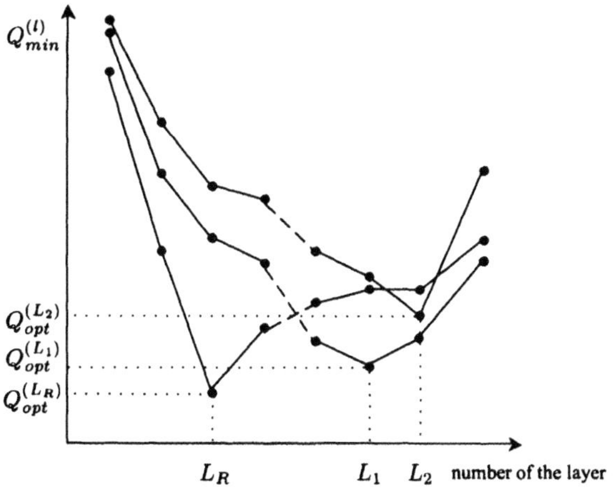

Fig. 3. Application of the optimal criterion for a multi-output network

The idea of this criterion is the determination of the quality index $Q_r(y_{r,n}^{(l)})$, for all N neurons included in the l layer, for each system output $y_1, \ldots, y_R$ independently:

$$Q_{r,min}^{(l)} = \min_{n=1,\ldots,N} Q_r(y_{r,n}^{(l)}) \text{ for } r = 1, \ldots, R. \tag{6}$$

The $Q_{r,min}^{(l)}$ represents the processing error of the best neuron in this layer for the output y_r.

The values $Q_r(y_{r,n}^{(l)})$ can be determined with the application of the defined evaluation criterion used in the selection process. The values $Q_{r,min}^{(l)}$ are calculated for each layer in the network. The *optimum criterion* is obtained when the following condition occurs:

$$Q_{opt}^{(L_r)} = \min_{l=1,\ldots,L_r} Q_{r,min}^{(l)} \quad \text{for} \quad r = 1,\ldots,R. \quad (7)$$

In other words, the synthesis of the network is completed when each of the calculated values $Q_{r,min}^{(l)}$ reaches the minimum.

The output y_r is connected to the output of the neuron for which $Q_{r,min}^{(l)}$ achieves the least value. The particular minimum could occur in different stages of network synthesis. This is why in the multi-output network, the outputs of the resulting structure are usually in different layers (Fig. 4).

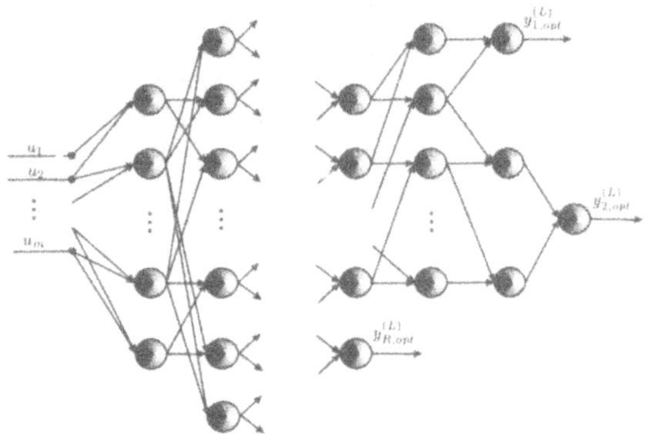

Fig. 4. Final structure of the MIMO GMDH network

The procedure of removing unnecessary neurons is the last stage of the synthesis of the GMDH neural network.

3 Simulation example

The objective of this section is to design an MIMO model of the actuator according to the approach described in the previous section based on the data gathered from a real plant. The data set is available on the web site the Research Training Network *Development and Application of Methods for Actuator Diagnosis in Industrial Control Systems* (DAMADICS) [1]. The RTN DAMADICS is focused on the diagnosis of valve plant actuators and looks towards real implementation methods for new actuator systems, and in particular, on the diagnosis of a 5-stage evaporation and boiler station process of the Lublin Sugar Factory S.A. in Poland. The element selected for modelling and fault detection, the third actuator, is a final control element, which interacts with the controlled process. The input of the actuator is the output of the process controller (flow or level controller) and the actuator modifies the position of the valve allowing a direct effect on the primary variable in order to follow the flow or level set-point. Each actuator is equipped with the similar measuring devices shown in Fig. 5.

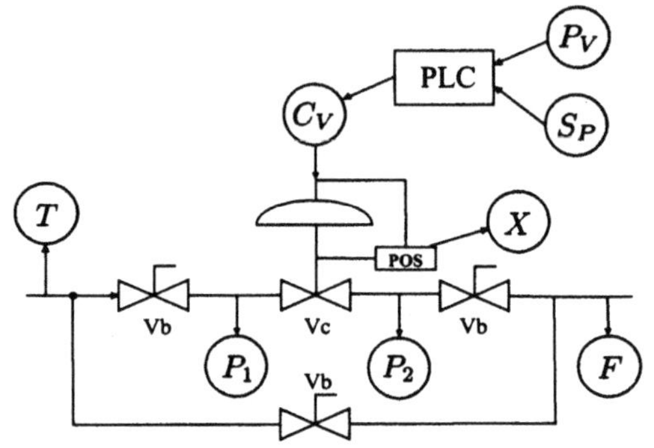

Fig. 5. Diagram of the actuator with measured variables

The actuator itself can be considered as a four-input and two-output system $(X, F) = r_{X,F}(C_V, P_1, P_2, T)$, where $r_{X,F}(\cdot)$ denotes the modelled relationship, and all process variables are described in Tab. I.

Table I. Specification of process variables

X	servomotor rod displacement
F	juice flow at the outlet of the valve
C_V	control value
P_1	pressure at the inlet of the valve
P_2	pressure at the outlet of the valve
T	juice temp. at the inlet of the valve

For the sake of comparison, the MISO $X = r_X(C_V, P_1, P_2, T)$ and $F = r_F(C_V, P_1, P_2, T)$ models were obtained. The data used for the identification and fault detection sets were collected on 17th November 2001. Below, in Tab. II., a detailed description of the introduced artificial faults is given.

Table II. Index of artificial faults introduced in the actuator

Fault	Tag sample	Fault description
no fault	1–20000	training data set
no fault	20001–30000	selection data set
F_{17}	53780–53794	unexpected pressure drop across the valve
F_{19}	55482–55517	flow rate sensor fault

The selection of best performing neurons in terms of their processing accuracy was realized with application of the soft selection method [4]. The evaluation criterion based on the regularity criterion [6] was defined where the identification data set was separated in two parts. Unfortunately, the data turned out to be sampled too fast. Thus, every 10-th value was picked resulting in the 2000-th elements training and 1000-th selection data sets. The dynamics in the GMDH network were realized with application of dynamic neurons with the *Infinite Impulse Response* (IIR) filter [5]. Such a neuron model consists of two submodules: the filter module and the activation module. The behaviour of the filter module is described by the following equation:

$$\begin{aligned}\tilde{y}(k) = &- a_1\tilde{y}(k-1) - \ldots - a_{n_d}\tilde{y}(k-n_d)\\ &+ v_{1,0}u_1(k) + v_{2,0}u_2(k) +, \ldots,\\ &+ v_{m,0}u_m(k)\\ &+ v_{1,1}u_1(k-1) + v_{2,1}u_2(k-1)\\ &+, \ldots, +v_{m,1}u_m(k-1)\\ &\ldots\\ &+ v_{1,n_d}u_1(k-n_d) + v_{2,n_d}u_2(k-n_d)\\ &+, \ldots, +v_{m,n_d}u_m(k-n_d),\end{aligned} \tag{8}$$

or, equivalently,

$$\begin{aligned}\tilde{y}(k) = &- a_1\tilde{y}(k-1) - \ldots - a_{n_d}\tilde{y}(k-n_d)\\ &+ \boldsymbol{v}_0^T\boldsymbol{u}(k) + \boldsymbol{v}_1^T\boldsymbol{u}(k-1) +, \ldots,\\ &+ \boldsymbol{v}_{n_d}^T\boldsymbol{u}(k-n_d),\end{aligned} \tag{9}$$

where $a_1, a_2, \ldots, a_{n_d}$ denotes the feedback filter parameters, n_d represent the filter order, $\boldsymbol{v}_0^T, \boldsymbol{v}_1^T, \ldots, \boldsymbol{v}_{n_d}^T$ is the input vector to the filter module, and m is the number of inputs to the dynamic neuron. The filter output is used as the input for the next module, namely, the activation module. The activation module can be described by

$$y(k) = \xi(\tilde{y}(k)). \tag{10}$$

In this way, each neuron in the network reproduces the output signal based on the past values of its inputs and outputs. The main objective is to estimate the parameters $a_1, a_2, \ldots, a_{n_d}$ and the parameter vectors $\boldsymbol{v}_i^T$, $i = 0, \ldots, n_d$. Using the fact that $\xi(\cdot)$ is an invertible activation function, the parameter estimation problem can be solved by means of the well-developed linear parameter estimation approaches [10]. The order of the IIR filters in each layer of the network was obtained with Lipschitz index based on so-called Lipschitz quotients [7], which in the multi-dimensional case (where $\varphi_1 = u_1(k), \ldots, \varphi_n = \tilde{y}(k-n_d)$) are defined as

$$l_{ij}^{(p)} = \frac{|y(i) - y(j)|}{\sqrt{(\varphi_1(i) - \varphi_1(j))^2 + (\varphi_n(i) - \varphi_n(j))^2}}, \tag{11}$$

for $i = 1, \ldots, n_t$, $j = 1, \ldots, n_t$ and $i \neq j$, where n_t is the number of samples in the data set and the superscript (p) in $l_{ij}^{(p)}$ stands for the number of inputs. The Lipschitz index can be defined as the maximum occurring Lipschitz quotient

$$l^{(n)} = \max_{i,j,(i\neq j)} (l_{ij}^{(n)}). \tag{12}$$

In the general case, the index is large if several inputs $\varphi_1, \ldots, \varphi_n$ are missing and is small otherwise. In other words, the correct order of the IIR filter can be detected when the index Lipschitz stops to decrease. The quality indexes for the training and selection sets, obtained with the application of the least square method, can be defined in the same way: $Q_{t,s} = \sum_{k=1}^{n_{t,s}}(y(k) - y_{opt}^{(L)}(k))^2 / \sum_{k=1}^{n_{t,s}} y(k)^2$, and have numerical values as given in Tab. III.

Table III. Quality indexes for the training and selection data sets

GMDH network	Q_t	Q_s
$X = r_{X,F}(C_V, P_1, P_2, T)$	0.0016	0.0015
$F = r_{X,F}(C_V, P_1, P_2, T)$	0.0031	0.0041
$X = r_X(C_V, P_1, P_2, T)$	0.0016	0.0016
$F = r_F(C_V, P_1, P_2, T)$	0.0027	0.0054

The responses of the system and the MIMO GMDH model obtained for the selection data set is given in Fig. 6.

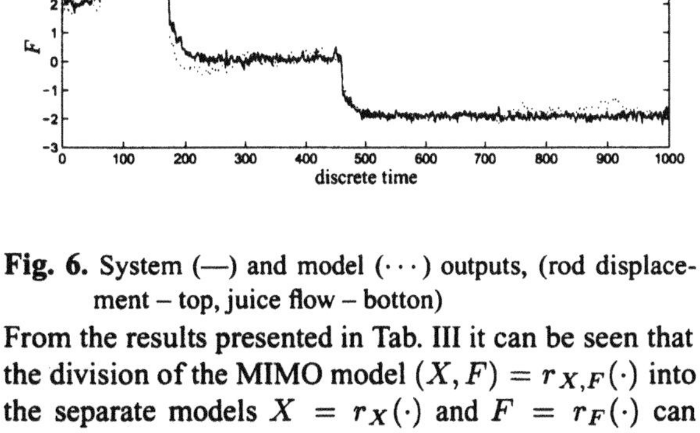

Fig. 6. System (—) and model (· · ·) outputs, (rod displacement – top, juice flow – botton)

From the results presented in Tab. III it can be seen that the division of the MIMO model $(X, F) = r_{X,F}(\cdot)$ into the separate models $X = r_X(\cdot)$ and $F = r_F(\cdot)$ can deteriorate modelling performance.

It follows from the fact that in real industrial systems outputs are usually correlated (Fig. 6), so during the synthesis of the next layers of the GMDH network, all selected potential outputs X and F from the previous layer should be used to generate each output X and F in the next layer.

Another objective of this work was to develop a residual generation scheme based on the presented identification framework. The residual signals for the faults F_{17} and F_{19} described in Tab. II. were determined by the comparison of the measured values and models outputs X and F respectively. The residual signals for the GMDH model are shown in Figs. 7-8.

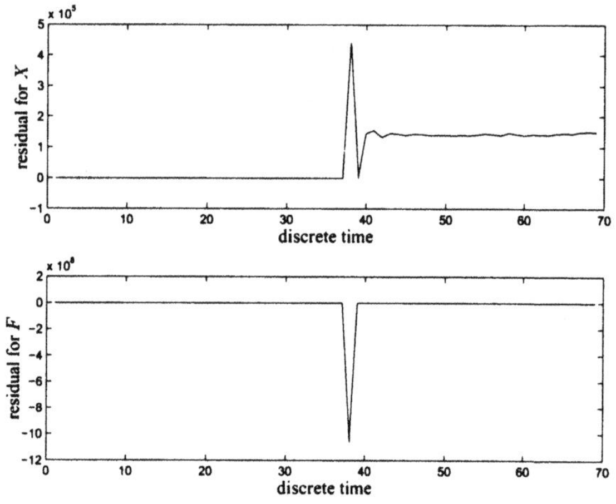

Fig. 7. Residual signals for the fault F_{17}

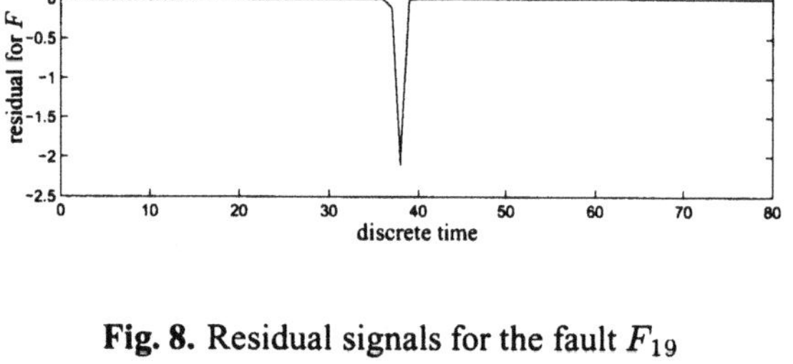

Fig. 8. Residual signals for the fault F_{19}

The occurrence of a fault is signalled by the deviation of the residual value from zero. For this reason the minimization of the identification error in the fault-free mode, as in the case of the proposed MIMO GMDH model, is of a great importance. As can be seen the fault is very easy to detect. e.g. using a simple threshold technique.

4 Conclusions

The objective of this paper was concerned with modelling an industrial process, especially with obtaining dynamical models directly from the observed data. To tackle this problem, the multi-dimensional GMDH neural network was employed. From the presented results it can be seen that the proposed non-linear MIMO model identification approach can be effectively applied to various multi-dimensional systems identification tasks. Finally, the proposed identification technique was applied to the identification of a chosen part of the evaporation station at the Lublin Sugar Factory, and the resulting model was used to design a residual generation scheme.

References

[1] DAMADICS (2002). Website of the Research Training Network *DAMADICS: Development and Application of Methods for Actuator Diagnosis in Industrial Control Systems*. http://diag.mchtr.pw.edu.pl/damadics/.

[2] Farlow, S. J. (1984). *Self-organizing Methods in Modelling – GMDH Type Algorithms*. Marcel Dekker, New York.

[3] Ivakhnenko, A.G., Muller, J.A. (1995). Self-organizing of nets of active neurons. *System Analysis Modelling Simulation*, Vol 20: pp. 93-106.

[4] Mrugalski, M., Arinton, E., Korbicz, J. (2002). *Methods of neuron selection in synthesis of GMDH neural networks*. Proc. 14th National Conf. on Automatic Control, June 24-27, Zielona Góra, Poland, pp. 845-850, (in Polish)

[5] Mrugalski, M., Witczak, M. (2002). Parameter estimation of dynamic GMDH neural networks with the bounded-error technique. *Journal of Applied Computer Science*, Vol 10, No 1: pp. 77-90.

[6] Mueller, J.E., Lemke, F. (2000). *Self-organising Data Maining*. Libri, Hamburg.

[7] Nelles, O. (2001). *Non-linear Systems Identification. From Classical Approaches to Neural Networks and Fuzzy Models*. Springer, Berlin.

[8] Nuck, D., Klawonn, F., Krause, R. (1997). *Foundations of Neuro-Fuzzy Systems*. John Wiley & Sons, Chichester.

[9] Patton, R.J., Frank, P., Clark, R.N. (2000). *Issues of Fault Diagnosis for Dynamic Systems*. Springer-Verlag, Berlin.

[10] Walter, E., Pronzato, L. (1997). *Identification of Parametric Models from Experimental Data*. Springer, Berlin.

Neural Network Control Approach for an Industrial Furnace

S. Martineau, E. Gaura, K.J. Burnham, O.C.L. Haas
Control Theory and Applications Centre
School of Mathematical and Information Sciences
Coventry University, Priory Street CV1 5FB, UK
Fax/Phone: +44 (0) 2476 888 052
Email: s.martineau@coventry.ac.uk, ctac@coventry.ac.uk

Abstract

The paper describes aspects of an on-going research programme in collaboration with AvestaPolarit (UK) Ltd, Sheffield, Corus Group plc, Rotherham and Eurotherm, Worthing. The work presented here involves the design and development of a neural network control strategy for continuously operated high temperature gas-fired industrial furnaces. At present, a bilinear control strategy is already applied routinely to the furnace. The results in this paper present the development of different neural network controllers compared to the bilinear controller.

1 Introduction

During the last decade, the Control Theory and Applications Centre (CTAC) of Coventry University has been involved, in collaboration with industry, in designing new control strategies to improve the temperature control of high temperature gas-fired industrial furnaces [1].

As for all practical systems, the furnaces exhibit nonlinear behaviour. Hence the required system performance, when use is made of a standard PID controller with fixed gains, is reduced; especially when the controller is required to operate over a region about the point of tuning. One solution to alleviate this problem is to use a bilinear controller, which has been developed recently to improve consistency in terms of performance over a range about the point of tuning [2].

The bilinear controller is nonlinear, involving products of system state and control. Whilst the bilinear controller offers improvement over a linear controller, further improvement in performance is sought and it was considered that neural network techniques may be good candidates for achieving this. The aim of this paper is to describe different structures of neural network controllers, i.e. Narma-L2 [3] and model predictive [4] controllers and compare their performances with the bilinear controller. In this context, computational time is addressed, together with implementation issues.

2 The furnace model and the bilinear controller

The system studied consists of three furnaces in series (labelled Furnace 1, 2 and 3) based at AvestaPolarit Ltd, Sheffield, UK and forms part of the annealing line used to process stainless steel strip [2]. The furnaces are designed to have high rates of gas circulation to provide uniform heating on the refractory walls. The walls then become radiant sources that provide heat to the strip. Each furnace is configured with three zones in series, each having its own temperature set point and associated controller. The furnace system is equipped with three Eurotherm T640 PID controllers, utilising one per furnace. Each controller has three individual PID loops covering the three zones in each furnace. At present the bilinear PID controller is implemented on Furnaces 2 and 3. A schematic illustration of the furnace system is given in Fig. 1.

The Bilinear controller which has been developed in collaboration with a number of industrial organisations in the UK, has shown energy savings of the order of 3% over the standard PID scheme, as well as an improvement in terms of product consistency [5, 6, 7, 8]. The bilinear controller is realised as a combination of a standard linear PID controller and a bilinear compensator. The resulting scheme is known as the BPID. Essentially the BPID is a four-term controller, which contains the existing three-term PID controller and an additional bilinear term, as illustrated in Fig. 2.

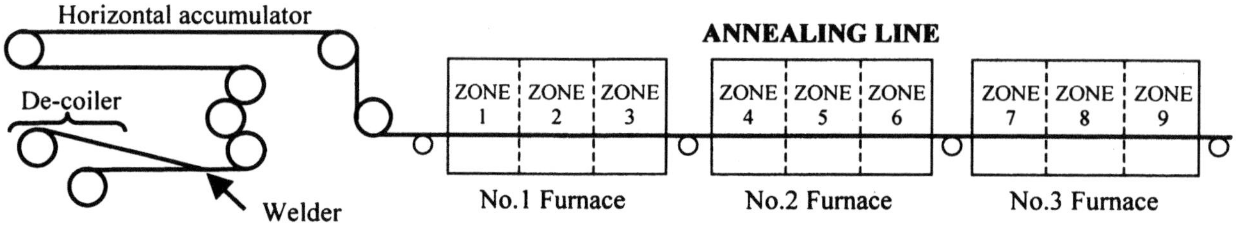

Fig. 1. Schematic illustration of furnace system based at AvestaPolarit.

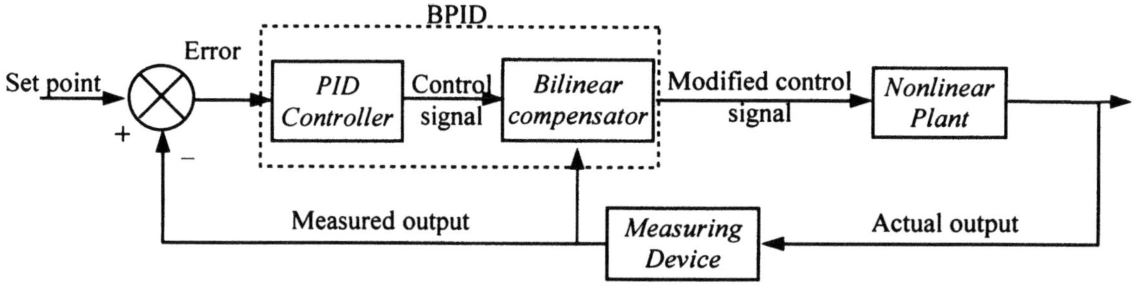

Fig. 2. Bilinear controller with PID structure for a nonlinear plant.

Before implementing the bilinear compensator in the Eurotherm T640, a mathematical model was used to simulate the continuously operated furnace under Matlab/Simulink. The same model has been used to design the neural network controller. The simulation model corresponds to a single zone.

3 Choice of a Neural Network controller

In an attempt to improve performance, different strategies to design a neural network controller were studied. The first approach considered was aimed at evaluating the complexity of a possible neural network controller, based on the existing BPID controller [7]. In this case, the neural network controller attempts to replicate the functionality of the existing nonlinear controller using collected simulation data. The multi-layer perceptron (MLP) architecture was chosen and the back-propagation method was used to train the neural network [9, 10].

The training sets were based on sampled data taken from the simulated system including the BPID controller (see Fig. 2). A set point has been applied to the simulated system and then the sampled signals (with error, delayed system output and delayed controller output as inputs) have been recorded for subsequent use in the training of the neural network controller. The best results with the neural network controller are obtained with a 3-input-1-output model and two hidden layers. The delayed controller and system outputs, as illustrated in Fig. 3, allow the neural network controller to capture the dynamics of the system.

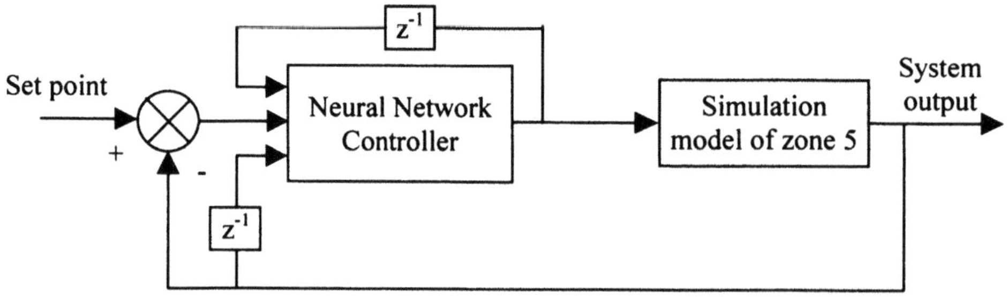

Fig. 3. Simulation model of Zone 5 with the neural network controller.

The integral absolute error (IAE) and the sum squared error (SSE) were used to compare the performance of neural network controller and the BPID controller. Neural networks of different architectures have been trained over 5000 epochs with different learning rates.

Fig. 4a shows the output of the closed loop system, in Fig. 3 when use is made of a MLP neural network with 3-9-7-1 architecture. This may be compared to the system output when use is made of the BPID controller (shown in Fig. 4a with a continuous time). The output in engineering units corresponds to a ratio of the actual temperature (i.e. 10 corresponds to 1000°C). The difference between the system output with the MLP 3-9-7-1 neural network controller and with the BPID controller is shown in Fig. 4b. It is clear that once the transients in the simulation have settled the MLP neural network controller is able to approximate the BPID controller, with a relative error of ±0.25%.

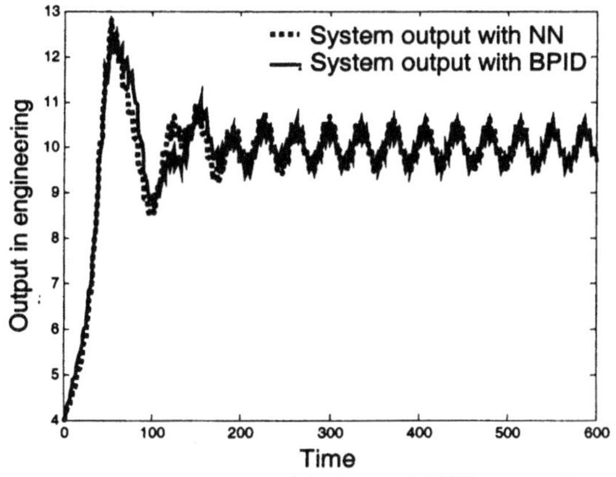

Fig. 4a. System output with NN and BPID controller.

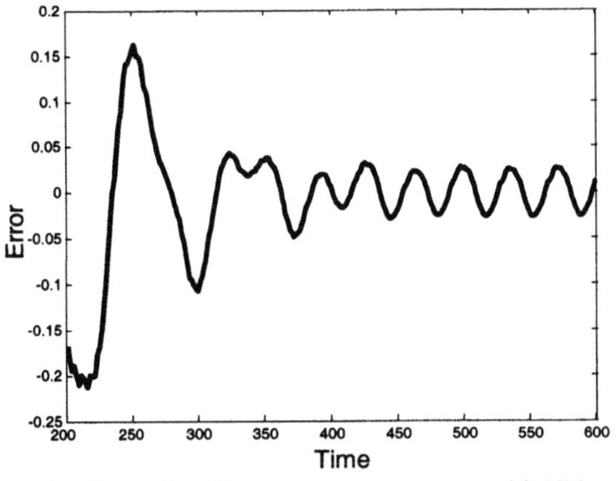

Fig. 4b. Error signal between system output with NN and BPID controller (from 200 to 600).

After evaluating the necessary complexity of a neural network controller able to approximate the bilinear controller, the development of independent neural network control architecture was the further step in the research. Design of a Narma-L2 controller [9] (adaptive control) and a neural network predictive controller [10] has led to interesting results which can be compared with the bilinear controller. Neural network toolboxes from Matlab were used to develop the controllers.

4 Development of a model predictive controller

High temperature furnaces have slow response characteristics for a given set point. The control action of the system depends on the opening of the valves on the burners. The control action required to reach a given set point, corresponding to a certain amount of gas, will be always be bounded such that the maximum opening of a valve is 100%. Furthermore, the slew rate of a valve will also limit the speed of response. These factors impose limitations on the possible variation and values of the control signal, which must be included in the simulation model. As the system is slow to react, using a predictive controller could be used to predict the future process outputs.

The control signal u' is chosen to minimize the quadratic performance criterion:

$$J = \sum_{j=N_1}^{N_2} (y_r(t+j) - y_m(t+j))^2 + \rho \sum_{j=1}^{N_u} (u'(t+j-1) - u'(t+j-2))^2$$

where N_1, N_2 and N_u define the horizons over which the tracking error and the control increments are evaluated. The u' variable is the tentative control signal, y_r is the desired response and y_m is the network model response. The value of ρ determines the contribution that the sum of the squares of the control increments has on the performance index.

A neural network model of the plant was constructed using data from the simulation model. Then through the optimization routine, the model predictive controller provides a control action to the system depending on the predictive horizon and the control weighting factor. Neural networks of different complexities were studied; the higher the number of units in the hidden layers, the larger the computational time but also the better the results. A model predictive controller with a longer prediction horizon and a small control-weighting factor provides good performance in terms of reduced error. However, the observation on the variation of the controller output provided an interesting result. Implementing such a controller on a real-time system would probably be prohibitive due to the fact that there are limitations on the incremental variation of the valves and the burners. Different options were studied to reduce this variation: changing the value of the weighting factor and/or introducing a percentage variation on the control signal. This limiter block, located immediately after the controller output, corresponds to a percentage of the previous value, e.g. if the value is set to 0.01, it means that the maximum variation between the previous value and the current one is ± 1%.

So far, better performances arise when use is made of a limiter block, i.e. a reduced overshoot and a reduction in the variation of the control signal are observed.

5 Development of a Narma-L2 controller

The Narma-L2 controller [9] requires less computational time compared to the model predictive controller. The controller is simply a rearrangement of the neural network plant model, which is trained offline, in batch form. The only online computation is a forward pass through the neural network controller, which transforms nonlinear system dynamics into linear dynamics by cancelling the nonlinearities. Several steps must be followed to apply the Narma-L2 controller within the system. The first step is to identify a neural network to represent the forward dynamics of the system. The nonlinear autoregressive moving average (NARMA) model is used to represent the system:

$$\hat{y}(k+d) = f[y(k), y(k-1), \ldots, y(k-n+1), \ldots, u(k-m+1)] + g[y(k), y(k-1), \ldots y(k-n+1), u(k-1), \ldots u(k-m+1)].u(k)$$

where $d \leq 2$. Using the NARMA-L2 model, the controller is obtained of the form:

$$u(k+1) = \frac{y_r(k+d) - f[Y,U]}{g[Y,U]}$$

where $\begin{cases} Y = [y(k), \ldots, y(k-n+1)] \\ U = [u(k), u(k-1), \ldots, u(k-n+1)] \end{cases}$

The performance of the Narma-L2 controller depends essentially on the identification of the system with a neural network. An accurate neural network model of the system provides good results in terms of set point tracking, hence reduced IAE values. However, as there is no factor to adjust the control weighting, use of a limiter on the control action appears to be necessary. The performances are hence reduced but the control signal variation is also reduced.

From Figure 5, it can be observed that the limiter block provides attenuation on the variation of the control signal. In steady state, it is found that the difference at the output is similar to that obtained with the BPID controller.

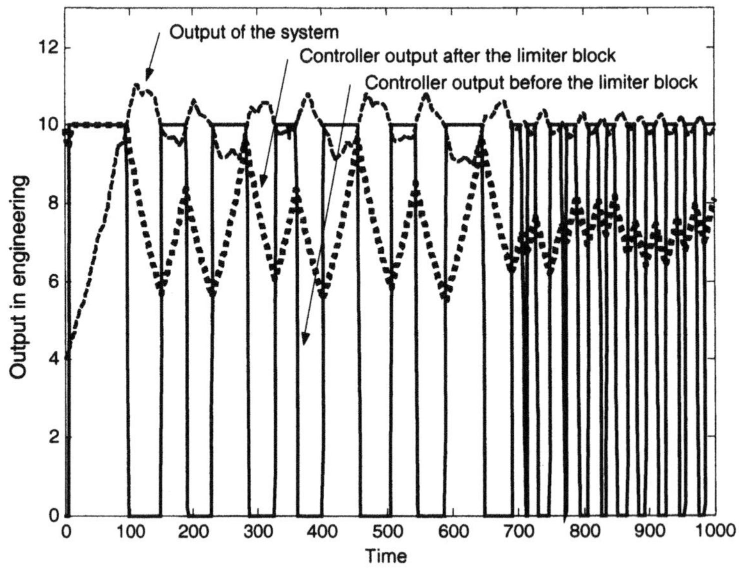

Fig. 5. Output of the system and controller output with a NARMA-L2 controller.

6 Conclusion

The paper has shown, firstly how a neural network may be trained to approximate an existing nonlinear controller, namely the BPID. The development of independent neural network control architectures, e.g. model predictive controller and Narma-L2 controller presented a further step in the research. In theory, predictive control, discussed here only, would probably provide benefits for a furnace system by reducing the use of gas. However, the model predictive controller requires larger computational time, and also requires the values for the weighting factor and the prediction horizons to be determined. It is believed that it could be difficult to justify the replacement of a conventional controller (PID controller) by such controller in a real industrial environment. Use of Narma-L2 controller, however, could provide a good compromise as the training of the neural network controller is done off-line. Performances in terms of IAE value, overshoot and time to reach the set point are very similar to those of the BPID controller.

Comparison studies between an existing BPID controller and potential neural network controllers are currently on-going. A further enhancement to the control strategy would be to consider a combined BPID neural network approach.

7 Acknowledgements

The authors wish to acknowledge the support of the Energy Technology Support Unit in pursing this work, as well as colleagues at AvestaPolarit Ltd, CORUS Group plc and Eurotherm Ltd, UK.

8 References

[1] Burnham K. J., Dunoyer A. and Marcroft S. (1999). Bilinear controller with PID structure, IEE Computing and Control Engineering Journal, v.10, n.2, p63-69.

[2] Minihan J.A. (2001). *Design of Bilinear Controllers for Industrial Plants*, PhD Thesis, Coventry University, Coventry, UK.

[3] Narendra, K.S. and Mukhopadhyay S. (1997). Adaptive Control Using Neural Networks and Approximate Models, IEEE Transactions on Neural Networks Vol. 8, p. 475-485.

[4] Hagan, M.T., De Jesus O., and Schultz R. (1999). Training Recurrent Networks for Filtering and Control, chap. 12 in Recurrent Neural Networks: Design and Applications, L. Medsker and L.C. Jain, Eds., CRC Press, p. 311-340.

[5] Martineau S., Minihan J.A., Burnham K.J., Eastell J.M., Rhine J.M., Marcroft S., Heeley A. and Andrews G. (2001a). Development of bilinear control strategies for industrial furnace applications, *International Gas Research Conference*, Amsterdam, Holland, CDROM.

[6] Martineau S., Burnham K.J., Minihan J.A., Marcroft S., Andrews G. and Heeley A. (2002). Application of a bilinear PID compensator to an industrial furnace, 15th IFAC World Congress on Automatic Control, Barcelona, Spain, CDROM.

[7] Martineau, S., Gaura, E., Burnham, K.J. and Haas, O.C.L. (2002). Neural network control approach for an industrial furnace, *Proceedings of the 14th International Conference on Systems Science*, Las Vegas, USA, p227-233.

[8] Martineau, S. (2002). Real-time results of a bilinear PID controller applied to an industrial furnace, *Proeedings of UKACC Control 2002, Postgraduate Symposium*, Sheffield, UK, p113-117.

[9] Haykin S. (1999). *Neural Networks, a comprehensive foundation, 2nd Edition*, Prentice-Hall International.

[10] Limanond, S. and Si, J. (1998). Neural-networks-based control design: an LMI approach, IEE Transactions on Neural Networks, v.9, No.6, p1422-1429.

An Inductive Inference Approach to Large Scale Text Categorisation

Catarina Silva [12] Bernardete Ribeiro[1]

[1]Departamento de Engenharia Informática, Centro de Informática e Sistemas, Universidade de Coimbra, Coimbra 3030, Portugal, email:{catarina,bribeiro}@dei.uc.pt [2]Escola Superior de Tecnologia e Gestão - IPLeiria - Portugal

Abstract

Automatic text categorisation of documents has received a resounding interest in last years due to the increased availability of documents in digital form and the commanding need to organize them. In this paper, our main focus is the development of tools that will enable very fast and accurate text classifiers in large scale databases. To pursue this objective, we start by introducing the main issues of text categorisation and present possible ways of handling them. Kernel based methods, such as, Support Vector Machines (SVMs), are learning methods with strong potential for solving the tasks involved in automatic text categorisation. The first results achieved with Reuters-21578 collection are reported and some points of possible improvements are identified.

1 Introduction

Text categorisation can be defined as the assignment of natural language texts to one or more pre defined categories, based on their content. Automatic text categorisation can play an important role in information management tasks, such as text retrieval, routing and filtering. The growth of information availability, supported by the omnipresent Internet, has led to the necessity of better organizing the overloaded information. In many contexts, as Yahoo or spam categorisation for instance, trained professionals are employed to categorise new items. This process is very time consuming and costly, thus limiting its applicability, especially for large or rapidly changing collections [1]. Consequently there is an increased interest in developing technologies for automatic text categorisation. In this paper we explore the potential of SVMs for text categorisation. Section 2 will examine the text categorisation problem, setting guidelines for problem formulation. Section 3 will tackle the solid mathematical foundations of SVMs, first introduced by Vapnik [2]. Section 4 will explore the application of SVMs to automatic text categorisation with some state of the art examples. Empirical results will be addressed in section 5 and in the last section some conclusions and future lines of project research are presented.

2 Text categorisation

The task of text categorisation is the classification of natural text or hypertext (documents for simplicity) into a fixed number of pre defined categories based on their content. Each document can be assigned to none, one or several categories, making it a multi-class and multi-label problem, that is usually treated as several separate binary classification problems. Using a category-pivoted approach, for each category a document either belongs or does not belong to it. On the other hand, using a document-pivoted approach, each document belongs or not to each one of the available categories [3]. Which approach to follow depends on the kind of application one is running. If the number of categories is *a priori* fixed and the objective is to categorise new documents then a category-pivoted categorisation should be used. If one is in a category discovering phase, a document-pivoted categorisation should be followed.

To accomplish automatic text categorisation, the set of documents, typically strings of characters, has firstly to be converted to an acceptable representation that the learning machine can handle. Then a learning procedure has to take place. These steps are the concern of the following two subsections.

2.1 Document Representation

The most common, simple and successful representation so far is the vector space model, also known as the *Bag of Words*. Each document is indexed with the *bag* of the terms occurring in it, i.e., a vector with one component for each term occurring in the whole collection, having as value the number of times the term occurred in the document. Each document is thus represented as a point in a vector space with one dimension for every term in the vocabulary, thus loosing word order information.

The number of times a term occurs in a document is usually referred to as *term frequency* or tf. The method described so far gives equal importance to all terms. It can be however more effective to weight terms according to their discriminative power within the document collection. This is usually done based on the idea that terms occurring in fewer documents are better selectors. The *document frequency* $df(t)$ of a term t is the number of documents in the collection in which the term occurs. The *inverse document frequency* or $idf(t)$ is (1):

$$idf(t) = \log\left(\frac{|D|}{df(t)}\right), \qquad (1)$$

where $|D|$ is the number of documents in the collection.

Vector components are weighted according to the *idf* of the corresponding term. Usually some monotonous function of the tf, such as the log or the square root, is used instead of the tf itself, to avoid giving more importance than appropriate to multiple occurrence of terms [3].

Using the *Bag of Words* approach, and fairly all document representation schemes, a high dimensional space is reached, where the number of features is much larger than the number of documents available for training. This fact is inevitable, but can be mitigated by the use of some dimension reduction techniques, such as stop words and stemming.

In the simplest approach, a term is any space-separated word in a document. However, there is a great number of non-informative words, such as articles, prepositions and conjunctions, called *stop words*. For this reason, a *stop-list* it is usually built with words that should be filtered in the document representation process. Words that are to be included in the *stop-list* are language and task dependent. For instance, the word "bank" can refer to a financial institution, a piece of furniture or the side of a river [7]. Besides this, only words that appear in more than a number of documents are deemed relevant.

Another commonly used method is *stemming*, where the word stem is derived from the occurrence of a word by removing case and inflection information. For example "computes", "computing" and "computer" are all mapped to the same stem "comput".

Yet another dimensionality reduction technique is Latent Semantic Indexing, whose aim is to handle problems of synonymous[1] and polysemous[2] words. This technique compresses vectors representing documents into other vectors of a lower-dimensional space whose dimensions are obtained as combination of the original dimensions by looking at their patterns of co-occurrence.

[1] Words that mean the same, but are written differently.

[2] Words that are written the same way but have different contextual meanings.

2.2 Learning phase

Having a suitable representation, the goal is to find a mapping from the representation of documents into the class of possible labels. A classifier is a function that maps an input vector attribute vector $\mathbf{x} = (x_1, x_2, ...x_n)$ to the confidence that the input belongs to a class - that is $f(\mathbf{x}) = confidence(class)$. In the case of text categorisation, attributes are words in the document and classes correspond to text categories.

The problem of inductive construction of a text classifier has been tackled in a variety of different ways, such as [4]:

- Rule-Based Approach
- k-Nearest Neighbour
- Linear Models (like Rochio's method)
- Neural Networks
- Probabilistic Models (like Bayesian classifiers)

3 Support Vector Machines

As already referred, SVMs have been introduced by Vapnik [2] and are a promising approach to many machine learning problems, among which text categorisation. Formulation of SVMs is based on the *Structural Risk Minimization* principle, and not on the traditional *Empirical Risk Minimization* principle.

The huge difference between *Empirical Risk Minimization* and *Structural Risk Minimization* is that the first aims at reducing the training error, making it unreliable to assure that there will be a small test error; while bounding the second will indubitably bound the real error obtained when facing new unseen examples.

3.1 Foundations

Considering we want to estimate a function $f : \mathbb{R}^N \rightarrow \{\pm 1\}$ using input-output training data $(\mathbf{x}_1, y_1), ...(\mathbf{x}_l, y_l) \in \mathbb{R}^N \times \{\pm 1\}$, such that f will correctly classify unseen examples $(\mathbf{x}, y)$, generated from the same underlying probability distribution $P(\mathbf{x}, y)$ as the training data.

The Empirical Risk (or training error) can be defined as (2):

$$R_{emp}[f] = \frac{1}{l}\sum_{i=1}^{l}\frac{1}{2}|f(\mathbf{x}_i) - y_i|. \qquad (2)$$

Putting no restrictions on f, it is possible to choose a function that classifies well training data, but does not generalize well on test or real data, i.e., f will not have minimized the real *Risk* (3).

$$R[f] = \int \frac{1}{2}|f(\mathbf{x}) - y|dP(\mathbf{x}, y). \qquad (3)$$

Vapnik and Chervonenkis [2] (in the denominated *VC theory*) prove that it is necessary to restrict the class of functions that f is chosen from to one that has the *capacity* that is suitable for the complexity underlying training data. VC theory provides bounds on the test error, circumventing the generalization problems presented earlier. Minimizing these bounds leads to the principle of *Structural Risk Minimization.* A function's capacity can take the form of VC *dimension*, defined as the largest number h of points that can be separated in all possible ways, using functions of the given class. If $h < l$ is the VC dimension of the class of functions that the learning machine can implement, then for all the functions of that class, with a probability of at least $1-\eta$, the bound (4):

$$R(\alpha) \leq R_{emp}(\alpha) + \phi\left(\frac{h}{l}, \frac{\log(\eta)}{l}\right) \qquad (4)$$

holds, where the *confidence term* ϕ is defined as (5):

$$\phi\left(\frac{h}{l}, \frac{\log(\eta)}{l}\right) = \sqrt{\frac{h(\log\frac{2l}{h} + 1) - \log(\frac{\eta}{4})}{l}} \qquad (5)$$

3.2 Support Vector Classification

Although text categorisation is indeed, as referred, a multi-class, multi-label problem, it can be broken into a number of two class problems without loss of generality. This means that instead of classifying each document into all available categories, for each pair $\{document, category\}$ we have a two class problem: the document either belongs or does not to the category.

Consider the example in figure 1. Although there are several linear classifiers that can separate the two presented classes, only one, the Optimal Separating Hyperplane, maximizes the margin, i.e., the distance between it and the nearest data point of each class, thus presenting better generalization potential.

The formula for the output of a linear SVM is $u = \mathbf{w} \times \mathbf{x} - b$, where $\mathbf{w}$ is the normal vector to the hyperplane and $\mathbf{x}$ is the input vector. Maximizing the margin can be seen as an optimization problem:

$$\begin{aligned} &minimize \quad \frac{1}{2}||\mathbf{w}||^2, \\ &subjected\ to\ y_i(\mathbf{w} \times \mathbf{x} - b) \geq 1, \forall i, \end{aligned} \qquad (6)$$

where $\mathbf{x}_i$ is the training example and y_i is the correct output of the SVM for the ith training example, as represented in figure 1.

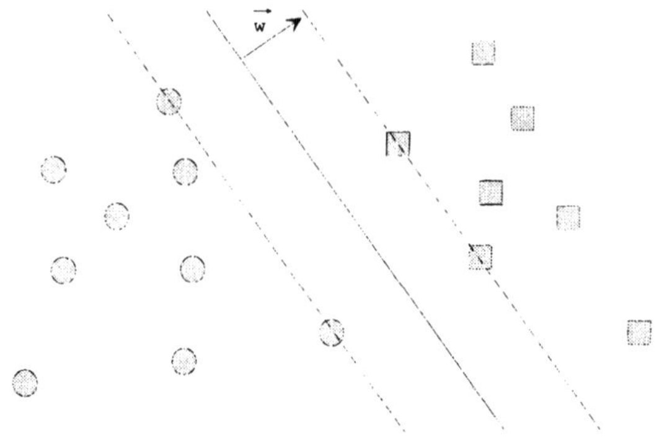

Fig. 1. Optimal Separating Hyperplane.

SVM are universal learners. In its basic form, shown so far, SVMs learn linear threshold functions. However, using an appropriate kernel function, they can be used to learn polynomial classifiers, radial-basis function networks and three layer sigmoid neural networks.

A note-worthy property of SVMs is that their ability to learn is independent of the dimensionality of the feature space. Complexity is based on the margin with which they separate the data, instead of the number of features, what will be very helpful in text classification matters.

4 Automatic Text categorisation with SVMs

There is a number of text categorisation properties that make it a suitable SVM application field [6]:

- High dimensional input space: fairly all document representation schemes have very high dimensional representation spaces (more that 10000 features). SVMs have, as mentioned in the last section, a complexity not proportional to the number of features.
- Few irrelevant features: the dimensionality reduction techniques referred in previous sections remove only a few non-informative words. It is easy to show by experiments that almost all features carry significant relevance to classification.
- Sparse document vectors: each vector representing a document has a very significant percentage of null values, since most words (all the words that appear in all the documents in the set) do not appear in each document.

5 Results

5.1 Data set

The Reuters-21578 collection was used (The collection is publicly available at: http:// kdd.ics.uci.edu/

databases/reuters21578/reuters21578.html). In this corpus there are about 12000 classified stories into more than 100 categories. Reuters-21578 is a financial corpus with news articles averaging 200 words each. Example categories are trade, earn or crude.

The ModApte split was used, using 75% of the articles (9603 items) for training and 25% (3299 items) for testing.

Table I presents the ten most frequent categories, comprising 75% of the training items.

Category	Train	Test
Earn	2715	1044
Acquisitions	1547	680
Money-fx	496	161
Grain	395	138
Crude	358	176
Trade	346	113
Interest	313	121
Ship	186	89
Wheat	194	66
Corn	164	52

Table I. Ten most frequent categories training and test documents

Reuters is a very heterogeneous corpus, since there are stories not assigned to any of the categories and stories assigned to 12 categories. Moreover there are categories with only one assigned document and others with thousands of assigned documents.

The documents pre-processing was carried out with the Lemur Toolkit for Language Modelling and Information Retrieval. The toolkit is being developed as part of the Lemur Project, a collaboration between the Computer Science Department at the University of Massachusetts and the School of Computer Science at Carnegie Mellon University. This software is available at http://www-2.cs.cmu.edu/ lemur/.

A list of stop words was used to filter the documents and Porter stemming was applied. The documents were represented by the frequency of the terms that occurred in each one and this representation was scaled to avoid result disturbance, but no further manipulations were carried out.

5.2 Performance *Criteria*

For evaluating the simulation results besides accuracy, two other performance *criteria* were used: Recall and Precision. Take note that all these measures are taken in the test set for each category.

Recall is the percentage of total documents for the given topic that are correctly classified (7).

$$recall = \frac{categories\ found\ and\ correct}{total\ categories\ correct} \qquad (7)$$

Precision is the percentage of predicted documents for the given topic that are correctly classified (8).

$$precision = \frac{categories\ found\ and\ correct}{total\ categories\ found} \qquad (8)$$

An alternative representation is the use of the number of true positives (f_{++}), false positives (f_{+-}) and false negatives (f_{-+}) as depicted is equations (9) and (10).

$$recall = \frac{f_{++}}{f_{++} + f_{-+}}, \qquad (9)$$

$$precision = \frac{f_{++}}{f_{++} + f_{+-}}. \qquad (10)$$

The point at which recall equals precision is called the *break-even* point and it is considered a good compromise. As the text categorisation multi-class problem has been sub-divided in several two-class problems, averaging has to be used to find total precision and recall values. There are two types of averaging: *micro-averaging* and *macro-averaging*. In *micro-averaging*, performance tables for each of the categories are added and recall and precision are computed. In *macro-averaging*, performance measures are computed separately for each category and the mean of the resulting performance is taken.

Table II presents the values reached for accuracy in the 10 referred categories.Using *macro-averaging* an encouraging, yet misleading, value of 97.31 is reached.

Category	Accuracy
Earn	95.92
Acquisions	94.93
Money-fx	96.13
Grain	97.96
Crude	97.04
Trade	97.64
Interest	97.15
Ship	98.45
Wheat	98.77
Corn	99.08

Table II. Accuracy values for the most significant categories.

Table III presents the values for precision and recall. Using again *macro-averaging* we found a precision 85.91 of and a recall of 62.54. While precision values remain encouraging there is yet room for improvements with recall, as can be confirmed in table IV.

Examining the results achieved, it is possible to say that there is an average significant number of false negatives, that will reduce recall as can be acknowledged by

Category	Precision	Recall
Earn	93.53	95.50
Acquisions	93.09	85.15
Money-fx	71.43	52.8
Grain	92.55	63.04
Crude	84.85	63.64
Trade	79.49	54.87
Interest	77.03	47.11
Ship	89.36	51.85
Wheat	84.44	57.58
Corn	93.33	53.85

Table III. Precision and recall values for the most significant categories.

Category	False Positives	False Negatives
Earn	69	47
Acquisions	43	101
Money-fx	34	76
Grain	7	51
Crude	20	64
Trade	16	51
Interest	17	64
Ship	7	28
Wheat	5	39
Corn	2	24

Table IV. False positives and false negatives for the most significant categories.

equation (9). As the data set is very large and the number of positive examples for each category is small (between 30% for Earn and 1,4% for Corn) a high accuracy can be reached betting always on the negative answer (a document does not belong to a category).

6 Conclusions

SVMs present good results in automatic text categorisation, since their objective is to maximize the margin between positive and negative examples for each class.

It remains to be studied how the document representation and the SVM parameters can positively influence specially the recall results, since there are still a significant number of false negatives.

Other work is also foreseen with new corpora for result validation.

Acknowledgements

CISUC - Center of Informatics and Systems of University of Coimbra and Project POSI/SRI/41234/2001 are gratefully acknowledged for partial financing support.

References

[1] S. Dumais, J. Platt, D. Heckerman, "Inductive Learning Algorithms and Representations for Text categorisation", in Proceedings of the ACM-CIKM98, pp. 148-155, 1998.

[2] V. Vapnik, "The Nature of Statistical Learning Theory", 2nd edition, Springer, 1999.

[3] F. Sebastiani, "A Tutorial on Automated Text categorisation", in Analia Amandi and Alejandro Zunino (eds.), Proceedings of ASAI-99, 1st Argentinian Symposium on Artificial Intelligence, Buenos Aires, AR, pp. 7-35, 1999.

[4] J. Kwok, "Automated Text categorisation Using Support Vector Machine", in Proceedings of the International Conference on Neural Information Processing (ICONIP'98), pp. 347-351, Kitakyushu, Japan, 1998.

[5] B. Schölkopf, C. Burges, A. Smola, "Advances in Kernel Methods - Introduction to Support vector Learning", MIT Press, pp. 1-15, 1999.

[6] T. Joachims, "Text categorisation with Support Vector Machines: Learning with Many Relevant Features", in Proceedings of the European Conference on Machine Learning (ECML), Springer, pp. 137-142, Berlin, 1998.

[7] T. Joachims, "Learning to Classify Text Using Support Vector Machines - Methods, Theory and Algorithms", in The Kluwer International Series in Engineering and Computer Science, Kluwer Academic Publishers, 2001.

[8] S. Gunn, "Support Vector Machines for Classification and Regression", Technical Report, Faculty of Engineering and Applied Science, Department of Electronics and Computer Science, 1998.

The accurate estimation of meteorological profiles employing ANNs

Tatiana Tambouratzis   Myrsini Gazela

Institute of Nuclear Technology – Radiation Protection, NCSR 'Demokritos'
Aghia Paraskevi 153 10, Athens, Greece
e-mail tatiana@ipta.demokritos.gr

Abstract The lack of meteorological measurements at a location of interest (target location) constitutes a problem that is crucial for the purposes of both weather forecasting and energy system design/validation. This paper constitutes a pilot study for the accurate estimation of meteorological values at a target location employing the meteorological measurements collected at a nearby location (reference location). Artificial neural networks (ANNs) are investigated and compared with traditional estimation methods. The significance of the improvement obtained via the ANN approach both over the traditional estimation methods as well as over simply considering the measurements at the reference location is demonstrated in a number of energy applications.

1. Introduction

The accurate knowledge of meteorological parameters (ambient temperature, solar radiation, relative humidity, wind speed and wind direction) at a location of interest (target location) is significant in terms of both meteorology and energy applications. For instance, the exact prediction of the current/near-future ambient temperature can prevent undesirable phenomena caused by ice on street surfaces, while it can also restrict the destruction of crops as well as the damage of sensitive and expensive instruments. Concerning energy applications, the accurate knowledge of the meteorological profile of a target location is necessary in order for a weather-driven system to be successfully designed, installed and profitably operated. The meteorological profile is commonly expressed via the Typical Meteorological Year. This comprises typical hourly values, spanning Julian Days 1 to 365, of ambient temperature, solar horizontal radiation, and – less importantly - relative humidity, wind speed and wind direction. The typical hourly values of the Typical Meteorological Year are derived from the hourly measurements collected long-term (over no less than 20 years) at the target location, after the extreme measurements relating to extraordinary weather conditions have been discarded [1]. Unfortunately, the evaluation of a Typical Meteorological Year at the target location is not only time consuming but also expensive, as it requires the installation of a weather station at the target location and the long-term collection of the meteorological parameters of interest. A commonly employed solution is to use the Typical Meteorological Year of a nearby reference location, at which data collection is performed anyway (e.g. at a weather station). Often, however, the reference and target locations have distinct meteorological profiles, whereby the available Typical Meteorological Year of the reference location may be inappropriate for calculating the long-term performance of a system to be installed at the target location. An alternative methodology is proposed here:

I. Perform short-term collection of meteorological data at the target location.
II. Create an estimator of the meteorological data of the target location employing (a) the data collected during step I, and (b) the corresponding data of a nearby reference location.
III. Construct an "approximate" Typical Meteorological Year of the target location employing the Typical Meteorological Year of the reference location as input to the estimator of step II.
IV. Evaluate the long-term performance of the weather-driven system to be installed at the target location using the "approximate" Typical Meteorological Year of step III.

This paper is focused on the estimation of ambient temperature (T, measured in ^{o}C) and solar horizontal radiation (I, measured in W/m^2) at a target location employing meteorological measurements at a nearby reference location. As mentioned before, T and I constitute the most salient meteorological parameters in terms of both meteorology and energy applications. ANN [2] estimators are developed and compared with traditional estimation methods. The improvement effectuated in the estimation of the weather parameters at the target location over simply assuming those of the reference location is determined by evaluating the performance of two solar hot water systems. Additionally, the possibility of employing the estimator of step II in order to construct an "approximate"

Typical Meteorological Year for the target location is examined.

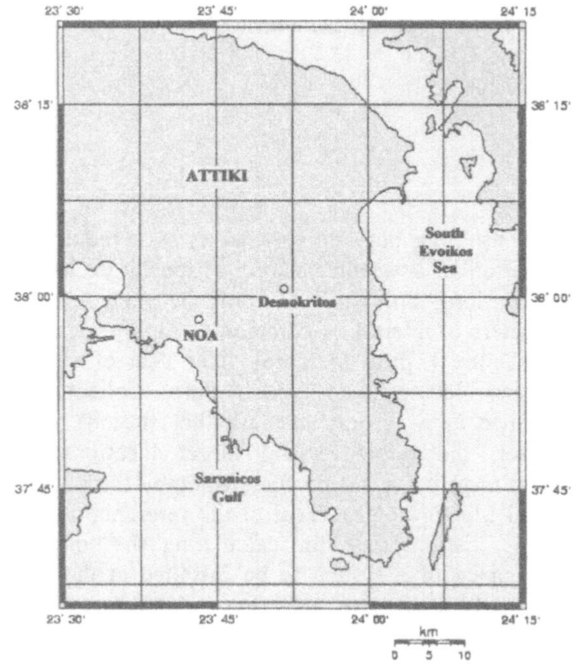

Fig. 1 The map of Attiki with the two weather stations.

2. Data

Fig. 1 illustrates part of Attiki, Greece, with the two weather stations located at (i) the National Observatory of Athens (NOA) and (ii) NCSR 'Demokritos' (DEM). The two locations are near each other (less than 15km distance as-the-crow-flies) but have distinct morphological – and, thus, also meteorological – profiles. The dataset employed here consists of hourly values of ambient temperature T and solar horizontal radiation I collected at the two stations during 1997 and 1998: the values of 1997 have been reserved for training the various estimation techniques, while those of 1998 have been employed for testing. In order to evaluate the prediction accuracy of the implemented estimation methods and to examine their influence on energy calculations, the improvement effectuated by the estimates over simply considering the measurements at the reference location has been employed. The standard error of estimates has been utilised for estimating the accuracy of each estimates implemented/employed.

In the following, T_{NOA} and T_{DEM} (both measured in °C) denote the hourly values of ambient temperature collected at the National Observatory of Athens and at NCSR 'Demokritos', respectively; I_{NOA} and I_{DEM} (both measured in W/m^2) denote the hourly values of solar horizontal radiation collected at the National Observatory of Athens and at NCSR 'Demokritos', respectively.

3. Parameter estimation

3.1 Estimation of T_{DEM}

Only T_{NOA} has been employed for the estimation of T_{DEM}. This is owing to the high value of the linear correlation coefficient between T_{NOA} and T_{DEM} (ρ=0.977) and the significantly lower values between T_{NOA} and the other meteorological variables at the reference location: $\rho(T_{NOA},H_{NOA})$=-0.58 for hourly relative humidity H; $\rho(T_{NOA},I_{NOA})$=0.53; $\rho(T_{NOA},WS_{NOA})$=-0.04 for hourly wind speed WS.

Observation of the (T_{NOA},T_{DEM}) pairs of the training set (Fig. 2) reveals distinct patterns of behaviour during the daylight and night hours, thereby prompting the application of problem decomposition into the daylight partition (t=7:00 to 18:00) and the night partition (t=19:00 to 6:00).

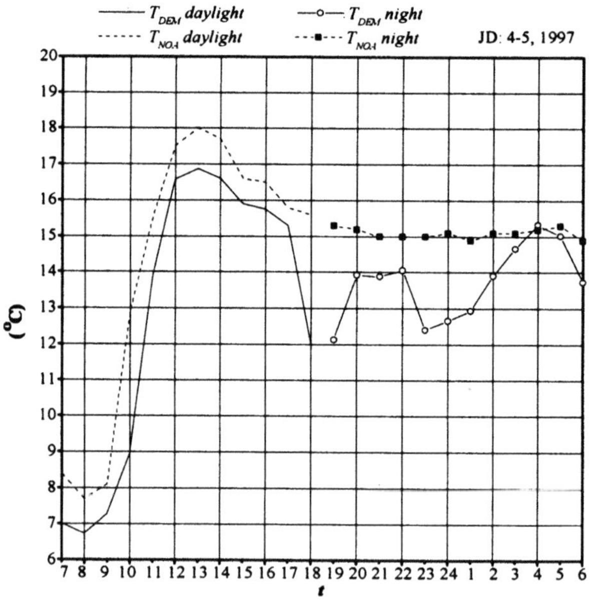

Fig. 2 The distinct patterns of behaviour of the T_{NOA} and T_{DEM} time series for the daylight and night partitions; example for Julian Day (JD) 4-5 of 1997 (training set).

The most accurate ANN approach for estimating T_{DEM} is composed of two 12-8-4-8-12 auto-associative ANNs (AANNs) trained via batch mode BP training [4], each dedicated to one partition: 12

consecutive daylight/night T_{NOA} values constitute the inputs to the AANNs, while the 12 corresponding consecutive daylight/night $\hat{T}_{DEM}$ values constitute the outputs. Training of both AANNs was terminated when the error of the training and validation sets stopped decreasing.

The most accurate alternative methods for estimating T_{DEM} are the 1st order memory-less linear model with problem decomposition, the 36th order linear model with memory and no problem decomposition and the non-linear model with problem decomposition.

Estimators	1997	1998
T_{NOA}	2.6	2.6
ANN	1.6	1.3
1st order linear model	1.8	1.4
36th order linear model	1.8	1.4
Non-linear model	1.7	1.6

Table 1. Standard error of estimates (SEEs) between T_{DEM} and T_{NOA} as well as the $\hat{T}_{DEM}$ given by the implemented estimation methods.

All of the implemented estimation methods effectuate an improvement over $\hat{T}_{DEM} = T_{NOA}$, with the ANN approach demonstrating the lowest standard error of estimates (Table 1). Furthermore, the maximum and mean differences $|\Delta T|$ between actual T_{DEM} values and the various $\hat{T}_{DEM}$ estimates have been found significantly higher for $\hat{T}_{DEM} = T_{NOA}$ than for any of the implemented estimation methods: an improvement (over $\hat{T}_{DEM} = T_{NOA}$) of around 6°C for the maximum differences and of 1.1°C for the average differences is demonstrated by the ANN approach, with smaller improvements (2-3°C and 1°C, respectively) effectuated by the other implemented estimation methods. An analogous improvement is observed throughout the relative cumulative frequencies of $|\Delta T|$; for instance, only 20% of the estimations for $\hat{T}_{DEM} = T_{NOA}$ but over 50% of the estimations for the implemented methods (with the ANN approach being, once more, superior) fall within $|\Delta T| \leq 1°C$.

The improvement in the estimation of all implemented estimation methods - and especially the ANN approach - over $\hat{T}_{DEM} = T_{NOA}$ is significant in terms of both meteorology and energy calculations:

(a) Since 0.2°C constitutes the acceptable thermometer resolution worldwide, estimations for which $|\Delta T| \leq 0.2°C$ are be assumed to be exact, i.e. $\hat{T}_{DEM} \cong T_{DEM}$. While less than 3.6% exact estimations are made by $\hat{T}_{DEM} = T_{NOA}$, this percentage rises to over 11.5% for all the implemented estimation methods (over 12.5% for the ANN approach).

(b) The precise estimation of ambient temperature at the target location - in cases where the ambient temperature is very near or below zero - can prevent the potential destruction of crops as well as the damage of sensitive and expensive instruments. When $T_{DEM} \leq 4°C$, the standard error of estimates of $\hat{T}_{DEM} = T_{NOA}$ and the ANN approach (which is superior to the other implemented estimation methods) is 3.3°C and 1.6 °C, respectively, demonstrating the accuracy of the ANN approach in estimating low ambient temperatures at the target location.

(c) The precise estimation of ambient temperature - in cases of high ambient temperatures - can prevent increasing demands on power supply as well as financial losses caused by the erroneous prediction of energy consumption (e.g. cooling energy loads). When $T_{DEM} \geq 33°C$, the standard error of estimates of $\hat{T}_{DEM} = T_{NOA}$ and the ANN approach (which is superior to the other implemented estimation methods) is 1.9°C and 1.7°C, respectively, illustrating a small, but potentially significant, superiority - in terms of energy consumption - of the ANN approach in estimating high ambient temperatures at the target location.

While the findings of Table 1 and point (a) concern the accuracy/exactness of the estimation, points (b) and (c) are especially important when short-term prediction of meteorological and energy parameters is desirable at the target location. In this case, forecast T_{NOA} values can be employed, the aim being to predict the corresponding T_{DEM} values as accurately as possible in order to avoid the problems discussed above.

3.2 Estimation of I_{DEM}

Only I_{NOA} has been employed for the estimation of I_{DEM}. This is owing to the high value of the linear correlation coefficient between I_{NOA} and I_{DEM} (ρ=0.941) and the significantly lower values between I_{NOA} and the other meteorological variables at the reference location: $\rho(I_{NOA},H_{NOA})$=-0.60; $\rho(I_{NOA},T_{NOA})$=0.53; $\rho(I_{NOA},WS_{NOA})$ =-0.18.

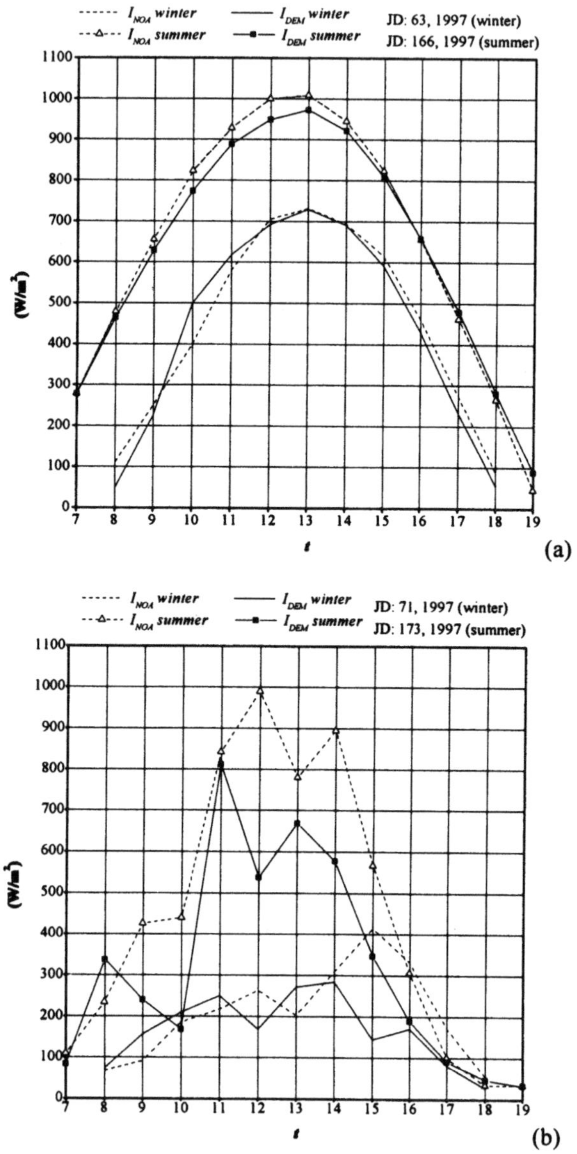

Fig. 3 (a) I_{NOA} and I_{DEM} curves during: a sunny day of winter and a sunny day of summer (a); a cloudy day of winter and a cloudy day of summer (b).

Solar horizontal radiation is the most important energy parameter, whose values and alterations determine the design and performance of solar systems (e.g. solar hot water systems and solar cells). Three ranges of I can be distinguished, which correspond to qualitative differences concerning solar system performance:

- Range 1: [0-~250]W/m^2. The system does not respond to alterations of I within this range, a phenomenon that is due to the thermal capacitance of the system.
- Range 2: [~250-~600]W/m^2. The system responds to changes of I within this range; the sensitivity of the system is related to the magnitude of I as well as the change in I.
- Range 3: [~600-~1100]W/m^2. The system becomes very sensitive to the values of I; even small alterations of I significantly influence system performance.

Partitioning of the I_{NOA} curves (consecutive values during a given Julian Day) has been performed both by season and according to sunny/cloudy weather conditions; the effects of both season and weather conditions on the characteristics of the I_{NOA} and I_{DEM} curves are illustrated in Fig. 3. The best partitioning has been accomplished using a 16 x 4 hexagonal SOM [3], with the following input features:

- The daily maximum I_{NOA} value. This gives an indication of the height of the I_{NOA} curve and, subsequently, of the season of the year.
- The maximum absolute difference between consecutive I_{NOA} values around noon. This indicates the shape of the upper part of the I_{NOA} curve.
- The average of the differences between $I_{NOA}(t+1)$ and $I_{NOA}(t)$ from sunrise until noon. This provides information about the shape of the leftmost part of I_{NOA} curve.
- The average of the differences between $I_{NOA}(t)$ and $I_{NOA}(t+1)$ from noon until sunset. This provides information about the shape of the rightmost part of I_{NOA} curve.

The SOM accomplishes clustering of the dataset into four partitions: the first partition represents the sunny days of summer (bell-shaped I_{NOA} curves), the second the intermittently cloudy days of spring and fall (curves with abrupt changes between consecutive I_{NOA} values), the third the sunny days of winter, the beginning of spring and the end of fall (almost bell-shaped I_{NOA} curves), and the fourth the cloudy days of winter (curves with low I_{NOA} values).

Estimators	Range 1	Range 2	Range 3
I_{NOA}	63.34	93.40	72.94
ANN	58.2	77.96	66.80
1st order linear model	59.16	83.87	71.02
4th order linear model	59.86	80.50	70.42

Table 2. Standard error of estimates (SEEs) for specific energy-oriented ranges of I_{DEM}.

Subsequently, one single-layer BP ANN [5] has been assigned to each partition for estimating I_{DEM}. Each BP ANN comprises three inputs and a single output. Inputs $I_{NOA}(t)$ and $I_{NOA}(t\text{-}1)$ represent the shortest history of the signal at the reference location that allows the accurate estimation of $I_{DEM}(t)$ at the target location, while t provides a further indication of the range within $I_{DEM}(t)$ is expected to lie in the given partition. The number of hidden nodes varies in the four BP ANNs (2, 4, 6 and 15, respectively), reflecting the complexity of the given partition. For instance, two hidden nodes are adequate for the first partition, where both I_{NOA} and I_{DEM} have almost similar bell-shaped curves; by contrast, in the fourth partition - where there is a marked variability between the I_{NOA} and I_{DEM} curves - 15 nodes are required. The sigmoid activation function and the batch mode of training have been employed for all BP ANNs. Training of the BP ANN assigned to the first partition was terminated when the error of the training and validation sets stopped decreasing; training of the other BP ANNs was terminated when the error of the validation sets started increasing.

The most accurate alternative methods for estimating I_{DEM} have been achieved via the 1st and 4th order memory-less linear models with problem decomposition.

As shown in Table 2, both the ANN approach and the alternative implemented estimation methods produce superior results for all the energy ranges of I over $\hat{I}_{DEM} = I_{NOA}$:

- For Range 1, all the implemented estimation methods effectuate an improvement in estimating I_{DEM} over $\hat{I}_{DEM} = I_{NOA}$, with the ANN approach producing slightly superior results.
- For Range 2, the ANN approach effectuates a significant improvement in I_{DEM} estimation over $\hat{I}_{DEM} = I_{NOA}$ (around 15.5W/m^2) with smaller – but still important - improvements of 9.5 and 13W/m^2 for the 1st and 4th order linear models, respectively; for this range, the improvement realized by the ANN approach is adequate to influence the behaviour of a solar system.
- For Range 3, the ANN approach effectuates an improvement of around 6W/m^2 over $\hat{I}_{DEM} = I_{NOA}$, compared with an improvement of 2 and 2.5W/m^2 for the 1st and 4th linear models; the reduction in the standard error of estimates realized by the ANN approach is again capable of modifying energy calculations.

4. Energy application

In order to establish the improvement effectuated in terms of energy calculations when estimating the meteorological parameters at the target location over simply employing those collected at the reference location, the behaviour of two solar hot water systems has been investigated; the first (smaller) system is suitable for domestic use, while the second (larger) system is more appropriate for industrial use. The Transient System Simulation Program (TRNSYS) [6] has been used for calculating the performance of the two systems. The inputs to TRNSYS constitute collected/estimated/evaluated meteorological parameters (T, measured in °C; I measured in W/m^2; I^d and I^t, both derived from I and measured in W/m^2) and the daily temperature of the cold water supply (T^{CW}, measured in °C). The outputs of TRNSYS comprise the daily average temperature inside the storage tank (T^{av}, measured in °C), the daily useful energy obtained by the collector (Q^u, measured in kJ), the daily energy removed from the tank (Q^{sup}, measured in kJ), the daily energy supplied by the auxiliary heater (Q^{aux}, measured in kJ) and the daily energy loss from the tank to the environment (Q^{loss}, measured in kJ); while T^{av}, Q^u, and Q^{sup} indicate the benefits from the utilization of a solar system, Q^{aux} and Q^{loss} represent the losses.

4.1 Energy calculations for 1997-1998

The short-term performance of the two systems has been simulated employing:

(i) Measured inputs at NCSR 'Demokritos' (DEM), representing the actual performance of a solar system installed at the target location; the TRNSYS outputs comprise T^{av}_{DEM}, Q^{u}_{DEM}, Q^{sup}_{DEM}, Q^{aux}_{DEM} and Q^{loss}_{DEM}.

(ii) Measured inputs at the National Observatory of Athens (NOA), representing the actual performance of a solar system installed at the reference location; the TRNSYS outputs comprise T^{av}_{NOA}, Q^{u}_{NOA}, Q^{sup}_{NOA}, Q^{aux}_{NOA} and Q^{loss}_{NOA}.

(iii) Estimated inputs at Demokritos (DEM) (via the ANN approaches described in section 3), representing the estimated performance of a solar system installed at the target location; the TRNSYS outputs comprise $\hat{T}^{av}_{DEM}$, $\hat{Q}^{u}_{DEM}$, $\hat{Q}^{sup}_{DEM}$, $\hat{Q}^{aux}_{DEM}$ and $\hat{Q}^{loss}_{DEM}$.

It has been found that the TRNSYS outputs are closer between simulations (i) and (iii) than between simulations (i) and (ii); this is true of both the domestic and the industrial system. Furthermore, the values of T^{av}, Q^u, and Q^{sup} (i.e. the calculated benefits) of simulation (ii) are consistently higher than those of simulation (ii), while those of Q^{aux} and Q^{loss} (i.e. the calculated losses) of simulation (ii) are consistently lower than those of simulation (i). This suggests that employing the meteorological parameters of the reference location for the design/validation of a solar system to be installed at the target location leads to an overestimation of the energy benefits and an underestimation of the energy losses. This can, in turn, prompt the design of a smaller system than necessary, which will be inadequate in satisfying the actual energy demands at the target location. By contrast, the more correct estimation of the benefits and losses of simulation (iii) - as witnessed by the smaller differences in T^{av}, Q^{u}, Q^{sup}, Q^{aux} and Q^{loss} as well as by the preference for a slight underestimation of the benefits and a slight overestimation of the losses - is advantageous as it will prompt the design of a solar system that will be appropriate in satisfying the actual energy demands at the target location.

The larger standard error of estimates of the TRNSYS outputs for the two systems between simulations (i) and (ii) than between simulations (i) and (iii) further indicates that the simulation process of both solar hot water systems is more accurate when estimations of T_{DEM} and I_{DEM} are employed than when the actual T_{NOA} and I_{NOA} values at the reference location are utilized; the more pronounced improvement observed for the industrial system is due to its larger size.

Finally, the percentages of the sums of the absolute differences for the daily outputs Q^{u}, Q^{sup}, Q^{aux} and Q^{loss} have been examined on a yearly basis (a measure commonly used for determining data consistency in energy applications and employed to distinguish between random and systematic errors). The absolute yearly differences between simulations (i) and (ii) support that the differentiation between the meteorological profiles of the two locations is systematic and, therefore, significant in terms of solar energy system performance. By contrast, the absolute yearly differences between simulations (i) and (iii) always suggest that the actual and estimated meteorological parameters at the target location do not differ significantly for the purposes of energy calculations.

4.2 Energy calculations employing Typical Meteorological Years

The Typical Meteorological Year of the reference location has been constructed [1]. Subsequently, the ANN estimators of T_{DEM} and I_{DEM} (section 3) have been employed for constructing an "approximate" Typical Meteorological Year for the target location. The meteorological profiles of the reference and target locations have, subsequently, been employed in order to assess the long-term (typical) performance of the two systems.

The TRNSYS output values for simulations (ii) and (iii) are lower than those calculated during 1997 and 1998. The observed value reduction is expected; it is – in fact – accounted for by the smoothing performed when constructing the Typical Meteorological Year at the reference location. This reduction notwithstanding, a comparison between the mean daily energy TRNSYS output values of simulations (ii) and (iii) demonstrates that their differences are always of the same nature (sign) and of a similar proportion for the Typical Meteorological Years as during 1997 and 1998. These findings indicate a consistency between the yearly and Typical Meteorological Year output energy parameters and, thus, support the superiority - in terms of energy calculations - of employing the ANN estimators of section 3 for constructing an "approximate" Typical Meteorological Year for the target location over simply using the Typical Meteorological Year of the reference location.

5. Conclusions

The accurate knowledge of the meteorological profile at a specific target location is (a) crucial in terms of meteorology, and (b) necessary in order for an energy system to be successfully designed, installed and profitably operated. Since the meteorological profile of the target location is rarely – if ever - known, it is common practice to employ the known meteorological profile of a nearby reference location instead. Unfortunately, even nearby locations may not share the same meteorological profile, whereby the "adoption" of the known meteorological profile of the reference location can lead to erroneous weather predictions and energy calculations at the target location.

A more viable alternative has been implemented here. The meteorological data collected short-term (over two years) at the target location have been employed in order to create an estimator of meteorological data at the target location from the corresponding data of a nearby reference location. ANN estimators have been investigated and found superior to traditional estimation methods in:

- Estimating meteorological parameter values at the target location from the corresponding measurements collected at the reference location.
- Constructing an "approximate" meteorological profile of the target location from the known meteorological profile of the reference location.

The ANN-estimated profile is significantly more accurate than the typical meteorological profile of the reference location in evaluating the benefits and losses of a solar energy system installed at the target location, thus prompting the design of systems that are more appropriate in satisfying the long-term energy demands at the target location.

References

[1] Hall I.J., Prairie R.R., Anderson H.E., Boes E.C. (1978), Generation of a typical meteorological year, *Report SAND-78-1096C, CONF-780639-1* (Sandia Laboratories).

[2] Haykin S. (1994), **Neural Networks – A comprehensive foundation**, (Prentice-Hall Inc., USA), pp. 59-72.

[3] Kohonen T. (1997), **Self-Organizing Maps, (2nd edn., Springer, Germany)**, pp. 85-144.

[4] Kramer M.A. (1992), Autoassociative neural networks, *Computers and Chemical Engineering* **16**(4), 313-328.

[5] Rumelhart D.E., Hinton G.E., Wiliams R.J. (1986), Learning representations by back-propagating errors, *Nature* **323**, pp. 533-536

[6] TRNYS (1994), A transient system simulation program, (Solar Energy Laboratory, University of Wisconsin, Madison).

Acknowledgments

This research was funded by the National Centre for Scientific Research 'Demokritos', project DEMOEREYNA No. 643.

Use of Neural Networks for Modelling and Fault Detection for the Intake Manifold of a SI Engine

Jocelyn A. F. Vinsonneau[1], Derek N. Shields[1], Paul King [2], Keith J. Burnham[1]

[1]Control Theory and Applications Centre, School of Mathematical and Information Sciences, Coventry University, Coventry CV1 5FB. U.K. Tel: +44 2476 888972 Email: j.vinsonneau@coventry.ac.uk [2]Jaguar Cars Ltd, Coventry, U.K.

Abstract

A Jaguar Car engine is used to provide data for modelling the throttle body, engine pumping and manifold body. Based on the gas law of the intake dynamics, input/output variables are identified and used to train a neural network. Various structures are compared and assessed. The best structure is then used for fault detection. A neural network observer is developed and error stability is assessed. Two fault scenarios are considered.

1 Introduction

On-board fault diagnosis in most modern cars is introduced to aid control and diagnose engine malfunctions. More recently, it has been used to meet increasingly stringent emissions standards set by the Environmental Protection Agency (EPA), in the context of a low emission vehicle (LEV) programme. The first LEV standards were adopted by the California's Air Resources Board (CARB) and operate from 1994 through to 2003. LEV II regulations, running from 2004 until 2010, require on-board diagnostics to be more sophisticated, detecting emission problems relating to any sensor or component of the engine [1].

A mean of detecting those malfunctions is to make use of fault detection algorithm, to diagnostic the system (ref. [2], [3] and [4]). Over the great variety of methods used in literature based on mathematical models of the monitored system, the most common is the model-based approach, where residuals are generated [5]. In the framework of Fault Diagnosis and Isolation (FDI), fault are detected by setting threshold on each residual signal. FDI was recently introduced in most modern cars to control and diagnose engine malfunctions (ref. [6], [7] and [8]). Also, neural networks have been proven to be useful for nonlinear system modelling in the automotive field (ref. [9], [10] and [11]). This paper is organized as follows: A model description will be given. In section 3, neural network modelling for the intake manifold will be developed. Section 4 will propose a fault detection observer using neural network. Section 5 is an application section.

2 Model Description

The purpose of the modelling phase is to predict the mass airflow and the manifold pressure through the throttle and the intake manifold of a spark ignition (SI) engine, respectively. This is required to be done for various atmospheric pressures and air temperatures, using mathematical expressions. The airflow rate through the intake manifold is controlled by a throttle plate. Fundamental work (ref. [12] and [13]) give equations to describe the throttle and intake manifold dynamics, which form part of a complete engine model.

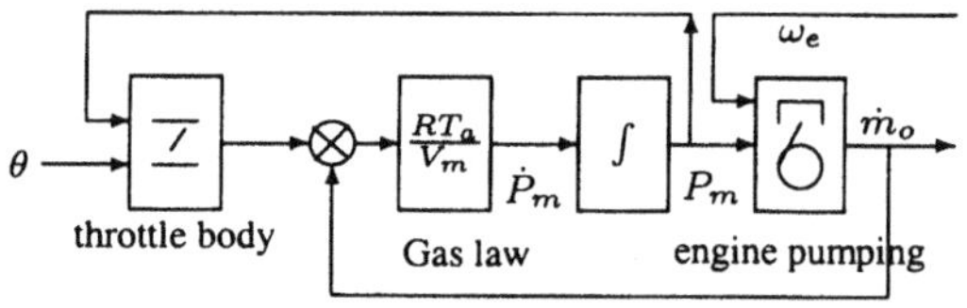

Fig. 1. Manifold diagram.

The basic throttle model assumes the air-gas is a one-dimensional, steady compressible flow of an ideal gas. From gas dynamics, the rate of air introduced into the intake manifold is expressed as the product of the function of the throttle plate angle and a function of both the manifold and atmospheric pressure, as described in Equation (1).

There are no accurate and simple physical models describing the flow from the manifold into the cylinder. The state equation for the manifold pressure is obtained by applying conservation of mass inside the volume of the manifold, between the throttle and cylinders. The ideal gas law is applied, assuming that the temperature and atmospheric pressure are homogeneous, as described in Equation (2). The full model is represented in Fig. (1). The following set of equations (1) to (3) define the model:

$$\dot{m}_{ai} = f_1(\theta)g_2(P_m) = f_2(\theta, P_m), \quad (1)$$

$$\dot{P}_m = \frac{RT_a}{V_m}(\dot{m}_{ai} - \dot{m}_o), \quad (2)$$

$$\dot{m}_o = h_2(\omega, P_m), \quad (3)$$

This can be rearranged into one equation

$$\dot{P}_m = \frac{RT_a}{V_m}(f_2(\theta, P_m) - h_2(\omega, P_m)), \quad (4)$$

3 Neural Network Modelling

Based on the above knowledge of the air intake system, it is proposed here to make use of neural network to model the dynamics equation. Some inputs (u) and outputs (y) (measured variables) are defined in Table I. Note in Table I that measured variable are scaled to 1. It

	name	description	unit
u_1	θ	throttle angle	[deg/50]
y_1	P_m	manifold pressure	[Bar/1]
u_2	ω	engine speed	[rad/s/150]

Table I. Definition of inputs/outputs

is assumed that the throttle angle do not exceed 50 degree, that the manifold pressure is always less that 1 bar and that the engine speed doesn't exceed 150 rad/s. The model described in Section 2 becomes:

$$\dot{y}_1(t) = G(u_1(t), u_2(t), y_1(t)), \quad (5)$$

where $G = \frac{RT_a}{V_m}$. Equation (5) is discretized using Euler giving:

$$\frac{y_{1,k+1} - y_{1,k}}{T} = G(u_{1,k}, u_{2,k}, y_{1,k}), \quad (6)$$

where T is the sampling time.

From equation (6), the neural network structure for the manifold pressure is proposed as

$$y_{1,k} = ANN(y_{1,k-1}, u_{1,k-1}, y_{2,k-1}), \quad (7)$$

and represented in Fig. 2.

3.1 Training Method (on real data)

A multi layer perceptron (MLP) is used, mainly because of its simplicity of implementation. The learning method is a back-propagation, using the *Levenberg-Marquardt approach* [14]. Typical training parameters are: epochs=1000, learning rate =0.5. Two comparison indices are defined, in order to assess the design of the neural networks: The integral absolute error (IAE, the lower the better is the model) and the coefficient of determination (R_T^2, the closer to unity the better).

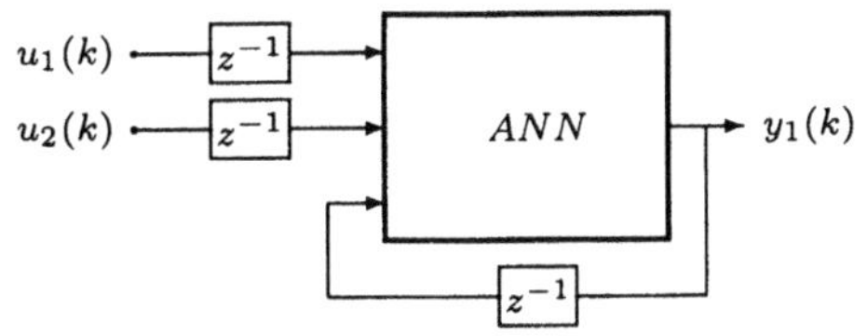

Fig. 2. Neural Network configuration for training.

Two data sets were taken for a general driving cycle, with normal atmospheric conditions, and temperature. The data for training was that defined on the last third of *data set a*. *Data set b* gave a means of verifying the model as it is untested, independent, data. Training Data are shown in Fig. (4-6).

3.2 Training Results

Table II shows a restricted comparison of network structure for ANN. The best neural structure is 351 and

ANN	**data set a**	
Network structure	IAE	R_T^2
331	0.0076	0.9893
341	0.0047	0.9922
351	0.0052	0.9922
361	0.3577	0.4546
3541	0.0384	0.2685

Table II. Structure comparison for ANN.

is optimum according to both indices.

4 Fault Detection

A fault scenario is now considered which is of particular importance to present engine design (Jaguar engines).

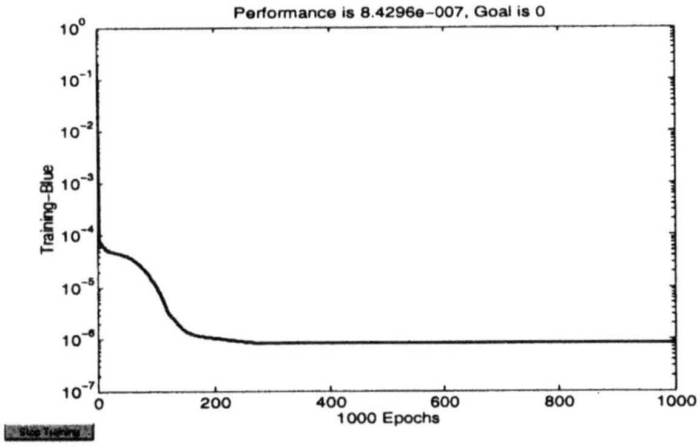

Fig. 3. Performance behaviour while training of Neural network.

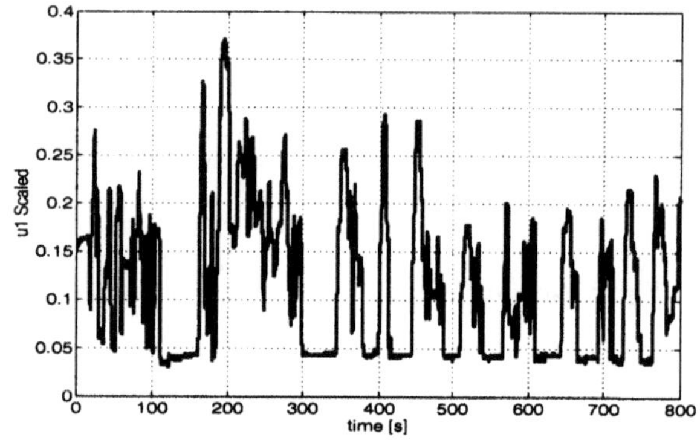

Fig. 4. Throttle angle scaled.

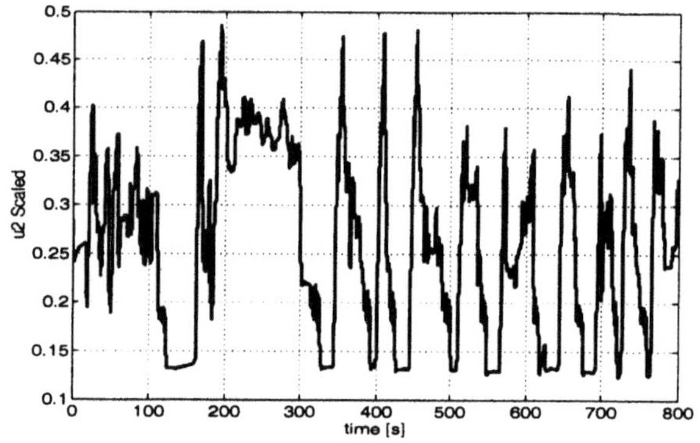

Fig. 5. Engine speed scaled.

Sensor degradation is tolerated but needs detecting, as engine pollution increases and engine performance decreases. Detecting the latter was deemed very important for limiting engine pollutants. Some input/ouput measurements are defined in Table III, where f^s is sensor additive faults and f^a a leakage in the manifold (actuator fault).

inputs	outputs
$u_1 = \theta$	$y_1 = P_m + f^s$
$u_2 = \omega$	

Table III. Inputs/outputs for FD.

4.1 Design of Nonlinear observer using neural Network

Here, a Neural Network observer is proposed, mimicking a continuous-time observer. Consider the following system,

$$\dot{x} = Ax + f(x,u) + f_k^a \tag{8}$$

$$y = Cx + f_k^s, \quad C = 1, \tag{9}$$

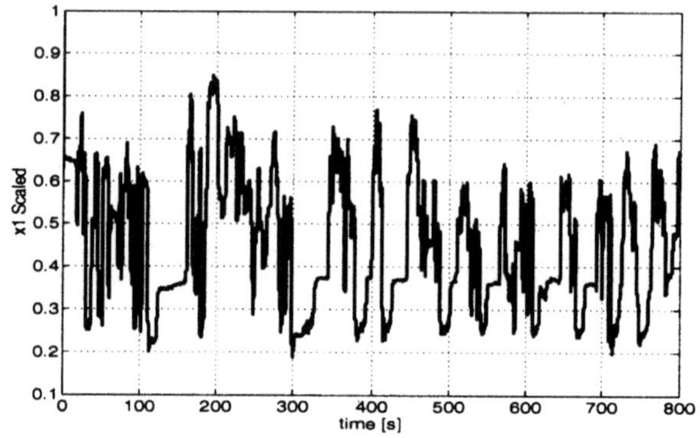

Fig. 6. Manifold pressure scaled.

and the observer in the form

$$\dot{\hat{x}} = \hat{A}(\hat{x}) + f(\hat{x},u) + K(x - \hat{x}), \tag{10}$$

where K is introduced to provide stability and convergence of the observer error.

From (8), a discrete model is obtained using the Euler method:

$$\frac{x_{k+1} - x_k}{T_s} = Ax_k + f(x_k, u_k) + f_k^a \tag{11}$$

$$x_{k+1} = T_s\Big[Ax_k + f(x_k, u_k) + f_k^s\Big] + x_k \tag{12}$$

where T_s is the sampling time. From (10), the corresponding discretized observer would be

$$\frac{\hat{x}_{k+1} - \hat{x}_k}{T_s} = A\hat{x}_k + f(\hat{x_k}, u_k) + K(x_k - \hat{x_k}) \tag{13}$$

$$\begin{aligned}\hat{x}_{k+1} = {} & T_s\Big[A\hat{x}_k + f(\hat{x_k}, u_k)\Big] \\ & + T_s K(x_k - \hat{x_k}) + \hat{x}_k\end{aligned} \tag{14}$$

Now replace $T_s[A\hat{x}_k + f(\hat{x_k}, u_k)]$ by the neural network developed and trained in Section 3.

The dynamic of the error $e_k = x_k - \hat{x_k}$ is on the from,

$$\begin{aligned}e_{k+1} = {} & x_{k+1} - \hat{x}_{k+1} && (15)\\ = {} & ANN(x_k + e, u_k) + T_s f_k^a - ANN(\hat{x_k}, u_k) \\ & - T_s K(x_k - \hat{x_k}) && (16)\\ = {} & \left[\frac{\partial ANN(\hat{x}_k, u_k)}{\partial \hat{x}} + 1 - T_s K\right] e_k \\ & + T_s f_k^a + \hat{H}(e_k, \hat{x}_k, u_k) && (17)\\ = {} & \lambda_k e_k + \hat{H}_k, && (18)\end{aligned}$$

where $\hat{H}(.) = \hat{H}_k$ for simplicity, and represents higher order terms in e_k, and where $\hat{H}(0, \hat{x}_k, u_k) = 0$.

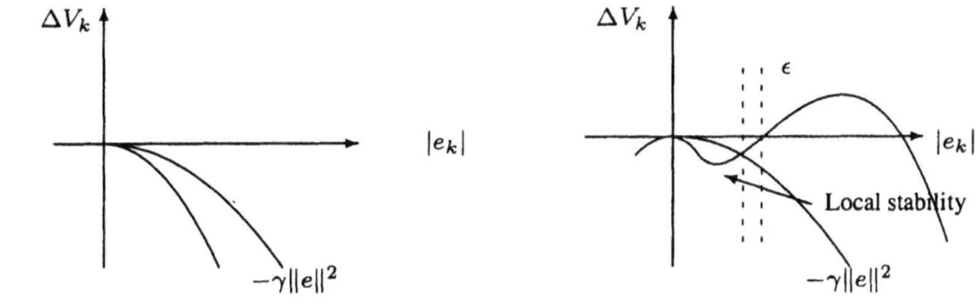

case a: $\forall\beta, \lambda \leq \frac{\sqrt{2}}{2}$

case b: $\forall\beta, \lambda > \frac{\sqrt{2}}{2}$

Fig. 7. Two cases studies of stability.

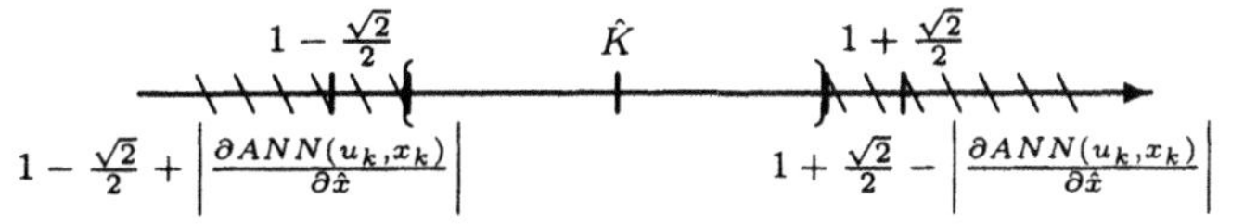

Fig. 8. Boundary condition on $\hat{K}$.

4.1.1 Estimation of a region of asymptotic stability. It is now assumed $f_k^a = 0$ for simplicity. Consider the candidate Lyapunov function, V_k. The aim here is to find some conditions for ΔV_k to be negative definite, guaranteeing local stability for (18):

$$V_k(e_k, k) = e_k^2 \tag{19}$$
$$\Delta V_k(e_k, k) = V_{k+1} - V_k \tag{20}$$
$$= e_{k+1}^2 - e_k^2 \tag{21}$$
$$= \left(\lambda_k e_k + \hat{H}_k\right)^2 - e_k^2 \tag{22}$$
$$= e_k^2(\lambda_k^2 - 1) + 2\lambda_k e_k \hat{H}_k + \hat{H}_k^2. \tag{23}$$

The following assumption is now made:

$$|\hat{H}_k| < \beta |e_k|^2 \tag{24}$$

It follows,

$$\Delta V_k \leq -e_k^2(1 - \lambda_k^2) + 2\lambda_k\beta|e_k|^3 + \beta^2|e_k|^4 \tag{25}$$
$$\Delta V_k \leq -e_k^2\left[(1 - \lambda_k^2) + 2\lambda_k\beta|e_k| + \beta^2|e_k|^2\right] \tag{26}$$

We want to show that $\Delta V_k(e_k, k) < -\gamma\|e_k\|^2$, according to Lyapunov theory.

The RHS of equation has 3 poles located at $p_1 = 0$ and $p_2, p_3 = \frac{2\lambda\beta \pm \delta}{2\beta^2}$, where $\delta^2 = 4\beta^2\lambda^2 + 4\beta^2(\lambda^2 - 1)$. If $\delta^2 < 0$, p_2, p_3 are not real. The critical value, found when $\delta = 0$ is $\lambda_k = \frac{\sqrt{2}}{2}$.

The observer error is locally asymptotically stable if $e_k < \epsilon$, with $\epsilon = min(p_2, p_3)$. Fig. (7) shows 2 cases for the study of stability. Case a is always stable ($\Delta V_k < 0, \forall e_k \neq 0$), and case b, show the boundary of local asymptotical stability for the condition $|\lambda_k| < \frac{\sqrt{2}}{2}$. If $\lambda_k = \frac{\sqrt{2}}{2}$, $\epsilon = \frac{\lambda}{\beta} - \frac{1}{\beta}\sqrt{2\lambda_k^2 - 1} = 0.54$.

A safe mode would be to define $|\lambda_k| < \frac{\sqrt{2}}{2} - \epsilon$. It follows

$$\Delta V_k \leq -e_k^2\left[(1 - \lambda_k^2) + 2\lambda_k\beta|e_k| + \beta^2|e_k|^2\right] \tag{27}$$
$$\leq -e_k^2\left[(1 - (\frac{\sqrt{2}}{2} - \epsilon)^2) + 2(\frac{\sqrt{2}}{2} - \epsilon)\beta|e_k| + \beta^2|e_k|^2\right] \tag{28}$$
$$\leq -e_k^2\left[(1 - \frac{1}{2} + \epsilon^2) + (\sqrt{2} - 2\epsilon)\beta|e_k| + \beta^2|e_k|^2\right] \tag{29}$$

Neglecting higher order terms (assuming they are small),

$$\Delta V_k \leq -e_k^2\left[\left(\epsilon - \frac{\sqrt{2}}{2}\right)^2\right] \tag{30}$$
$$\Delta V_k \leq -\gamma\|e_k\|^2 \quad \gamma \geq 0 \tag{31}$$

For $\epsilon = 0.5417$ and $\lambda_k = 0.707$, then $\gamma = 0.027$. It is therefore safe to define λ such as

$$\lambda_k = \left|\left|\frac{\partial ANN(\hat{x}_k, u_k)}{\partial \hat{x}}\right| + 1 - T_s K\right| < \frac{\sqrt{2}}{2} \tag{32}$$

Two solutions exist for the above equations ($\hat{K} = T_s K$).

$$\begin{cases} \left|\frac{\partial ANN(\hat{x}_k,u_k)}{\partial \hat{x}}\right| + 1 - \hat{K} < \frac{\sqrt{2}}{2} \\ \left|\frac{\partial ANN(\hat{x}_k,u_k)}{\partial \hat{x}}\right| - 1 + \hat{K} < \frac{\sqrt{2}}{2} \end{cases} \tag{33}$$

$$\hat{K} > 0.293 + \left|\frac{\partial ANN(\hat{x}_k, u_k)}{\partial \hat{x}}\right|, \tag{34}$$

$$K < 1.707 - \left|\frac{\partial ANN(\hat{x}_k, u_k)}{\partial \hat{x}}\right|. \tag{35}$$

The aim is now to estimate the value of $\left|\frac{\partial ANN(\hat{x}_k,u_k)}{\partial \hat{x}}\right|$. In Fig. (9), it is shown the maximal amplitude of the partial differential with respect to the state x is 0.1, through data set a.

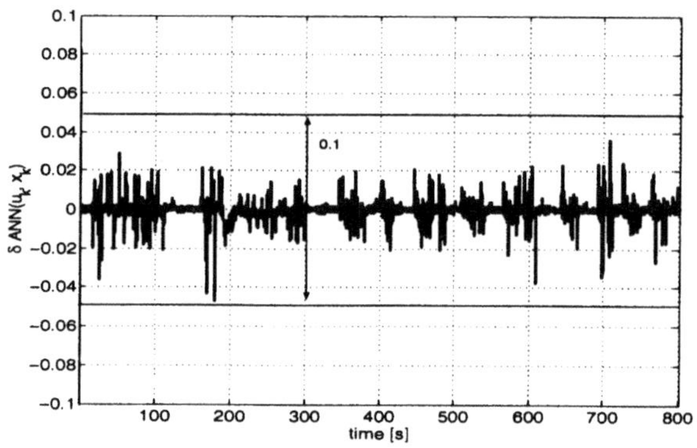

Fig. 9. Partial differentiation of ANN.

The feedback gain $\hat{K}$ is therefore bounded $0.4 < \hat{K} < 1.6$. A safe value in the middle of the range is chosen, $\hat{K} = 1$.

4.2 Implementation of the residual

The residual is defined as:

$$r_k = W[y_k - \hat{x}_k] \tag{36}$$

$$= W[x_k + f_k^s - \hat{x}_k] \tag{37}$$

$$= W e_k + W f_k^s \tag{38}$$

where W is a low pass filter defined as

$$r_{f,k} = \frac{1-a}{1-az^{-1}} r_k, \quad |a| < 1 \tag{39}$$

where a the pole of the filter to be tuned. Fig. (10) shows a residual generation based on a neural network observer.

In equation (38), the sensor fault affects directly the residual. The actuator fault, f_k^a, affects the residual through the error e_k, as shown in equation (18). Here, the sensor fault and actuator fault cannot be isolated. The threshold ϵ is designed, such that,

$$\|r_{f,k}\| < \|\epsilon\|, \{f_k^s \wedge f_k^a\} = 0, \textit{ (no-faulty)} \tag{40}$$

$$\|r_{f,k}\| > \|\epsilon\|, \{f_k^s \vee f_k^a\} \neq 0, \textit{ (faulty)} \tag{41}$$

5 Results

The Jaguar Car engine considered is a model "XJ8 Saloon", normally aspirated 4 litre, V8 engine. Two data set a and b were taken for a general driving cycle and with normal atmospheric conditions, and temperatures. The sampling time was T=0.05s. Shown in Fig. (4-6) are: the throttle angle scaled in Fig. (4), the engine speed scaled in Fig. (5), the manifold pressure scaled in Fig. (6). The observer was implemented in a Matlab/Simulink environment, using the neural network toolbox.

A leak, an actuator fault, f^a, corresponding to a hole ($\emptyset 3mm$) in the manifold, was simulated in a Simulink environment, where the Simulink model was taken as the real car. Modified data were saved to file. f^a was applied for $t > 500s$. The sensor fault f^s was of multiplicative type, with an amplitude of 10%. $f^s$was applied on y_1 for $t \in [50 : 150]$. The successful design of the residual is shown in Fig. (11), where faults are detected in both scenarios. It is easy to see that the residual is dying off after the initial condition. In Fig. (12), the fault detection strategy is applied on *data set b*. The residual detects also the sensor fault and the leak. The leak (3mm) is detected from t=500s onwards. However, it is not clear in Fig. (12), at time $t = 800s$, if the leak is detected or not. Clearly, for a smaller leak, the detection would be difficult.

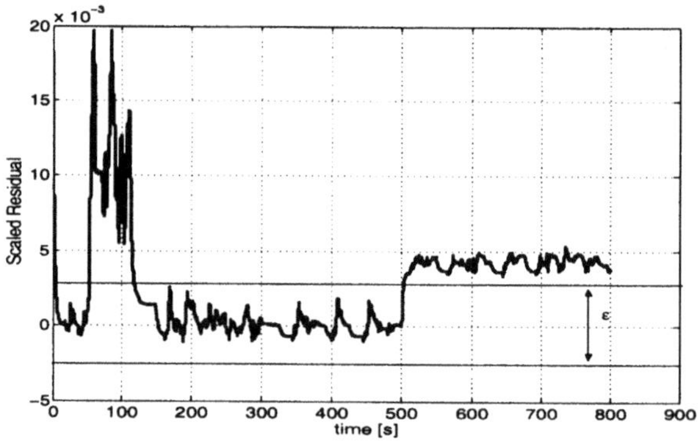

Fig. 11. Filtered residual on *data set a*.

6 Conclusion

The modelling of the intake manifold of a Jaguar Car is considered. Neural network structures are developed and compared. The best one is chosen to create a neural network observer, with feedback. A residual is implemented from the observer and system output. Stability

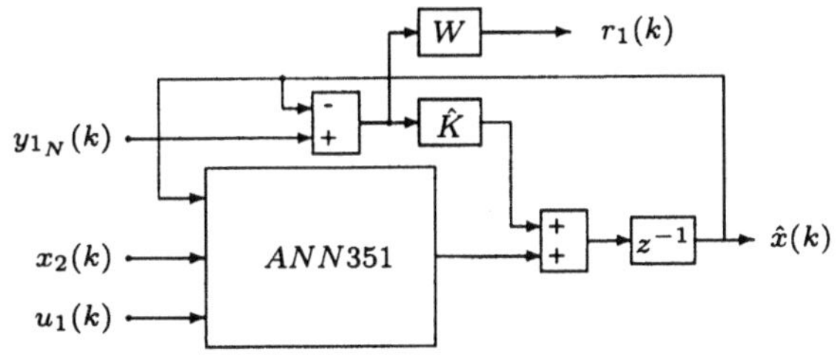

Fig. 10. Stable neural network observer for the manifold pressure.

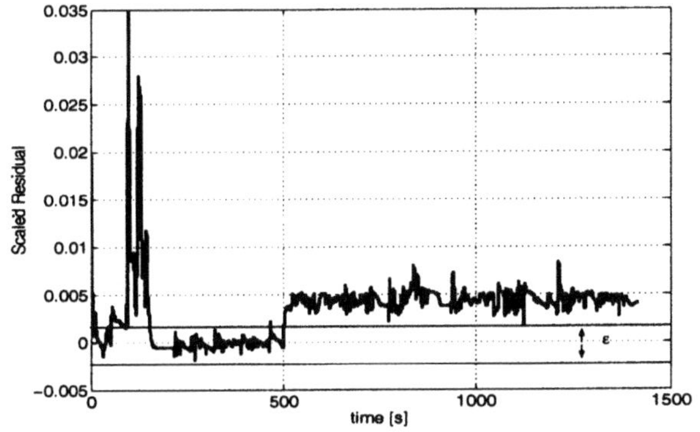

Fig. 12. Filtered residual on untrained *data set b.*

is assessed. Two fault scenarios are considered: a leak in the manifold and a fault in the sensor measuring the manifold pressure. The observer is able to detect both faults.

References

[1] Jeffrey A. Cook, Jing Sun, and J. W. Grizzle. Opportunities in automotive powertrain control applications. *Proceedings of the 2002 IEEE International Conference on Control Applications*, 2002.

[2] R. J. Patton and J. Chen. Observer based fault detection and isolation: Robustness and applications. *Control Engineering Practice*, 5(5):671–682, 1997.

[3] P. M. Frank and X. Ding. Survey of robust residual generation and evaluation methods in observer-based fault detection systems. *Journal of Process Control*, 7(6):pp. 403–424, 1997.

[4] D. N. Shields and S. Ashton. A fault detection observer method for nonlinear systems. In *IFAC Fault Detection, Supervision and Safety for Technical Processes*, pages 226–231, Budapest, Hungary, 2000.

[5] P. M. Frank, S. X. Ding, and B. Köppen-Seliger. Current developments in the theory of fdi. In *IFAC Fault Detection, Supervision and Safety for Technical Processes*, pages 467–471, Budapest, Hungary, 2000.

[6] Y.-W. Kim, G. Rizzoni, and V. Utkin. Automotive engine diagnosis and control via nonlinear estimation. *IEEE Control Systems*, pages 84–99, 1998.

[7] G. Rizzoni, Y. Kim, and A. Soliman. Estimation problems in engine control and diagnosis. In *IFAC Fault Detection, Supervision and Safety for Technical Processes*, pages 124–130, Budapest, Hungary, 2000.

[8] R. Isermann. Diagnosis methods for electronic controlled vehicles. *Vehicle Systen Dynamics*, 36(2-3):77–117, 2001.

[9] Tan Yonghong and Saif Mehrdad. Non linear dynamics modelling of automotive engines using neurals network. *Proceedings IEEE International Conference on Control Applications*, 6:408–410, 1997.

[10] M. Ayeb, D. Lichtenthler, T. Winsel, and H. J. Theuerkauf. Si engine modeling using neurals networks. *SAE Electronics Engine Controls*, SP-1357(Paper No. 980790), 1998.

[11] J. A. F. Vinsonneau, D. N. Shields, P. J. King, and K. J. Burnham. A comparison of two si engine modelling techniques with application to fault detection. In *UKACC Control 2002*, volume 1, 10 September 2002, Sheffield, UK., 2002. United Kingdom Automatic Control Concil.

[12] U. Kienche and L. Nielsen. *Automotive Control System - For Engine, Driveline and Vehicle.* Springer-Verlag, Berlin, Germany, first edition, 2000.

[13] W. Pulkrabek Willard. *Engineering Fundamentals of the Internal Combustion Engine.* Prentice-Hall, London, first edition, 1997.

[14] Nigel C. Steele. Neural networks - summary paper. *Workshop on Computer Architecture, Technology and Formalism for Computer Science, Research and Applications, Naple, Italy.*, 1992.

RBFG Neural Networks for Insulation Thermal Ageing Prediction

L.Mokhnache (*), A.Boubakeur (**)

(*) University of Batna- Faculty of Engineering- Dept. of Electrical Engineering, Batna, ALGERIA
(**) Ecole Nationale Polytechnique- High Voltage Laboratory, BP 182 EL-Harrach, Algiers, ALGERIA

Abstract :This paper presents RBFG neural networks trained by ROM and applied in prediction. Two techniques of Gaussian centers disposition are presented: regular trellis in a one dimension and an original method combining the regular trellis in a two dimensions and the K-mean clustering method. The second technique presents the best prediction.

1 Introduction

Prediction is an important subject for power systems economy point of view. It allows solving economical problems of energy and making maintenance in a simplest way.
The prediction is an interpolation of present values. Previous values curves are used. For this task the RBF neural networks are found to be the best.
In this paper the prediction of solid and liquid insulation thermal ageing has been done using RBF neural networks.

As ageing experiments are costly and time consuming, the aim of this paper is to reduce the ageing experiment time and predict thermal ageing stress for the PVC used in cables insulation and the transformer oil named BORAK22.

Having the database of the full thermal ageing interval, we have trained our net in the aim to predict insulation characteristics at longest intervals. In this way, we have trained the net within an interval less than the used in experiments, to predict insulation characteristics corresponding to the full experimental interval. This had permitted us to compare predicted and experimental properties.
The used ANNs are Radial Basis Function Gaussian networks (RBFG) trained by a Random Optimization Method (ROM). The main advantage of this kind of ANN is its independence of the mathematical model. Also, with this method the error decreases, avoiding a possible divergence. Two RBFG nets trained by ROM have been used. In both of them the training is done on a set of samples having the form (y_i, y_{i+1}) where y is the network output. The network is optimized by always having the same number of trained data values.
To learn its input and output mapping of a system Algorithms of RBF networks contain two sub-classes of algorithms namely 'ref 1':

-Algorithms to select centers
-Algorithms to determine the weights
The choice of centers is too important to the performance of the RBF network. Two centers distributions have been used. The first is in regular trellis represented in one dimension and the second is an original combination of the cited distribution in two dimensions and the K-mean clustering method. We have used one input representing the characteristic to predict the ageing time is included implicitly in the centers representation.

2 Deterioration Of The Insulation

Insulation in oil-filled power apparatus degrades with time of service. Such degradation may be caused by water absorption from the air, impurities in the oil, etc. Most of the degradation is caused by oxidation of the oil, i.e. air contact with the heated oil in the apparatus. Besides, insoluble materials formed by chemical reactions may deposit on the ferrous core, coil, wall and bottom of the tank in the form of sludge. This sludge, if formed, will lower the oil voltage breakdown. Even if it is deposited, it will reduce transformer cooling and may contribute to the heating of oil, resulting in further degradation of the oil 'ref2'. Ageing experiments have been done with 80-100-120C°.

It has been shown 'ref3' that the temperature affects the properties of PVC. The degradation of the physical and electrical properties is sensitive to the temperature and can cause electrical failures. Several test pieces of PVC submitted to voltage, have been exposed to several ageing temperatures (80, 100, 120°C and 140°C).

3 The RBFG Neural Networks

The main idea of the RBF networks was introduced by Powell 'ref4', and it states that any function f(x) can be approximated by an interpolation composed by the sum of p core functions:

$$f(x) = \sum_{i=1}^{p} w_j \Phi(|x - \xi_i|) \qquad (1)$$

ξ_i are nodes of interpolation for i = 1-n, called centers.
w_j are synaptic weights interconnecting neurons to the output and Φ is the core function.

Φ insures the continuity in the nodes. It has a radial

symmetry and its contribution changes in function of the input positions relative to the centers and it must produce local responses. In this work Gaussian function is selected.

The RBFG networks constitute efficient models for the function generalization. The use of Gaussian functions allows benefiting from its local characteristic to facilitate the training and improve the generalization. The most important advantage of this technique is that the network has always one hidden layer.

So, the RBF networks satisfy naturally the following constraints:

- Universal approximation ability.
- Cover ability.
- Local generalization.

To build a predictive neural net model, we have chosen Radial Basis Function Gaussian network (RBFG). The architecture consists of three ordered layers: an input layer with n units, a hidden layer with m units and an output layer with a single node. Each node in the hidden layer has a radial symmetric response around a node parameter vector called center. The output layer is a linear combiner with connection weights 'refl,4'. Giving a set of input and output data x_i, y_i i=1,2...n, the state of the hidden unit j will be denoted by:

$$\rho_j(x_i) = \exp\left(-\frac{1}{2}\sum_{i=1}^{n}\frac{(x_i - c_{ij})^2}{\sigma^2_{ij}}\right) \qquad (2)$$

Where c_{ij} : i=1,...n. and j=1,...,m are the RBFG centers σ_{ij} define the widths of the Gaussians.

The network response y^* is given by:

$$y^* = \frac{\sum_{j=1}^{n} w_j \rho_j(u)}{\sum_{j=1}^{n} \rho_j(u)} \qquad (4)$$

$\left(\sum_{j=1}^{n} \rho_j(u)\right)$ is a normalizing factor and w_j is the weight connection between the hidden unit j and the output unit.

This network is different from the non-normalized one in which the neuron receives only a local data for the calculation of its activation. But the introduction of a normalization factor gives each neuron its own data and also information about the other neurons.

3.1 Covering rates

The RBFG centers are vectors of dimension n obtained from the training data.

Training with ROM, the first technique used in centers disposition consists of an arrangement of the centers in a regular trellis in a one dimension in order to cover uniformly the data input space. The covering rate is given by:

$$\tau = \exp\left(-\frac{1}{2}\frac{\left(\frac{\delta}{2}\right)^2}{\sigma^2}\right) \qquad (3)$$

Where δ denotes the distance between two neighboring centers. This disposition is presented in Fig. 1.

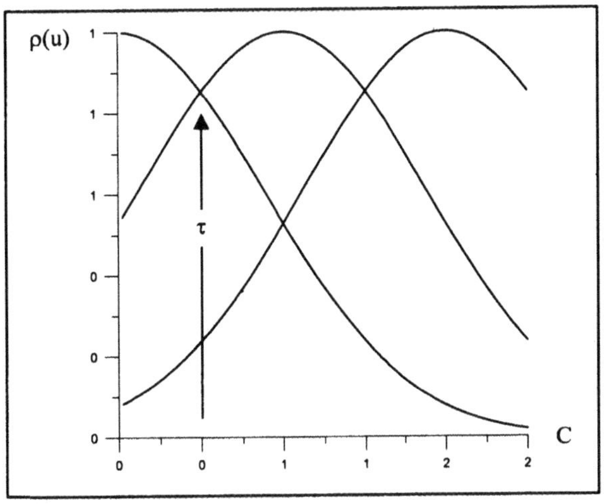

Fig.1 Input space covering in one dimension

If the network contains two inputs, the trellis covering ensures the same radii for all the Gaussians. But the number of hidden units is equal to the square of the cardinal of C, where C is the Gaussians centers set (Fig.2) 'refl'.

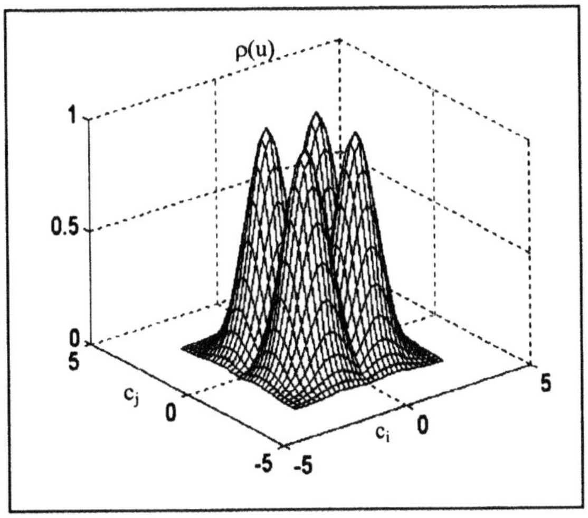

Fig.2 Input space covering in two dimensions

From equation (1), it can be seen that all the RBF functions indicate a change of contributions in the approximation as they move radially away from the center.

The RBF network effectively constructs a linear function space, which depends on the position of the

known data points according to an arbitrary distance measure. Alternatively, it can be seen as a method to perform a hypersurface reconstruction. This approach has been widely accepted to yield good multivariable approximation where other techniques fail 'ref1'.
The size of the RBF given in equation (1) depends on the number of data points or input vector. In many practical situations, this may not be feasible as the data set is large. Moreover this data is often corrupted by noise. Thus to train a network to interpolate strictly, these points will produce a complex, if not erroneous network.
To this end a close approximation is sought at the interpolating points. This gives rise to an RBF network with data points (x_i, i=1,n) and centers (c_j, j=1,m) where m<<n.
The choice of centers is crucial to the performance of the RBF network. It should be chosen to sample the input domain. The approach is to find a subset in centers to suitably sample the input domain such that the RBF network can be more sensitive to more than just one pattern.
In this work we have used one input for both of the two centers dispositions.
The first disposition is in regular trellis in one dimension (Fig.1) given in equation (3).
The second distribution is an original method, it is the combination of the equation (3) and a part of the K-mean clustering method.

3.1.1 K-mean clustering method

This method defines a set of cluster centers and partition of the training data into subsets. Each cluster center is associated with one of the hidden node (M). The algorithm find a local minimum in $E_{K\text{-}min}$, the total squared Euclidean distances between the training points assigned to each cluster and the cluster centers c_i:

$$E_{K-mean} = \sum_{m=1}^{M} \sum_{k=1}^{K} B_{mk} \|x_k - c_m\|^2 \qquad (4)$$

Where B_{mk} is the cluster membership function.

- In the proposed centers distribution method, equation (3) is applied where δ is calculated in two dimensions (y, t) (t is the ageing time) multiplied by a coefficient as B_{mk} . This coefficient changes from an input set to another ($0< B_{mk} <1$) if the new input set represents another cluster of inputs.

3.2 RBFG Trained by the ROM

Random optimization is based on random search technique and was used with success in several optimization problems. It is more suitable in some cases where the error function has several minimums. Notice that ROM converges to a minimum with a probability equal to 1 in a compact set. The idea of this algorithm is to attribute to the weights of the net a set of a white noise sequence and then compute the network output. If the target output is less than the previous error, the new weights are kept; otherwise, the previous weights are kept until the desired error is obtained.

The architecture of this network is the same as the RBFG trained by Back-Propagation 'ref5'. The difference is in the adjustment of the weights. In addition, ROM algorithm does not need to adjust c_{ij} and σ_{ij} because it is independent of the minimized criterion.

The ROM developed by Solis and Wetts depends on the Gaussian variation vectors, i.e. a sequence with a big variation generates a big error change, but no oscillations. However, sometimes a big variation gives a stagnation of the error surface 'ref1'. Here, the training is started with a big variance value, which is reduced progressively.
The ROM algorithm is given by 'ref6' in figure 2.

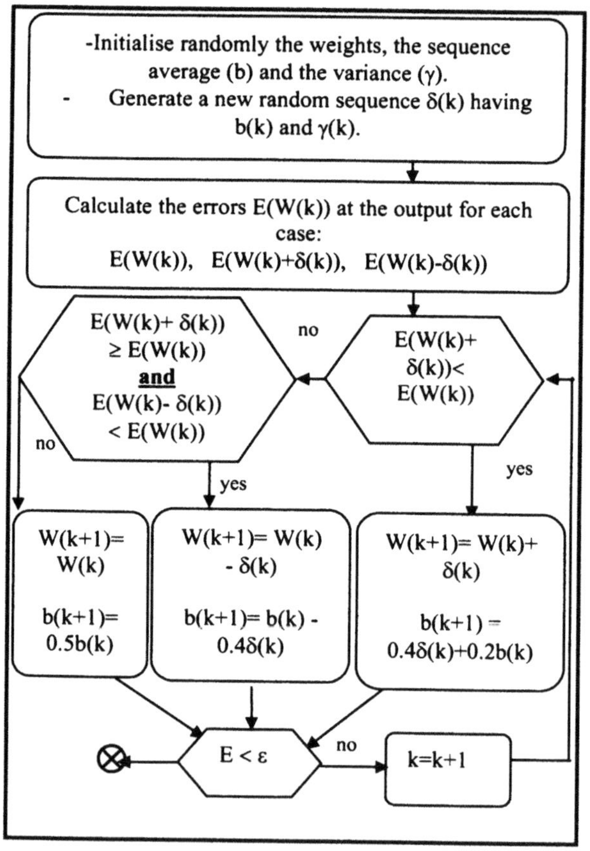

Fig.3 Algorithm of the RBFG net trained by the ROM

4 Prediction Phase

To predict a future value y_{n+1} of a set of measured data y_i, i=1,...,n, the algorithm will be trained on a set of samples having the form (y_i,y_{i+1}), i=1,...n-1. After this training, the weights of the net are updated so as when the network receives the value y_i, its response will be y_{i+1}. Now, a new set of data is obtained with n+1 values y_i, i=1,...,n+1. The training is repeated from the beginning in order to predict the value y_{n+2}, the new set of data will contain n+2 values y_i, i=1,...,n+2, and this procedure is repeated until the prediction of all values is obtained.

In order to improve the prediction quality, after each future value prediction, the first value is omitted from the set of data. For example, y_1 is omitted after the prediction of y_{n+1}, and y_2 is omitted after the prediction of y_{n+2} etc. In this way, the same number of values trains the network.

5 Results and Discussion

Results in figures 4 and 5 given for PVC are obtained for T = 100°C and a learning time of 1600h. We have trained the net on an interval, which is, less than the experimental one and predicted the property at the experimental interval. This will allow us to validate our prediction. In the generalized phase, the net is trained on the full experimental interval to predict properties on longest periods.

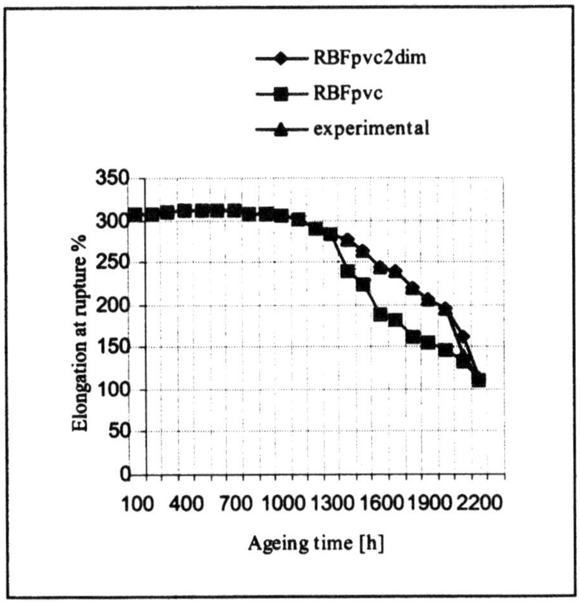

Fig.4 Determination of the PVC life duration from the elongation at rupture

We note that the centers distribution affects greatly the prediction quality. The second centers disposition (in two dimensions represented on the figures RBFpvc2dim) has improved the prediction comparing with results with a regular trellis in one dimension.

In the second distribution of centers, one input was used but in the calculation of δ the time is introduced directly as if two inputs were used. The coefficient B_{mk} ($0< B_{mk} <1$) change from a training to another. It changes when the input vector represents another cluster of input data. That is the cause why we have named it the cluster coefficient.

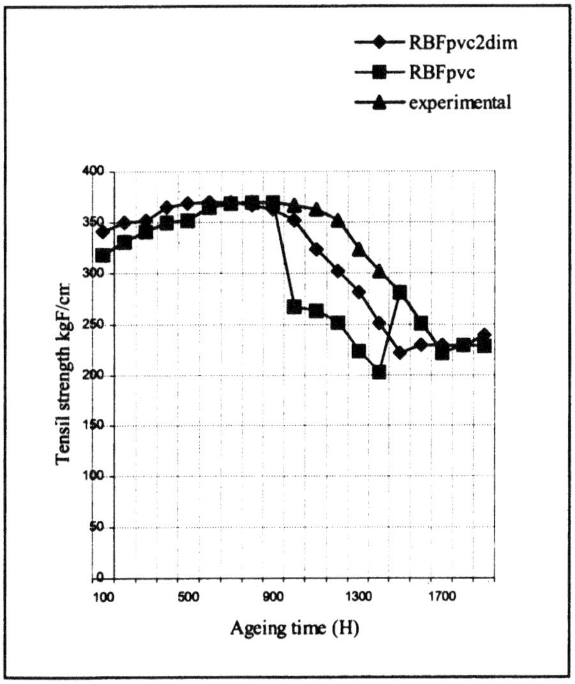

Fig.5 Determination of the PVC life duration from the tensile strength

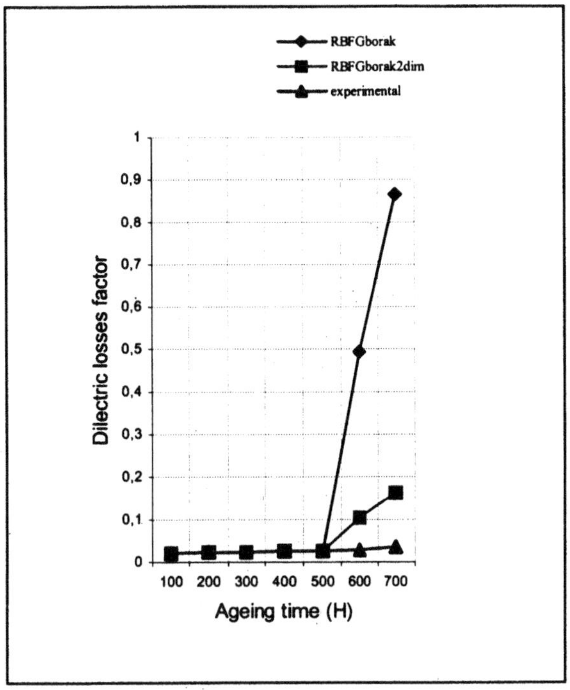

Fig.6 Variation of the dielectric loss factor of the BORAK22

Figure 6 gives the variation of the BORAK22 dielectric loss factor in function of the ageing time using a 100 time step and A 500 h learning times.

We note that with the first centers disposition, the prediction follows the real experimental values in the beginning of the prediction and goes away from in advancing in the prediction. The second network presents the best prediction.

In other results given for the regular trellis centers disposition 'ref7', we have used three time steps (150, 200, 250) and we have noted clearly that the training time and the time step influence the network training, but there is not a linear dependence between them. We have noted that there is an optimum training time and an optimum time step giving the best results but not together. The same was done with the PVC.

6 Conclusion

- It would be very economical to use ANNs in investigations on HV insulation. We may reduce the laboratory test and let the network predict insulation properties.

- The choice of the RBFG net is argued by the fact that it solves the problem of selecting the hidden layers number. The optimization of a network is not an easy task; it can take a long time as it can diverge completely. The RBFG local characteristic avoids this problem.

- Two principal RBFG nets trained by ROM were investigated; they have presented their ability to adjust the weights without determining any gradient.

- The choice of Gaussian centers is a crucial to the performance of the RBF network.

- The second method of centers distribution is an original tentative to combine two known techniques: the two dimensions regular trellis and the K-mean clustering method.

- The simulation results give for the new distribution centers the best prediction performances.

7 References

[1] Khemaissia, S., Morris, A.S. (1994) Review of networks and choice of radial basis function networks for system identification. Technologies Avancées N°6: 55-85.

[2] Vanegas, O, Mizuno, Y., Naito, K. (1997) Diagnostic of oil-insulated power apparatus by using neural networks simulation, IEEE transactions on dielectrics and electrical insulation Vol.4, N° 3: 290-299

[3] Nedjar, M. (1991) Influence of thermal ageing of polymers used in HV cables, Master thesis, University of Tizi-Ouzou, Algeria.

[4] Poli R., Jones, R.D. (1994) A neural net model for prediction. Journal of the American Statistical Association Vol. 89, N° 425.

[5] Mokhnache L., Boubakeur A., Feliachi A. (2002) Prediction of Thermal Ageing in Transformer Oil and High Voltage PVC Cables Using Artificial Neural Networks. IEE Proceeding Science Measurement and Technology, final acceptance for publication.

[6] Sollis, F.J., Wetts, R.J.B. (1981) Minimization by random search techniques. Mathematics of Operations Research, The Institute of Management Sciences, Vol.6. N°1: 19-29.

[7] Mokhnache L., .Boubakeur A. (2002) The Use of some Paradigms of Neural Networks in Prediction of Dielectric properties for High Voltage liquid-solid and gas Insulations. IEEE international symposium on electrical insulation proceedings, ISEI'2002 :306-309. April 2002, Boston Massachusetts, USA.

Meteorological data mining employing Self-Organising Maps

Tatiana Tambouratzis[1] George Tambouratzis[2]

[1] Institute of Nuclear Technology – Radiation Protection, NCSR 'Demokritos'
Aghia Paraskevi 153 10, Athens, Greece
e-mail tatiana@ipta.demokritos.gr

[2]Institute for Language and Speech Processing
Artemidos 6 & Epidavrou, Paradissos Amaroussiou 151 25, Athens, Greece
e-mail: giorg_t@ilsp.gr

Abstract A data mining application is described, which is focused on the analysis of high-dimensional meteorological data collected long-term (over 43 years) at 130 weather stations in Greece. A hybrid clustering method (combining artificial neural networks and statistical-based techniques) has been employed for grouping the data. The proposed method has been found effective in clustering the stations and partitioning Greece into areas with distinct meteorological profiles: stations in the same area have similar meteorological profiles, while those classified in different areas have distinct meteorological profiles. Additionally, the most salient parameters per area have been determined, whereby the parameters characterizing the different meteorological profiles of Greece have been uncovered.

1. Introduction

Data mining aims at analysing high-dimensional data in order to determine their characteristics, i.e. expose the natural groups that exist in the pattern space (classification) as well as extract the salient information/characteristics inherent in the data (dimensionality reduction). To this end, a variety of techniques have been developed including statistical-based, neural network-based and hybrid methods.

The task studied in this piece of research involves the analysis of high-dimensional meteorological data collected long-term (over a period of 43 years) at 130 weather stations covering Greece (Fig. 1). Analysis has been performed by a hybrid method that combines the application of the self-organising map (SOM) [1] with statistical clustering [2,3]. The experimental results indicate that the proposed hybrid method is effective in clustering the stations and partitioning Greece into areas with distinct meteorological profiles such that stations in the same area have similar meteorological profiles, while stations classified in different areas have distinct meteorological profiles. Additionally, the most salient parameters per area have been determined, uncovering the parameters (and their ranges) that characterize each distinct meteorological profile in Greece.

This paper is organised as follows: the principles of the SOM and its extensions are presented in section 2; the data mining/clustering/ classification task is introduced in section 3, while the experimental results are given in section 4; section 5 concludes the paper.

Fig. 1 Geographical map of Greece with the locations of the 130 meteorological weather stations.

2. The self-organising map

The origins of the SOM [1] are traceable to the study of biological neural networks in humans and animals, which has revealed that neurons become organised during learning via a self-organisation process in such a manner that neighbouring neurons become sensitised to similar stimuli. Accordingly, the SOM consists of a set of nodes organised into a regular structure (most frequently two-dimensional) and arranged in a rectangular or hexagonal lattice.

The SOM self-organisation process is based on an unsupervised adaptation law, which provides a global ordering of the input patterns through competition between the SOM nodes. The weights of the nodes are initialised either randomly or with pre-defined values that introduce a partial order in the map. During training, each input pattern is broadcast throughout the SOM and the node whose weights best match the pattern is determined. This node together with its neighbours are updated via the following adaptation rule:

$$\begin{aligned} m_i(t+1) &= m_i(t) + a(t)[x(t) - m_i(t)], && \forall i \in N_c(t) \\ m_i(t+1) &= m_i(t), && \forall i \notin N_c(t) \end{aligned} \tag{1}$$

where $m_i(t)$ is the weight of node i at time t, $x(t)$ is the input pattern at time t, $a(t)$ is the gain factor at time t, and $N_C(t)$ is a function denoting the nodes which belong to the neighbourhood of the best matching node at time t. Both the neighbourhood function $N_C(t)$ and the adaptation gain $a(t)$ decrease as training progresses. Since neighbouring nodes are adapted towards the input patterns in a similar matter, they learn to respond alike to each input.

SOM training is divided in two phases:

- *Rough training*, during which the neighbourhood comprises several nodes in an effort to achieve a gradual ordering of the nodes.
- *Fine-tuning*, during which the neighbourhood comprises a very small number of nodes, allowing the weights of each node to be fine-tuned to their optimal values.

A number of extensions to the SOM model have been proposed. These include:

(i) Map structures that evolve during training [4]. This approach results in maps with a non-uniform structure, but customised to accommodate the characteristics of the input pattern space and, thus, to accurately represent the input data.

(ii) Hierarchical, multi-layer map structures [5]. Each node of the first (or a higher) SOM layer corresponds to a SOM of the second (or a lower) layer; the inputs to this SOM correspond exactly to the patterns assigned to the node of the first (or higher layer), thereby providing a detailed representation and ordered classification of the data.

(iii) Hybrid methods [2]. The data are initially processed by a standard SOM. The resulting codebooks are then grouped using a statistical clustering technique in order to expose the main clusters of the input pattern space.

3. The meteorological data mining/ clustering/classification task

3.1 The meterorological data

The National Meteorological Service (E.M.Y) of Greece maintains a network of 130 stations covering the Greek territory (Fig. 1). The data collected at these stations [6] are mainly utilized for studying the weather patterns of different areas and providing the foundation for weather forecasts. They comprise different parameters including: the monthly and yearly maximum, minimum and average temperature, barometric pressure and relative humidity; the monthly and yearly average number of sunny and cloudy hours per day; the monthly and yearly average levels of precipitation and snow; the monthly and yearly average wind speed.

The data employed here span a period of 43 years (from 1955 to 1997); a single value (equal to the numeric average over all years) per parameter has been utilised. The target is to determine whether the 28 pooled parameter values per station can provide a reliable foundation for clustering the stations and partitioning Greece into areas with distinct meteorological profiles, as well as to analyse the meteorological characteristics of the Hellenic region and extract the most salient parameters per area.

3.2 The hybrid method

Two-dimensional SOMs with hexagonal lattice neighbourhoods have been employed for clustering the 28-dimensional data of section 3.1; the map sizes range from 2x3 to 20x30 nodes. To initialise the SOM nodes, the weight values have been set randomly. A relatively lengthy training procedure has been employed, namely 1,000 and 10,000 iterations of the entire training set during the rough-training and fine-tuning phases, respectively. To simulate the SOM network, the SOMTOOLBOX for the MATLAB environment has been used [7].

Following the assignment of the stations to the SOM nodes, the topological properties of the formed map have been investigated by determining the similarity of the codebooks of the SOM nodes. Clustering has been performed combining:

(I) The unified distance matrix (U-matrix) [8], which constitutes a tool for visualising the distances between neighbouring nodes and identifying the cluster structure within the trained map. High values of the U-matrix indicate a cluster border, while areas of low values indicate regions within the clusters.

(II) Ward's method via the squared Euclidean distance [3]. The created dendrogram indicates the gradual grouping of the codebooks into clusters as well as the distance at which clustering is achieved. The SPSS statistical package has been used.

4. Experimental results

In order to investigate as well as validate the performance of the proposed hybrid method described in section 3.2, the 130 stations were uniformly normalized and subsequently split into two sets, the training set (120 stations) and the test set (the 10 remaining stations).

While the training set was employed for the assignment of the 120 stations into the SOM nodes and the assignment of the codebooks of the SOM nodes into the resulting clusters (areas with similar meteorological profiles), the test set examined the generalization capability of the hybrid method, i.e. whether data from unknown stations would be classified in a manner consistent with the characteristics of the best matching SOM node and/or the corresponding area. Twenty different training (and test) sets were created, each time randomly selecting the 120 stations of the training set. The results reported here have been restricted to SOMs of size 2x5. This choice has been made as the limited map size produces meaningful clustering of the data. Provided that none of the ten stations of the test set were outliers, the proposed method was proved consistent both in classifying the stations of the training set into groups of stations characterised by similar meteorological profiles and in successfully classifying novel stations. The operation of the hybrid method is detailed in the following, which are further illustrated via two representative as well as complementary examples (Figs. 2-3).

4.1. Comments concerning SOM training

Figs. 2(a) and 3(a) depict the classification accomplished by the SOM nodes, where each station has been annotated with a square marker inheriting its colour from the colour of the corresponding SOM node. Evidence of ordering as well as clustering is apparent. For instance, all stations located at islands of the Ionian sea are assigned to either one or two neighbouring SOM nodes; similarly, all the stations corresponding to Northern Greece are assigned to three neighbouring SOM nodes, which are situated far from the nodes dedicated to the islands of the Aegean.

This demonstrates the considerable difference in the meteorological profiles of the stations of Northern Greece as compared to these of the Aegean islands. The relative distances of the codebooks, as corroborated by the U-matrix and the dendrogram, consistently verified the clustering of the SOM nodes into three areas with distinct meteorological profiles (Figs. 2(b) and 3(b)) corresponding to:

(a) The islands of the Ionian and Aegean sea as well as the northern part of Crete.

(b) The southern and/or coastal (sometimes including some islands) part of Greece and the southern part of Crete.

(c) Northern Greece (the provinces of Macedonia, Epiros and Thraki) as well as mountainous areas of Southern Greece.

The nodes corresponding to each area were immediate neighbours in the map, so that area (a) occupied one end of the map, area (b) the central portion of the map and area (c) the other end of the map.

4.2. Comments concerning SOM testing

The stations corresponding to the test set are indicated by dots in Figs. 2-3. As can be seen, these stations were correctly assigned to a SOM node as well as an area (indicated by their colour, which is similar to that of neighbouring stations that were used during the SOM training).

An indicative example is provided in Fig. 3. Crete constitutes the sole island containing stations from all three areas, a fact that is due to Crete being the largest Greek island and having mountains exceeding 1,000 meters in altitude as well as extensive coasts and planes situated near sea level. One of the stations of the test set corresponds to one of the only two stations in Crete that are assigned to area (c). Although its classification is different to that of the stations in the vicinity, it is given correctly owing to its similarity – in terms of parameter values, i.e. meteorological profiles - to the other stations belonging to the same area. This can be verified by

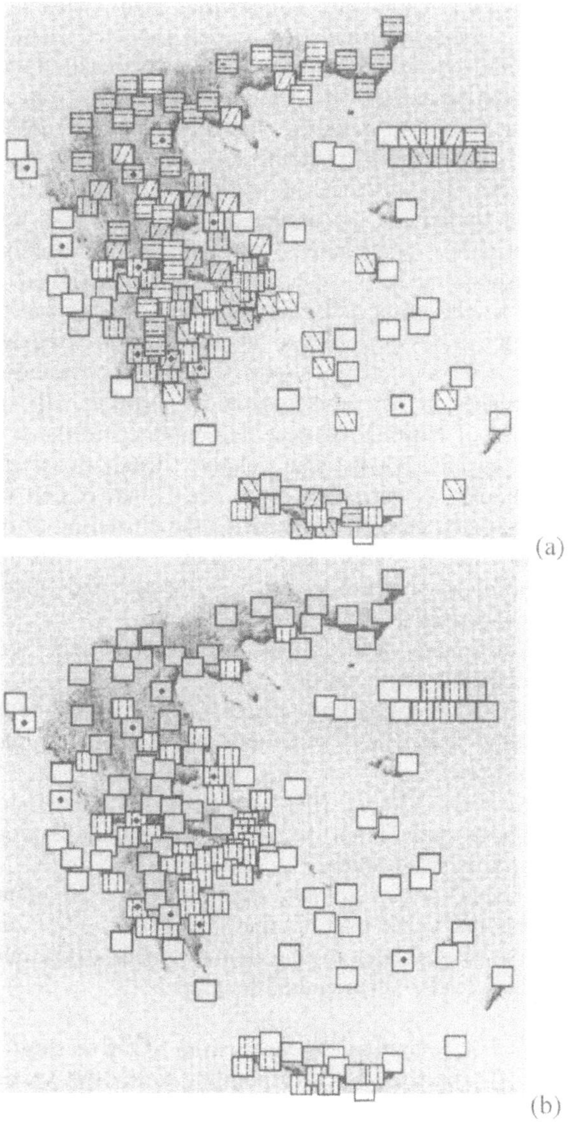

Fig. 2 Assignment of the weather stations to the ten nodes of the 2x5 SOM map (a) and the three areas obtained by clustering (b). Stations-belonging to the test set are denoted by a dot within the corresponding square marker.

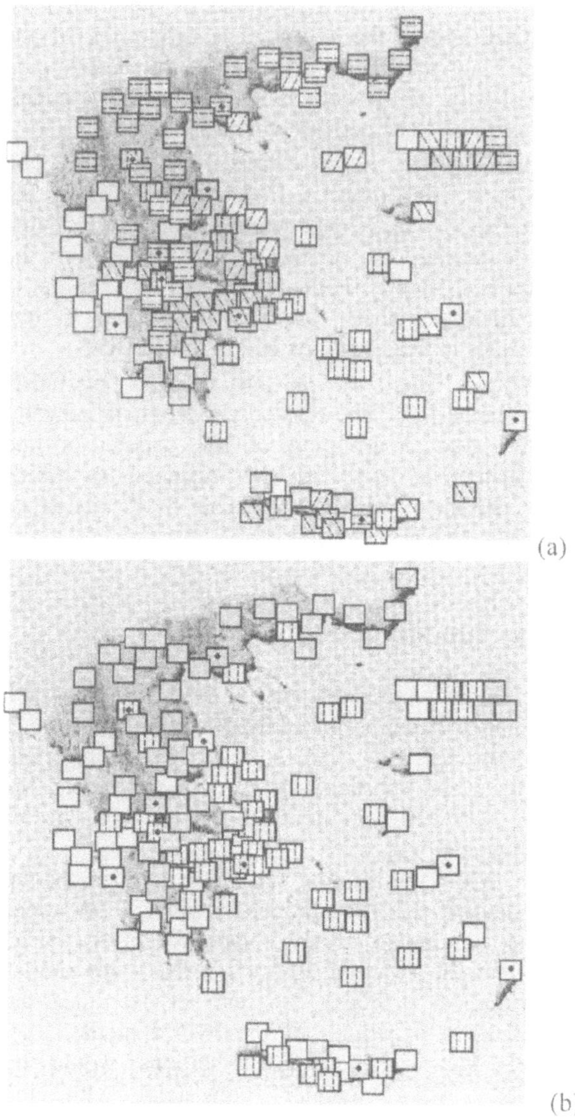

Fig. 3 Assignment of the weather stations to the ten nodes of the 2x5 SOM map (a) and the three areas obtained by clustering (b). Stations belonging to the test set are denoted by a dot within the corresponding square marker.

examining the classification of this station in Fig. 2, where the station belongs to the training set.

Finally, the most salient parameters over the 130 stations as well as per area have been determined by examining the parameter values of the codebooks as well as their variation over neighbouring nodes. It has been found that parameters related to temperature, wind speed, ice and snowfall are the most significant in the classification of the stations into the SOM nodes and the three areas, while parameters related to humidity, cloud cover, wind direction, rain and stormy conditions are not. Furthermore, moderate temperature values as well as higher values of humidity, precipitation and the chance of fog characterize the *moderate & humid* meteorological profile of area (a), lower temperature values and stronger winds characterize the *cold & windy* meteorological profile of area (c), with area (b) lying between these two extremes with characteristics closer to those of area (c).

5. Conclusions

A hybrid method combining the SOM with statistical clustering has been implemented for performing data mining. The data analysed here constitute high-dimensional meteorological measurements collected long-term (over a period spanning 43 years) at 130 weather stations in Greece. The SOM, followed by a statistical clustering method applied on the SOM codebook vectors, has been found capable of clustering the stations and partitioning Greece into areas with distinct meteorological profiles such that stations in the same area have similar meteorological profiles, while stations classified in different areas have distinct meteorological profiles; additionally, the most salient parameters per area have been determined. It has been found that the stations are consistently divided into three areas corresponding to (a) the region of the Greek islands (temperate & humid weather), (b) southern Greece (intermediate weather) and (c) Northern and inland Greece (cold & windy weather).

References

[1] Kohonen, T. (1997) Self-Organising Maps (2nd. Edition.) Springer-Verlag, Berlin.
[2] Vesanto, J. & Alhoniemi, E. (2000) Clustering of the Self-Organising Map. IEEE Transactions on Neural Networks, Vol. 11, No. 3, pp. 586-600.
[3] Ward, J.H. (1963) Hierarchical Grouping to Optimise an Objective Function. Journal of the American Statistical Association, Vol. 58, pp. 236-244.
[4] Fritzke, B. (1994) Growing Cell Structures - A Self-Organising Network for Unsupervised and Supervised Learning. Neural Networks, Vol. 7, No. 9, pp. 1441-1460.
[5] Merkl, D. (1999) Document Classification with Self-Organising Maps. In Kohonen Maps, Oja, E. & Kaski, S. (eds.), pp. 183-192. Elsevier, Amsterdam.
[6] Kornaros, G. (1999) Climatic Data of the Stations of the Hellenic National Meteorological Service (period 1955-1997), Vols. 1 & 2. National Meteorological Service (E.M.Y.), Athens, Greece (in Greek).
[7] Vesanto, J., Himberg, J., Alhoniemi, E. & Parhankangas, J. (2000). "SOM Toolbox for Matlab 5". Report A57. SOM Toolbox Team, Helsinki University of Technology, Finland (available at http://www.cis.hut.fi/projects/somtoolbox).
[8] Ultsch, A. (1993) Self-Organised Feature Maps for Monitoring and Knowledge Acquisition of a Chemical Process. Proceedings of the ICANN-93 Conference, Gielen, S. & Kappen, B. (eds.), pp. 864-867, Amsterdam, The Netherlands. Springer-Verlag.

Artificial Intelligence methods for large microsensor arrays : feasibility and design

E. Gaura, R. M. Newman, N. Steele
Coventry University, Priory St, Coventry, CV1 5FB, UK
email: e.gaura@coventry.ac.uk

Abstract

Recently several authors have turned their attention to the design problems of very large arrays of intelligent sensors, such as would be suitable for inclusion as an integral part of a 'smart structure'. Such systems have architectural and data throughput advantages, but to be used effectively need to be designed so as to be 'controllerless' and 'decentralised'. This paper explores some of these design issues, and illustrates how artificial intelligence (in particular Neural Networks techniques) could be effectively incorporated within the design of large intelligent microsystems with the scope of enhancing their functionality. Micromachined acceleration sensors are considered as a case study.

The paper opens with a discussion on the ever-increasing appropriateness of local sensor health monitoring, fault diagnosis and measurement confidence indices. The use of Artificial Intelligence techniques is suggested for implementing on-chip sensor diagnosis. Design issues for large, intelligent arrays of microsensors are brought forward.

1 Introduction

As man-made dynamical systems become increasingly complex, there is an ever-present need to ensure their safe and reliable operation. These requirements extend beyond the normally accepted safety-critical systems (e.g. nuclear reactors, chemical plants, and aircraft) to new systems such as autonomous vehicles and rapid transport systems. Early detection of faults and/or malfunctions in industrial processes and systems can help reduce downtimes and the incidence of catastrophic events. Sensors are essential components of any process or system which makes use of automatic control.

It follows that an important aspect of any process/system fault diagnosis strategy is to attempt to determine their state of functionality. With current fabrication processes it is feasible to integrate a micromachined sensor along with a substantial processing capacity onto a single chip making the intelligent sensor a realistic and very economical possibility. In fact, such a device could be sufficiently inexpensive to make possible the building of systems using much greater numbers of sensors than considered before, given the open ended processing capacity of such systems.

ANNs have been successfully used in a variety of applications for complex data analysis and feature extraction [1]. In the context of the proposed discussion, some of their main advantages are:

- the majority of intensive computation takes place during the training process.
- due to the self-learning capability of neural networks, no detailed a priori information regarding the system under analysis or design is required.
- the massive parallelism of neural networks offers a very fast and robust multiprocessing technique when implemented using neural chips or other types of parallel hardware.

Once the ANN is trained for a particular task, operation consists of propagating the data through the mapping produced by the ANN, thereby making possible real-time self-diagnosis, self-validation and monitoring. Acceleration sensors provide a good example for the discussion here, as their lack of accessible internal signals makes the tasks of diagnosis and validation particularly challenging.

2 Intelligent Sensor Arrays - Motivation

In 1994 the one of the authors patented an 'intelligent vibration sensor' [2]. The device was essentially an accelerometer with integrated analog and digital processing capability as shown in Figure 1, which allowed a large proportion of the diagnostic functions of a complete vibration monitoring system to be localised at the accelerometer.

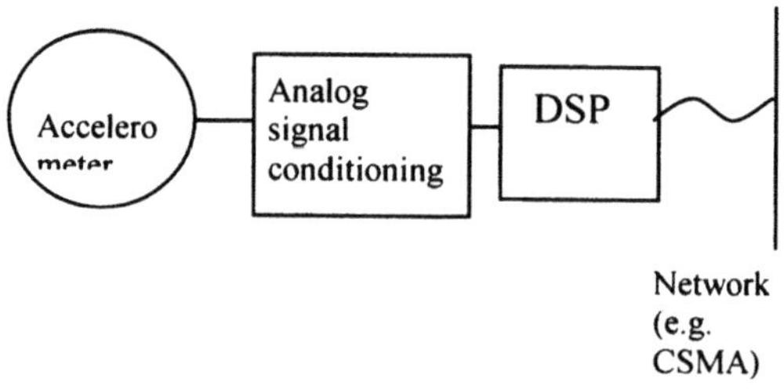

Figure 1: Intelligent Vibration Sensor

This development was a response to one of the major problems encountered with design of such systems at the time. The number of accelerometers typically used in a system had escalated steadily, causing severe data processing bottlenecks. An array of intelligent sensors provided a parallel machine, the processing capacity of which expanded to meet the number of sensors used. By localising signal processing and reduction the data communication bandwidth was also reduced.

In the light of technological developments, the original intelligent vibration sensor architecture could be modified to include three further communications links and a dedicated communications processor (Figure 2).

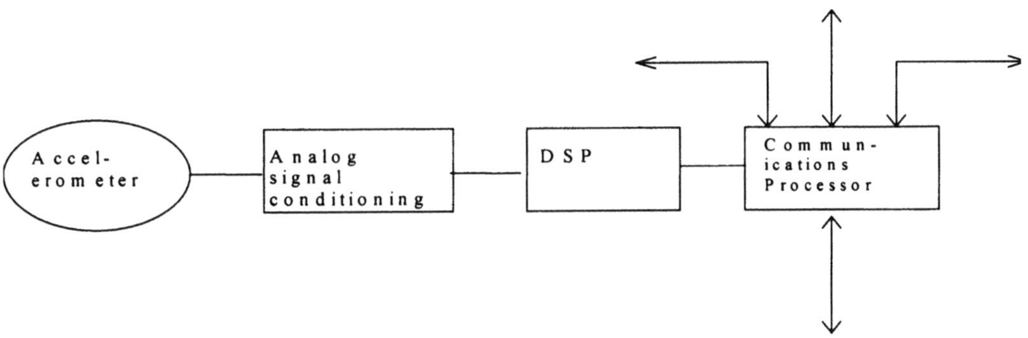

Figure 2: Intelligent Array Vibration Sensor (IAVS)

The device then becomes an 'intelligent array vibration sensor', which can be connected together in a square, two dimensional array of sensors of any size required (Figure 3), which could be integrated permanently into the fabric of a machine or vehicle under test, for example, the main constraints being those of power consumption and the ability to calibrate sensors and handle their failure. Previous work by Gaura [1] shows how the individual sensors might be calibrated and linearised using neural network methods.

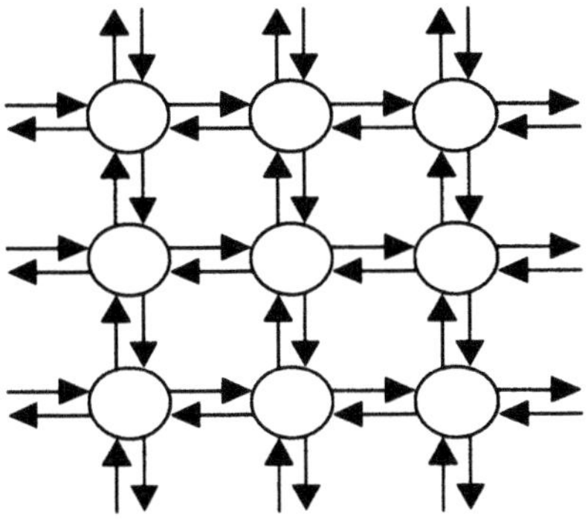

Figure 3: Two dimensional array of sensors

Such an array may be used in a number of ways.

- The large number of processors in the array may be used to enhance reliability by allowing for redundancy. Means would need to be developed for automatic detection of faults and allocation of tasks to adjacent sensors. This possibility is dealt with later in the paper.
- Different analysis tasks may be distributed to different processors within the array, allowing a single system to perform several analyses simultaneously.
- Recognition and analysis algorithms might be distributed to the array as a whole, using the complete system as a 'pattern recognition' device for different spatial vibration patterns. Such patterns are often diagnostic of particular mechanical faults, and could be detected by such means.

Systems are likely to use all three of these possibilities, providing them with specialized processing and an element of redundancy, providing powerful diagnostic procedures with graceful degradation in the case of individual sensor failure.

Several authors have discussed the issues of designing and using systems comprised of very large arrays of intelligent sensors[3,4,5]. If an array of sensors is to be built which can function independently of a master controller, then decentralized means of handling the functions of that controller must be found. One such function is detection of faulty sensors, and their removal from the array [6]. One convenient way to model such a system is as a collection of concurrent processes. This can lead simply to programmed solutions from an appropriate formal specification, in an appropriate programming language [7].

A means for handling this was developed by the authors, described in such a formalism π-calculus [8]. It has been shown that a simple, fail safe protocol (in that sensors which simply fail to work are switched out of the array automatically), it is possible for the array to reconfigure itself in a distributed manner, with no central control over the process.

It follows that sensor self-diagnosis could positively contribute to enhancing the measurement capabilities of an array described above.

3 Sensor Self Diagnosis

In most applications, accurate and reliable sensor readings are vital for good overall system performance [1,9]. Despite advances in fabrication technologies, sensors generally exhibit imperfections (for acceleration sensors, for example, common imperfections are: offset, drift, non-linearity and noise) and the magnitude of these imperfections is found to vary both from sensor to sensor and with time [1].

Localized, real-time sensor self-validation, self-diagnosis technology will enable improved automatic process/system monitoring and control, removing this overhead from a central process controller. Early detection of small, incipient (rather difficult to detect) sensor faults can be achieved and therefore downtime can be reduced and catastrophes can be avoided. More robust, fault-tolerant control can be designed on the above basis, to accommodate/compensate for soft sensor failures (i.e., recoverable sensor failures that leave no permanent damage). Finally, easy identification of sensors which have suffered hard failures (irrecoverable sensor failures) can be achieved.

In safety-critical systems, (e.g. civilian aircraft) typical solutions for overcoming the sensor fault problem are the use of hardware redundancy or majority voting [9]. Such a 'collective' measurement validation technique is often prohibitive. For example, one of the problems associated with systems comprising large number of accelerometers, is that of data analysis bottlenecks. The data analysis generally includes: (1) Fourier analysis; and (2) statistical methods for determining outliers and inconsistent measurements, both of which are computationally expensive processes.

Another approach to validate measurements obtained from sensors and to diagnose sensor faults is to use mathematical/knowledge-based modelling of the system under measurement, to identify inconsistencies in measurement data. Such a system/process specific approach has two major drawbacks:

- detailed mathematical models of the system/process are required; these are generally extraordinarily complicated to construct and may have significant errors (although some success has been recently reported on using ANN techniques for the modeling and fault detection of such systems);
- the validation algorithms are 'tuned' for each system/process. Retuning, due to a slight modification of the system or the introduction of the same techniques to a new application, is extremely expensive and demands a high degree of expertise.

Considerable research effort has been expended on sensor self-validation techniques. No definitive results have been found in the literature referring to self-validating and diagnosis of acceleration sensors, although some interest in the field was shown by NASA, PNRL and Berkeley University.

When considering the sensor at macrosystem level, there is a variety of fault detection methods which could be applied for implementing the self diagnosis task, including:

- analytical methods – the simplest being the threshold analyzer that performs limit checking on the sensor signals;
- signal processing techniques – signature analysis and spectrum analysis;
- statistical - mean and variance analysis.

Initial studies carried out by the authors have established that, initially, a set of design requirements should be drawn to enable unique sensor failure diagnosis and measurement validation with minimal sensor requirements (no hardware redundancy is necessary). The approach taken will be to exploit the information content of readily available signals: the sensor output signal and contextual information gathered from the sensor's working environment. The above capabilities could be incorporated into a validation and diagnosis module (VDM), which, associated with the sensor, should be able to detect, in real-time, several common sensor faults and failures and issue specific warnings and provide confidence indices for each validated measurement value.

Prerequisites for this design activity are:

- identification (based on experimentally obtained sensor signatures) of features which characterise several common sensor faults.
- determination of the nature of additional, application related information, which can be used in conjunction with the sensor output for fault diagnosis and measurement validation purposes.
- assessment of the feasibility and applicability of using ANN techniques for qualitatively and quantitatively representing the information gathered above.

The remainder of this paper presents an initial analysis of some of these prerequisites, using acceleration sensors as a working example.

4 Acceleration Sensors – Example

In the development of a reliable signal-based diagnosis and validation strategy, there is a need to consider not only sensor failures (soft and hard) but also situations which can give rise to faulty (false) measurement data, in the specific application where the sensor is used. For the chosen case study, a set of fault signatures for acceleration sensors could be obtained under laboratory conditions for several different scenarios.

To gain an initial idea of the nature of these signatures we developed a number of scenarios relating to the type of information regarding different types of sensor fault condition that might be obtained from the data available to the processor of the smart sensor and perform a simple analysis using electrical equivalent circuits.

Case 1. The sensor being not rigidly attached to the structure under test. We can imagine a simplified version of this case as illustrated in Figure 4. The test structure is represented by a beam, which will have some characteristic vibration frequency. The improperly mounted sensor is mounted on this beam by a compliance or spring. This is in itself a resonant system, with some characteristic resonance given proportional to the mass of the sensor and the stiffness of the spring. We can produce an electrical analogue of this system, as shown in Figure 5.

So long as the resonant frequency of the poorly mounted sensor is widely separated from that of the beam, then the symptoms of the mounting fault can be clearly differentiated. Since a micromachined sensor is likely to be very light, and even a poor mounting quite stiff, it can be expected that the symptomatic resonance of the sensor mounting will be very high. Thus, at first sight it appears that a properly designed sensor should be capable of self-diagnosis of such a fault by itself.

However, we should also consider the way in which faults in the structure are likely to manifest themselves. Imagine that the beam shown above suffers an incomplete fracture, as shown below in Figure 5. This results in two beams, connected by a compliance, giving an electrical analogue shown here. . This is similar to that for the loosely mounted sensor, with the exception of the coupling compliance Now there are two beam resonances of concern, but in most cases it is likely that they will be at a much lower frequency than the symptomatic resonance of the sensor. The only case in which there might be some scope for misidentification is if the sensor is located close to the end of the beam and the fracture is located close to the sensor, resulting in a very high resonant frequency for the piece of the beam close to the sensor. If this frequency is in the range of that expected from a loose sensor, then the difference between the two signatures is the compliance linking the two resonant systems which occurs if the fault is in the beam, rather than the sensor mounting. The effect of this compliance is shown in the frequency plot in Figure 6. It is possible that a suitably designed diagnostic could detect such a characteristic, but the presence of information from adjacent sensors as to the magnitude of such a resonance would certainly make diagnosis easier.

Case 2. An extreme case of the above is if the sensor detaches completely from the structure under test, in which case no resonances will be detected. Such a condition may occur simply because there is no stimulus to cause resonance. Information from other sensors in the locality can be used to identify whether this is the case, a detached sensor being indicated by a single sensor detecting no resonances surrounded by ones which do. Complete loss of function of a sensor, for instance due to loss of power supply, would cause similar symptoms, except that there would not even be residual noise detected.

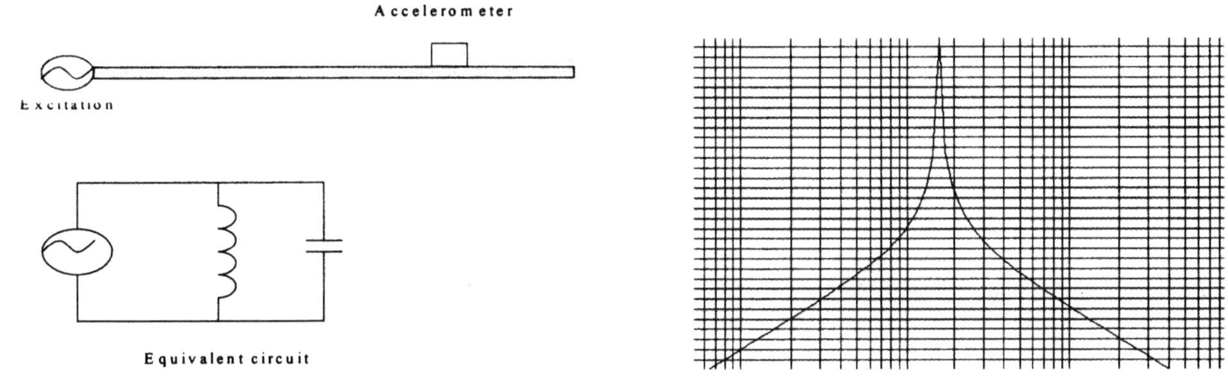

Figure 4: Beam model of structure under test, equivalent circuit and resonance

Case 3. Another related case is if the sensor housing suffers some structural damage. Such a problem may include a number of cases, the simplest of which is a small part of the housing becoming partially detached. Once again this takes the form of a small mass attached to the sensor by a compliance. The difference from the case discussed earlier is that the mass is still smaller and so the resonance will be higher, thus the discussion above applies, although the two symptoms should be more easily separated.

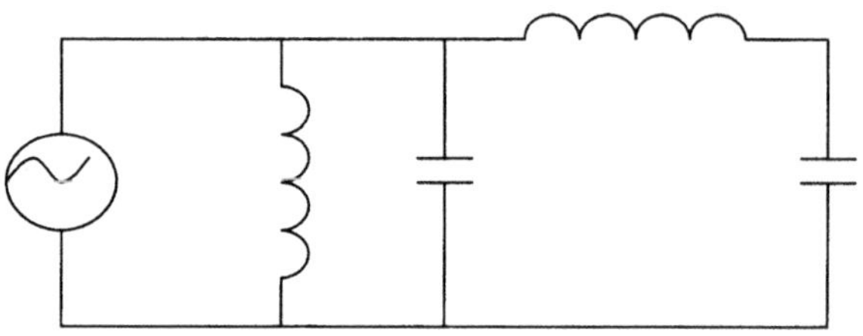

Equivalent circuit

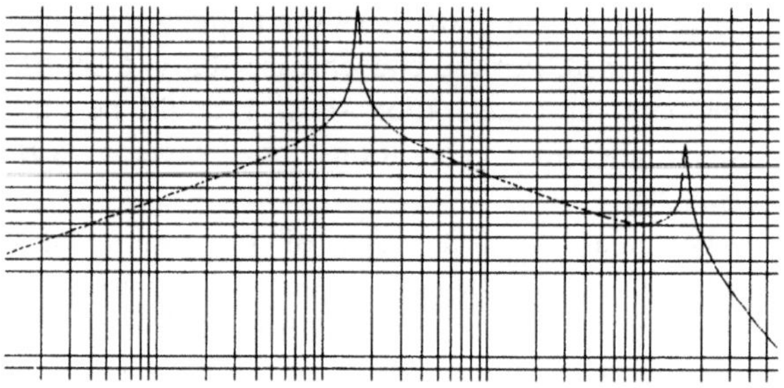

Figure 5: Detached accelerometer model and resonance

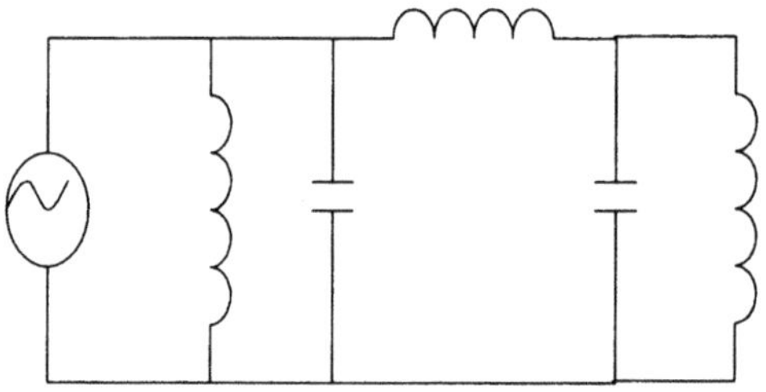

Equivalent circuit

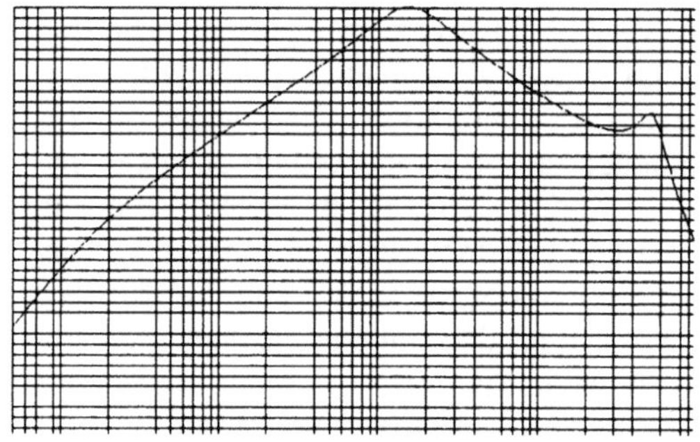

Figure 6: Beam fractured near accelerometer, model and resonance

Case 4. Changes in ambient conditions can be detected at the sensor, if properly equipped and compensated for appropriately.

Case 5. Detection of parameter drift of an individual sensor, or of internal damage to its structure would cause would depend on detection of variation of the output of that sensor from its neighbors, once again in a way which could not be confused with symptoms of the structure under test. The nature of these variations would need to be characterized to be detected.

The faulty and healthy sensor signatures will be analyzed by an ANN, with a view to extracting their characteristic features.

A set of rules based on physical principles can be deduced for the data expected from these sensors. (For example, acceleration measurements from sensors situated at adjacent locations along a cantilever will only be permitted to be different within certain limits imposed by the expected accelerations to which the object is subjected.) The application-related reasoning should be kept to a minimum, in order to maintain the overall generality of the method. In designing the experiment it will be necessary to ensure that the range of accelerations used for developing the rules encompass the accelerations used for fault identification.

A typical node in a sensor array configuration which includes the ANN processing is depicted in Figure 7, with the Neural Network component detailed in Figure 8. For the 5 fault cases above, a possible NN design (based on a feed-forward topology) would have 5 input signals (one from the sensor to which the net belongs plus four coming from its neighbors) and 4 binary outputs (one for the Healthy/Non-healthy diagnostic and the remaining 3 for a code of the specific fault). Various strategies for training the network are currently investigated (the network inputs could be time-domain signals or frequency domain ones, for example). The window size for the inputs history for accurate diagnosis is seen as one of the critical network design factors.

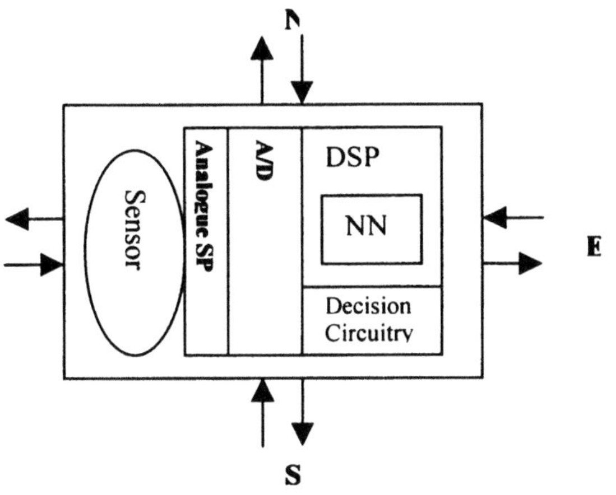

Figure 7:Internal structure of self diagnostic sensor

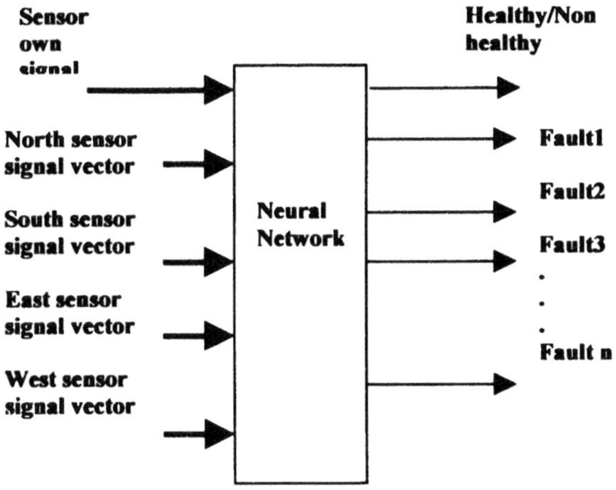

Figure8: Diagnosis for an individual sensor

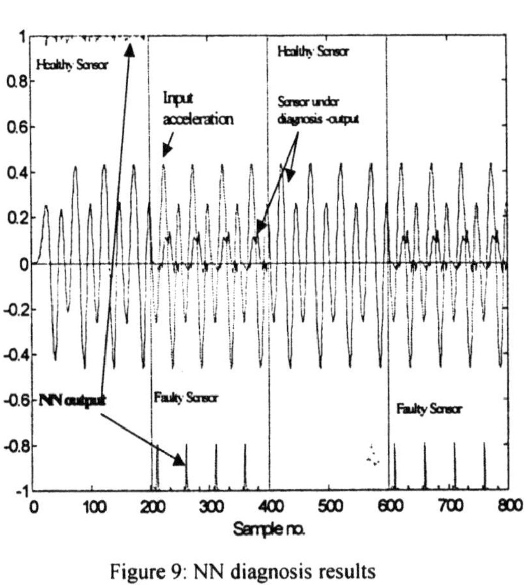

Figure 9: NN diagnosis results

Following on from the discussion in section 3, it becomes clear that, the sensor self-validation on the basis of its output only is an impossible task. A set of rules based on physical principles can be deduced for the data expected from neighboring sensors. For the study presented here, it is assumed that we have two neighboring sensors, S1 and S2, situated in close proximity of each other, on a vibrating beam. The diagnosis system under design is associated with sensor S1 and the contextual information is provided by sensor S2. The input acceleration for the sensors ($a1(\omega t)=a2(\omega t+\varphi)$) is a filtered white noise signal, with a frequency range of 0-100Hz, sampled with a sampling rate of 4kHz. One sensor fault only was considered at this stage, corresponding to Case1 in the previous section. The scaled input acceleration for the sensor being diagnosed, healthy sensor output and faulty sensor output are shown in Figure 9. The diagnosis module to be designed will consist of an ANN, whose task is to identify the healthy/faulty response of sensor S1 in the following situations: S1, S2 – healthy; S1 healthy, S2 faulty; S1 faulty, S2 healthy.

The ANN proposed is of a tap-delayed feed-forward type, with two hidden layers, trained by dynamic backpropagation (with a momentum term and a variable learning rate). Use is made of tap-delayed-lines in order to incorporate the dynamic behavior of the sensor into the model. The present and the past values of the sensors S1 and S2 outputs form the input vector to the neural network. The network output represents the Healthy/Faulty condition of S1 at any instant of time. Since no feedback loop exists in the model, static error backpropagation (BKP) can be used to adjust the network parameters.

Based on these considerations, the electrical equivalent circuits in Figure 1 and 2 were simulated in SPICE in order to gather the input-output network training data. The ANN was trained and tested using Matlab. The best network performance (in terms of least false alarms and highest correct diagnosis rate) was obtained with a 8x31x17x1 network architecture. The network performance on a test set (produced under the same conditions as the training set) is shown in Figure 9. The continuous line represents the correct diagnosis expected from the sensor, with +0.99 identifying the Healthy condition and –0.99 identifying the Faulty condition. The network correctly assesses the sensor's health throughout the test.

It has therefore been possible to design a working self-diagnosis module for the acceleration sensor considered, on the basis of its own output signal and a minimum of contextual information, non-application based, provided by one neighboring sensor. The results obtained encourage the continuation of this line of research towards multiple faults diagnosis. It should be noted here that a single neighbor as context might not be sufficient for such a task.

5 Conclusions

The paper discussed the suitability and feasibility of enhancing the reliability of microsensors by adding an on-chip self-diagnosis capability. The approach taken is based on Artificial Intelligence techniques and sensors with no accessible internal signals are taken as an example. Some common acceleration sensor faults are considered and an indication is given of the manner in which these faults can be detected and isolated, either on an individual sensor basis or based on cooperative work within a sensor network. The design requirements for such self-diagnosable measurement systems are set and further practical implementation issues are raised.

A Self-diagnosis sensor module was designed which is able to successfully assess its state of health in respect to one fault condition. Work is in progress to enhance the module's ability to detect several other fault conditions.

References

[1] Gaura, E. Neural Network Techniques for the Control and Identification of Acceleration Sensors. PhD Thesis, Coventry Univ., UK, 2000

[2] Newman, R.M. and Robinson, B. (1994)Intelligent Vibration Sensor, U.K Patent No GB 2251071, Grant Date 3 Aug 1994

[3] M. Chu, H. Haussecker, F. Zhao, (2002) Scalable information-driven sensor querying and routing for ad hoc heterogeneous sensor networks. Xerox Palo Alto Research Center Technical Report P2001-10113, May 2001.

[4] Estrin, D., Govindan, R., Heidemann, J., Kumar, S., (1999) Next Century Challenges: Scalable Coordination in Sensor Networks. Proc. ACM International Conference on Mobile Computing and Networking, (Mobicon 1999) pp. 263-270.

[5] Wang, A, Chandraskan, W (2002) Energy Efficient DSPs for Wireless sensor networks, IEEE Signal Processing magazine, July 2002, pp. 68-78.

[6] Jaikaeo, C., Srisathapornphat, C., Shen, C.-C., (2001) Diagnosis of Sensor Networks, Proc IEEE International Conference on Communications, Helsinki, Finland, June 11--14, 2001.

[7] Newman, R. M. (1998) The ClassiC Programming Language and Design of Synchronous Concurrent Object Oriented Languages, Journal of Systems Architecture, Elsevier Scientific, September 1998.

[8] Newman, R. M.(2003) Using very large arrays of intelligent sensors, to appear in Proc. Modelling and Simulation of Microsystems 2003, San Francisco, CA February 2003.

[9] NASA TechTracS. System Fault Detection and Accommodation, Technology Opportunity Sheet, http://technology.larc.nasa.gov/scripts/nls_ax.dll/twDispTOPSItem(111;TOP3-2;0;1), 2000

A new approach for on-line visual encoding and recognition of handwriting script by using neural network system

Basma JOUINI, Monji KHERALLAH, Adel M. ALIMI
REGIM: Research Group on Intelligent Machines,
University of Sfax, ENIS, Department of Electrical Engineering.
BP W - 3038, Sfax, Tunisia
Tel. +216-74-274-088, Fax. +216-74-275-595
Emails: { Monji.Kherallah,Basma Jouini, Adel.Alimi }@ ieee.org

Abstract:

In this paper we have developed a prototype of lecture support system using PDA's, This lecture support system concern the fourth students. The extraction of parameters is based on the visual encoding gives 72% as a extraction rate. The originality of the on-line recognition method is in the vector of input of the neuronal network system. We developed database contains 50 000 Arabic words that 70% are used for learning the neural network system and 30% are used for testing the recognition system. The recognition rate obtained is 90%.

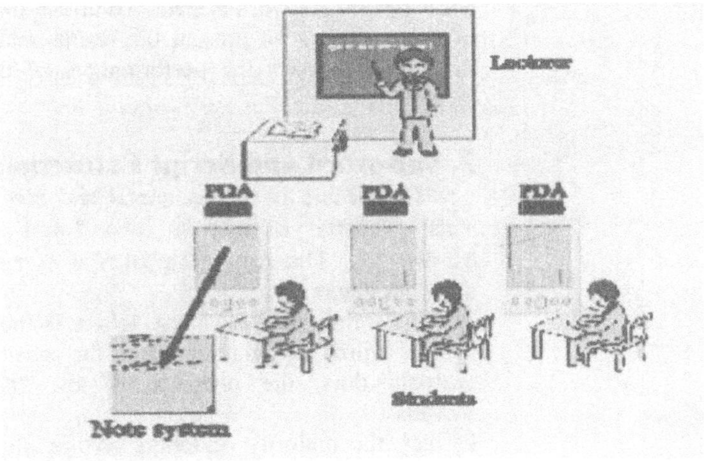

Fig.1. *Configuration of lecture support system using PDA's*

Fig.2. *Personal Digital Assistant*

1. Introduction:

In the last few years a number of pen-based computers and personal digital assistants (PDA's) have been released (Fig.1., Fig.2.). Most of the PDA's have difficulty to recognize handwritten text. Since handwriting is one of the most familiar communication media, pen-based interfaces combined with automatic handwriting recognition offer a very easy and natural input method. We notice the great increase of interest in the development of on-line handwriting recognition systems [1,2,3,4,5,6,7].

On line recognition refers to the recognition mode in which the machine recognizes the handwriting while the user writes on the surface of a digitizing tablet with an electronic pen. The digitizing tablet captures the dynamic information about handwriting, such as number of stokes, stroke order, writing speed, etc, all in real time. Off-line recognition, by contrast, is performed after the handwriting has been completed and its image has been scanned in. Thus, dynamic information is no longer available.

Another advantage of on-line recognition over off-line recognition is that there is interaction between the user and the machine. The user can, thus, correct any recognition error immediately as it occurs.

In this paper we have applied a prototype of a lecture

support using PDA's as shown in figure.1. It concerns the fourth year students. This prototype of lecture refers to the recognition mode in which the PDA recognizes the handwriting while the student writes on the surface with an electronic pen.
Most lecture support systems consist of many PCs in a computer exercise room and use special software for lectures [17].
This work presents 3 sections. In the first we introduce a new approach for on-line encoding the
handwriting script. The second section develops the on-line recognition system using the MLP neural Network. At the end, we give the results and *discussion.*
The recognition module is accomplished in three stages: (Fig.3)

1. Data preprocessing
2. Feature extraction
3. Recognition system

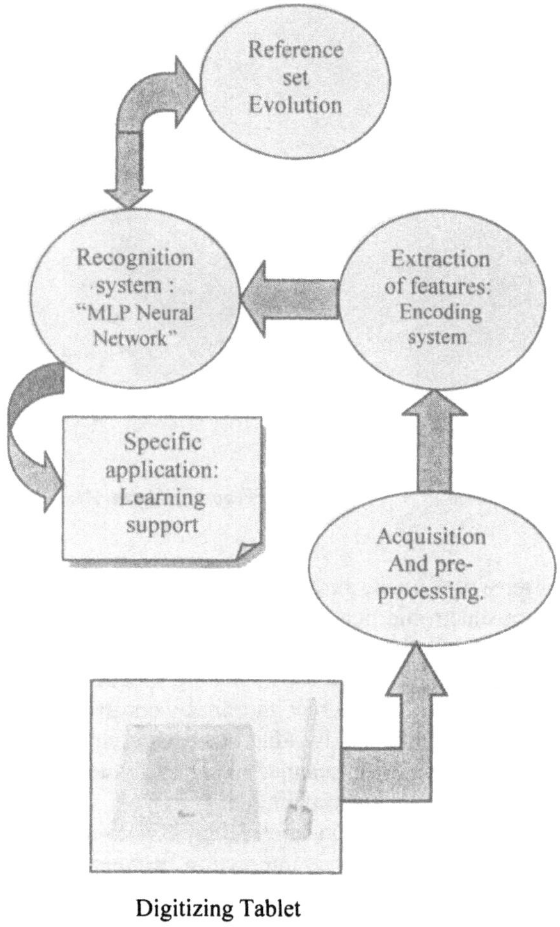

Fig.3. *Different steps of the recognition module of on-line Handwritten cursive word*

Feature extraction is a very crucial step, as the success of a recognition system is often attributed to a good feature extraction method.
In recent years, some researches have made a lot of efforts in the field of the cognitive psychology and the linguistics to conceive the model of perception and reading by man.[8,9,10,11,12,13,14,15].
First, we present an approach to on-line encoding the Arabic handwritten word. It consists of representing a word from its original sequence of 2D coordinating it to sequence of visual codes. In fact, our approach is based on the psychology of the visual perception to extract primitive description of the word, which is useful in the stage of on-line handwritten recognition. For this reason, some algorithms are developed to extract the visual codes from the cursive script. The extraction experiment has been conducted with 50 Arabic words of only one written style. The extraction rate of this experiment is 72 %.
After that, we developed a data base containing 50000 Arabic words that are used for learning and testing the MLP Neural Network system. To affirm the usefulness of these stages, we present the results and discussion that tend to prove the performances of the different algorithms.

2. Sub-word and Script Extraction

The Arabic is a consonantal and cursive writing. This property is met in two forms printed or handwritten. The Arabic alphabet is composed of 28 different letters.
The difference between these letters is their positions in the word, the number and the position of the diacritic dots, the presence of the "hamza" and Vowels.
In fact, the majority of letters change slightly in the shape of their character according to their position in the word (initial, medium or final) if they are joined to another letter either if they are isolated.

2.1 Diacritical symbols influence:

Generally, an Arabic sub word consists of a single main stroke written without lifting the pen, sometimes combined with secondary stroke for which the pen has to be lifted in order to complete sub-word. Accordingly, any Arabic sub-word has exactly one main stroke and zero or more secondary stroke. Some Arabic characters have the same shape, however, they are distinguished from each other by the addition of dots in different positions relative to the main stroke. Some Arabic characters use special marks to modify the character accent. When diacritical symbols (dots, specials marks) are used, they appear above or below the characters and they are drawn as isolated entities as

shown in fig. 4. Diacritical symbols are positioned at a certain distance from the character. In fact, this makes some difficulties in separating the border of a text line. Indeed, diacritical symbols can generate some redundant separate lines [13, 14].
We count 15 among the 28 letters of the alphabet, which contain dots. Some letters present a zigzag shape said 'Hamza'. It is considered as accent "vowel" in the Arabic alphabet.

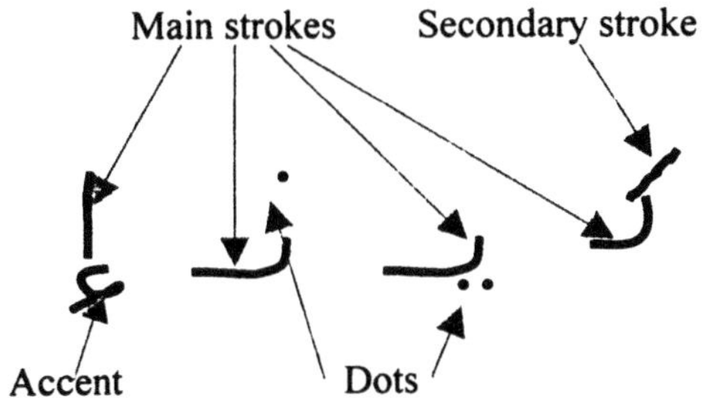

Fig.4. *Characters with diacritics and secondary Stroke*

2.2. Vowels:

The number of these vowels is seven. They are defined as being auxiliary entities that accompany letter to clarify their pronunciation. They take place above and below the letters.
In the majority of cases, the Arabic writing does not use vowels; the sense of the word is often determined by the context of the sentence.

2.3. Primitive extraction

Thanks to the analogy with man to recognize the handwriting, the researchers were able to resolve the problem of primitive extraction of the cursive word [8, 9, 10].
The fact to raise perceptive information such as 'visual indications ' allows us to facilitate the development of the recognition system of the handwriting.
The primitive must be relevant: (little variants to the changes of the same class and discriminate with great variants inter-classes). The word has generally two main zones: the median zone and the striking zone that will be described afterward. Every zone is composed of its own visual indications as it is presented in the following tables. Table (1, 2).

3. Proposed method

According to the harmonic oscillator description of the muscle action involved in handwriting production, cursive handwriting generation can be viewed as a sequential modulation of two coupled oscillations; One in vertical direction and one in the horizontal direction. In this context, it is natural to characterize the writing as an ordered sequence of "upward" and "downward" strokes. However, it has

Table.1. *Visual features of the striking zone and Freeman codes associated*

Right oblique shaft	**Hod**		F5, F5, ...,
Leg	**Ja**		F5, .., F6, ..,F7..
Pocket	**Po**		F6,.,F4,.,F2,F2,
Alif	**Al**		F6, F6, ...
Shaft	**Ha**		F2, ..., F2, ...
Left oblique shaft	**Hog**		F7, ...
Dot up	**Ph**		
Down dot	**Pb**		

Table.2. *Visual features of the median zone and Freeman codes associated*

Curve right opening	**Cod**		F4,..,F5,..,F6 .F7.F0.
Curve down opening	**Cob**		F4,..,F5,..,F6 F6.F7.F0.
Ha	**Haa**		F7, ..., F5, ...
Valley	**Va**		F4, .., .F3,...
Ain	**Ain**		F3,F0 ..., F5, ...
Occlusion	**Oc**		F4 F3,..F2, F1, F0, F7, F6, F5, F4
Sad	**Sad**		F4, ...F4, ...F2, ...F0, F6, F4
Space intra-tracing	**#**	**#**	

been previously suggested that downward strokes that represent the median zone in a word are more important than upward strokes because they are always part of the letters [16].
Our method consists of encoding the handwritten word by a string of visual indices. So we can deduct clear information of word.
The first step of encoding system is detection and extraction of the median zone. This phase allows us to localize, the majority of visual lines such as occlusions and the valleys.
The second step is to make the Freeman's codes in order to generate easily a string of codes describing the different entities which form the word, and we use these codes to test related entities by deducting the visual code afterward which is a translation of Freeman code in to visual indices. The visual indication is a sequence of Freeman's codes from which we quote for example ' Ja ' of the character ' Ra ' which is coded as follows (F7, F7, F7... F5, F4,), what gives us an approximate form of letter.
The different steps of our algorithm are described as follows: (Fig.5.)

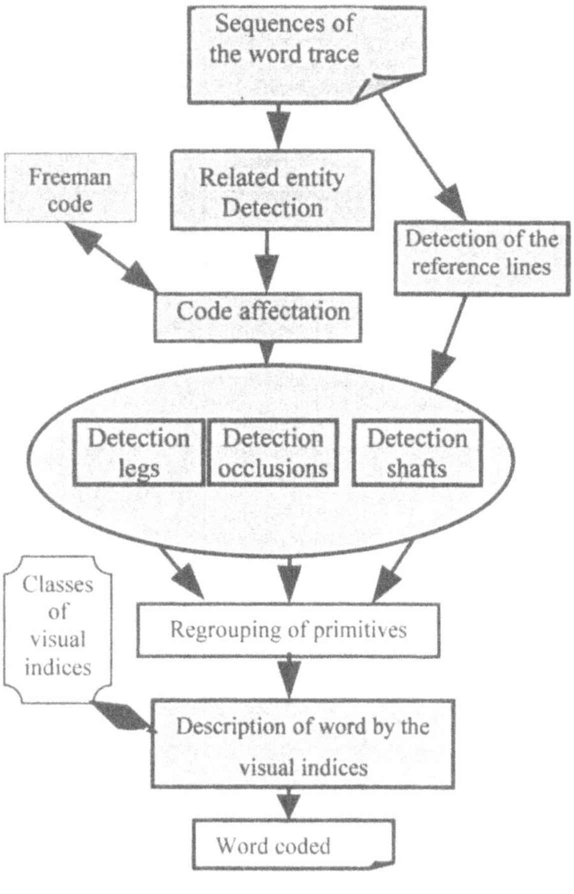

Fig.5. *Different steps of encoding the handwritten words*

- Detection of the primitive: detection of occlusions, legs and handles...
- Grouping of the primitive: grouping of primitive discovered in a single entity describing word.
- Description of the word by visual indices.

3.1 Segmentation

In the case of the on-line handwriting, the decomposition of words in related entities called pseudo-words is directly done. Points acquired among the put and the levying of the electronic pen form a related entity. (Fig.6.)

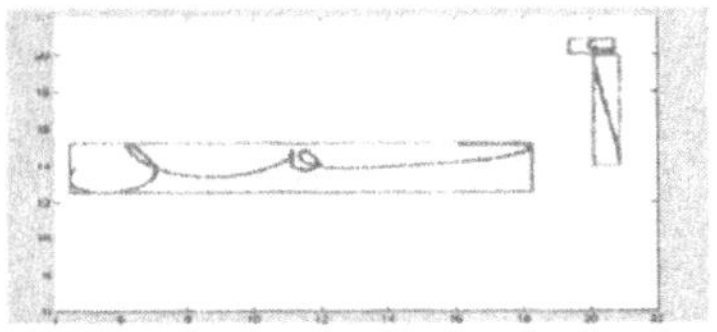

Fig .6. *Segmentation of the word in pseudo word*

3.2 Reference lines detection

Base-line plays an essential role on Arabic writing. Most characters connect each other on the base line. The base line detection is done by detecting the peak in the horizontal density histogram of the text line. Others two lines represent two limits up and down the median zone. (Fig.7)

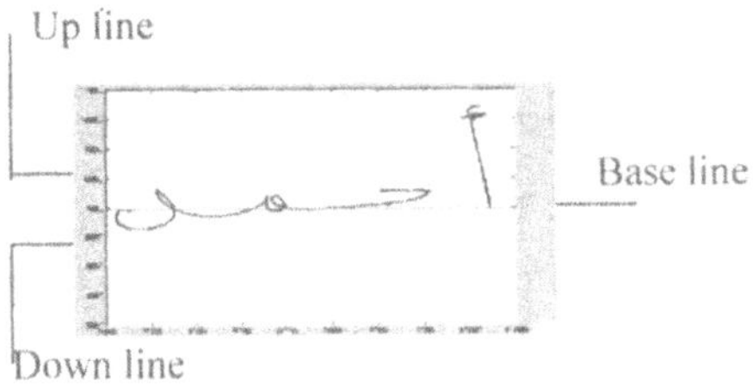

Fig. 7. *The significant lines in word*

3.3 Freeman's code Affectation

This stage represents the base step of our analysis, in fact, we are interested in the generation of Freeman's code to characterize every entity related to part, and this is to be able to generate the visual code of each of the related entities of the word.
The importance of this phase is represented by the generation of the form of the word. These codes are {F0, F1, F2, F3, F4, F5, F6, F7}.
They are affected by making out a will on the values of hillsides and their orientation with regard to the direction of the writing between two points successive of the same entity extracted from the word. (Fig.8, Fig.9)

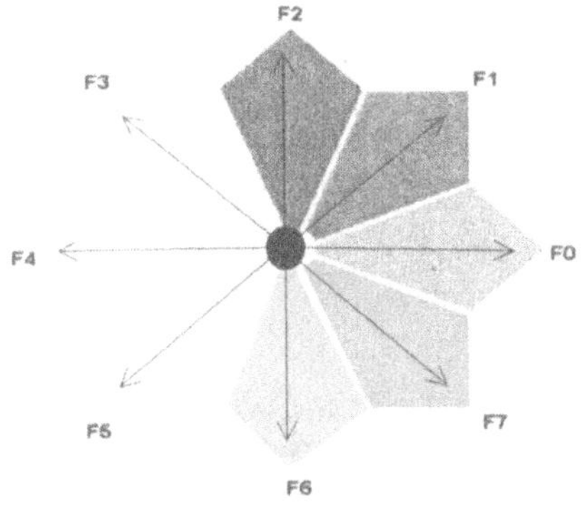

Fig.8. *The eight Freeman codes*

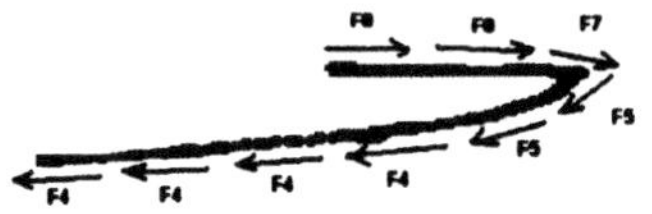

Fig.9. *Example of Freeman's code affectation* (letter ' HA ')

3.4 Word encoding

In this stage, various codes allocated to various visual indications are regrouped to obtain the coding of the complete word. (Fig.10)
For example: Coding the word (الكريم)

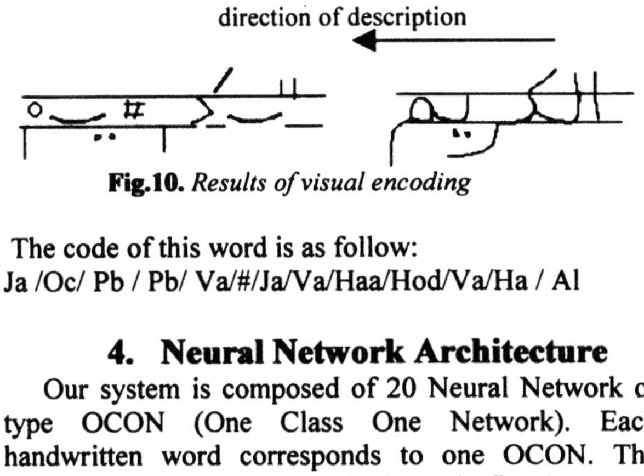

Fig.10. *Results of visual encoding*

The code of this word is as follow:
Ja /Oc/ Pb / Pb/ Va/#/Ja/Va/Haa/Hod/Va/Ha / Al

4. Neural Network Architecture

Our system is composed of 20 Neural Network of type OCON (One Class One Network). Each handwritten word corresponds to one OCON. The global recognition processes is shown in fig.12.
A handwritten word is composed of n primitives, 1< n <16. The input vector is composed of 16 values. We attribute the number 1 for the existing primitive and 0 if the primitive doesn't exist without considering the emplacement of primitives. (Fig.11)

1	0	1	. . . .	0	0	1

Fig.11. *The input vector*

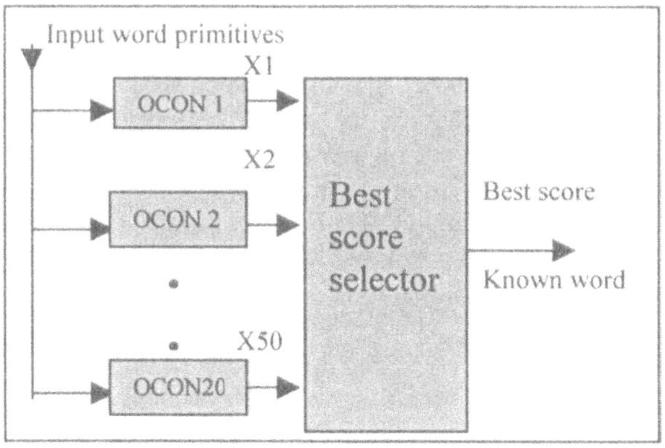

Fig.12. *Global recognition process using* 50"OCON"

The architecture of each OCON neural network is shown in fig.13.
Each neural network was trained by the standard back propagation algorithm. The back propagation learning parameters of interest are: (μ: The learning rate; alpha: the momentum factor and r: the range of the random initial weight) are adjusted so as to trade off speed and accuracy. In the learning phase, 35 000 words were used. 15 000 words were used to test the system.

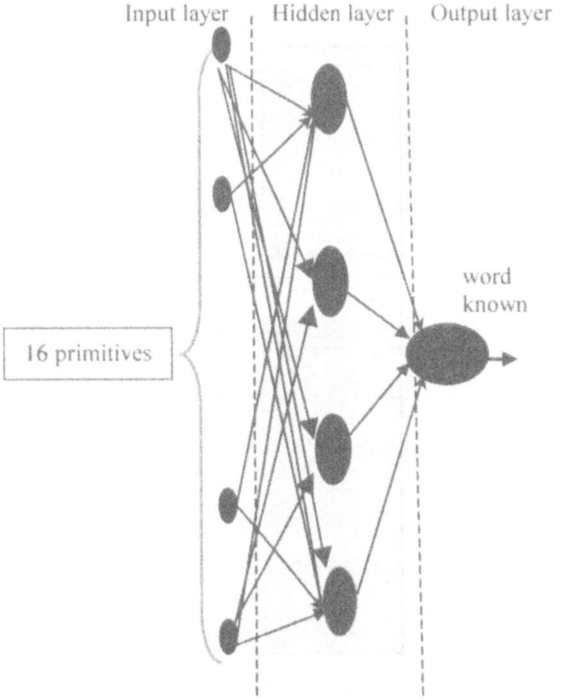

Fig.13. *Architecture of an OCON used for the recognition of a handwritten word*

5. Results and discussion

The performance of the visual extraction indications is rather delicate and remains subjective. The rate of success of the visual coding does not depend on stage of the affectation of the Freeman codes but it is the result of the success of all the stages of process of the visual coding going from the detection of related entities and reference lines until grouping the primitive and description of the word by visual indices.

Table.3. *Results of primitive extraction*

Primitive	*Extraction rate*
Right oblique shaft: Hod	*90,84%*
Leg : Ja	*80,73*
Pocket: Po	*75,03%*
Alif: Al	*95,65%*
Shaft : Ha	*85,44%*
left oblique shaft: Hog	*90%*
High point: Ph	*96,66%*
Point bas: Pb	*95,62%*
Right curve opening : Cod	*76,66%*
Ha : Haa	*46,73%*
Vallet : Va	*80,33%*
Ain : Ain	*73,33%*
Occlusion : Oc	*85%*
Sad : Sad	*50,42%*

In our work we suppose that words to code are normalized and filtered and the base line of the handwritten word is aligned. Experimental tests are realized on a base of 50words acquired by a tablet of type WACOM4 with a frequency of 119.5Hz and a resolution of 500 lines / centimeter.

The following table presents the rate of extraction for every type of primitive with regard to a human observation.
We notice that the rate of the primitive extraction in the median zone is weaker than those of the striking zone because the median zone presents primitive very similar as ' Sad ', ' Oc ' and 'Ain'.

Conclusion

This study concerns the development of a prototype of lecture support system using PDA's. This prototype concerns the fourth year students.
We treated at the first of this study the visual encoding of the handwritten Arabic in real time.
Visual encoding method is based itself on the psychology of perception and temporal information. It contains two great parts: the detection of related entities and reference line then the affectation of Freeman's codes.
The second section of this paper propose a MLP Neural Network method for the on line recognition of the word. The last section of this paper gives the results and discussion. In fact, we tested our algorithms on a base of 50 words and the extraction rate of visual encoding is 72 %. The neural network system achieves a recognition rate up to 90%.
We can exploit this work to develop a prototype of lectures system of the fourth year students. The class room became interactive between the lecturer and students as shown in figure 1.

References

[1] Alimi M.A. and Ghorbel O.A.,"Comparative study of Two Algorithms for the Recognition of On-Line Arabic Handwritten Characters", *Proc. First Int.* Conf. Electronics, Circuits & systems: ICECS'94, Cairo, Egypt, Dec., pp. 973-977, 1994.

[2] Alimi M.A., "A Neuro-Fuzzy Approach to recognize Arabic Handwritten Characters ", Proc. *Int. Conf. Neural Networks*: ICNN'97, Huston, Tx, USA, June,(1997).

[3] Alimi A. M , "Evolutionary Computation for the Recognition of On-Line Cursive Handwriting», IETE Journal of Research, Special Issue on "Evolutionary Computation in Engineering Sciences" edited by S.K. Pal et al., in Press,(2002).

[4] Kherallah M., Njah S., Alimi M.A. and Derbel N., "Recognition of on-line handwritten digits by neural networks using circular and beta approaches", *Proc.Int.* Conf. *IEEE'02, Trans. On man machine interface, Hammamet, Tunisia, 2002.*

[5] Al-Emami S., Usher M., "On-line recognition of

handwritten Arabic characters". *IEEE Trans. On Pattern analysis and machine intellegence. Vol. 12. N° 7.1990.*

[6] Saadallah S., Yacu S. G., "Design of an Arabic character processing and transmission of the Arabic language". *Kuwait 1985.*

[7] Tappert C.C., Suen C.Y and Wakahara T., "The state of the art in on-line handwriting recognition ". *IEEE Trans. on Pattern Analysis and machine intelligence*, Vol.12, no. 8, pp. 787-808, 1990.

[8] Miled H., " Stratégies de résolution en reconnaissance de l'écriture semi-cursive : application aux mots manuscrits arabes", *phd, Université de Rouen U.F.R. de science et techniques, 1998.*

[9] Coté M., " utilisation d'un modèle d'accés lexical et de concepts perceptive pour la reconnaissance d'images de mots cursifs ", *phd, National school of*

Télécommunication, Paris, (1997).

[10] Coté M., Cherie M., Leconet E and Suen C.Y,. "Building a perception Based Model for Reading Cursive Script", *ICDAR, VOL I, pp 898-901, 1995.*

[11] Maddouri S.S,. " Modèle perceptif neuronal à vision globale-locale pour la reconnaissance de mots arabes omni-scripteurs ", *phd, National school Engineering of Tunis, Tunis 2001*

[12] Foorster K.,. "Computaionnel modelling and elementary processs analysis in visual word recognition". *Journal of Experimental Psychology,. "Human Perception and Performance", Special Section, Modeling Visual Word Recognition, 20, no 6:1292-1310, 1994.*

[13] Saadallah S., Yacu S G., " Design of a year Arabic character processing and transmission of the Arabic language ". Kuwait on 1985.

[14] Zahour A., Taconet B and Ramdane S., "Arabic Handwritten Text-Line Extraction", *Proc. Int. Conf. ICDAR'2001.*

[15] Elgammal A M and Ismail M A., "A Graph-Based Segmentation and Feature Extraction Framework for Arabic Text Recognition", *Proc. Int. Conf. ICDAR'2001.*

[16] Giovanni S., "Large Vocabulary Recognition of-line Handwritten Cursive Words". *PhD., Department of Computer Science of the state University of New York at Buffalo for the degree of Doctor of philosophy. August, 1995.*

[17] Munemori J.,Yoshino T and Yoshida A., " Movable Lecture Support System Using PDAs", *Proc. Int.* Conf. *IEEE'02, Trans. On man machine interface, Hammamet, Tunisia, 2002.*

An Immune System-Based Genetic Algorithm to Deal with Dynamic Environments: Diversity and Memory

Anabela Simões[1,2], Ernesto Costa[2]
[1] Dept. of Informatics and Systems Engineering, Coimbra Polytechnic, Quinta da Nora, 3030 Coimbra, Portugal
[2] Centre for Informatics and Systems of the Univ. of Coimbra, Pinhal de Marrocos, 3030 Coimbra, Portugal
E-mail: abs@isec.pt; ernesto@dei.uc.pt

Abstract. The standard Genetic Algorithm has several limitations when dealing with dynamic environments. The most harmful limitation as to do with the tendency for the large majority of the members of a population to convergence prematurely to a particular region of the search space, making thus difficult for the GA to find other solutions when changes in the environment occur. Several approaches have been tested to overcome this limitation by introducing diversity in the population or through the incorporation of memory in order to help the algorithm when situations of the past can be observed in future situations. In this paper, we propose a GA inspired in the immune system ideas in order to deal with dynamic environments. This algorithm combines the two aspects mentioned above: diversity and memory and we will show that our algorithm is also more adaptable and accurate than the other algorithms proposed in the literature.

1 Introduction

When using evolutionary techniques in order to cope with dynamic environments it is necessary to overcome some limitations inherent to traditional evolutionary algorithms. In fact, the long-term success of any biologic evolutionary system is assured by maintaining diversity of individuals within a population. Genetic diversity allows a population to adapt to modifications in the environment. In the classical GA it is difficult to maintain diversity because the algorithm assigns exponentially increasing number of trials to the observed best parts of the search space [9]. Consequently, the GA has strong convergence properties. When dealing with dynamic environments strong convergence can be problematic, because the GA is unable to respond effectively to modifications in the environment. To avoid premature convergence of all individuals of the population towards the optimum, several approaches have been used: hypermutation [1], Random Immigrants [10], new genetic operators [15]. Other approaches tested with non-stationary environments used the incorporation of memory mechanisms in the GA in order to help it, when some situation saw in the past is observed again [2], [3], [11]. In this work we propose a new algorithm, inspired in some ideas present in the natural immune system (IS), and we will test its performance in a classical dynamic optimization problem. The proposed GA will be referred as ISGA (Immune System-Based Genetic Algorithm). The application of ideas of the immune system in dynamic environments is not completely new and other approaches can be found in [6] and [7].

The IS exist to protect us from the dysfunction of our own cells and against the action of exogenous infectious microorganisms. The IS has two important characteristics. First, it is able to respond to infinity many pathogens by its capability to build also infinity many diverse agents adapted to kill each type of pathogen. The mechanism used combine gene libraries with a process of clonal selection, which includes a phase of somatic hypermutation. Second, the IS is able to build up a memory of previous pathogens and how to fight them, so the next time they invade the body the response will be faster and stronger.

In our work we will incorporate these two main ideas of the IS in a standard GA. We tested the ISGA with a biologically inspired genetic operator called transformation already used with success in problems dealing with dynamic environments [17], against a standard GA enhanced with memory (MGA) using one-point, two-point or uniform crossover. We will see that transformation is able to promote high levels of diversity and consequently the results were better than the ones obtained with crossover. The results also show that the proposed ISGA has a typical behavior of primary and secondary response: as the time goes on, the reaction to the changes becomes faster and stronger. We also compared ISGA with other three approaches: the GA used only with transformation (ETGA) [16], the Hypermutation GA (HMGA) proposed by Cobb [1] and the Random Immigrants GA (RIGA) proposed by Grefenstette [10].

2 The Immune system

The IS is a complex, distributed and multi-layered system, which includes cells, molecules and organs that constitute an identification mechanism capable of recognize and eliminate foreign molecules called antigens.

The human body maintains a large number of immune cells. Some belong to the innate IS, e.g. the macrophages, while others are part of the acquired, or adaptive IS and are called lymphocytes. There are mainly two types of lymphocytes, the **B-cells** and the **T-cells**, which cooperate but play different roles in the immune response. The B-cells can be further decomposed into plasma B-cells and memory B-cells. The same happens with the T-cells, which can be partitioned into helper T-cells and killer T-cells. The main functions of the B-cells are the production and secretion of antibodies as a response to exogenous organisms. Each B-cell produces a specific antibody,

which can recognize and bind to a specific pathogen. In order to do their job correctly the B-cells replicates by a process called **clonal selection**. This process is similar to the evolution of a population by means of a genetic algorithm using only mutation: only those cells that have high affinity with an antigen proliferate. Therefore, the B cells that have antibodies, which bind to the pathogen, are selected and cloned. Nevertheless, during cloning some variations may occur due to a process of somatic **hypermutation**. This may increase the affinity between the antibody and the antigen, making the B-cell more adapted to bind to the antigen.
When an antigen enters an organism for the first time, only a few number of B-cells can recognize it. Those cells are then stimulated to produce antibodies specific to the antigen (*primary response*). But, the IS can remember patterns that have been seen previously. This is possible because some cells, including the B-cells, become memory cells, which persist in the circulation and are capable of recognizing enemies when and if they attack again. Therefore, the next time the same pathogen invades the organism, the response of the IS will be much faster and strong (*secondary response*). We may say then the IS has **learning** capabilities based on the **memory** cells.
If each antibody recognizes only a specific antigen, how is possible that a huge number of pathogen can be recognized by the IS? The answer to this question remains in the **diversity** of the antibodies. A heavy and a light chain form each antibody molecule, and a variable and a constant region create each one of these chains. These variable and constant regions of the light chain are created by the concatenation of modular chunks of genes aggregated in **gene libraries** called V, J and C regions (see Fig. 1). The heavy chain is formed by an analogous process [4].

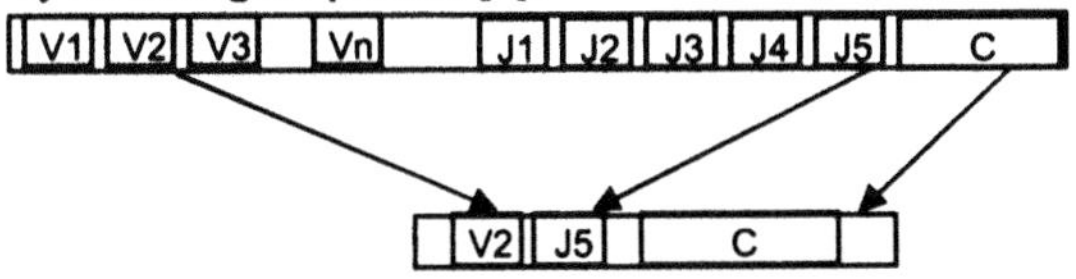

Fig. 1. Concatenation of genetic chunks to create an antibody molecule

It will be those ideas of gene libraries, clonal selection (including somatic hypermutation) and memory B-cells that we will translate and include into the standard GA. They are responsible for promoting diversity and ascribing memory capabilities to our modified GA.

3 Computational Implementation

As we said, in this work it is not our intention to mimic the functioning of the IS. We only used some of the main characteristics of the IS in order to enhance the GA so it will be more efficient when dealing with dynamic environments. In the next sections we will describe the mapping between those aspects of the IS we will use and the modified GA.

3.1 The immune system genetic algorithm

Our starting point is to view the environment as an antigen, and the changes in the environment as the appearance of a different antigen. We will have a process that detects whenever a change occurs. In practice, this will be achieved when degradation in the average fitness of the population is observed. The detecting mechanism is equivalent to the role of the helper T-cells.
But let's see how the GA behaves when the environment is stationary. First of all in the ISGA there are two populations. The first one is a set of individuals (the antibodies of plasma B-cells) that evolve in a process similar to clonal selection: the individuals that have the best match to the optimum (antigen) are selected and cloned. During the cloning phase every individual has a chance of being modified by a mechanism that remembers the somatic hypermutation of B-cells. This mechanism is based in a biologically inspired event called **transformation** and will be explained later. The second population is formed by a collection of individuals, which were the best ones at different moments in the past when they belong to the first population (they are the antibodies of memory B-cells). Each individual of the second population has attached a value which corresponds to the average of the first population's fitness is some particular situation, when that individual had good performance.
When a change occurs and is detected, it probably exists in the domain of the memory B-cells one that has proximity with the new conditions. This proximity is measured by the average of plasma B-cells population's fitness and the value attached to the memory B-cells. The most suitable memory B-cell is then activated, cloned and reintroduced in the population of plasma B-cells, replacing the worst ones. As we see the two populations communicate with each other.
At the beginning of the simulation, we randomly create a set of M libraries, each one containing N gene segments of size L. The libraries are kept constant during the entire evolutionary process (Fig 2).

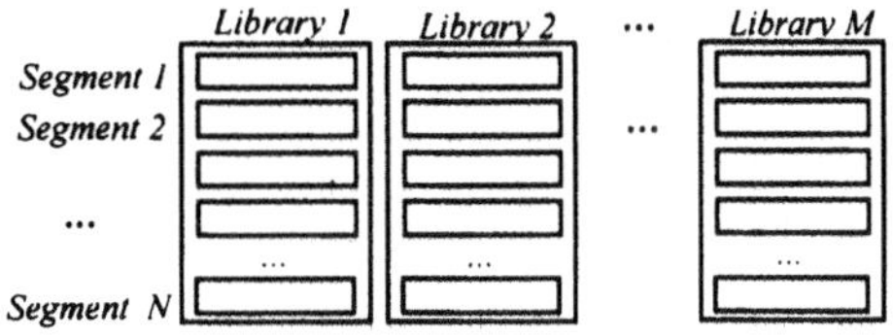

Fig. 2. Organization of the gene segments in libraries

These gene libraries will be used for the creation of the first memory B-cells and in the process of somatic hypermutation. In our experiments we used three libraries with eight segments of size four each and one library with eight segments of size five. The concatenation of one element of each library creates a binary string of size 17 (the length of our B-cells). The number and size of these gene libraries is problem dependent. This idea of gene libraries was also used by [12].

3.2 Transformation

Somatic hypermutation is implemented by a biologically inspired mechanism called transformation. Biological transformation consists in the transfer of small pieces of extra cellular DNA between organisms. These strains of DNA, or gene segments, are extracted from the environment and added to recipient cells [14]. The process of using transformation can be described as follows: the ISGA starts with an initial population of individuals (plasma B-cells) and an initial pool of gene libraries both created at random. In each generation we select individuals to be transformed and we modify them using one gene segment of one gene library chosen at random. To transform each individual we now choose, also randomly, a point in the selected individual. The gene segment is incorporated in the genome of the individual, replacing the genes after the transformation point, previously selected. Form more details see [15].

4 Experimental Setup

In order to test de adequacy of ISGA to dynamic environments we made two types of experiments. In the first one, we compared ISGA with the standard GA enhanced by the memory B-cells (MGA), to see if transformation is a more effective, diversity preserving mechanism; in the second one, we compared ISGA also with three other algorithms: the Triggered Hypermutation GA (HMGA), the Random Immigrants GA (RIGA) and the enhanced transformation GA (ETGA) proposed by [1], [10] and [15], respectively. Now the goal was to see if the double effect of diversity and memory could give better results. In all cases we used the dynamic version of the 0/1 Knapsack Problem (DKP).

4.1 The 0/1 dynamic knapsack problem

The well-known single-objective 0/1 knapsack problem is defined as follows: given a set of n items, each with a weight $W[i]$ and a profit $P[i]$, with $i = 1, ..., n$, the goal is to determine which items to include in the knapsack so that the total weight is less than some given limit (C) and the total profit is as large as possible. In the classical 0/1 knapsack problem, the capacity of the bag is kept constant during the entire run. In the dynamic knapsack problem (DKP) the weight limit can change over time between different values.

We used as a test function a 17-object 0/1 knapsack problem with oscillating weight constraint, proposed by [8]. The vectors of values and weights used for the knapsack problem are exactly the same as that used by the authors. The penalty function for the infeasible solutions is defined by: $Pen=K(\Delta W)^2$, where ΔW is the amount which the solution exceeds the weight constraint and K=20. A solution is considered infeasible if the sum of the weights of the items exceeds the knapsack capacity.

Goldberg and Smith used the DKP to compare the performance of a haploid GA and a diploid GA with fixed dominance map and a diploid GA with a triallelic dominance map. In [10] it is referred that their experimentation used variation of the knapsack capacity between two different values every 15 generations. This problem was already used by other authors to test evolutionary approaches in dynamic environments [11], [13].

In this work we enlarged the number of case studies: we used three types of changes in the capacity of the knapsack: periodic changes between two values (C1=104 and C2=60) and between three values (C1=60, C2=104 and C3=80) and non-periodic changes between 3 different capacities (C1=60, C2=80 and C3=104). Each trial allowed 10 cycles with cycle lengths of 30, 100, 200 and 300 generations.

When the changes in the environment are non-periodic we run the algorithms during 2000 generations and selected randomly several moments of change. In these moments the capacity of the knapsack was altered to a different value chosen among the same three values used in the periodic situation: 60, 80 and 104.

4.2 The parameters of the algorithms

The set of parameters used by the different algorithms were the following.

The three standard GA enhanced with memory (MGA) used 70% of crossover rate and 0.1% of mutation rate.

The ISGA used only transformation with a probability of 90%. Moreover the gene segment length is equal to 4 or 5, depending on the selected gene library. A detailed explanation of the reason for these values is given in [16].

The HMGA uses a baseline mutation rate (usually very low) when the algorithm is stable and increases the mutation rate (to a high value) whenever there is degradation in the performance of the time-averaged best performance [1]. We implemented this mechanism with a baseline mutation rate of 0.1% and whenever degradation is observed we increased the mutation rate to 10%, 20% or 30%. The best results were achieved by a hypermutation rate of 10% (the results presented refer to this value).

The RIGA replaces a fraction of a standard GA's population each generation, as determined by the replacement rate, with randomly generated values. This mechanism views the GA's population as always having a small flux of immigrants that wander in and out of the population from one generation to the next. This strategy effectively concentrates mutation in a subpopulation while maintaining a traditionally low (i.e., 0.001) mutation rate in the remainder of the population [10]. We tested the Random Immigrants GA with a replacement rate of 10%, 20% and 30%. The best results were achieved with the value 10% (the results presented refer to this value) Both HMGA and RIGA were run with one-point crossover with a probability of 70%.

The ETGA was run with the set of parameters obtained by the empirical study presented in [16]. The chosen

values were: no mutation rate, transformation rate equal to 90%, replacement rate of 50% and gene segment length of size 5.
All the algorithms used populations with 100 individuals and were repeated 30 times. In the ISGA and MGA we used an additional population of 20 memory B-cells. The results reported in the next section are the average values of the 30 runs.

4.3. Performance Measures

In order to evaluate the performance of the five approaches (seven algorithms) solving the dynamic 0/1 KP, we used two well known measures, usually employed in non-stationary problems. Those measures are the **accuracy** and the **adaptability**. They are based on a measure proposed by De Jong [5], the **off-line performance**, but evaluate the difference between the value of the current best individual and the optimum value, instead of evaluating just the value of the best individual. Accuracy (Acc) is the difference between the value of the current best individual in the population of "just before change" generation and the optimum value averaged over the entire cycle. Accuracy measures the capacity to recover to the new optimum before a new modification occurs. Adaptability (Ada) is the difference between the value of the current best individual of each generation and the optimum value averaged over the entire cycle. Adaptability measures the speed of the recovery.
The smaller measured values for accuracy and adaptability the better results. If the accuracy reaches a zero value it means that the algorithm found the optimum every time before a change occurs. If adaptability is equal to zero it means that the best individual in the population was at the optimum for all generations, i.e., the optimum was never lost by the algorithm.
These two measures can be mathematically defined by:

$$Acc = \frac{1}{K}\sum_{i=1}^{K} Err_{i,n-1} \qquad (1)$$

$$Ada = \frac{1}{K}\sum_{i=1}^{K}\left[\frac{1}{r}\sum_{j=0}^{r-1} Err_{i,j}\right] \qquad (2)$$

Where **K** is the number of changes during the run; **r** is the number of generations between two consecutive changes; $\mathbf{Err_{i,j}}$ is the difference between the value of the current best individual in the population of jth generation after the last change ($j \in [0, r-1]$) and the optimum value for the fitness after the ith change ($i \in [0, K-1]$).

5 Results

In this section, due to the constraints of space, we will present part of the obtained results. First we compare the ISGA with the MGA with one point crossover (MGA-Cx1). After that, we present the performance measures obtained with the proposed approach (ISGA), MGA and the other approaches already used by us (ETGA) and by other authors (HMGA and RIGA).

5.1 Transformation versus Crossover: the problem of diversity

As described in the paper, the proposed approach based on IS ideas incorporated a sort of memory mechanism in the GA in order to improve its efficiency when dealing with dynamic environments. Nevertheless, this mechanism of memory was not completely successful if we cannot preserve the diversity in the population. In fact, the MGA with one point crossover couldn't reach the best solution when changes occur. On the other hand ISGA, as the time passes, is always improving until reach the best solution. This behavior that can be seen in Figure 3 corresponds to the primary response and secondary response of the IS. Figures 3 and 4 show the behavior of the two algorithms when changes occur between to values every 15 generations.

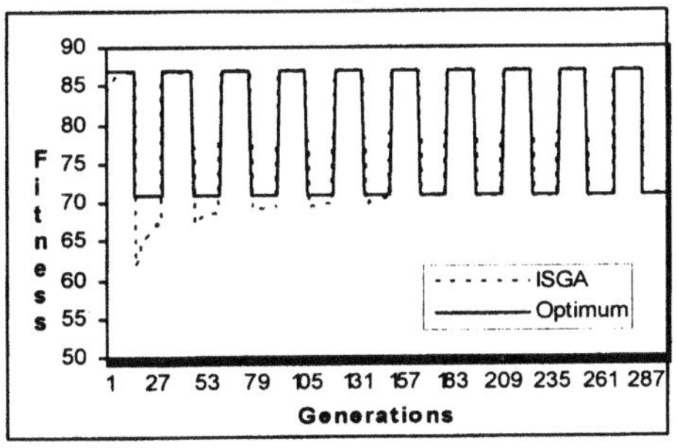

Fig. 3. ISGA: changes between 2 values, cycle=30

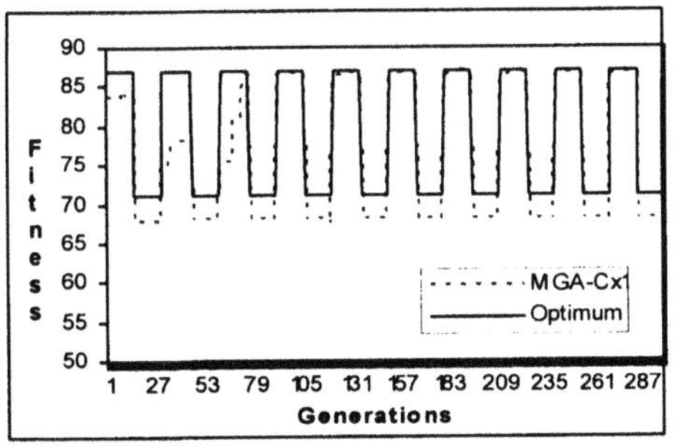

Fig. 4. MGA-Cx1: changes between 2 values cycle=30

For larger cycles, with changes every 150 generations, ISGA after the first change is able to recognize all the changes in the environment and the adaptation is very rapid. MGA-Cx1 had a poor performance because it memorizes one of the values but can't improve it until reach the best solution.
The behavior of ISGA and MGA-Cx1 when changes occur between three different values in periodic intervals is very similar to the previous cases. ISGA performs much better than MGA-Cx1. Figures 5 and 6 show the obtained results for cycles of 30 generations.
For non-periodic changes between three different values, the results were not as good as the ones obtained in the case of periodic changes. Nevertheless, ISGA was also better than MGA-Cx1. In this case, MGA-Cx1 had some situations where it was not able to find the new solution when a change occurred. Figures 7 and 8 show the achieved solutions.

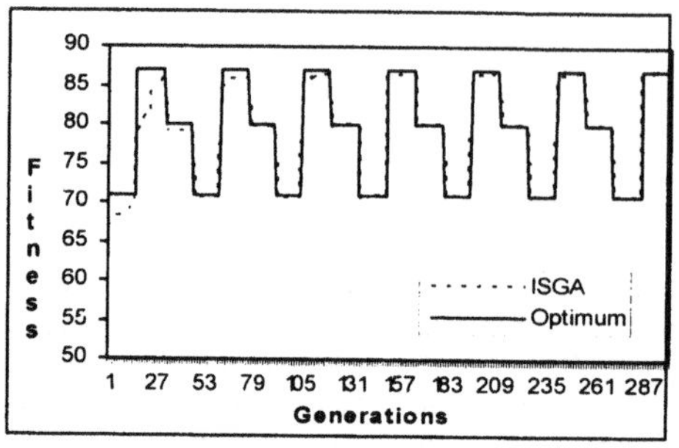

Fig. 5. ISGA: changes between 3 values, cycle=30

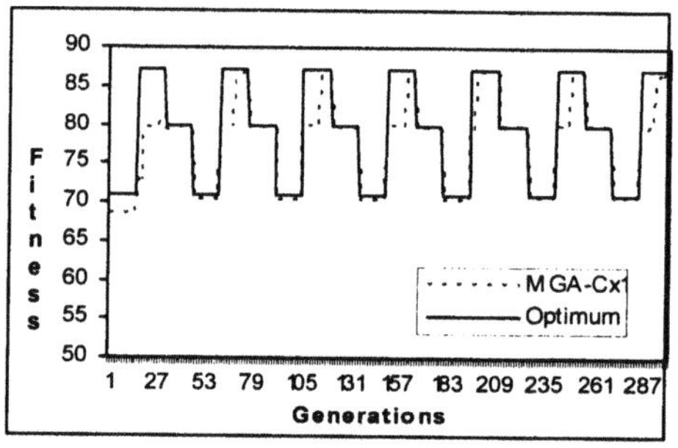

Fig. 6. MGA-Cx1: changes between 3 values, cycle=30

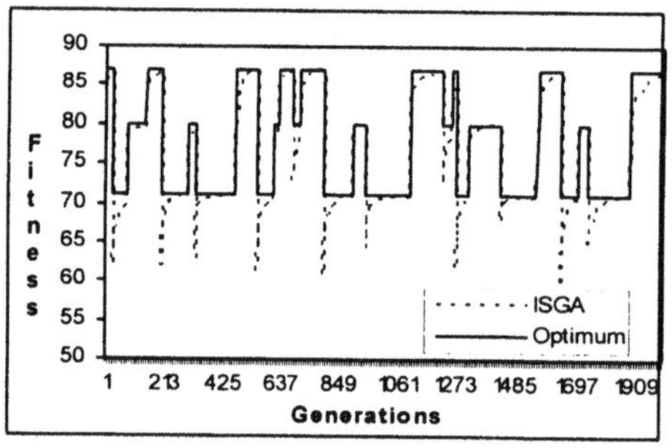

Fig. 7. ISGA: changes between 3 values, non-periodic

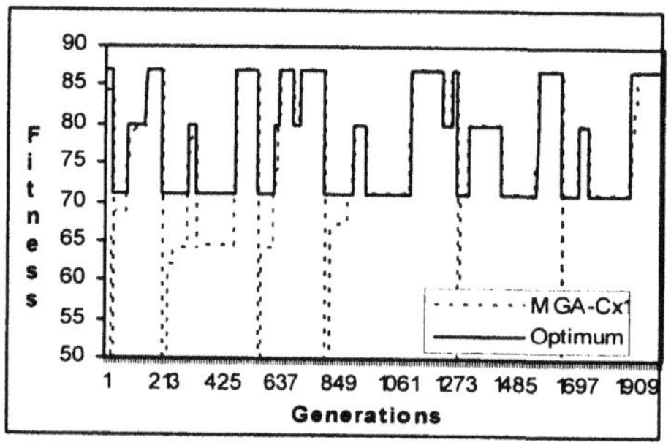

Fig. 8. MGA-Cx1: changes between 3 values, non-periodic

The main reason for the worst performance of MGA-Cx1 seems to be the fact that the population's diversity is loss and the memory mechanism by itself isn't enough to readapt the solutions to the new conditions of the environment. We used a standard measure of diversity defined by (3) and we can see that transformation can preserve diversity at very high.

$$Div(Pop) = \frac{1}{LP(P-1)} \sum_{i=1}^{P} \sum_{j=1}^{P} HD(p_i, p_j) \qquad (3)$$

where **L** is the length of the chromosome, **P**, the population size, $\mathbf{p_i}$, the ith individual in the population and **HD** the hamming distance.

Figure 9 shows the values measured for ISGA and MGA-Cx1 in the case of periodic changes between two values every 15 generations. As we can see, the MGA-Cx1 with mutation a rate equal to 0.1% cannot preserve the diverity in the population..

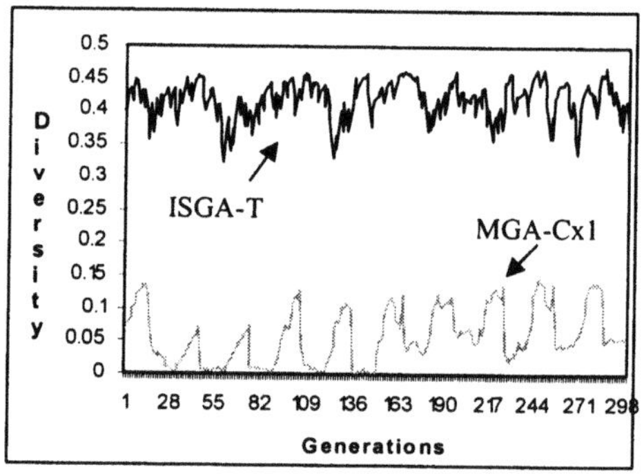

Fig. 9. Diversity in the population preserved by ISGA-T and MGA-Cx1

5.2 Immune system GA: comparing with other approaches

To evaluate the ISGA more deeply, we test it against other algorithms. We re-implemented the well-known HMGA and RIGA. We also used the GA with transformation, but no memory, enhanced by the appropriated choice for the parameters as studied in [16] (ETGA). To compare the results obtained by the proposed approach (ISGA) and also MGA (mutation rate of 0.1%), with these well known algorithms we used the measures introduced in section 4.4. Table 1 shows the accuracy and adaptability values measured during all the generations. As we can see, the ISGA was the approach that obtained, in general, the best results for periodic or non-periodic changes, with smaller or larger cycle lengths.

Moreover, in order to see how the accuracy and adaptability were in the end of the generational process (to see the effect of primary and secondary response) we present in table 2 the values obtained in the last cycle of each experiment. As we can see, ISGA reached always a zero value. This means that the new optimum was always achieved after the change occurs. MGA-Cx in some situations also achieved similar values.

Table 1. Accuracy and Adaptability obtained with ETGA, HMGA, RIGA, ISGA and MGA during the entire generational process

			ETGA	HM 10%	RI 10%	ISGA	MGA-Cx1	MGA-Cx2	MGA-CxU
Periodic 2 values	Cycle = 30	Acc	2.20	1.25	1.78	**0.58**	2.22	4.20	3.68
		Ada	5.01	3.26	4.37	**0.78**	2.73	4.24	3.70
	Cycle = 100	Acc	0.29	0.23	0.59	**0.06**	4.14	3.64	3.94
		Ada	2.35	0.98	2.58	**0.16**	4.24	3.83	4.08
	Cycle = 200	Acc	0.05	0.23	0.50	**0.06**	0.49	3.64	4.14
		Ada	1.11	0.61	1.68	**0.17**	1.16	3.70	4.14
	Cycle = 300	Acc	0.02	0.23	0.45	**0.04**	3.28	0.79	0.14
		Ada	0.73	0.52	1.27	**0.15**	3.43	1.05	0.21
Periodic 3 values	Cycle = 30	Acc	1.45	0.70	1.08	**0.38**	0.72	1.28	2.16
		Ada	3.53	1.84	2.74	**0.68**	2.02	1.85	3.59
	Cycle = 100	Acc	0.18	0.36	0.22	**0.09**	0.57	6.68	1.10
		Ada	1.30	2.18	1.16	**0.31**	1.25	7.52	4.22
	Cycle = 200	Acc	**0.02**	0.42	0.15	0.04	3.64	1.48	0.28
		Ada	0.61	1.39	0.65	**0.19**	4.59	4.45	1.59
	Cycle = 300	Acc	**0.01**	0.49	0.22	0.05	0.58	1.34	0.16
		Ada	0.42	1.35	0.58	**0.18**	1.79	3.38	1.04
NonPeriodic 3 values	-	Acc	0.77	0.54	0.93	**0.53**	1.27	1.27	0.98
		Ada	2.02	1.13	2.22	2.29	5.71	5.26	4.03

Table 2. Accuracy and Adaptability obtained with ETGA, HMGA, RIGA, ISGA and MGA in the last cycle

			ETGA	HM 10%	RI 10%	ISGA	MGA-Cx1	MGA-Cx2	MGA-CxU
Periodic 2 values	Cycle = 30	Acc	1.90	0.73	0.47	**0.00**	3.00	4.00	4.00
		Ada	4.01	2.25	0.76	**0.00**	3.00	4.00	4.00
	Cycle = 100	Acc	0.97	0.20	0.83	**0.00**	4.00	4.00	4.00
		Ada	3.40	0.72	2.18	**0.00**	4.00	4.00	4.00
	Cycle = 200	Acc	0.23	0.20	0.73	**0.00**	0.13	4.00	4.00
		Ada	1.29	0.69	1.77	**0.00**	0.13	4.00	4.00
	Cycle = 300	Acc	0.10	0.40	0.50	**0.00**	**0.00**	0.27	**0.00**
		Ada	0.81	0.62	1.14	**0.00**	**0.00**	0.27	**0.00**
Periodic 3 values	Cycle = 30	Acc	1.73	0.60	1.93	**0.00**	0.30	**0.00**	**0.00**
		Ada	5.72	2.40	5.61	**0.00**	4.24	1.63	4.73
	Cycle = 100	Acc	0.07	0.20	0.47	**0.00**	**0.00**	5.00	3.00
		Ada	1.78	0.61	2.32	**0.00**	0.84	6.64	4.16
	Cycle = 200	Acc	0.00	0.30	0.10	**0.00**	4.00	**0.00**	**0.00**
		Ada	1.03	0.54	1.00	**0.00**	4.50	0.98	0.57
	Cycle = 300	Acc	0.00	0.20	0.40	**0.00**	**0.00**	**0.00**	**0.00**
		Ada	0.44	0.38	0.98	**0.00**	1.40	0.96	0.85
NonPeriodic 3 values	-	Acc	0.60	0.10	0.70	**0.00**	**0.00**	**0.00**	**0.00**
		Ada	2.03	0.46	2.12	**0.00**	**0.00**	**0.00**	**0.00**

6 Conclusions

In this paper we proposed an immune system-based GA to deal with dynamic environment. We tested it in different situations, namely periodic and non-periodic changes and different cycle lengths. The main characteristic of this ISGA is the ability to remember past situations with faster and stronger reactions obtained as time goes on, e.g. a kind of secondary response typical of the natural immune system. Moreover, using transformation, as a hypermutation operator, ISGA is able to keep the diversity in the population.

On the overall, when we compare ISGA with the other algorithms (MGA, ETGA, RIGA and HMGA), it always achieved the best accuracy and adaptability values.

As the results are very promising, the next step in our work is to make ISGA more close to the natural IS (for instance, we want to start with a plasma B-cells population created using also the gene libraries), and to test it in other problems involving dynamic environments.

References

1. H. Cobb (1990). *An Investigation into the Use of Hypermutation as an Adaptive Operator in Genetic Algorithms Having Continuous, Time-Dependent Nonstationary Environments.* Technical Report AIC-90-001.
2. J. Branke (1999). *Memory Enhanced Evolutionary Algorithm for Changing Optimization Problems.* In Proceedings of the 1999 Congress on Evolutionary Computation, pp. 1875-1881, IEEE.
3. J. Branke (2002). *Evolutionary Optimization in Dynamic Environments.* Kluwer Academic Publishers, 2002.
4. D. P. Clark, L. D. Clark (1997). *Molecular Biology made simple and fun.* Cache River Press.
5. K. A. De Jong (1975). Analysis of the Behavior of a Class of Genetic Adaptive Systems. Ph.D. Dissertation, Department of Computer and Communication Science, University of Michigan.
6. A. Gaspar, P. Collard (1999). *From GAs to Artificial Immune Systems: Improving Adaptation in Time Dependent Optimization.* Proc. of the 1999 Congress on Evolutionary Computation, pp. 1859-1866, IEEE.
7. A. Gaspar, P. Collard (2000). *Immune approaches to experience acquisition in time dependent optimization.* In A. S. Wu (ed.), GECCO'2000 Workshop Proceedings.
8. D. E. Goldberg, R. E. Smith (1987). *Nonstationary Function Optimization using Genetic Algorithms with Dominance and Diploidy.* In J. J. Grefenstette (ed.), Proceedings of the 2nd International Conference on Genetic Algorithms, pp. 59-68. Laurence Erlbaum Associates.
9. D. E. Goldberg (1989). *Genetic Algorithms in Search, Optimization and Machine* Learning. Addison-Wesley Publishing Company, Inc.
10. J. J. Grefenstette (1992). *Genetic Algorithms for Changing Environments.* In R. Maenner, B. Manderick (eds.), Parallel. Problem Solving from Nature 2, pp. 137-144. North Holland.
11. A. Hadad, C. Eick (1997). *Supporting Poliploidy in Genetic Algorithms using Dominance Vectors.* In P. Angeline et al. (eds.) Proc. of the 6th International Conf. on Ev. Programming, vol. 1213 of LNCS. Springer.
12. R. R. Hightower, S. Forrest, A. S. Perelson (1995). *The Evolution of Emergent Organization in Immune System Gene Libraries.* In Proc. of the 6th Int. Conf. on Genetic Algorithms, pp. 344-350, Morgan Kaufmann.
13. K. P. Ng, K. C. Wong (1995). *A New Diploid Scheme and Dominance Change Mechanism for Non-stationary Function Optimization.* In Proc. of the 6th Int. Conf. on Genetic Algorithms, pp. 159-166. Morgan Kaufmann.
14. P. J. Russell (1998). *Genetics.* 5th edition, Addison-Wesley.
15. A. Simões, E. Costa (2001). *On Biologically Inspired Genetic Operators: Transformation in the Standard Genetic Algorithm.* In Lee Spector et. Al (eds). Proceedings of the Genetic and Evolutionary Computation Conference (GECCO'2001), pp. 584-591, Morgan Kaufmann.
16. A. Simões, E. Costa (2003a). *Improving the Genetic Algorithm's Performance when Using Transformation.* Proceedings of ICANNGA 2003.
17. A. Simões, E. Costa (2003b). *A Comparative Study Using GAs to Deal with Dynamic Environments.* Proceedings of ICANNGA 2003.

Improving the Genetic Algorithm's Performance when Using Transformation

Anabela Simões[1,2], Ernesto Costa[2]
[1] Dept. of Informatics and Systems Engineering, Coimbra Polytechnic, Quinta da Nora, 3030 Coimbra, Portugal
[2] Centre for Informatics and Systems of the Univ. of Coimbra, Pinhal de Marrocos, 3030 Coimbra, Portugal
E-mail: abs@isec.pt; ernesto@dei.uc.pt

Abstract. Transformation is a biologically inspired genetic operator that, when incorporated in the standard Genetic Algorithm can promote diversity in the population. Previous work using this genetic operator in the domain of function c ptimization and combinatorial optimization showed that the premature convergence of the population is avoideu. Furthermore, the solutions obtained were, in general, superior to the solutions achieved by the GA with standard 1-point, 2-point and uniform crossover. In this paper we present an extensive empirical study carried to determine the best parameter setting to use with transformation in order to enhance the GA's performance. These parameters include the gene segment length, the replacement rate (percentage of individuals of the previous population used to update the gene segment pool), and the mutation and transformation rates.

1 Introduction

One of the key aspects related with the performance of Evolutionary Algorithms (EA) is the diversity of the evolved population. Loss of diversity may imply the premature convergence of the EA to local optima. The solutions converge to a region of the search space and the EA cannot continue exploring other promising areas. It is also true that a homogeneous population is less adapted to dynamic changes in the environment.

In EA, the diversity of a population is mostly achieved by the application of the genetic operators. The traditional genetic operators used in EA are crossover and mutation. Crossover is used to explore different promising areas of the search space, allowing the convergence to the optima (exploration). Mutation is used to "shake" the GA avoiding eventual traps in local optima (exploitation) [1]. Nevertheless, the premature convergence isn't always avoided with the application of these mechanisms. For instance, when faced with problems involving a dynamic environment, the traditional EAs aren't suitable to solve them. The population converges to the desired optimum of the problem, but when this optimum changes by some reason, the EA has troubles in readapting to the new solution.

In order to promote diversity in the individuals of the population other mechanisms have been used. Recently, Simões and Costa proposed a biologically inspired genetic operator called transformation that promotes diversity and avoids premature convergence [5]. The mechanism mimics the biological process and consists in the capacity of the individuals to absorb fragments of DNA from the environment. These gene segments are then reintegrated in the individuals' genome. The proposed mechanism was tested with several classes of problems, namely function optimization, the static and dynamic 0/1 knapsack problem ([5], [6]). The results showed that this new genetic operator preserves the diversity in the population during all the generations and the EA with transformation achieved, in general, better solutions than the standard GA (SGA). Nevertheless, the implementation of transformation used a parameter setting chosen based on some preliminary empirical studies. For instance, the choice of the gene segment length was always random. In this paper we report the results of an extensive empirical study that allows the user to choose the proper parameter setting in order to enhance the GA's performance when using transformation. The studied parameters are the gene segment length, the percentage of gene segments that are updated using the genetic information of the individuals of previous population (replacement rate), and the mutation and transformation rates.

The remaining paper is organized in the following manner. In section 2, we briefly explain the biological and computational functioning of the transformation mechanism. Section 3, briefly refers to previous work about transformation. In section 4, we detail the characteristics of the experimental study, including the test problems, the parameters and the experimental plan. Section 5 shows the results and states the main conclusions regarding the correct choice of the parameters. Finally, we present the relevant conclusions of the work.

2 Transformation

In nature, some bacteria can absorb fragments of DNA from the surrounding environment and reintegrate this DNA in its genetic material. These bacteria are called competent and the reintegration of this DNA fragments may confer some evolutionary advantages to them. The DNA fragments or gene segments proceed mostly from dead bacteria [3].

Simões and Costa proposed a computational implementation of the transformation mechanism and used it as the main genetic operator in the EA, instead of crossover [5].

At the beginning of the EA a population of individuals and a gene segment pool are created randomly. Then, using a roulette-wheel selection method some individuals are chosen to be transformed. Those individuals form the "mating" pool. The transformation was applied with a fixed probability of 70% and works as follows: select one individual from the "mating" pool and a gene segment from the gene segment pool. It is also chosen randomly a transformation point in the selected individual. The genes of the gene segment replace the genes in the selected individual, after the transformation point. After that, the gene pool is updated using the genetic information of the individuals of the previous population. These individuals are used to create 70% of the new segments, being the remaining created at random. These percentages were fixed to all the executed experiments. Furthermore, the length of each segment was also randomly chosen. The GA modified with this new genetic operator will be referred as Transformed-based GA (TGA)
Figure 1 describes the mechanism. For more details see [5]

3 Previous work

The genetic operator was tested in two domains: function optimization and combinatorial optimization. The chosen functions were four well-known functions: Ackley, Griewangk, Rastrigin and Schwefel and the 0/1 Knapsack problem (0/1 KP) for combinatorial optimization. The goals were to minimize the four functions (which minimum value is zero) and to maximize the value of the objects selected to a knapsack of limited capacity. Several instances of the 0/1 KP were implemented, namely with different number of items: 50, 100, 250 and 500 ([2]).
We compared the quality of the solutions found with transformation against the traditional crossover operators: 1-point, 2-point and uniform crossover. The obtained results are summarized in the tables 1 and 2 ([5]).
In Table 1 we show the results of the function minimization problems. As we can see transformation allowed the GA to achieve much better results. The principal conclusion to support these results was the great diversity that transformation introduces in the population avoiding its convergence, and allowing the evolution during a long period.
Nevertheless, either the GA or the TGA were able to achieve the optimum solution.

```
Transformed-Based GA (TGA)
Begin
1. P(t)=Generate initial population
2. S(t)=Generate initial gene segment pool
3. While (NOT stop_condition)
   a. Evaluate P(t)
   b. Select individuals from P(t)
   c. P'(t)=Transform individuals
   d. P''(t)=Mutate new individuals
   e. Update S(t) using segments from individuals from
      P(t) and new random segments
   f. Replace P(t) with P''(t)
End
```

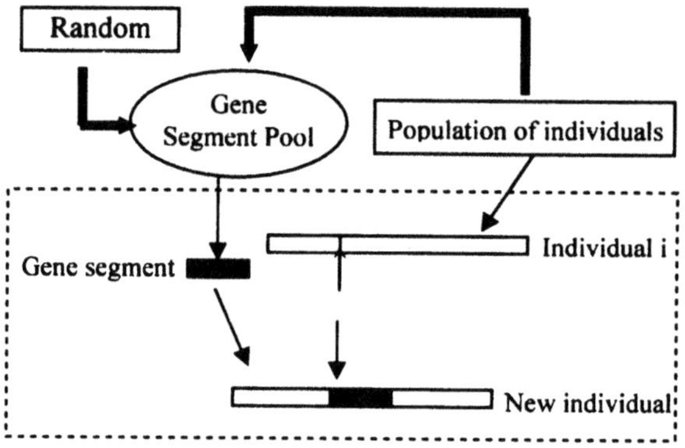

Fig. 1. Computational Transformation

In Table 2 we show the results for the 0/1 KP. In this case the Uniform Crossover was the genetic operator that achieved higher performance.
In order to enhance the transformation performance and achieve better solutions, in particular in the 0/1 KP problem, we performed an extensive parametric study to observe the influence that parameters, such as the gene segment's length, the transformation rate, mutation rate or replacement rate, could have in the obtained results.
The parametric study will be detailed in the next section.

Table 1. Results obtained by a SGA (using 1-point, 2-point and Uniform Crossover) and the GA using Transformation in the function optimization domain

	Genetic Operator											
	One-point Crossover			*Two-point Crossover*			*Uniform Crossover*			*Transformation*		
Nº evals	*50000*	*100000*	*200000*	*50000*	*100000*	*200000*	*50000*	*100000*	*200000*	*50000*	*100000*	*200000*
Function Rastrigin	23.464	16.502	11.599	23.164	15.879	10.739	31.079	22.791	16.676	38.401	18.828	**6.540**
Griewangk	0.006	0.001	0.001	0.004	0.002	0.001	0.007	0.004	0.003	0.010	0.003	**0.001**
Schwefel	23.611	2.565	0.409	36.140	13.178	5.484	53.705	11.166	0.521	36.212	0.475	**0.077**
Ackley	2.921	1.034	0.278	2.699	0.922	0.119	2.946	1.416	0.429	3.128	0.300	**0.002**

Table 2. Results obtained by a SGA (using 1-point, 2-point and Uniform Crossover) and the GA using Transformation solving the 0/1 Knapsack Problem

	Genetic Operator												
	One-point Crossover			*Two-point Crossover*			*Uniform Crossover*			*Transformation*			
P. Size→	*50*	*100*	*200*	*50*	*100*	*200*	*50*	*100*	*200*	*20*	*50*	*100*	*200*
Nº of Items 50	207.17	195.20	185.40	206.87	199.13	187.77	**207.20**	199.80	189.13	197.30	197.80	197.23	207.17
100	447.93	410.03	373.67	449.63	415.93	381.73	**453.07**	402.90	377.67	413.00	408.40	409.70	447.93
250	976.40	844.33	780.93	980.93	862.43	785.27	**981.40**	824.33	774.93	838.50	834.87	835.17	976.40
500	1914.43	1671.53	1535.50	1936.87	1696.20	1537.03	**1955.30**	1622.27	1533.30	1666.20	1669.07	1661.03	1914.43

4 Experimental setting

The GA was implemented with roulette-wheel selection with elite of size 2 and population size of 200 individuals in the minimization of the functions and with population of size 50 and 100 in the maximization of the 0/1 KP. Instead of the classical crossover operator we used transformation. The remaining parameters were change to study their influence in the GA's performance when using this mechanism.

We started analyzing the effect of the gene segment length in both problem domains. The length of the chromosome differs from problem to problem, and so the variation of the segment length was dependent on that.

In previous work the length of each segment was always chosen in a random way. So, we could have in the segment pool segments with several lengths. In this study, we fixed all the remaining parameters and changed the gene segment size from 5 to a maximum value depending on the chromosome length, but the segment was fixed for each experiment. The variation of this value was made in jumps of five units, i.e, 5, 10, 15, ... until the maximum allowed value. All the experiments were repeated over 30 runs.

Table 3 shows the chromosome length for each problem and the interval of variation for the gene segment's length.

As referred before, the replacement rate is the percentage of individuals of the previous generation that contribute for the modification of the gene segment pool at each generation.

Table 3. Interval of variation for the gene segment length

Problem	Chromosome Length	Interval of variation for the gene segment length
Ackley's Function	480	[5, 125]
Griewangk's Func.	160	[5, 125]
Rastrigin's Func.	280	[5, 125]
Schwefel's Func.	160	[5, 125]
0/1 KP – 50 Items	50	[5, 50]
0/1 KP – 100 Items	100	[5, 100]
0/1 KP – 250 Items	250	[5,125]
0/1 KP –500 Items	500	[5,125]

In our original work we used 70% of the individuals of the previous population to update the gene segment pool at each generation. In order to determine the influence of this parameter, we changed the replacement rate from 0% to 100%, in jumps of 5 units.

The third parameter to be analyzed was the mutation rate. We changed this rate among six values: 0%, 1%, 5%, 10%, 15% and 20%. Our previous work used 0.1% of mutation rate.

Finally, we altered the transformation rate from 0% to 100% in jumps of 10 units.

5 Results

This section will present some of the results obtained. We will show representative graphics obtained in the function optimization domain and in the 0/1 KP. To illustrate the results obtained in the function optimization domain we choose the Ackley's test function. For the 0/1 KP we show the results achieved with 100 items.

5.1 Segment length

The variation of this parameter had great influence in the obtained results. Smaller sizes allowed, always, better results. Figure 2 shows the variation of the best solutions found minimizing the Ackley's test function. The gene segment's length varied from 5 to 125. As we can see, segments of length equal to 5, 10 and 15 were the best choice for this parameter.

Figure 3, shows the results obtained in the maximization of the 0/1 KP, with 100 items. Once more, smaller segments allowed best results. In this case, segments of length 5 or 10 were the appropriated choice.

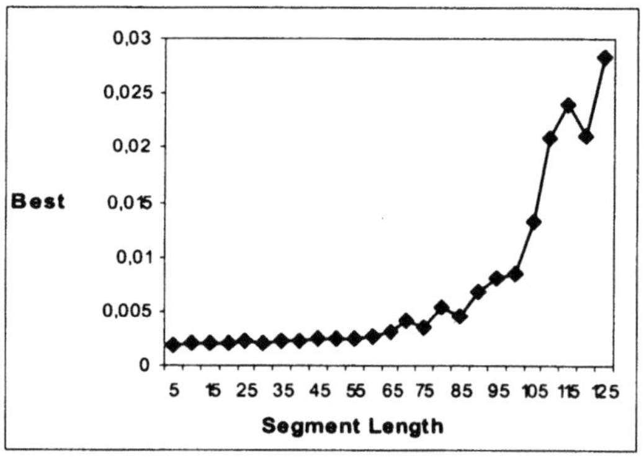

Fig. 2. Influence of the Gene Segment's Length in the Best Solution Found (minimization of Ackley's test function)

The main reason for the degradation of the results using larger segments appears to be the disruption introduced in the population. As we can see, smaller segments can also preserve the diversity of the population, but the levels are not as high as in the case of larger segments.

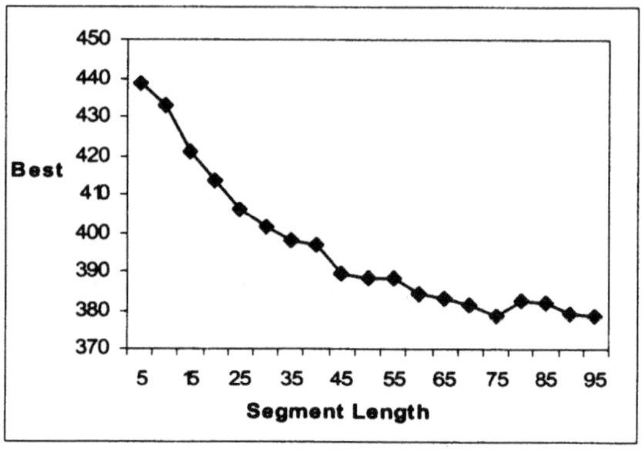

Fig. 3. Influence of the Gene Segment's Length in the Best Solution Found (maximization of the 0/1 KP – 100 items)

In order to explain these results we measure the diversity of the population using a standard measure defined by:

$$Div(Pop) = \frac{1}{LP(P-1)} \sum_{i=1}^{P} \sum_{j=1}^{P} HD(p_i, p_j)$$

where **L** is the length of the chromosome, **P**, the population size, $\mathbf{p_i}$, the ith individual in the population and **HD** the hamming distance.

Figures 4 and 5 illustrate how the diversity of the population evolved during the entire process, using gene segments of length 5 and 125 for the Ackley's function and 5 and 100 for the 0/1 KP, respectively.

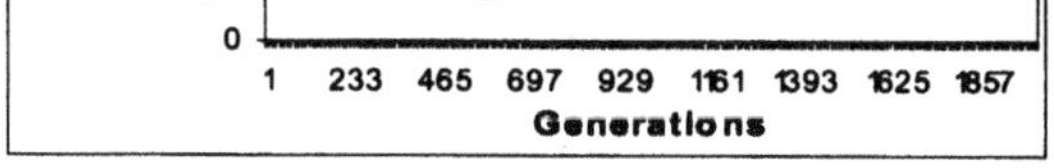

Fig 4. Population's Diversity using Gene segments of size 5 and 125 (Ackley's Function)

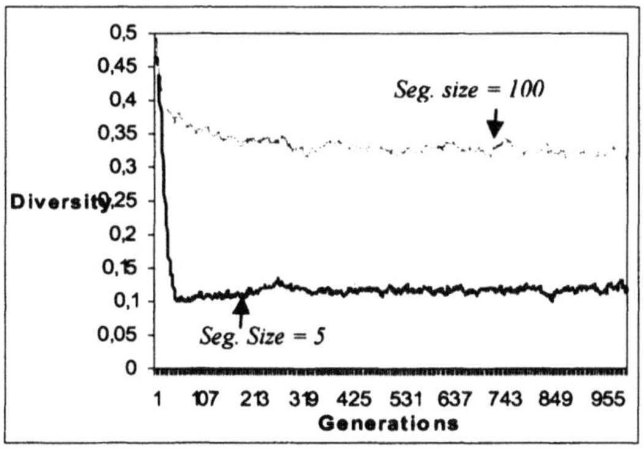

Fig. 5. Population's Diversity using Gene segments of size 5 and 100 (0/1 KP – 100 items)

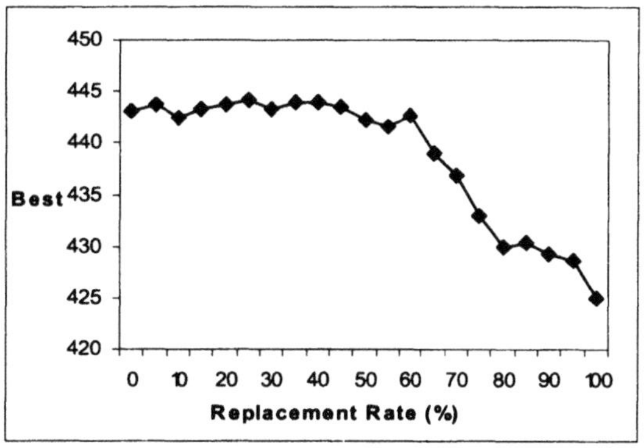

Fig. 7. Influence of the Replacement Rate in the Best Solution Found (maximization of the 0/1 KP – 100 items)

5.2 Replacement rate

The replacement rate was changed from 0% (all the gene segments are created at random) to 100% (all the segments proceed from individuals of previous generations). The results obtained in the two domains were quite different: when minimizing the test functions, replacement rates superior to 70% allowed the best results. In the 0/1 KP, the best solutions were found when using replacement rates inferior to 60%. Figures 6 and 7 show the behavior of the GA minimizing the Ackley's test function and maximizing the 0/1 KP, respectively.

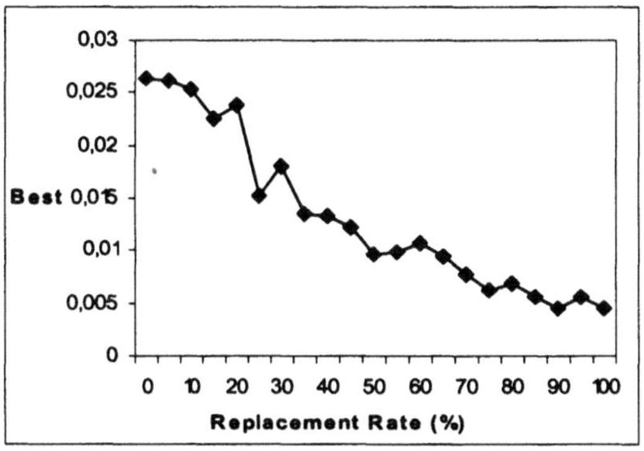

Fig. 6. Influence of the Replacement Rate in the Best Solution Found (minimization of Ackley's test function)

5.3 Mutation rate

The mutation rate was changed from 0% to 20% (0%, 1%, 5%, 10%, 15% and 20%). In both problems mutation rate equal to 0% or 1% allowed the GA to achieve the best performances. As we increase the mutation rate, the results become worst. Figures 8 and 9 show the results.

The observed results can be understand if we analyze the population's diversity when using 0% and 20% of mutation rate. If we observe figures 10 and 11 we can see that with 20% mutation, the diversity in the population is very high and consequently it was introduced a great degree of disruption. Transformation without any mutation was able to maintain the population's diversity within a reasonable value and lead the GA to the best solutions.

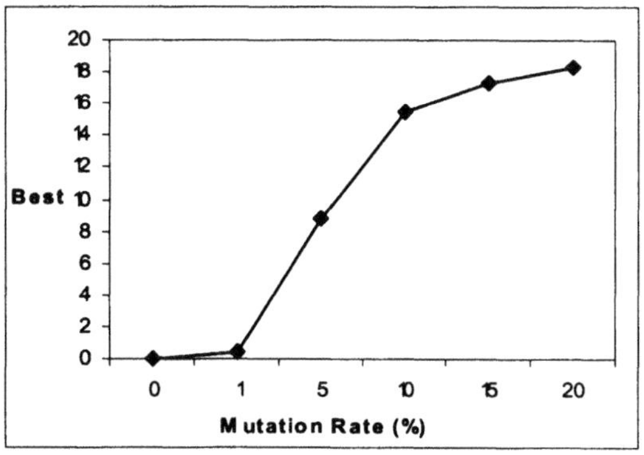

Fig. 8. Influence of the Mutation Rate in the Best Solution Found (minimization of Ackley's test function)

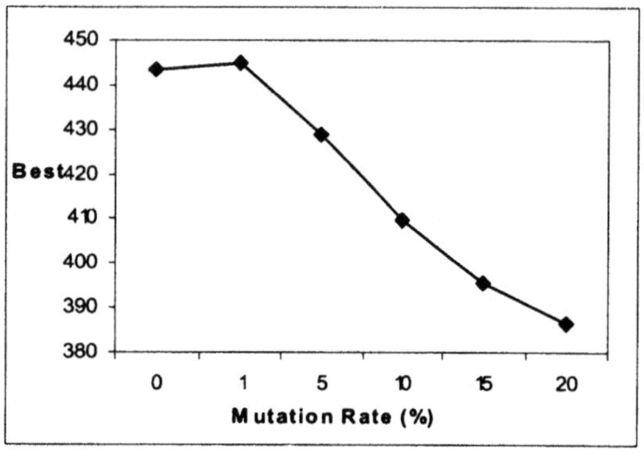

Fig. 9. Influence of the Mutation Rate in the Best Solution Found (maximization of the 0/1 KP – 100 items)

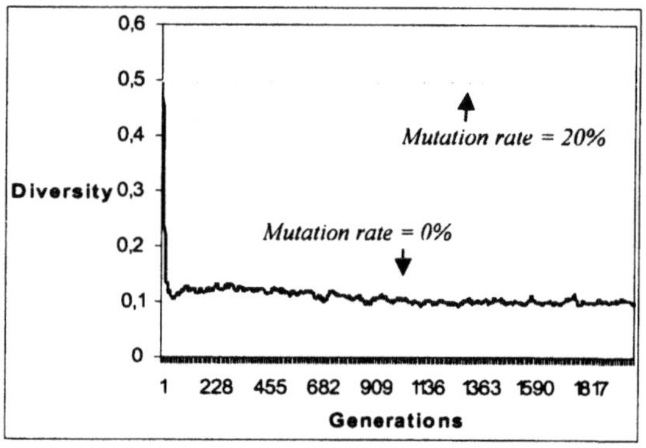

Fig. 10. Population's Diversity using Mutation rate = 0% and 20% (Ackley's Function)

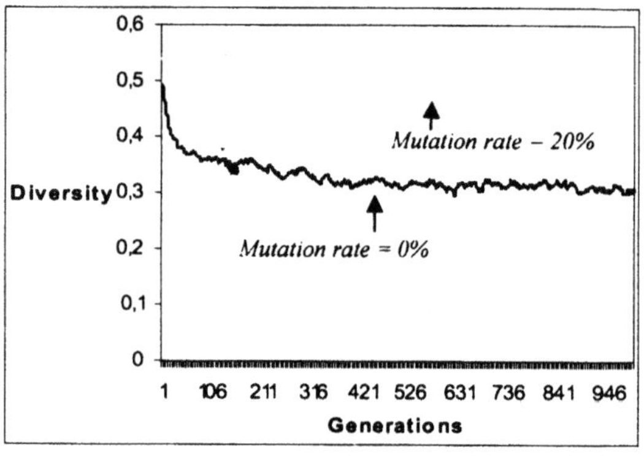

Fig. 11. Population's Diversity using Mutation rate = 0% and 20% (0/1 KP – 100 items)

5.4 Transformation rate

Transformation rate was changed from 0% to 100%. It is obvious that a rate equal to 0% didn't allow the evolution towards the global optimum, because no mutation was used. The best choice for this parameter, for both problems was a value superior to 60%.

Figures 12 and 13 show the influence of the transformation rate in the minimization of Ackley's function and in the maximization of the 0/1 KP with 100 items.

The transformation rate that leads the GA to the minimum values of the functions was 90%. In the case of the 0/1 KP, for all the instances, values of 90% or 100% of transformation rate allowed the GA to achieve the best solutions.

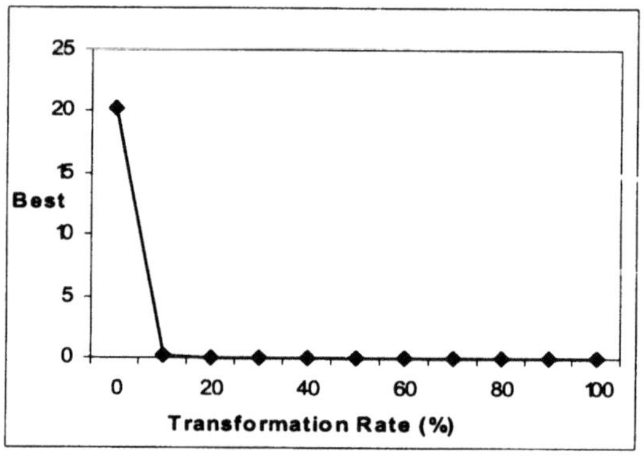

Fig. 12. Influence of the Transformation Rate in the Best Solution Found (Ackley's test function)

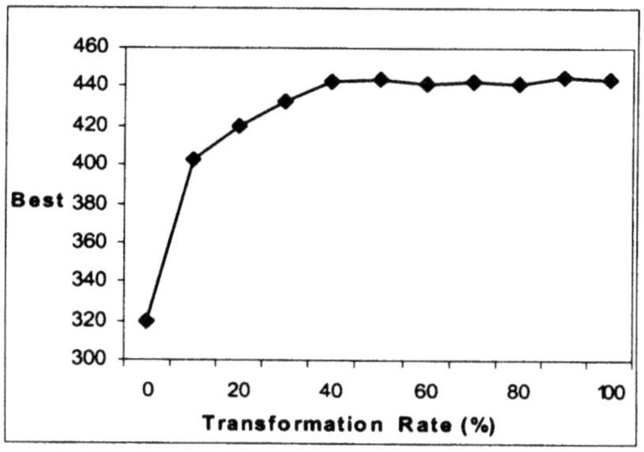

Fig. 13. Influence of the Transformation Rate in the Best Solution Found (maximization of the 0/1 KP – 100 items)

5.5 Global influence of the parameters

After the study we can conclude that the choice of parameters is an important aspect when using transformation in the GA. All the parameters influenced the results in an expressive manner.

We run the GA with a new set of parameters and we compared the obtained results with the solutions found by the GA with the initial parameter setting. Table 4 reports the results for the function optimization domain. As we can see, choosing the appropriate values for the parameters the quality of the results was significantly increased.

Table 5 shows the results obtained maximizing the 0/1 KP.

Table 4. Comparing Previous Results with the Results Obtained with the Appropriate Parameter Setting (Function Optimization)

Parametric Study				**Random Choice of Parameters**			
Segment length=5 Replacement Rate=90% Transformation Rate= 70% Mutation Rate=0.0%				Segment length=random Replacement Rate=90% Transformation Rate= 70% Mutation Rate=0.1%			
N° evals->	*50000*	*100000*	*200000*	*N° evals->*	*50000*	*100000*	*200000*
Ackley	2.678	0.044	**0.002**	Ackley	3.128	0.300	0.002
Griewangk	0.001	0.000	**0.000**	Griewangk	0.010	0.003	0.001
Rastrigin	8.290	0.821	**0.001**	Rastrigin	38.401	18.828	6.540
Schwefel	0.147	0.031	**0.008**	Schwefel	36.212	0.475	0.077

As we can see, choosing the GA parameters with some criteria, the results obtained in the minimization of the four test functions were quite better than the results achieved in our initial work, where the parameters were chosen based only in a small set of experiments.

Table 5. Comparing Previous Results with the Results Obtained with the Appropriate Parameter Setting (0/1 KP)

Parametric Study			**Random Choice of Parameters**		
Segment length=5 Replacement Rate=50% Transf. Rate= 90% Mutation Rate=0.0%			Segment length=random Replacement Rate=90% Transf. Rate= 70% Mutation Rate=0.1%		
Pop size->	*50*	*100*	*Pop size->*	*50*	*100*
50 items	204.60	**204.90**	50 items	197.30	197.80
100 items	442.50	**444.47**	100 items	413.00	408.40
250 items	**955.20**	954.60	250 items	838.50	834.87
500 items	**1926.87**	1910.00	500 items	1666.20	1669.07

In the 0/1 KP problem, the results were also improved. If we compare the new results with the ones obtained by the TGA, we can see that choosing the correct values for the parameters, the GA was able to reach higher values.

6 Conclusions

We used a new biologically inspired genetic operator called transformation in the GA in two different problem domains: function optimization and combinatorial optimization.

Previous work using this genetic operator showed that it is capable of preserve the population diversity during the entire evolutionary process. Nevertheless, the choice of the parameters to run the GA with transformation was always made without any strongly supported criteria.

In this paper we performed a parametric study to enhance the GA's performance when using transformation. In this study we varied four parameters: the gene segment length, the replacement rate, the mutation rate and the transformation rate. The results showed that the choice of these parameters influenced the results. In fact, using an appropriated parameter setting the GA achieved much better solutions than the ones obtained with the first set of parameters.

All the parameters had great influence in the obtained results.

As we increase the size of the segments the results become worst. Studies involving the population's diversity indicate that larger segments introduce more disruption in the individuals of the population.

Concerning the mutation rate, values of 0% allowed the ETGA to achieve the best solutions. In fact, the algorithm is able of continue evolving only with the application of transformation, and no mutation is necessary.

The replacement rate, i.e., the percentage of individuals that contribute in the generation of the gene segment pool of the next generation, must be chose in the appropriated interval: in the minimization of the test functions, replacement rates superior to 80% were the correct choice: in the maximization of the 0/1 KP, the appropriate choice was a value between 50% to 100%.

The correct choice for the transformation rate was a value in the interval 60% to 100%.

Combining all those parameters and choosing the correct values we obtained better results than the ones achieved previously by the TGA.

Acknowledgments

This work was partially financed by the Portuguese Ministry of Science and Technology under the Program POSI.

References

1. D. E. Goldberg (1989). *Genetic Algorithms in Search, Optimization and Machine Learning.* Addison-Wesley Publishing Company, Inc, 1989.
2. Z. Michalewicz (1999). *Genetic Algorithms + Data Structures = Evolution Programs.* 3rd Edition Springer-Verlag, 1999.
3. P. J. Russell (1998). *Genetics.* 5th edition, Addison-Wesley, 1998.
4. A. Simões, E. Costa (2001a). *Using Biological Inspiration to Deal with Dynamic Environments.* Proceedings of the Seventh International Conference on Soft Computing (MENDEL'01), pp. 7-12, Brno, Czech Republic, 6-8 June, Brno University of Technology, 2001.
5. A. Simões, E. Costa (2001b). *On Biologically Inspired Genetic Operators: Using Transformation in the Standard Genetic Algorithm.* Proceedings of the Genetic and Evolutionary Computation Conference (GECCO'2001), pp. 584-591, San Francisco, USA, 7-11 July, Morgan Kaufmann Publishers, 2001.

An ant colony algorithm for multiple sequence alignment in bioinformatics

Jonathan Moss and Colin G. Johnson *

*Computing Laboratory University of Kent at Canterbury Canterbury, Kent, CT2 7NF, England.
C.G.Johnson@ukc.ac.uk

Abstract

This paper describes a the application of *ant colony optimization* algorithms, which draw inspiration from the way ants organize themselves in searching for food, to the well-known bioinformatics problem of aligning several protein sequences.

1 Introduction

Swarm intelligence methods are computational techniques inspired by animals such as social insects acting together to solve complex problems. The main application of these techniques has been to combinatorial optimization problems. This paper discusses work-in-progress on the application of swarm intelligence ideas to a bioinformatics problem, *viz.* aligning multiple protein sequences which are believed to be related.

The paper begins with a brief survey of swarm intelligence and the multiple sequence alignment problem. The application of one to the other is then described, and some preliminary results are given both on synthetic problems and on real-world data.

2 Ant colony optimization and swarm intelligence

Ant colonies are able to organize their foraging behaviour in a seemingly efficient way without any centralized control [7]. This self-organizing structure is carried out via *stigmergic* communication, i.e. communication by changing the environment, in this case by laying down pheromone trails.

Initially ants have no idea of where food is in the environment, so they wander randomly, leaving a pheromone trail. When an ant finds food it wanders back to the nest. Initially these paths will be arbitrary, but when an ant follows a shorter path it will be able to follow that path more often within the same time period than an ant following a longer path, so there is a *positive reinforcement* process whereby the shorter paths get stronger.

A simple version of this is illustrated in figure 1, where ants have two possible routes from a nest to

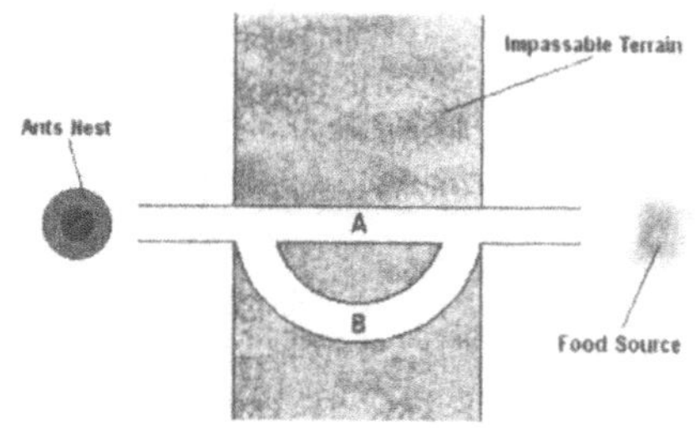

Fig. 1. A simple ant foraging problem.

a food source. If two ants set out at the same time, one taking route A and one route B, which is twice as long, then the ant taking A will have travelled back and forth between the food source twice in the same time that the other ant has travelled back and forth once. Therefore there will be a stronger pheromone trail on route A compared to route B. This idea can be effectively scaled up to solving route finding problems such as the TSP, with performance as good as or better than existing heuristics [2, 3].

3 Multiple sequence alignment

Proteins are complex molecules which consist of a long chain of *amino acids*. The sequence of these amino acids along the chain is specified by transscribing and translating the DNA sequence in the cell. These chains fold up into a complex three dimensional structure. Proteins are the basic building blocks of living organisms; most of the body is built from proteins of various kinds, and proteins are used to carry signals around the body and carry out the various actions which an organism needs to do to survive.

The fact that proteins consist of long linear sequences of simple subcomponents means that we can store this information easily on the computer. Finding relationships between such sequences is an important part of the subject known as *bioinformatics* [1, 5, 8].

Over the course of evolutionary history proteins

become modified as organisms evolve. Nonetheless enough commonality remains so that proteins with a common evolutionary history can be identified. There are a number of reasons for being interested in this, e.g.:

1. This information can be used to support the reconstruction of phylogenetic trees by giving an indication of how much time has passed since present organisms branched off from a common ancestor.

2. If several proteins have commonalities at the sequence level, this may correlate with commonalities in their three-dimensional structure, so this may contribute to the ongoing work on predicting three-dimensional structures of proteins.

3. Certain "families" of proteins are commonly found together. If two organisms have a strong alignment between certain proteins which belong to one of these families, then it is likely that the other proteins in the family will also be present [1].

There are a number of ways in which protein sequences can change. Firstly one amino acid can be substituted for another. In particular amino acids that have similar properties are more likely to be substituted, so alignment methods tend to incorporate measures of substitutability based either on data about known substitutions or biochemical properties. Secondly amino acids can be inserted or deleted from the sequence. Therefore one of the requirements for an alignment algorithm is to be able to include gaps in the sequence to enable a sequence which has lost or gained amino acids to be lined up against another sequence which hasn't. A number of methods have been applied in multiple sequence alignment such as hidden Markov models and dynamic programming (a good survey is [5]).

4 Applying ant colony optimization to multiple sequence alignment

We have developed a system, called *AntAlign* (summarized in figure 2), which applies the ant colony optimization techniques to the multiple sequence alignment problem.

The main idea of the system that ants take a subsequence and move in an interval associated with each sequence, strengthening a pheromone trail when a close match is found to a sequence at that position in other sequences. As the algorithm runs larger fragments of sequence are picked up by the ants. This is designed to encourage the removal of extraneous gaps later on in the process. The overall architecture of the program is indicated in figure 2.

Cycles	The number of cycles executed.
PopPerTime	The number of `Ants` generated per cycle.
StartLen	The length of the `Ant`'s subsequence in the first cycle.
EndLen	The length of the `Ant`'s subsequence in the final cycle.
Evap	The rate of trail evaporation.
Intensity	The intensity of the trail.
Drift	The distance `Ants` are allowed to drift from the current consensus and still score a match.
RndChance	The chance that an `Ant` will choose a random path rather than a matched trail.

Table I. Parameters for the algorithm.

The first type of object used to build the system is a "trail manager" (called an `ITrail`) which will organize the pheromone sequences. There is one `ITrail` for each sequence, and an additional `ITrail` in the system which will manage the emerging consensus sequence. The consensus sequence contains the strongest match so far at each position.

Objects of this `ITrail` class play a number of roles. The main role is in storing the pheromone trails which are created as the ants move along the sequences looking for a good match. Associated with each `ITrail` is an interval in which the pheromone trails will be placed, i.e. a number of locations (longer than the sequence itself) into which pheromone can be placed. The strength of these pheromone trails will eventually determine the sequence.

The `ITrail` object also manages the three main functions related to the pheromone trails (`PTrails`). Firstly when a new `Ant` is created the `ITrails` object compares the subsequence the `Ant` is carrying to those associated with trails already stored in the `ITrails` object to determine the possible paths the `Ant` could take. The second function it provides is the adding of `PTrails` into its interval, or enhancing existing `PTrails` if a matching `PTrail` is already present. The final role the `ITrail` plays is applying evaporation to the trail strengths at the end of each cycle.

The main driving program is contained in the `AntSystem` class. To start a run of the program a number of parameters are specified (table I). The

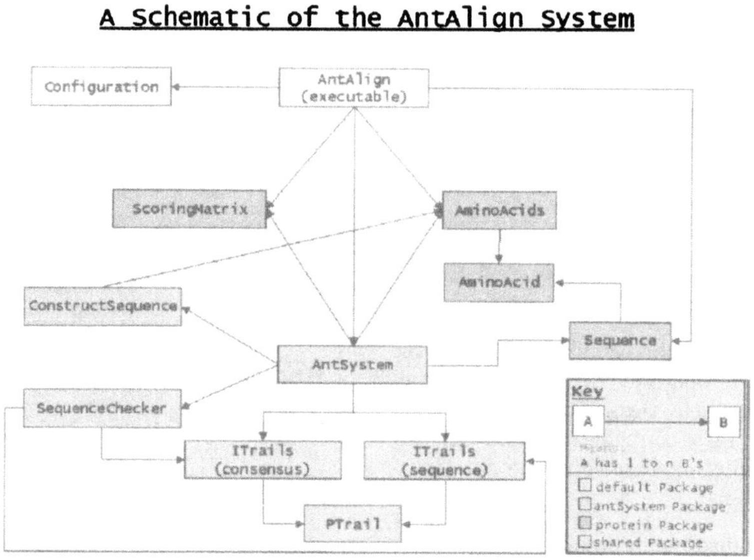

Fig. 2. An overview of the components of the *AntAlign* system and their interactions.

following pseudocode indicates the main algorithm used:

```
LOOP until number == cyclesRequired
  cargoLength := startingLength
      + (cycleNumber*rateOfLengthIncrease)
  LOOP though each sequence
    Generate number of ants required
          from current sequence
    LOOP through Ants
      Get matches of current Ant to
          consensus PTrails
      Determine match probabilities
      Determine path taken
      Generate alignmentScore for
          the path taken
      IF (alignmentScore>0.0)
        Put new PTrail into consensus trail
        Put new PTrail into current
            sequence trail
      ENDIF
    ENDLOOP
  ENDLOOP
ENDLOOP
```

The first few lines of the algorithm simply sets up loops through cycles, sequences and ants. During each iteration around the inner loop each **Ant** takes a random subsequence from its sequence and attempts to place it within the interval. Details of the core steps of the algorithm are as follows:

Get matches of Ants to consensus Ptrails. An important decision in creating the algorithm was that an **Ant** would have read access only to the pheromone trails related to the consensus sequence, not to the other sequences directly (it would be interesting future work to compare this with a variation where ants associated with a sequence have access to all other sequences). A match is determined to have happened if the subsequence carried by the **Ant** and the offset of the ant from the start of the **ITrail** interval is close to the offset of the **PTrail**. *Close* here means within the distance specified by the **Drift** parameter to the left or right.

Determine match probabilities. This determines the percentage probability of following each of the paths in the set $P = \{p_1, \ldots, p_n\}$ determined by the previous step in the algorithm, or another path chosen at random from all possible paths. The probability of a path p_i being chosen is

$$\frac{t_i}{\sum_{i=1}^{n} t_i} \times (100 - \textbf{RndChance})$$

where t_i is the strength of trail p_i and the **RndChance** is the parameter giving the percentage chance of a random path being followed instead of the path determined in the previous step of the algorithm.

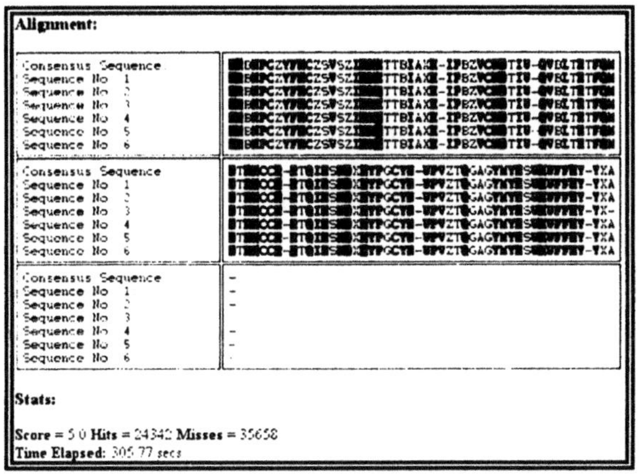

Fig. 3. Alignment of identical sequences

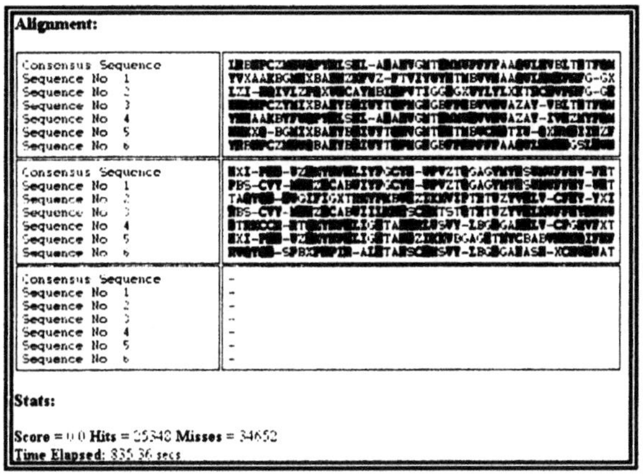

Fig. 4. Alignment of random sequences

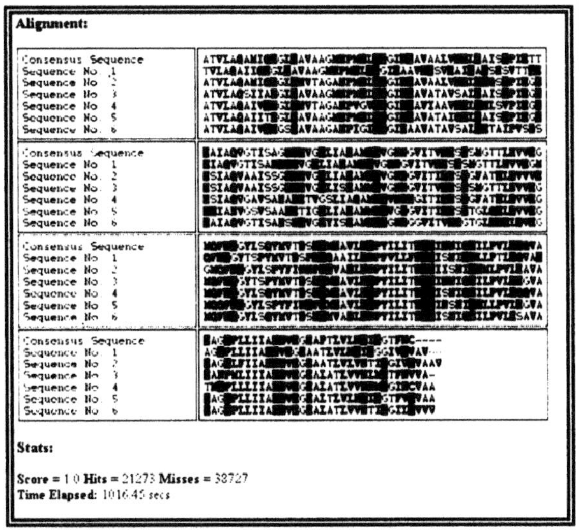

Fig. 5. Alignment of highly conserved sequences

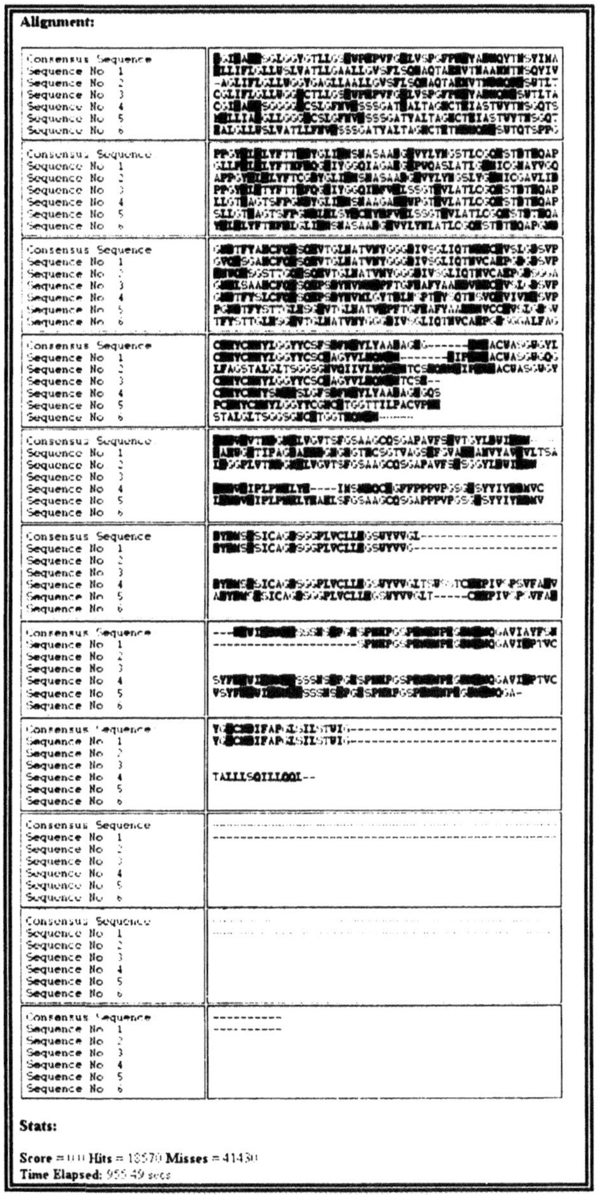

Fig. 6. Alignment of less highly conserved sequences

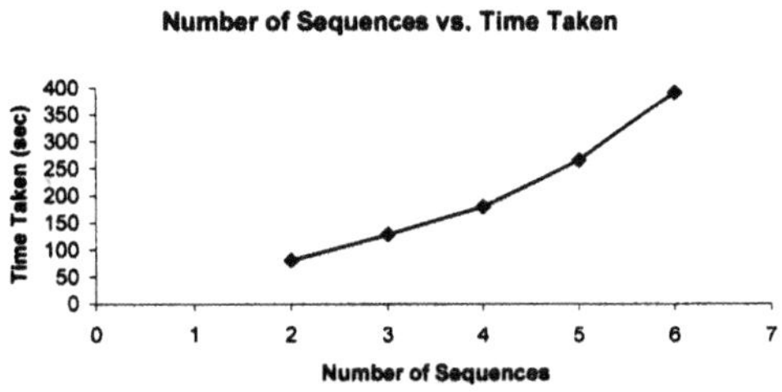

Fig. 7. Time taken vs. number of sequences

Note: Colour originals of all figures on the author's web page

Determine path taken. A path is chosen according to the probabilities calculated in the previous step of the algorithm.

Generate alignment score for path taken. A score is calculated based on the match between the current consensus sequence and the path taken by the ant. This is based on the well-known BLOSUM-62 matrix [6] which was calculated from data about probabilities of protein substitution from a large protein sequence database.

Because the subsequences which each `Ant` carries are generated at random, there is a chance that parts of the sequence are missed out or misaligned. To combat this a repair algorithm (the *SequenceChecker*) is invoked every 10 cycles, which tidies up inconsistencies. However as can be seen in the results below this is not ideal. Current work is attempting to improve the repair algorithm, and a new version of the algorithm is being implemented which divides the sequence completely between ants which are not allowed to move out of sequence order.

5 Results

Firstly we show the alignment of six copies of the same sequence (figure 3). With a small exception, these are lined up perfectly by the algorithm. The one exception is at the end of the third sequence, where an amino acid has been missed out. This demonstrates the need for improvements to the repair algorithm.

Secondly we take six randomly generated sequences (figure 4). It is unsurprising that these sequences are all bunched close to each other; there is minimal commonality between them which could form the basis for spreading out the sequences.

The third experiment worked with real data. This consisted of six variations on the chaperonin protein Cpn60 taken from six different bacterial species. The results from this are illustrated in figure 5. As would be expected from a highly conserved sequence there is little spread, the representation is rather compact. However several small improvements were not discovered by the algorithm, e.g. sequence number 1 lags behind a much stronger position by one position for much of the sequence.

The final experiment uses a less highly conserved set of sequences, consisting of six serine proteases from various species. These were chosen because they have small regions which are active in the function of the protein and much of the remainder has drifted over evolutionary timescales. These results are shown in figure 6. A particular point of interest is towards the end of the sequences, where a large gap has been left in the first sequence to make it line up strongly against the 4th and 5th sequences.

There is some evidence that the algorithm scales well as the number of sequences being aligned is increased. Results showing the time taken to align various numbers of sequences are given in figure 7. Adding a new sequence increases the time taken by approximately a factor of 1.5.

6 Conclusions and future work

This paper has presented the current state of work on this project. One important improvement which needs to be made is to improve the repair algorithm or replace the random assignment of subsequence payloads to `Ants` with a more consistent method. Another important area of work is carrying out a more formal comparison with other algorithms, including both heuristics developed specifically for multiple sequence alignment and metaheuristics adapted for this problem.

References

[1] T.K. Attwood and D.J. Parry-Smith. *Introduction to Bioinformatics.* Addison Wesley Longman, 1999.

[2] Eric Bonabeau, Marco Dorigo, and Guy Theraulaz. *Swarm Intelligence.* Oxford University Press, 1999.

[3] M. Dorigo, V. Maniezzo, and A. Colorni. The ant system: Optimization by a colony of cooperating agents. *IEEE Transactions on Systems, Man and Cybernetics—Part B*, 26(1):29–41, 1996.

[4] Marco Dorigo, Gianni Di Caro, and Michael Sampels, editors. *Ant Algorithms: Third International Workshop, ANTS 2002.* Springer, 2002.

[5] R. Durbin, S. Eddy, A. Krogh, and G. Mitchison. *Biological Sequence Analysis.* Cambridge University Press, 1998.

[6] S. Henikoff and J.G. Henikoff. Amino acid substitution matrices from protein blocks. *Proceedings of the National Academy of Sciences USA*, 89:10915–10919, 1992.

[7] B. Hölldobler and E.O. Wilson. *The Ants.* Springer, 1990.

[8] Arthur M. Lesk. *Introduction to Bioinformatics.* Oxford University Press, 2002.

The RBF-Gene Model

Guillaume BESLON, Carole KNIBBE, Hédi SOULA, Jean-Michel FAYARD[1]

[1]INSA Lyon, 69621 Villeurbanne, France, E-mail: gbeslon@prisma.insa-lyon.fr

Abstract

We present here the Radial Basis Function Gene Model as a new approach of evolutionary computation. This model enables us to relax the "locus constraint" that limits classical genetic algorithms. Both the principles and first results are presented, showing the great interest of this model.

1 Introduction

Genetic algorithms and evolutionnary computation, although they are proved to be powerful parametric optimization tools [4], suffer from a strong limitation: the genome structure has a strong influence on the exploration process. Indeed, neighbouring genetic region have a huge probability to evoluate concurrently (i.e. if one region is conserved/deleted in a phylogenetic branch, the other one will *probably* be conserved/deleted). On the opposite, far-off regions have few chances to be conserved together, *even though they are inter-dependant.* Since the genetic structure is generaly chosen by hand with no (or few) knowledge about the inter-dependencies, counter-productive structures can be introduced in the genome that prevent the evolutionnary process from converging in practice[1].

At a first glance, this limitation is due to the very simplified evolution operators we use during the "reproduction" process. The crossing-over operator, in particular, is generaly mono-point while a two-points crossing-over would partly overcome the structural constraints. However, it would be more interesting to *exploit* the interaction between genetic structure and evolution rather than *supress* it (although supressing it is better than suffering from it). That's why it has been proposed to introduced a *locus* operator that would enable the genome structure to evoluate during (and by) the evolution process. The main idea is that favorable structures will be over-selected because they are less liable to deleterious crossing-over. However, in classical genetic algorithm, the whole traduction process relies on the genome structure. Thus, the locus operator is very difficult to put into practice: It is meaningless to cross two genomes with different structures.

To our point of view, the main limitation of genetic algorithms actually comes from the oversimplistic relation between genotypes and phenotypes: In a genetic algorithm, this relation is predefined and constant over the entire population. The function of each "nucleotide" is given by its absolute position in the genome. In fact, in contrast to living beings, genetic algorithms suffer from the absence of intermediary coding levels (i.e. proteins in living beings). Such an intermediary level would enable genomes with different structures (i.e. coding for different proteins at different locus) to cross one with the other, giving a child that - although it may not be "viable" - actually is meaningful.

In this paper we introduce the Radial Basis Function Gene (RBF-Gene) Model, a new approach of evolutionary computation which is directly inspired of the *structure* and *structural evolution* of prokaryotic genomes [7] [8]. We will first present an "algorithmic" view of the model. Then, its perspectives and main interests will be explained. In the last section, we will present our first experiments with this model, showing that the genome structure actually evolve when individuals' fitness increase.

2 The RBF-Gene Model

The general principles of the RBF-Gene Model are similar to genetic algorithms: A randomly generated population evoluates thanks to an evaluation-selection-reproduction cycle. Actually, the main contribution of our model rests on the traduction process, i.e. on the mecanisms used to obtain a phenotype from a given genotype.

2.1 The traduction process in the RBF-Gene Model

"Classical" genetic algorithms directly convert the genotype into a phenotype thanks to a fixed, manualy defined, *traduction* process. As argued in the previous section, this process is strongly dependant on the *locus* of the genes in the coding sequence (figure 1).

Our objective, with this new model, is to relax the "locus constraint": We want identical sequences to have identical influence on the phenotype.

The RBF-Gene Model is based on three simple ideas directly inspired by the biology:

1. An individual is directly considered as an n-dimensional function (when, in classical genetic

[1]fortunately, it will still converge *in theory*.

algorithms, an individual is a set of parameters),

2. An individual is *made of* basic blocks, each of them representing an elementary function. These blocks can be compared to the *proteomic* level in biology. In the RBF-Gene Model, "proteins" are made of kernel functions (typically Radial-Basis Function, e.g. Gaussian functions) wich can be individually defined by a reduced parameter set. The phenotype (i.e. the "individual function") is then obtained by linear combination of a variable number of kernels,

3. The genome contains coding sequences and non-coding sequences; The later has no influence on the phenotype while the former directly contributes to it by coding for *one* kernel. Moreover, the kernels are encoded by an artificial "genetic code" (similar to biological "codons"). Coding and non-coding sequences or simply differenciated thanks to two genetic sequences: The *start* sequence (start codon) and the *stop* sequence (stop codon[2]). In between, the genetic code enables us to compute the kernel parameters (for gaussian kernels, we would have "Center" codons, "Width" codons and "weight" codons)..

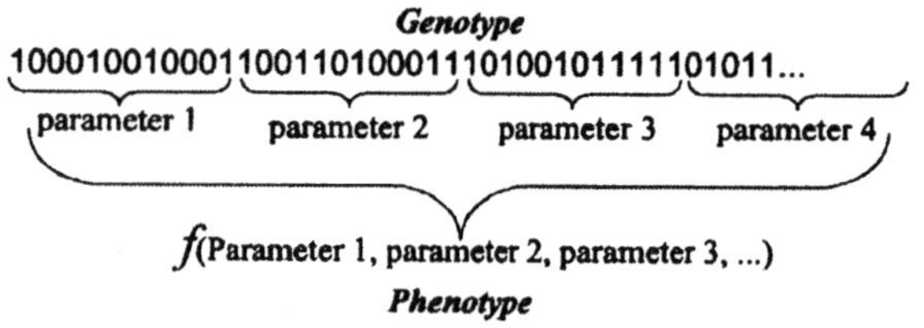

Figure 1. Overview of the traduction process in a genetic algorithm.

The figure 2 summarizes the principles of the RBF-Gene Model for an $R \rightarrow R$ individual function and a binary genetic code: Given a genetic sequence, the algorithm reads the sequence from the beginning to the end. Each time it finds a "start" codon, it begins the traduction process up to the next "stop" codon[3]. Between the start and the stop, all the coding sequence is analysed and the codons are used to compute the three parameters of the kernel thanks to the genetic code (table 1). In our model, the genetic code is used to compute the parameters by means of a binary coding scheme (negative power of two, see figure 2) but other coding scheme may be used[4]. One can note that, according to this code, a coding sequence can give rise to a "degenerated kernel" which will have no influence on the phenotype (e.g. "flat" kernels).

Codon	Function	Value
000	Start	-
001	Stop	-
100	Center (M)	0
101	Center (M)	1
010	Width (L)	0
011	Width (L)	1
110	Weigth (H)	0
111	Weigth (H)	1

Table I. Binary genetic code for an $R \rightarrow R$ individual function

When the kernel parameters have been computed by the algorithm, the genome reading process is resumed up to the next start codon. Of course, a new start codon can be found *inside* a coding sequence, either on the same reading frame (if a new "000" codon is found) or, more likely, on another reading frame. For instance, if an M0 codon ("100") is followed by an L1 codon ("011") on the first readin frame ("[100][011]"), then a new kernel begins ("1[000]11") on the second reading frame.

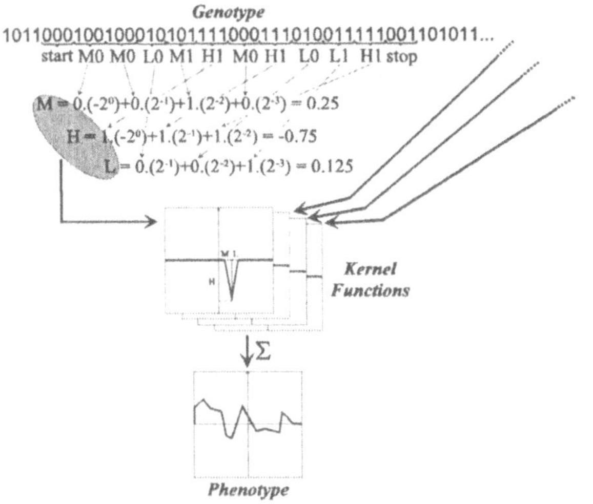

Figure 2. Overview of the traduction process in the RBF-Gene model.

At the end of the traduction process, an individual is represented by a set of Radial-Basis kernel functions and the resulting phenotype simply can be obtain by adding them all. Thus, the RBF-Gene model can be seen as a genetic paradigm to obtain and adapt structures that can

[2] In DNA, the start and stop codons begins and ends the coding sequence. In biology there are one start codon (AUG) and three different stop codons (UAG, UAA and UGA).

[3] In our model, the start and the stop codon should be on the same "reading frame" (with three bits condons it means that the distance between the start and the stop must be a multiple of three).

[4] Depending on the application, the genetic code can be completely different: Either the codon length, the "nucleotide" alphabet ou the coding scheme can be adapted without any influence on the global behaviour of the algorithm.

be compared to well known connectionist Radial-Basis Function Networks [6].

2.2 Interest of the RBF-Gene Model

As said previously, the RBF-Gene Model only differs from genetic algorithms in the traduction process. However, since the locus of a sequence has no direct influence on the phenotype, this novel approach of transciption enable us to cross two individuals having completely different genetic structure. Moreover, the suppression of the "locus influence" enables us to enlarge the set of mutation operators available for the reproduction process: While the "only" mutation operator in classical genetic algorithm is the "switch" operator (substitution of one nucleotide by its complement), we now can use operators that modify the length of the genome. Thus, simple InDel (Insertion/Deletion) operators can be used but it also become possible to use operators that will widely modify the genome structure. In particular, the RBF-Gene Model allows the use of sequences duplication operators (or sequences deletion). Besides, these operators are known to have a strong influence on the evolution of prokaryotic organisms: Sequences duplication enables an organism to increase its number of genes (in the RBF-Gene model, to increase the number of kernels) by duplication of an existing one followed by a drift of one (or both) of the now redundant sequence[5] [1].

As long as reproduction is concerned, the RBF-Gene Model also enables us to built more efficient crossing-over operators. Actually, prokaryotic organisms (e.g. *E. Coli*) can exchange genetic materials [3]. However, this particular form of crossing-over doesn't respect directly the position in the genome (i.e. the two parts of the genome may not be adjacent). On the opposite, they can cross at any position in the genome if the two sequences can be at least partly aligned (i.e. have strong similarities) at the crossing point [2]. Combined with sequences duplication, the specific form of crossing-over may help the genome to reorganise itself. Obviously, such a crossing-over operator is impossible to implement in a classical genetic algorithm; The RBF-Gene Model is perfectly suited for that.

3 Experiments

In order to validate and study our model, we have developped a simple prototype. It uses triangle kernels to identify a simple $R \rightarrow R$ functions (see figures 4 and 5). The population is composed of 100 individuals with genomes of (initaly) 1000 nucleotides. The population evolves thanks to a classical evaluation-selection-reproduction process. In this first experiment, we have used three mutation operators ("switch", "insert", "delete"), with probabilities of 0.001 for each nucleotide. Thus, the genome length is obviously not constant.

The selection process is based on a fitness function f where :

$$f = \int_{-1}^{1} |O(x) - I(x)| dx \qquad [1]$$

where $O(x)$ is the objective function and $I(x)$ is the individual function[5].

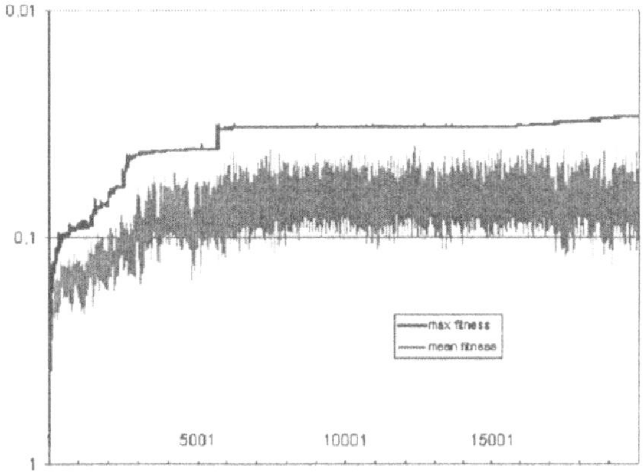

Figure 3. Evolution of the individual fitness (semi-log axis).

The figure 3 shows the evolution of the fitness (average and best individual) during the 18000 first generations. Figures 4 and 5 show the phenotypic functions of the best individual at the first generation and at the last generation. Obviously, the algorithm converges (the best individual has a fitness of approximately 0.027). However, more than this, the interesting point is the evolution of the genome structure: Figures 6 and 7 show the genome and the "proteome" (i.e. the kernels) of these two individuals.

We can see that the number of kernels has decreased and that the kernels have individually grown. Moreover, most of the degenerated kernel (around 85% of the kernels in the first generation) have disappeared in the last generation. This show that the RBF-Gene Model is able to modify the structure of the genome and to adapt it to its domain of application.

In order to study the evolution of the genome structure, we have plot the variation of the kernel set during adaptation process (figure 8). It shows that the number of kernels quickly decreases during the first generations while the average size of functional kernels (i.e. non-degenerated kernels) increases.

[5]Note that the fitness decreases to zero when the individuals improve themselves.

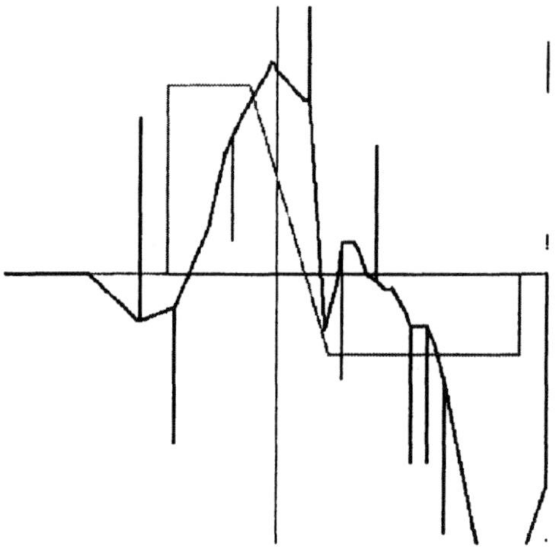

Figure 4. Objective function (in grey) and phenotypic function (in black) at the first generation.

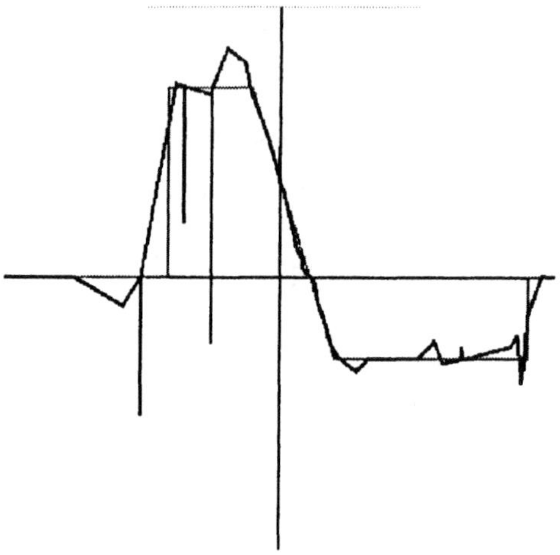

Figure 5. Phenotypic function (last generation).

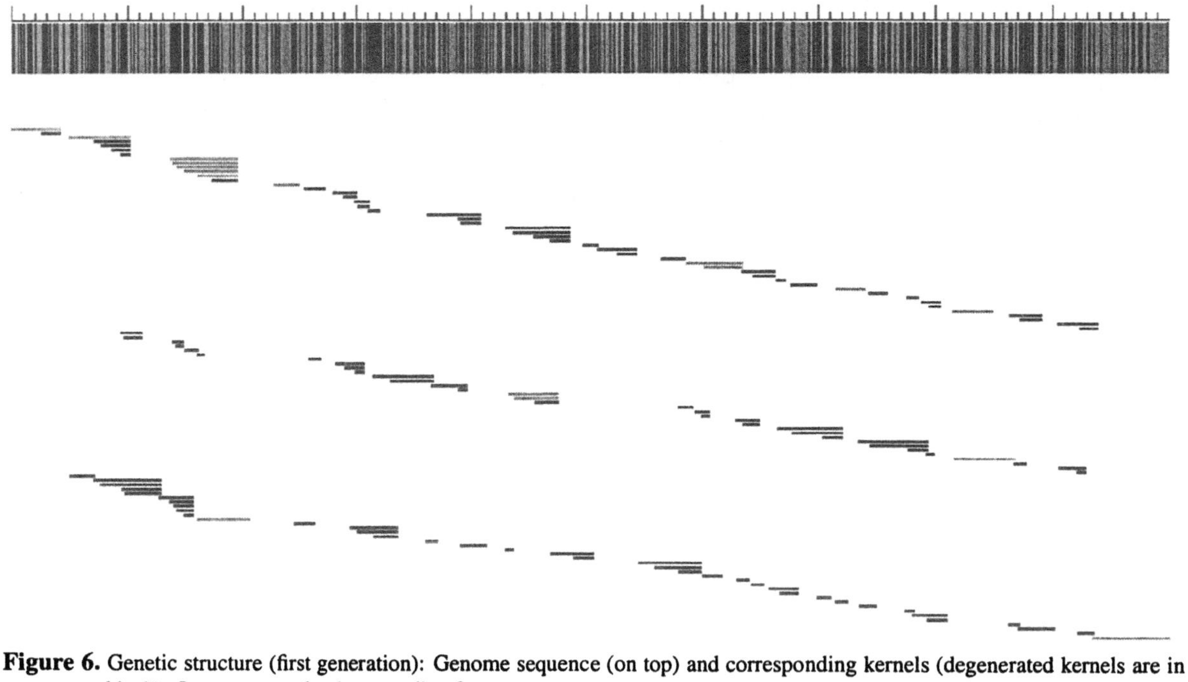

Figure 6. Genetic structure (first generation): Genome sequence (on top) and corresponding kernels (degenerated kernels are in black). One can note the three reading frames.

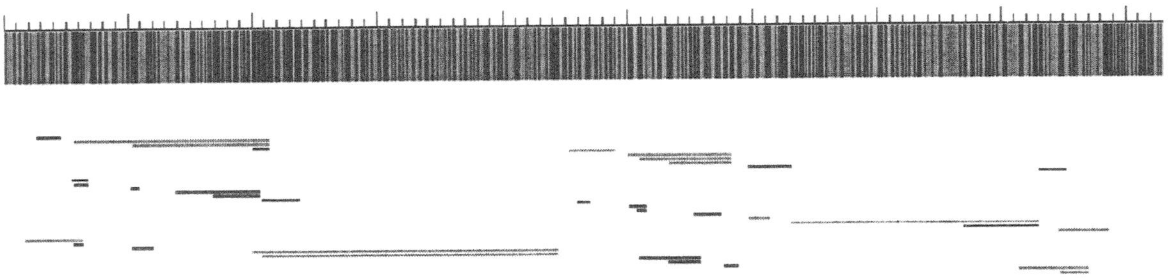

Figure 7. Genetic structure (last generation). One can note that most of the degenerated kernels have disappeared during the evolution process.

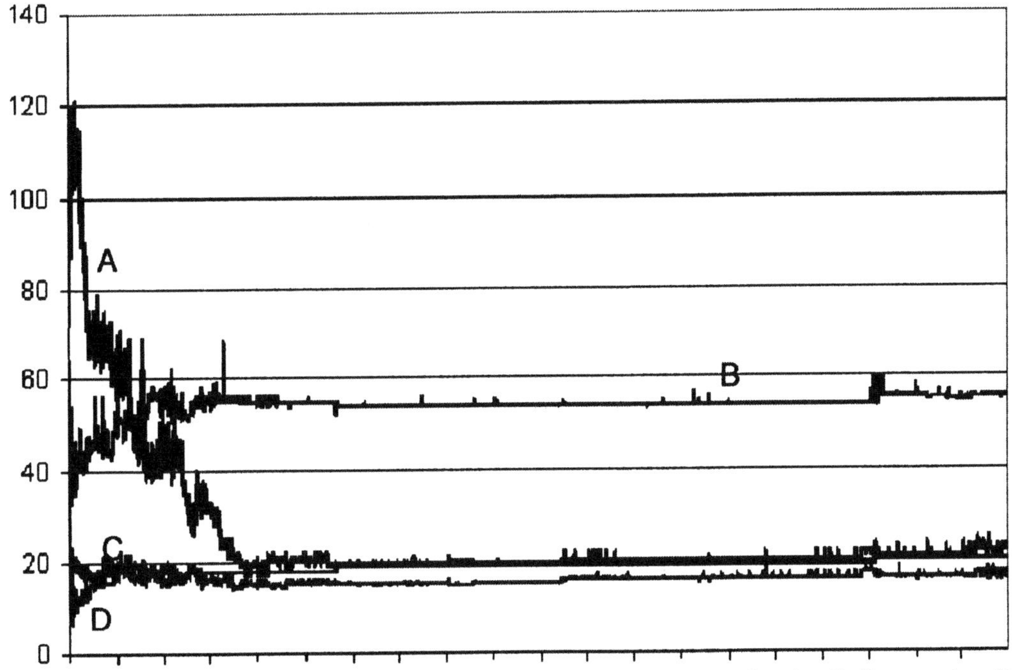

Figure 8. Evolution of the genetic structure (A: number of kernels (functional and non-functional), B: mean size of functional kernels, C: mean size of kernels (functional and non-functional), D: number of functional kernels).

4 Conclusion and perspectives

The RBF-Gene Model, although it is actually under study, appears as a promizing approach to enable artificial evolutionary algorithm to adapt both the coded function and the coding frame. Moreover, since it enable us to use innovating mutation/crossing-over operators, it should be studied further in order to express its entire potentialities. That's why we are studying further the model. Figure 9 shows, for instance, that, with larger genomes (5000 nucleotides), there is a strong correlation between the fitness evolution and the number/size of the kernel functions. Can this correlation explain some characteristics of prokaryotic genomes ? Going on our fructuous collaboration with biologists, we are looking at using the RBF-Model Gene not only as a tool but also as a model.

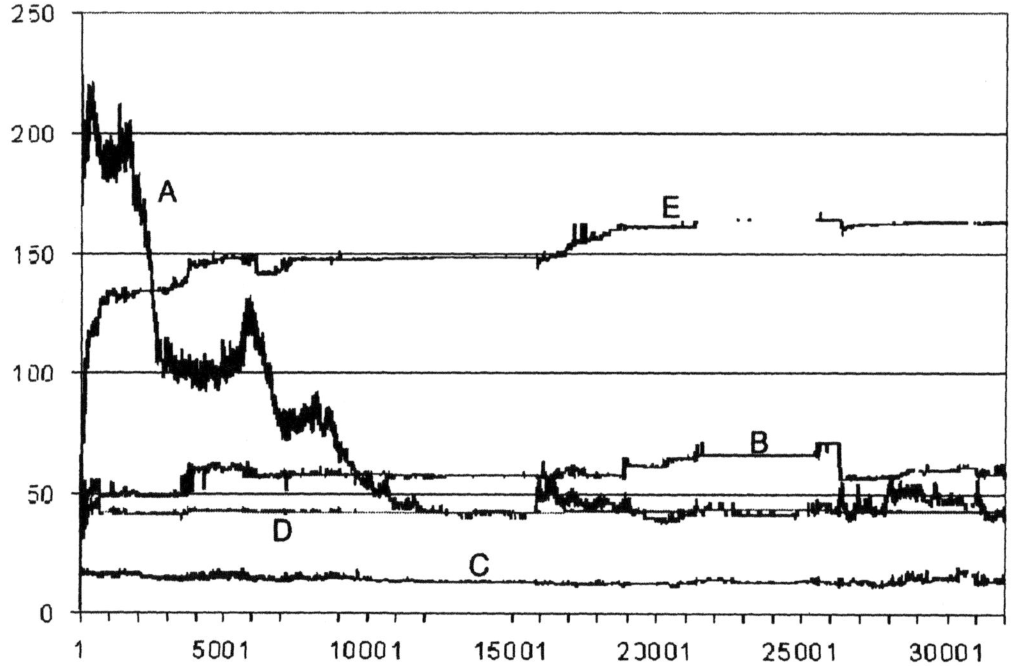

Figure 9. Experiments with a larger genome (A: number of kernels (functional and non-functional), B: mean size of functional kernels, C: mean size of kernels (functional and non-functional), D: number of functional kernels, E: best fitness (semilog scale).

References

[1] E. Betran and M. Long. Expension of genome coding regions by acquisition of new genes. *Genetica*, 115:65–80, 2002.

[2] E.A. Birge. *Bacterial and Bacteriophage Genetics (4th Edition)*. Springer-Verlag, 2000.

[3] J. Davison. Genetic exchange between bacteria in the environment. *Plasmid*, 42(2):73–91, 1999.

[4] D.E. Goldberg. *Genetic Algorithms in search, optimization and machine learning*. Addison-Wesley, 1989.

[5] D. Graur and H. Li. *Fundamentals of Molecular Evolution (2nd Edition)*. Sinauer Associates Inc., 1999.

[6] S.S. Haykin. *Neural Networks, A comprehensive foundation*. Prentice-Hall, 1999.

[7] E.V. Koonin and M.Y. Galperin. Prokaryotic genomes: The emerging paradigm of genome-based microbiology. *Current Opinin in Genetics and Developement*, 7(6):757–763, 1997.

[8] S. Krawiec and M. Riley. Organization of the bacterial chromosome. *Microbiological Review*, 54(4):502–539, 1990.

Evolving spiking neurons nets to control an animat

Hédi SOULA, Guillaume BESLON, Joël FAVREL [1]

[1]PRISMa Lab., INSA de Lyon 69621 Villeurbanne, France, e-mail: hed@prisma.insa-lyon.fr

Abstract

In this paper we propose to exploit the spiking neuron model to control an animat. Our basic assumption is that those neurons are well designed to exhibit both learning abilities and the modular structure of reactive control. Evolving the net of neurons is therefore a mean to study the complex interactions of spikes to produce a simple behaviour of predation and homing in simulation.

1 Introduction

As Behaviour-Based control emphazised it, modular organization appears as one of the fundamental features in animats architectures. However, we don't know how to learn this modular organisation (e.g with neural networks). So it usually has to be designed *a priori* by a designer. On the other hand, artificial neural networks exhibit learning abilities, but "robot/agent problem are often defined in terms of abstract goals rather than specific input-output pairs" which is unfortunatly the traditional key in ANN learning.

Thus when designing Neuro-Controller architecture, the need to combine both approaches lead toward modular neural networks. Some recent advances have been made in this field [13] and in some aspects, modules can be viewed as an artificial implementation of biological neural assemblies.

The aim of this paper is at firts to study a way to combine both behaviour-based and neural architecture for robot control. Then to return into biology to understand the underlying basis of such a solution. And finally to use the "*new*" generation of artificial neural networks (the spiking neural networks) as a step toward implementing our approach.

2 The structure/learning dilemma

As S. Nolfi emphasized it, modular systems are generally based on a "distal" decomposition of the global policy while effective control must be based on a "proximal" approach [10]. We call this the structure/learning dilemma because we have to allow a certain kind of structuration of the controller to emerge, while learning, from the relationship between the controller, the animat and its environment.

This means that the intermediate level (i.e the modularity itself) cannot be "hard-coded" as usual neural modules are: It has to be "soft-coded" inside the global neural network. Far from being independant modules communicating through specific links (as they are used in pattern recognition for instance), behavioral modules are to be functional groups of neurons (neural assemblies) [17] that are able to act and react together in order to perform *different* behaviors in a *single* network. Thus the abstract goals of an animat are coded via these groups of functional neurons. Moreover they are to emerge (to be *learned*) on-line (while the animat explores its environment).

3 Biological basis

Most of the biological neurons communicate by sending pulses to other neurons via the axons. A pulse is a small current charge that occurs when the voltage across the membrane exceeds a threshold. This pulse is known as a spike. After emitting a spike, the neuron enters a refractory period where it can no longer spike no matter how strong is the input stimulation.

In the last decade, many advances in neurobiology have changed our views about the neural treatment of information. Indeed, the major assumptions were based on the "firing rates" models of neural nets. They are the widespread model among the neural networks community. What matters in these models is the average number of spikes per neuron over a (rather) large time window.

But comtemporary neurobiology suggests now that the temporal factor (i.e. the exact time of a spike) is used as a coding factor in real brains. Thus, temporal synchronisations between far neurons have been detected in most of cerebral areas ([7, 14, 16]). At the same time, other authors ([8, 12, 11]) have showed that the exact time of firing of a neuron can code precisely the nature of stimulus received by this neuron. It is then useless to wait for more spikes to transmit the neural information to the remaining of the net[1]. Moreover, even lacking a specific definition, we can see neuron assemblies as spatio-temporal patterns in such a dynamical approach ([3, 4]).

Finally, a biological brain is structured *strongly* (cerebral areas) but most importantly *weakly* (functional

[1]for a review of all these arguments see [9]

groups of neurons without explicite definition). If the first structure is a constraint a priori, the latter however emerges from the evolution of the biological system inside its environment. The basis of this emergence is the specialisation of a certain number of neurons through learning. Indeed, nothing can no longer distinguish two neurons involved in such a specialisation. Only the contextual activity (with other neurons) and temporal (when) allow us to make a difference between them. Thus, behaviours express themselves with inter-modules relations while conserving an uniform global structure.

4 The spiking neurons

As discussed in the previous section, we need to introduce time inside the neuron. Indeed, time is no longer an abstract external concept. In an animat, the neural network is involved in a sensory-motor loop. Thus, any actions induced by a sensorial occurence may lead to modify the animat relation with the environment, itself modifying the sensorial occurence. Consequently, the sensorial modification may occur *before* the spikes have all been generated.

This fundamental feature is ignored in a classical neural networks since the update rule is independent of the animat's time (and all the neurons are updated at the same time). Thus, the time of firing is independent on how strong is the input stimulation. In previous works we proposed a simple model which preserved the *order* of the firing sequence. Therefore we called it *scheduled mode* [1, 15]. However, in spiking neurons model, we can keep track of not only the *order* of the firing sequence but also the *delays* between the spikes. Then, as long as we are able to reinforce such a spatio-temporal pattern, we have the potential to specialize temporally the neurons involved in a behavioural sequence. Morevover, since this is only a subset of all neurons, we can use this as a tool to make neural assemblies emergence possible. That's why we now turn toward spiking neurons model.

There are many possible and existing spiking neurons models. We used one of the simplest of them : the leaky integrate and fire model [5]. A neuron i is defined with four state variables (v, θ, γ, r). $v_i(t)$ is the voltage of the membrane at time t. θ_i is the threshold of the neuron, γ_i is the leak and finally r_i is the absolute refractory period. The voltage is calculated with the equation :

$$v_i(t) = v_{rest} + \sum_{\text{all presynaptic neurons } j} \gamma^{t-t_j} w_j \qquad (1)$$

where t_j is the date of firing of the presynaptic (input) neuron j and w_j is the synaptic weight between neuron j and i, t is the current time. v_{rest} is the voltage at rest and is constant. Each time we have a spike (when the neuron fires) we add to the current voltage of all the postsynaptic neurons the weight. The voltage tends to decrease exponentionnaly toward its resting potential, hence the γ^{t-t_j}. When the potential exceeds the threshold, the neuron fires. We then update the state of the neuron accordingly :

$$v_i(t) = v_{rest} \text{if} v_i(t) > \theta \qquad (2)$$

It means that the potential is reset to its resting value when the neuron has fired. The absolute refractory period is simulated by setting the threshold to infinity as follows :

$$\theta(t) = \begin{cases} \theta & \text{if } t - t_i > r, \\ \infty & \text{if } t - t_i < r, \end{cases} \qquad (3)$$

For sake of simplicity, we neglect the *relative* refractory period (i.e a decreasing function to the normal threshold).

As easily seen, this is possibly the simplest way to code spiking neurons. As a matter of fact, for computational purpose, we split the time into short period ($\Delta t's$) and repeat the above equations sequentially. In our case, the Δt are constant and equivalent to 1ms. We neglect propagation time through neurons as we update the input immediatly after the firing. In order to simulate noise, we added a random value to the voltage when scanned for firing. Consequently equation 2 becomes :

$$v_i(t) = v_{rest} \text{if } v_i(t) + \sigma(0, \beta)) > \theta \qquad (4)$$

where $\sigma(0, \beta)$ is a gaussian distribution of average 0 and standard deviation β.

5 Application : predation and homing

We wanted to test these ideas with a task of predation and homing. Despite we are in simulation we authorized only realistic sensors and motors in a realistic environment (i.e not a grid world).

5.1 The task

The animat had to learn a "food and nest" task (i.e moving to take the food from one point and bringing it back to its nest). There are obstacles in the environment. The animat can interact with them (i.e collision and shadowing). To take the prey, the animat has to step over it. The prey is capturedd and a new prey is created immediatly. To manage to capture new preys, the animat has to go back to its nest (a circle) and the brought prey is destroyed. In the simulation, the prey can be immobile or moving. The figure 1 shows the simulated environment.

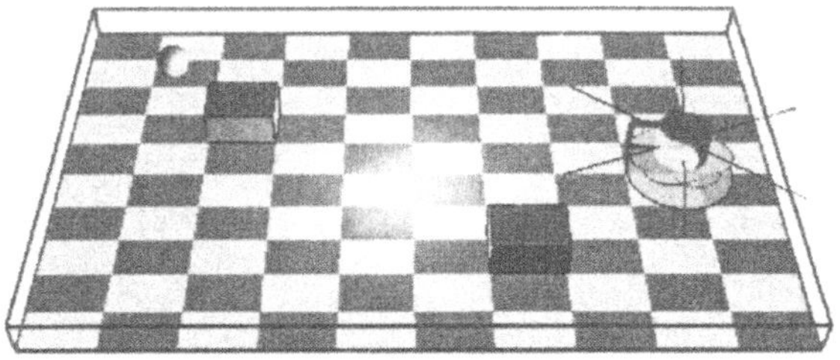

Figure 1. The simulated environment.

The animat is round shaped with six perception sectors. Each of them contains a binary proximity sensor and a target detector that indicates which target faces this sector (figure 2.A). Two binary inputs indicate to the animat which target (food or nest) it should reach. It means it can distinguish the nest and the prey but it can actually "see" them only with a 60° of precision. Moreover, the animat is aware of an obstacle only within its own radius range (the circles around the animat in figure 2.A). The prey and the nest are not considered as obstacles. The animat has two propulsion wheels (Differential propulsion, figure 2.B).

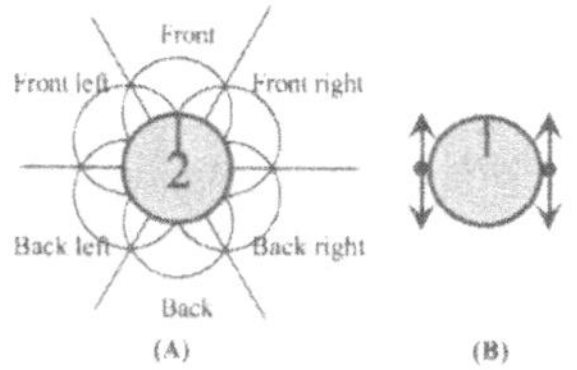

Figure 2. Agent's perceptions (A) and actions (B).

5.2 The architecture

The controller is a spiking neural network consisting of 20 input neurons, 10 hidden ones and 4 for output. The input consist of 6 neurons of contact, 6 neurons for the food, 6 more for the nest and finally two neurons for the target. The input layer is connected to the hidden layer. The latter is itself totaly connected and connects with the output layer. There are many ways to transform the boolean input into spikes. We chose to define two frequencies (one low and one high) to separate the two values. The firing of an input neurons is a probability proportionnal to the frequency of the corresponding input. At each time step we check for firing with the adequate probability. In our simulation the frequencies for 0 ("no signal") and 1 ("signal") were respectively 2 Hz and 10 Hz.

The motors output are calculated in an easier way. Each motor (left and right) is coded with 2 neurons. Each of these codes respectively for forward and backward motor speed. We measure the firing rates over a time window (of 20 ms here) to code the speed of each motor.

5.3 Adaptation process

Our main aim is to study the network. We then try to set artificially the synaptic weights. We are fully aware that our approach cannot be fully validated without actual learning. However we use a simple genetic algorithm to evolve the neural controller as a start of a broader approach. Our aim is first to be sure that the network can do the task with only 10 neurons. Moreover, this approach enables us to estimate parameters influence on the animat and on the network. Particularly we think the influence of the refractory period and the influence of noise will be both fundamental (the noise for example will be used to allow exploration which is a fundamental feature in a learning process).

Now let us describe the adaptation process. Each population is composed of 100 genotypes. Each genotype codes for the synaptic weights of the neural net. The other states variables are constant through the adaptation process and are set at its start. We used $\theta = 0.1, \gamma = 0.9$, $r = 4ms$ and $v_{rest} = -0.5$. The synaptic weights are chosen between -1.0 and 1.0 in the initial genotype set.

At each generation, we evaluate at each generation all the animats during 100000 step time (i.e $100s$). However in order to quicken computations, we divide the trial session into sections of 15000 time steps during which the animat must have a number of capture or homing greater than the section number unless it dies. All the animats in a generation are evaluated according to a fitness function which depends on the number of capture and return and on the average distance made after the last capture (or return).

The first 20 individuals are selected and reproduced to obtain the next generation. During the reproduction process, all the genotypes are altered through mutation and crossover.

5.4 Results

5.4.1 The task. We ran the adaptation process in two steps and over 500 generation. One evolution process was conducted with a non-mobile prey and the same after a certain number of generation with a moving (up and down) prey. The figure 3 shows the results of these experiments.

It displays the average and maximum number of prey taken and return made (the average is the lowest plot of course). We shifted toward mobile prey at the generation 317 when it was estimated that the task was learned. The 200 first generations are used to learn the task (i.e one capture and one return). The "hunting" behaviour is then consolidated during the second period; hence the shape of the fitness curve. Indeed the higherslope means that the selection pressure is directed toward optimization of

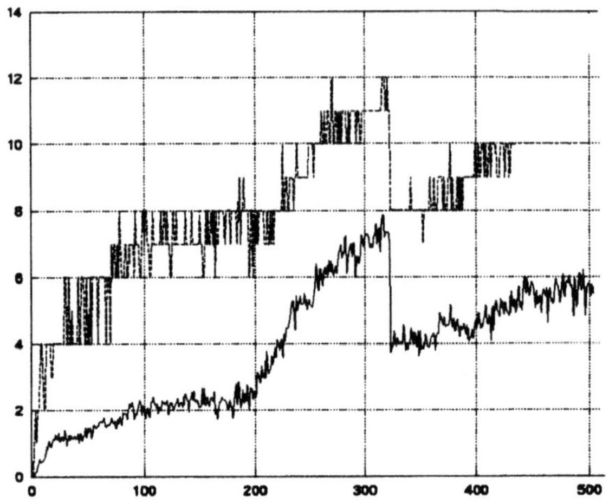

Figure 3. Average and Maximum fitness

the task.

We display a screenshot of a typical move of the robot on figure 4. The circle with an arrow is the animat and the arrow indicates the direction of the animat. The squares are obstacles. The great circle on the right is the nest and the black disk is the prey. The prey is moving up and down changing direction after colliding the walls. The animat showed here didn't touch the walls, managed to get a prey, to get back into its nest, turning around the obstacle and again getting the prey. As shown it managed to follow the wall (the bottom wall). So it worked

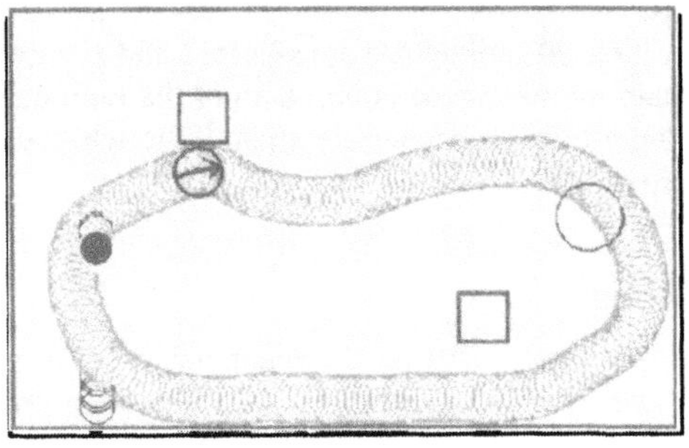

5.4.2 The refractory period. In order to begin our study of the temporal aspect of the neural nets, we looked at the refractory period. We took one animat created by the adaptation process and used different refractory period to test the net.

We start to decrease the value of the refractory period down to zero. The impact on the behaviours is shown in figure 5. The standard value (4 ms) is displayed at the bottom (figure 5.E). The pictures labeled "A", "B", "C"

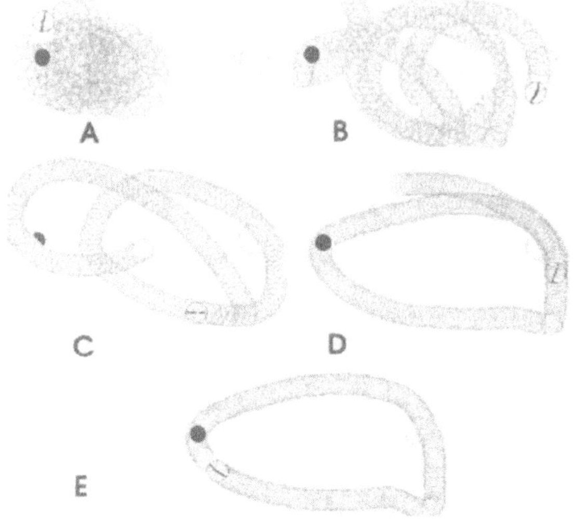

Figure 5. The effect of refractory period on behaviour

and "D" are showing the behaviours with 0 ms, 1ms, 2ms and 3ms respectively for the refractory period for the same net as "E". We could expect a higher average speed (because the network can generate more spikes) but not this type of degradation of the behaviours. It seems that the refractory period plays an important role in the spikes synchronizations. Some neurons, in the original case, are "waiting" to spike and this wait is used by the networks to code for the behaviour. On the other hand, we expected to change nothing by increasing the refractory period except the average speed (always because of the motor coding). As we expected, the behaviours were not qualitatively changed and the average speed of the animat decreased.

This type of analysis shows us just a glimpse of the potential of spiking neurons net even if its full study remains to be done.

6 Conclusion and future works

Spiking neuron nets are a terrible challenge for robot control and designing spiking neural nets able to display simple cognitive function and autonomous learning abilities is still an open problem [6]. Artificial evolution seems a good start to study the complex dynamics underlying this temporalized approach. In order to obtain robots that can, during learning, develop their own structure based on temporal synchronization of neurons, we had to follow that way.

The step forward will be the designing of a learning algorithm which respects the principles evoked previously (i.e that allows the *emergence through living* of such modularities). As already mentioned, we did previously, with another model, succeed in designing such learning algorithm for a very simple version of spiking neurons (*scheduled mode* [2, 15]). We showed some evident structures emerging. But this model lacked both the temporal and computational power of spikes and thus did only partially validate our approach of learning. On the other side, it shows the promising possibilities of doing it with spiking neurons.

References

[1] G. Beslon. *Contrôle Sensori-Moteur par Réseaux Neuromimétiques Modulaires*. PhD Thesis, INSA de Lyon, 1995.

[2] G. Beslon. A neural model for animats brain. In *Proc of ICANNGA 2001, Prague*, pages 352–355, 2001.

[3] E. Daucé. *Adaptation Dynamique et Apprentissage dans les Réseaux de Neurones Récurrents Aléatoires*. PhD Thesis, ENSAE, 2000.

[4] E. Daucé and M. Quoy. Random recurrent neural networks for autonomous system design. In *proc. of the sixth Int. Conf. on Simulation of Adaptive Behavior*, 2000.

[5] E. Domany, J.L.V. Van Hemmen, and K. Schulten (Eds). *Model of Neural Networks II; Temporal Aspects of Coding and Information Processing in Biological Systems*. Springer-Verlag, 1994.

[6] D. Floreano and C. Mattiussi. Evolution of spiking neural controllers for autonomous vision-based robots. In T. Gomi (ed.), editor, *Evolutionary Robotics*. Berlin: Springer-Verlag., 2001.

[7] C.M. Gray, P. Köning, A.K. Engel, and W. Singer. Oscillatory responses in cat visual cortex exhibit intercolumnar synchronization which reflects global stimulus properties. *Nature*, 338(6213):334–337, 1989.

[8] J.J. Hopfield. Pattern recognition computation using action potential timing for stimulus representation. *Nature*, 376(6535):33–36, jul 1995.

[9] W. Maas and C.M Bishop. *Pulsed Neural Networks*. MIT Press, 2001.

[10] S. Nolfi. Using emergent modularity to develop control systems for mobile robots. *Adaptive Behavior*, 5(3/4):343–363, 1997.

[11] R. Van Rullen, J. Gautrais, A. Delorme, and S. Thorpe. Face processing using one spike per neuron. *Neuron. Biosystems*, 48(1-3):229–239, 1998.

[12] M. Samuelides, S. Thorpe, and E. Veneau. Implementing hebbian learning in a rank-based neural network. In *Proc. of ICANN'97, Lausanne*, pages 145–150, 1997.

[13] A.J.C. Sharkey. *Combining Artificial Neural Nets; Ensemble and Modular Multi-Net Systems*. Springer-Verlag, 1999.

[14] W. Singer. Synchronization of cortical activity and its putative role in information processing and learning. *Annual Review of Physiology*, (55):349–374, 1993.

[15] H. Soula, G. Beslon, and J. Favrel. Controlling an animat with a self-organized modular neural network. In *Proc. of EWLR'2001, Prague*, pages 39–46, 2001.

[16] F. Varela, J-P. Lachaux, E. Rodriguez, and J. Martinerie. The brainweb: Phase synchronization an large-scale integration. *Nature Neuroscience Review*, 2:229–239, 2001.

[17] T. Ziemke. On parts and wholes of adaptive behavior: Functional modularity and diachronic structure in reccurrent neural robot controllers. In *Proc. of Int. Conf. on Simulation of Adaptive Behavior, Paris*, pages 115–124, 2000.

Synapsing Variable Length Crossover: Biologically Inspired Crossover for Variable Length Genomes

Ben Hutt and Kevin Warwick
Department of Cybernetics, University of Reading, Whiteknights, Reading, Berkshire, RG6 6AY, UK.

Abstract

Biological Crossover occurs during the early stages of meiosis. During this process the chromosomes undergoing crossover are synapsed together at a number of homogenous sequence sections, it is within such synapsed sections that crossover occurs. The SVLC (Synapsing Variable Length Crossover) Algorithm recurrently synapses homogenous genetic sequences together in order of length. The genomes are considered to be flexible with crossover only being permitted within the synapsed sections. Consequently, common sequences are automatically preserved with only the genetic differences being exchanged, independent of the length of such differences. In addition to providing a rationale for variable length crossover it also provides a genotypic similarity metric for variable length genomes enabling standard niche formation techniques to be utilised. In a simple variable length test problem the SVLC algorithm outperforms current variable length crossover techniques.

1. Introduction

Traditional Genetic Algorithms operate on a population of fixed length genomes, which are viewed as a set of potential solutions to a problem. Typically, the genome is regarded as a set of variables, which are to be optimised by the GA in order to solve a given problem. In other words the goal of the GA is to find an optimum within a given, fixed parameter space. If a large number of parameters are required to be optimised then the search space a GA must traverse in order to find the optimum can be immense. Considering the length of biological genomes a classical GA is unlikely to be a good model of the long term process of biological evolution.

The human genome, for example, is vast and consists of approximately three billion base pairs specifying some hundred thousand genes [1]. In fact, if the DNA from a single human cell were to be unravelled and stretched out it would be nearly two metres long. It would clearly be preposterous to think that evolution had started with strings of this length and merely optimised them over time. Instead it is thought that life started off rather more simply with complexity arising over time. First, single celled organisms evolved, these then led to multi-cellular organisms, which in turn led to the first plants and animals. Complexity and genome length has been able to grow gradually over several billion years through the course of natural evolution. It is worth noting that biological evolution does not necessarily cause an increase in complexity or genome length both can change or remain static over time, all that matters is how good an organism is at passing on its genes to subsequent generations. Also, a longer genome does not necessarily lead to increased complexity. It is only *possible* for complexity to increase, whether this occurs or not is likely to be dependent on the organism in question and the environmental conditions.

2. Variable Length Genomes

A number of studies have investigated variable length genomes in the context of Genetic Algorithms. In all of the methods described below, a position independent genotype encoding must be used. This can be achieved by either encoding the locus in a fixed length representation or by letting certain genetic sequences acts as markers, that effectively identify the sections encoding problem data – see [2]. Other evolutionary methods such as Koza's Genetic Programming [3] and certain Evolution Strategies [4] also use variable length representations. However, the investigation presented in this paper is limited to the context of Genetic Algorithms as other non-GA based variable length evolutionary methods differ in fundamental ways to those used by Genetic Algorithms. For example, Genetic Programming techniques evolve Lisp S-expression programs, whilst Evolution Strategies use adaptive mutation rates and rather different selection strategies, making them exceedingly difficult to compare to any GA based methods in a consistent manner.

2.1 Messy Genetic Algorithms

One of the first studies to utilise a variable length representation was the messy GA [1]. In a messy GA the traditional crossover operator is replaced by the cut and splice operators. First the cut operator is applied on each parent genome, a point is selected at random on each genome cutting each genome into two strings, forming four strings. Then the splice operator is applied to rejoin the strings in a random order.

Whilst messy GAs use a variable length representation they are in fact still based on a fixed length scheme. The reason for this is that genes in a messy GA contain both a value and a tag that specifies the position or locus of that value in a fixed length genome. The messy GA allows for both over specification and under specification of the fixed length genome. In the case of over specification, there are multiple specified values for a specific locus. In this case an ad hoc rule is used to decide which value gets to be used at the specified locus – a popular method is to give precedence to the last gene in the genome. An easy way of implementing this is to allow specified values to overwrite those contained in the fixed length genome, thereby giving precedence to the last value specified. In the case of under specification, some loci in the fixed length genome do not have a specified value. For such loci with missing values the value is taken from a universal template in order to fill in the gap – this template would usually contain a sensible initial or default value for that particular variable.

2.2 The SAGA Cross

In [5] the concept of the SAGA cross is introduced. The idea of this algorithm is to maximise the similarity between sections that are exchanged during recombination thereby ensuring that a sensible and meaningful crossover occurs between the parent genomes. The SAGA cross is intended to be used within the general SAGA framework.

In a fixed length GA, crossover is trivial as the crossover point on both parents is at the same position - the offspring of such a cross will be of exactly the same fixed length as that of the parents thereby ensuring that each individual has a full complement of genes. Within the SAGA framework the evolving population is relatively converged - meaning that different values at the same locus are highly likely to reflect different values of the same phenotypic variable.

The question is how crossover should be applied to variable length genomes in order to ensure that resulting offspring have a full complement of genes and to ensure that only the values of like phenotypic variables are exchanged.

The SAGA cross is based on the sequence similarity of the two parent genomes to be crossed – the metric used is the Longest Common Sub Sequence. The Longest Common Sub Sequence is the longest uninterrupted matching sequence of symbols found between two strings of arbitrary length. Meaning is therefore given by the context of a gene within the genome. If a gene has the same context it is likely to be a different allele of the same gene.

The SAGA algorithm works as follows: first a random crossover point is chosen on the first genome, then the algorithm tests every possible crossover point on the second genome. For each potential crossover point the algorithm calculates the sum of the LCSS on both the left and right portions of the genomes. Only the crossover point(s) with the highest score are eligible as a crossover point for the second genome. If there are multiple eligible crossover points then one of them will be selected at random.

2.3 Virtual Virus

A slightly more biologically plausible crossover is the homologous crossover used by the virtual virus (VIV) project [6]. Like the SAGA cross the VIV crossover algorithm is based on the sequence similarity between parent genomes. The major difference from the SAGA cross is that crossover can only occur in similar sections. In VIV the probability of crossover is controlled by the degree of local similarity between the parent genomes. This local similarity is determined by the number of matches between parents within a specified fixed size window.

The VIV algorithm works as follows: as in the SAGA crossover algorithm a random crossover point is initially chosen on one of the parent genome. The algorithm then compares a window of bases from the selected point with all possible windows of the same size on the other parent genome. The window position that achieves the greatest number of matches is then recorded. The genomes are then crossed within the matched window with a probability based on the similarity of the best matching window.

2.4 Overview of Current Methods

All the above methods provide a rationale to perform Variable Length Genetic Crossover, messy GAs use a simple cut and splice implementation – however this totally ignores any sequence similarity between parent genomes and worse still, messy GAs are fundamentally based on an underlying fixed length representation.

Both the SAGA and VIV crossover methods are based on the similarity between the two parent genomes. Both select a random point on one genome and then search the entirety of the other parent genome in order to find the optimal crossover point(s).

The SAGA cross however explicitly attempts to preserve a complete genetic sequence using the parent genomes as a guide and is, in this respect, far superior to the VIV cross.

It is important to note that in all these variable length methods the genomes are considered to be inflexible, rigid arrays of data. In addition the selection of the initial crossover points are entirely random – this is unlikely to be the case in biological crossover.

3. Biological Crossover Revisited

In animals genetic recombination by crossover occurs during the production of sperm or egg cells (*gametes*) this process is called *meiosis*.

3.1 Meiosis

Before meiosis begins the chromosomes within the cell have already been doubled, forming pairs of homologous dyads. Each dyad consists of two sister chromatids held together by a single shared kinetochore.

First, each pair of homologous dyads align lengthwise with one other and form synapses between themselves resulting in a tetrad. The two homologous dyads are held together at one or more chiasmata. The chromatids attached by chiasmata then slip apart and the homologous dyads separate from each other.

3.2 Crossover

Chiasmata are in fact the remnants of where non-sister chromatids swapped sections whilst they were joined together during the early stages of meiosis. The process of crossover effectively exchanges adjacent strands of DNA. However, this exchange must be performed in such a way that not even a single nucleotide is lost or gained during the exchange. The exact mechanism of crossover is unknown but a plausible mechanism that has significant support is as follows [7]. During the process of crossover one non-sister chromatid is cut through on both strands and an enzyme eats away at these exposed ends leaving two single stranded tails. One of these tails then invades the double helix of the other non-sister chromatid displacing the strands. The invading strand then aligns itself with a complementary sequence of nucleotides with which it can pair; the complementary tail also pairs up but with the other displaced strand. DNA synthesis then fills in any gaps using the second unbroken chromatid as a template. Then all the strands are cut, exchanged and rejoined, effectively exchanging the arms of the non-sister chromatids, this process is illustrated in Figure 1.

4. Inspiration from Biological Crossover

It is worth noting that crossover can only occur in regions that are homogenous due to the double stranded structure of DNA, consequently each strand of DNA can only pair with a complementary strand. Figure 1 only shows a single chiasma however multiple chiasmata are commonly found.

We can therefore safely assume that, in general, crossover does not occur between regions that have a dissimilar sequence and that it is highly likely to occur between regions that have identical or a highly similar sequence. More importantly we should consider the regions between chiasmata – it is these sections that are likely to contain any significant genetic differences between genomes and it is these sections that are actually exchanged by crossover. Further, we should consider the fact that DNA is a flexible molecule and so the sections actually exchanged may be of differing length.

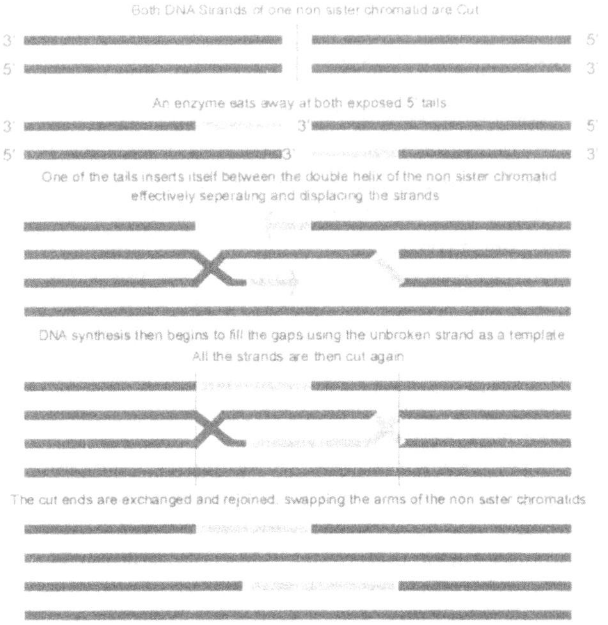

Figure 1 A plausible mechanism for crossover.

Thus, by restricting crossover to homogenous sequences the common sequence similarity is preserved using both parent genomes as a template and only the differences between the parent genomes are exchanged independent of the length of those differences. Intuitively, it must be the differences between genetic sequences that cause any variation in phenotypic fitness. Thus by preserving the similarity and exchanging differences an efficient crossover is performed, therefore only genetic variations that may cause phenotypic fitness variations are recombined.

5. SVLC Algorithm

In order to perform such a crossover we must first have a method with which to compare the two genetic sequences for similarity in order to find regions, which are to be subsequently synapsed together. The metric used is the same as that used by Harvey's SAGA crossover algorithm – the Longest Common Sub Sequence. In addition Harvey also points us towards an efficient algorithm for finding the LCSS. A modified version of this algorithm is used to recursively find the LCSS in order to find any similarities between genomes with priority given to longer common sub sequences. Full C++ source code to perform this crossover is available by e-mailing the author at B.D.Hutt@rdg.ac.uk.

The rationale behind the algorithm is as follows, given two variable length parent strings the position and length of the LCSS between the two strings is located.

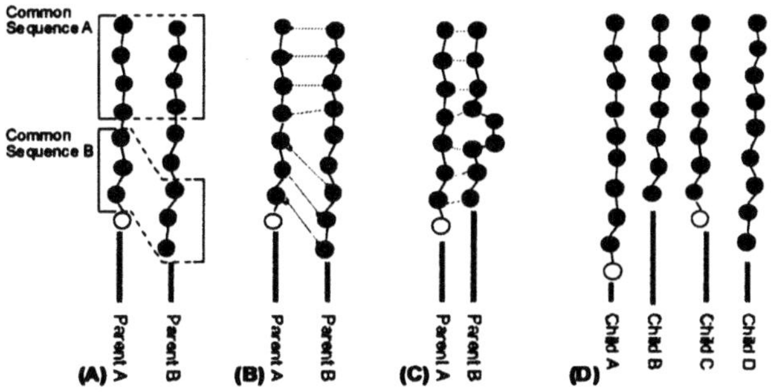

Figure 2 Example of Synapsing Variable Length Crossover. (A) Two similar parent genomes are shown; The common sequences on the parent genomes are labelled. (B) First regions that are identical are identified. (C) The genome is thought of as being flexible so that identical sections can be aligned - crossover is only permitted within these identical regions. (D) The implication of this is that the similarity between parent genomes is always preserved – it is only the differences between parent genomes that are exchanged.

The position and length of the LCSS is recorded and the beginning and end of the LCSS on *both* strings is then used to create two sub problems, finding the LCSS between the two sub strings to the left of the current LCSS and the two sub strings to the right of the current LCSS. This is repeated recursively only dropping out when the LCSS is shorter than a predefined limit. This process produces a list of matched or synapsed segments between the two parent strings. This list is then collated in order to form a set of possible crossover points between the two genomes.

Crossover can then be performed by choosing one or more crossover points at random from this set of possible crossover points in order to produce the offspring.

This process is illustrated in Figure 2, the two parent genomes are effectively synapsed together at their points of similarity, with longer common sub-sequences being given priority. This aligns the chromosomes in a sensible fashion with identical regions being synapsed together. Crossover is then only permitted within the synapsed regions producing offspring that inherit the entirety of the common sub sequences of the parents. Any sequence differences between the parent genomes may or may not be included in the offspring dependent on the crossover point(s) selected.

5.1 Computational Requirements

Since the SVLC (Synapsing Variable Length Crossover) algorithm recursively eliminates the LCSS from each string it will always be somewhat less computationally efficient than the SAGA crossover algorithm. The SAGA crossover algorithm chooses a one-point crossover between parent strings of length m and n, this is done extremely efficiently, being of the order $O(mn)$ having the additional advantage of being time independent of the sequence similarity between parent genomes [5].

The SVLC algorithm is however highly dependent on the similarity between the parent strings. At worst case the SVLC algorithm will always be less than order $O(mn^2)$ however this is a worst case and things are exceedingly unlikely to be this bad, especially when we consider that evolving populations within a SAGA framework are considered to be relatively homogenous [8]. In the best case the algorithm will be as efficient as the SAGA cross and be of order $O(mn)$, this is when both parent genomes are identical or entirely dissimilar. Since we can expect the population to be relatively converged we can also expect that the order will be much closer to order $O(mn)$ rather than $O(mn^2)$.

Due to the fact that SVLC becomes increasingly computationally expensive the longer a genome grows, it is intended for use primarily when fitness evaluation is also computationally expensive, such as is the case when evolving neural network controllers for robots. If fitness evaluation is not computationally expensive and a large number of generations are required and the resulting genomes are likely to be long, then another variable or fixed length method should be considered.

5.2 Similarity Metric

In addition to locating crossover points the SVLC algorithm can be used to give an effective metric with which it is possible to compare the similarity of two variable length strings. This is given by:

$$H = \frac{2\sum_{i=0}^{\psi-1} L_i}{m+n}$$

Where, L_i is the length of the ith synapsed section, ψ is the total number of synapsed sections and m and n are the lengths of the two parent genomes. H therefore gives us a hamming distance like measure of the similarity or homogeneity of the two genomes and varies between *0.0*, for genomes that are totally dissimilar and *1.0* for identical genomes. This immediately provides a method by which standard fixed length genetic algorithm niche formation techniques can be directly applied to variable length genetic algorithms.

6. Initial Results

In order to test the performance of the SVLC algorithm it was tested on a simple variable length problem. The task was to produce a piecewise linear approximation of a non-linear target function $T(x)$ given by:

$$T(x) = 0.5(1+\sin(20.\pi.x)) \text{ where } 0 \le x \le 1$$

Genotypes are interpreted as a set of points between which lines are drawn in order to create the actual

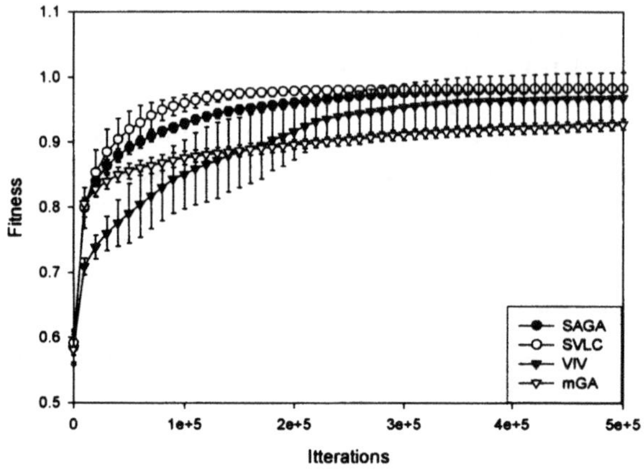

Figure 3: Comparison of various crossover algorithms with a length penalty in place.

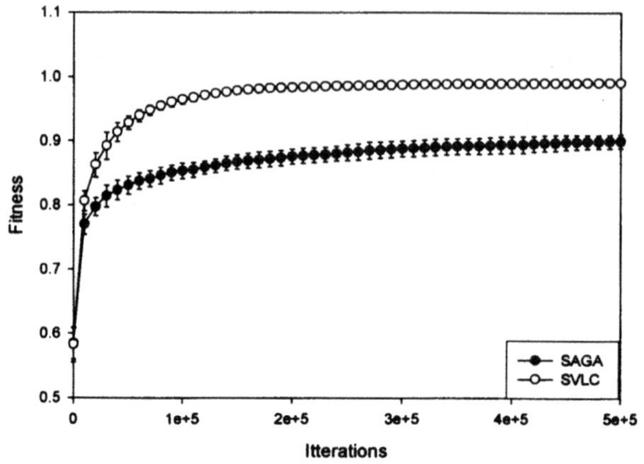

Figure 4: Comparison of SAGA and SVLC crossover algorithms with no length penalty in place.

phenotype *P(x)*. The fitness *F* is calculated using an estimate of the root-mean square error being given by:

$$F = 1 - \sqrt{\frac{1}{N}\sum_{x=0}^{1}\left(T(x) - P(x)\right)^2} - \mu.L$$

The RMS error is estimated by the sum over N discrete parts, with the constant μ effectively scaling the genome length L to create a weak genome length penalty. In all cases Harvey's microbial GA [9] was used with a tournament size of 2 and an initial genome length of 32 with each crossover algorithm being tested over 10 runs of $5x10^5$ iterations.

The first set of tests are shown in Figure 3 with the length penalty μ set to 10^{-6}. As can be seen both the SAGA and SVLC crossover algorithms rapidly obtain excellent approximations of the target function. The VIV crossover algorithm takes far longer to obtain a good solution and the messy GA struggles to obtain a good solution. Figure 4 shows a similar test run for both the SAGA and SVLC crossover algorithms but with the length penalty μ set to 0. Surprisingly this cripples the performance of the SAGA crossover to a similar level to that of the messy GA in the previous run whilst the SVLC algorithm remains unaffected. Without a length penalty in place it is possible for fitness neutral sequences to increase in length. It therefore seems likely that the SAGA crossover can be misled by long neutral sequences forming the LCSS. In such circumstances the SAGA algorithm will select crossover points based on neutral portions of the genome. SVLC does not suffer from this deficiency since it recursively synapses common sequences together restricting crossover to such sections. These are preliminary results and further testing on a wider range of variable length test problems is required to better evaluate the performance of the SVLC algorithm.

7. Conclusion

The SVLC algorithm offers a biologically inspired rationale for variable length crossover. It also provides a similarity metric creating the possibility of using niche formation techniques. In the simple test problem it outperforms current variable length techniques.

References

[1] Goldberg, D., K. Deb, B. Korb (1989). Messy genetic algorithms: motivation, analysis, and first results. In Complex Systems 3: 493-530.

[2] Fullmer, B., Miikkulainen, R. (1991). Using Marker-Based Genetic Encoding of Neural Networks to evolve Finite-State Behaviour. In Varela, F., Bourgine, P. (eds.), Proceedings of the First European Conference on Artificial Life. MIT Press.

[3] Koza, J. (1990). A paradigm for genetically breeding computer programs to solve problems. Technical Report STAN-CS-90-1314, Department of Computer Science, Stanford University.

[4] Lee, C., Antonsson, E.. (2000). Variable Length Genomes for Evolutionary Algorithms. In Proceedings of the Genetic and Evolutionary Computation Conference, Morgan Kauffman, p. 806.

[5] Harvey, I. (1992). "The SAGA cross: the mechanics of crossover for variable-length genetic algorithms". In Männer, R., Manderick, B. (eds.), Parallel Problem Solving from Nature 2: 269-278.

[6] Burke, D., De Jong, K., Grefenstette, J., Ramsey, C., Wu, A. (1998). Putting More Genetics into Genetic Algorithms. In Whitley, D. (ed.), Evolutionary Computation 6(4): 387-410.

[7] Kimball, J. (1994). Biology. McGraw Hill Education.

[8] Harvey, I. (1992). Species adaptation genetic algorithms: a basis for a continuing SAGA. In Varela, F., Bourgine, P. (eds.), Proceedings of the First European Conference on Artificial Life, MIT Press, pp. 346-354.

[9] Harvey, I. (2001). Artificial Evolution: A Continuing SAGA. In Gomi, T. (ed.), Evolutionary Robotics: From Intelligent Robots to Artificial Life, Proc. of 8th Intl. Symposium on Evolutionary Robotics. Springer-Verlag Lecture Notes in Computer Science LNCS 2217 2001.

A Comparative Study Using Genetic Algorithms to Deal with Dynamic Environments

Anabela Simões[1,2], Ernesto Costa[2]
[1] Dept. of Informatics and Systems Engineering, Coimbra Polytechnic, Quinta da Nora, 3030 Coimbra, Portugal
[2] Centre for Informatics and Systems of the Univ. of Coimbra, Pinhal de Marrocos, 3030 Coimbra, Portugal
E-mail: abs@isec.pt; ernesto@dei.uc.pt

Abstract. One of the approaches used in Evolutionary Algorithms (EAs) for problems in which the environment changes from time to time is to use techniques that preserve the diversity in population. We have tested and compared several algorithms that try to keep the population as diverse as possible. One of those approaches applies a new biologically inspired genetic operator called transformation, previously used with success in static optimization problems. We tested two EAs using transformation and two other classical approaches: random immigrants and hypermutation. The comparative study was made using the dynamic 0/1 Knapsack optimization problem. Depending on the characteristics of the dynamic changes, the best results were obtained with transformation or with hypermutation.

1 Introduction

Evolutionary Algorithms (EAs) are often used to solve problems involving a stationary environment and so, the fitness function does not change over the time. When the environment changes over time, resulting in modifications of the fitness function from one cycle to another, we say that we are in the presence of a dynamic environment.

Traditional EAs are not suitable to solve problems facing a dynamic environment, because the population quickly converges to an optimum and when changes do occur, it is very difficult to readapt the solutions to the new conditions. In general, this is a consequence of the lost of the population's diversity.

Several approaches have been proposed to deal with dynamic environments: approaches based on using an explicit memory ([12], [11], [10],[2]), approaches based on promoting diversity ([8], [4], [15]), or hybrid approaches combining these two aspects.

Solutions that appeal to an explicit memory, does not seem to be the best approach in cases where we can not anticipate the dynamics of the changes (for instance, non-periodic changes).

In this work we present a comparative study, using three different approaches, all based on promoting diversity in the population, to solve the dynamic 0/1 Knapsack problem (0/1 KP) proposed by [6].

In particular, we will test the effectiveness of these approaches using the technique of the introduction of random immigrants [8], the application of hypermutation [3] and the introduction of a biologically inspired genetic operator called transformation proposed by [14].

In this paper we will use two forms of transformation: the original proposal, based on the application of random parameters, and the enhanced version, proposed by [16] that uses a set of optimized parameter setting obtained by an extensive empirical study.

This paper is organized in the following manner. First, in section 2, we briefly explain the transformation mechanism. We will describe its biological functioning and the proposed computational implementation. Section 3, details the characteristics of the experimental environment. In section 4, we report the results using the three approaches. Finally, we present the main conclusions of the work.

2 Transformation

2.1 Biological transformation

Transformation is a process that modify certain bacteria (and occasionally other cells as well) which, when grow in the presence of killed cells, take up foreign DNA from that cells and acquire characters encoded by it [13].

2.2 Previous work on computational transformation

We incorporate transformation into the standard genetic algorithm as a new genetic operator that replaces crossover. This modified GA is briefly described in Figure 1. The foreign DNA fragments, consisting of binary strings of different lengths, will form a gene segment pool and will be used to transform the individuals of the population.

```
1. Generate Initial Population
   Generate Initial Gene Segment Pool
2. DO
 2.1. Evaluate Population
 2.2. Select Individuals
 2.3. Transform Individuals
 2.4. Replace Population with New Individuals
 2.5. Create New Gene Segment Pool
WHILE (NOT Stop_Condition)
```

Fig. 1. The GA using Transformation

The GA starts with an initial population of individuals and an initial pool of gene segments, both created at random. In each generation, we select individuals to be transformed and we modify them using the gene segments in the segment pool. After that, the segment pool is changed, using the old population to create part of the new segments with the remaining being created at random. The segments that each individual will take up from the "surrounding environment" will proceed, mostly, from the individuals existing in the previous generation. In the used experimental setup, we changed the segment pool every generation.

After selecting individuals to a mating pool, we use the transformation mechanism to produce new individuals. In this case, there is no sexual reproduction among the individuals of the population. Each individual will generate a new one through the process of transformation. We can consider this process a form of asexual reproduction.

To transform an individual we execute the following steps: we select a segment from the segment pool and we randomly choose a point of transformation in the selected individual. The segment is incorporated in the genome of the individual, replacing the genes after the transformation point, previously selected. This corresponds to the biological process where the gene segments when integrated in the recipient's cell DNA, replace some genes in its chromosome. For more details about transformation see [14].

2.3 The basic version of Transformation-Based Genetic Algorithm (TGA)

The first application of transformation used a set of parameters chosen without any particular criterion. The gene segment lengths were always defined randomly, the mutation rate was set to 0.1%, the transformation rate was 70% and the replacement rate, i.e., the percentage of individuals of the previous generation that contributes for the update of the gene segment pool at the present generation, was set to 70%.

The GA using this primary form of transformation will be referred as Transformation-based Genetic Algorithm (TGA) and was used in the domains of function optimization and combinatorial optimization (static 0/1 KP and dynamic 0/1 KP). Besides the choice of the parameters has been done without any previous study, the obtained results were very promising. In fact, in the domain of function optimization, the TGA achieved much better results than the Standard GA (SGA) using the classical crossover operators.

One of the main conclusions of these studies was the ability of transformation to preserve the diversity in the population during the entire course of computation.

One of the main drawbacks of EA to solve dynamic problems is the fact of, as they converge for the optimum, the population's diversity is lost and the algorithm cannot continue exploring different areas of the search space. So, the use of transformation in non-stationary problems, seem to be a good idea. We used the TGA in the dynamic 0/1 KP but the obtained results were not compared with another approaches. The results obtained were very promising and are reported in [15].

2.4 The enhanced version of Transformation-Based Genetic Algorithm (ETGA)

Analyzing the results obtained by the TGA, it was obvious that if the values of the parameters were adjusted carefully, the algorithm performance could be improved. In order to conclude about those intuitions, we carried out an extensive parametric study to obtain the correct choice of parameters when using transformation in the GA [16].

In that study we analyzed four parameters: the gene segment length, the replacement rate, and the mutation and transformation rates.

The main conclusion was that, the random choice for the length of the segments that was made in the TGA deteriorated its performance in a very expressive manner. We fixed the gene segment length in values from 5 to a maximum depending on the chromosome length and it was clear that as we increase the segment length the results became worst. In fact, larger segments introduced a great degree of disruption.

The choice for the replacement rate also influenced the obtained solutions. For this parameter, in the case of function optimization, the best choice was a value superior to 60%, in the case of the 0/1 KP, the best choice was a value inferior to 50%.

Analyzing the effects of mutation, we concluded that when using transformation, no mutation in necessary. In fact, using mutation, even with a small rate, the results become worst.

Finally, the transformation rate must be chosen in the interval 50% to 100%.

The GA using this improved version of transformation will be denoted as Enhanced Transformation–based Genetic Algorithm (ETGA).

Table 1 reports, for each problem domain, the intervals of each parameters and the best choice in the case of the instances tested.

Table 1. Parameter Choice when using Transformation

Problem	Parameters	Interval	Best choice
Function optimization	Seg. length	[5, 15]	5
	Replac. rate	[70%, 100%]	90%
	Mut. rate	[0%, 1%]	0%
	Transf. rate	[50%, 100%]	90%
Static 0/1 KP	Seg. length	[5, 10]	5
	Replac. rate	[0%, 50%]	40%
	Mut. rate	[0%, 1%]	0%
	Transf. rate	[50%, 100%]	90%

3 Experimental Setup

In this section we will explain the characteristics of the dynamic 0/1 KP (DKP), the characteristics of the four algorithms and the performance measures used to compare the studied approaches.

3.1 The Zero/One knapsack problem

The well-known single-objective 0/1 knapsack problem is defined as follows: given a set of n items, each with a weight $W[i]$ and a profit $P[i]$, with $i = 1, ..., n$, the goal is to determine which items to include in the knapsack so that the total weight is less than some given limit (C) and the total profit is as large as possible.

More formally, given a set of weights $W[i]$, profits $P[i]$ ($i = 1 ... n$) and capacity C, the task is to find a binary vector x = {x[1], ...,x[n]}, such that:

$$\sum_{i=1}^{n} x[i]W[i] \leq C$$

and for which

$$\rho(x) = \sum_{i=1}^{n} x[i]P[i]$$

is maximum.

The knapsack problem is an example of an integer linear problem that has NP-hard complexity.

In the classical 0/1 knapsack problem, the capacity of the bag is kept constant during the entire run. In the DKP the weight limit can change over time between different values.

3.2 The 0/1 dynamic knapsack problem

We used as a test function a 17-object 0/1 knapsack problem with oscillating weight constraint, proposed by [6]. The vectors of values and weights used for the knapsack problem are exactly the same as that used by the authors. The penalty function for the infeasible solutions is defined by: $Pen=K(\Delta W)^2$, where ΔW is the amount which the solution exceeds the weight constraint and K=20. A solution is considered infeasible if the sum of the weights of the items exceeds the knapsack capacity.

Goldberg and Smith used the DKP to compare the performance of a haploid GA and a diploid GA with fixed dominance map and a diploid GA with a triallelic dominance map. In [7] it is referred that their experimentation used variation of the knapsack capacity between two different values every 15 generations.

In this work we enlarged the number of case studies: we used three types of changes in the capacity of the knapsack: periodic changes between two values (C1=104 and C2=60) and between three values (C1=60, C2=104 and C3=80) and non-periodic changes between 3 different capacities (C1=60, C2=80 and C3=104). In each of the periodic experiments we started with a total capacity C1 and after half a cycle the constraint was switched to C2. When using 3 values, after a complete cycle the capacity is changed to the third value C3. Each trial allowed 10 cycles with cycle lengths of 30, 100, 200 and 300 generations.

When the changes in the environment are non-periodic we run the modified GA during 2000 generations and selected randomly several moments of change. In these moments the capacity of the knapsack was altered to a different value chosen among the same three values used in the periodic situation: 60, 80 and 104. The moments when a change occurred and the new chosen knapsack capacity were randomly generated at the beginning of the first run and kept constant for all trials.

3.3 The parameters of the algorithms

In order to compare the efficiency of the GA using transformation we used two well known techniques also based on promoting diversity in the population: the Triggered Hypermutation GA (HMGA) and the Random Immigrants GA (RIGA). Cobb and Grefenstette proposed those techniques and already used them for problems dealing with dynamic environments.

The GA with hypermutation uses a baseline mutation rate (usually very low) when the algorithm is stable and increases the mutation rate (to a high value) whenever there is a degradation in the performance of the time-averaged best performance [3]. We implemented this mechanism with a baseline mutation rate of 0.1% and whenever degradation is observed we increased the mutation rate to 10%, 20% or 30%. The best results were achieved by a hypermutation rate of 10%.

The Random Immigrants mechanism replaces a fraction of a SGA's population each generation, as determined by the replacement rate, with randomly generated values. This mechanism views the GA's population as always having a small flux of immigrants that wander in and out of the population from one generation to the next. This strategy effectively concentrates mutation in a subpopulation while maintaining a traditionally low (i.e., 0.001) mutation rate in the remainder of the population [8]. We tested the Random Immigrants GA with a replacement rate of 10%, 20% and 30%. The best results were achieved with the value 10%.

Both HMGA and RIGA were run with one-point crossover with a probability of 70%.

The TGA used the parameters proposed in the original work [14]: mutation rate of 0.1%, transformation rate equal to 70%, replacement rate of 70% and the gene segment lengths were defined randomly.

The ETGA was run with the set of parameters obtained by the empirical study presented by [16]. The chosen values were: no mutation rate, transformation rate equal to 90%, replacement rate of 40% and gene segment length of size 5.

All the algorithms used population of size 100 individuals and were repeated 30 times. The results reported in the next section are the average values of the 30 runs.

3.4 Performance measures

In order to evaluate the performance of the four approaches solving the dynamic 0/1 KP, we used two well known measures, usually employed in non-stationary problems. Those measures are the accuracy and the adaptability. They are based on a measure proposed by De Jong [5], the off-line performance, but evaluate the difference between the value of the current best individual and the optimum value, instead of evaluating just the value of the best individual.

Accuracy (Acc) is the difference between the value of the current best individual in the population of "just before change" generation and the optimum value averaged over the entire cycle. Accuracy measures the capacity to recover to the new optimum before a new modification occurs.

Adaptability (Ada) is the difference between the value of the current best individual of each generation and the optimum value averaged over the entire cycle. Adaptability measures the speed of the recovery.

The smaller measured values for accuracy and adaptability the better results. If the accuracy reaches a zero value it means that the algorithm found the optimum every time before a change occurs. If adaptability is equal to zero it means that the best individual in the population was at the optimum for all generations, i.e., the optimum was never lost by the algorithm.

These two measures can be mathematically defined by:

$$Acc = \frac{1}{K}\sum_{i=1}^{K} Err_{i,n-1} \; ; \; Ada = \frac{1}{K}\sum_{i=1}^{K}\left[\frac{1}{r}\sum_{j=0}^{r-1} Err_{i,j}\right]$$

where:

K – Number of changes during the run;

r – Number of generations between two consecutive changes;

$Err_{i,j}$ – Difference between the value of the current best individual in the population of jth generation after the last change ($j \in [0, r-1]$) and the optimum value for the fitness after the ith change ($i \in [0, K-1]$).

4 Results

In this section we report the results obtained with the four studied approaches.

4.1 Periodic changes between two values

Changing the capacity of the knapsack between two values, all the algorithms detected the changes, but the readaptation of each one to the new solutions were not quite the same. Figure 2 shows the behavior of the four algorithms in the case of cycle length equal to 100 (changes every 50 generations).

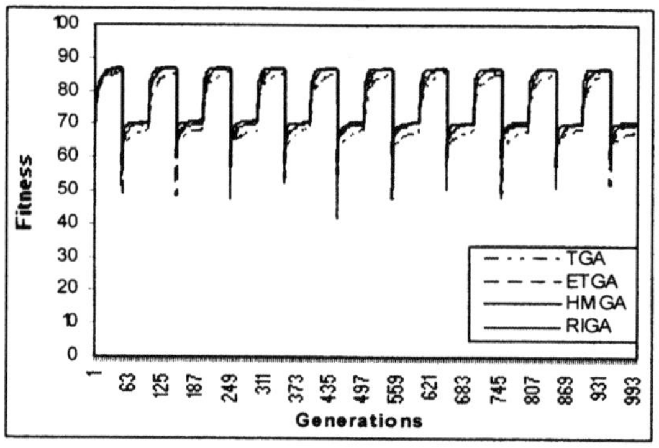

Fig. 2. Performance of the Algorithms with Cycle Length = 100 in the DKP with Periodic Changes between Two Values

The cycle length had major influence in the obtained results. Table 2 and Figure 3 show the accuracy measured in the four approaches. As we can see, TGA obtained very poor performance that improves as the cycle length is increased. ETGA obtained the best accuracy for larger cycle lengths and HMGA, behave better in situation where the changes were more abrupt, especially when the modifications occur every 15 generations.

Table 2. Accuracy Obtained in Experiments with Periodic Changes between Two Different Values

	TGA	ETGA	HMGA	RIGA
Cycle = 30	5.43	2.20	**1.25**	1.78
Cycle = 100	2.34	0.29	**0.23**	0.59
Cycle = 200	1.23	**0.05**	0.23	0.50
Cycle = 300	0.84	**0.02**	0.23	0.45

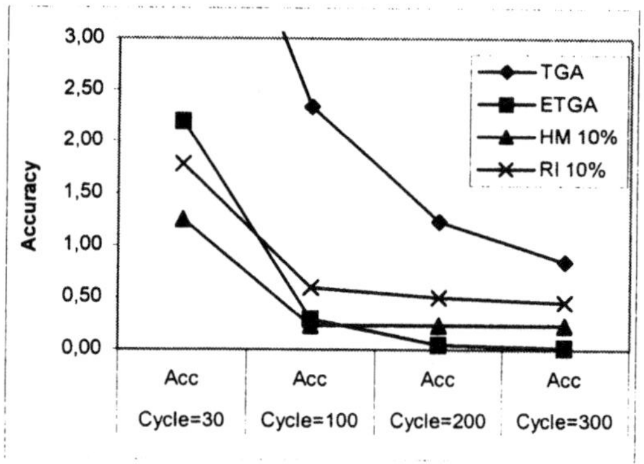

Fig. 3. Accuracy solving the Dynamic 0/1 KP with Periodic Changes between Two Values

Concerning the adaptability of the algorithms, the HMGA, in this particular case of changes between two values, had better performance. This means that the HMGA had faster adaptation to the new optimum value, every time a change was observed. Table 3 and Figure 4 report the results.

Table 3. Adaptability Obtained in Experiments with Periodic Changes between Two Different Values

	TGA	ETGA	HMGA	RIGA
Cycle = 30	8.22	5.01	**3.26**	4.37
Cycle = 100	4.67	2.35	**0.98**	2.58
Cycle = 200	3.22	1.11	**0.61**	1.68
Cycle = 300	2.49	0.73	**0.52**	1.27

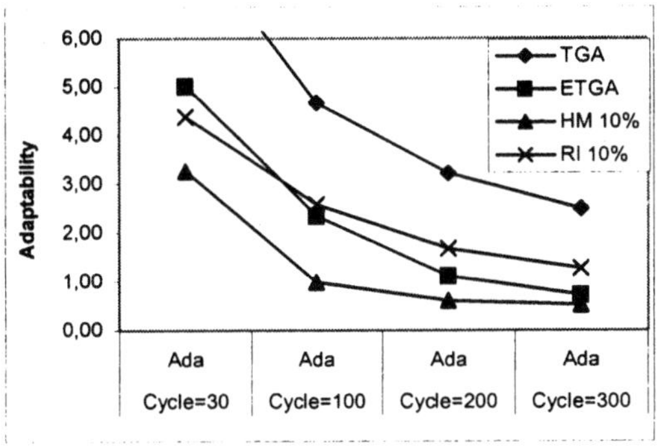

Fig. 4. Adaptability solving the Dynamic 0/1 KP with Periodic Changes between Two Values

4.2 Periodic changes between three values

The performance of the algorithms when the changes were between three different values was slightly different from the previous case, particularly in the adaptability. Figure 5 shows the behavior of the four algorithms with cycles of length 100.

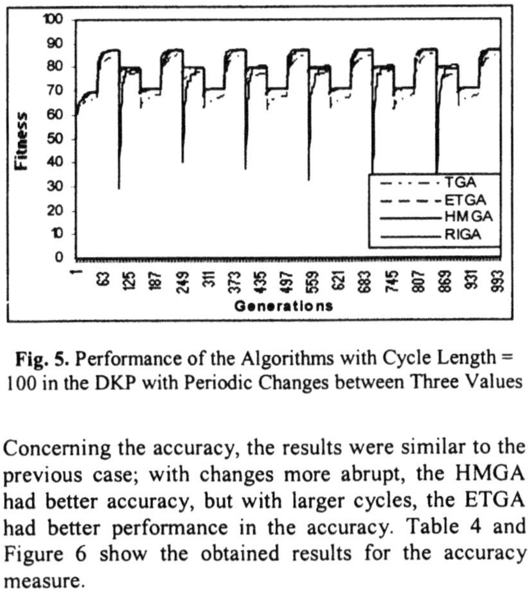

Fig. 5. Performance of the Algorithms with Cycle Length = 100 in the DKP with Periodic Changes between Three Values

Concerning the accuracy, the results were similar to the previous case; with changes more abrupt, the HMGA had better accuracy, but with larger cycles, the ETGA had better performance in the accuracy. Table 4 and Figure 6 show the obtained results for the accuracy measure.

Table 4. Accuracy Obtained in Experiments with Periodic Changes between Three Different Values

	TGA	ETGA	HMGA	RIGA
Cycle = 30	5.27	1.45	**0.70**	1.08
Cycle = 100	2.53	**0.18**	0.36	0.22
Cycle = 200	1.27	**0.02**	0.42	0.15
Cycle = 300	0.78	**0.01**	0.49	0.22

Table 5. Adaptability Obtained in Experiments with Periodic Changes between Three Different Values

	TGA	ETGA	HMGA	RIGA
Cycle=30	7.63	3.53	**1.84**	2.74
Cycle=100	4.59	**1.30**	2.18	1.16
Cycle=200	3.10	**0.61**	1.39	0.65
Cycle=300	2.40	**0.42**	1.35	0.58

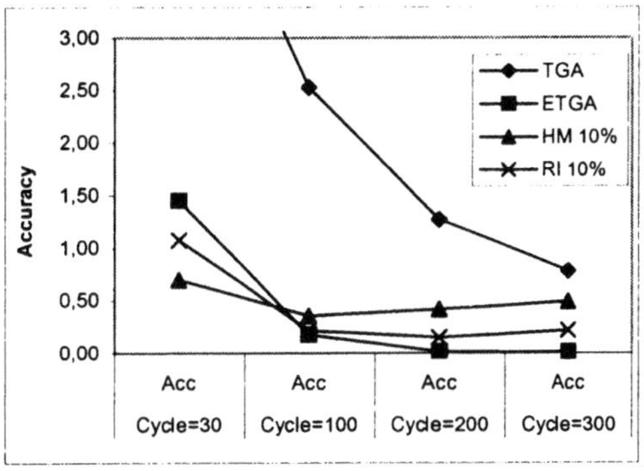

Fig. 6. Accuracy solving the Dynamic 0/1 KP with Periodic Changes between Three Values

Analyzing the adaptability of the algorithms, in this case, ETGA obtained better accuracy than HMGA with larger cycle lengths. With cycles of size 30, HMGA achieved better accuracy, just like in the case of changes between two values.

The TGA and RIGA were, once again the worst algorithms.

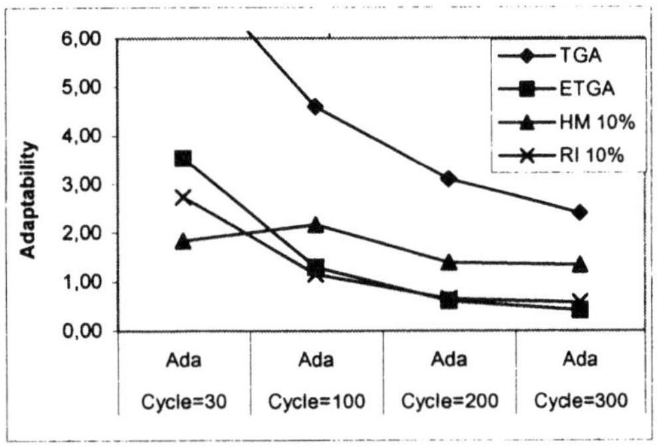

Fig. 7. Adaptability solving the Dynamic 0/1 KP with Periodic Changes between Three Values

In order to understand why HMGA with larger cycles had worst performance than ETGA we measured the population's diversity during the entire computation process. As Figure 8 illustrates, with cycles of 300 generations, when the end of each cycle is reached, the population's diversity in the HMGA achieves very low values. This may be the reason for the high values of accuracy measured. On the other hand, ETGA can preserve the population's diversity at higher levels, and consequently, the algorithm's accuracy was better.

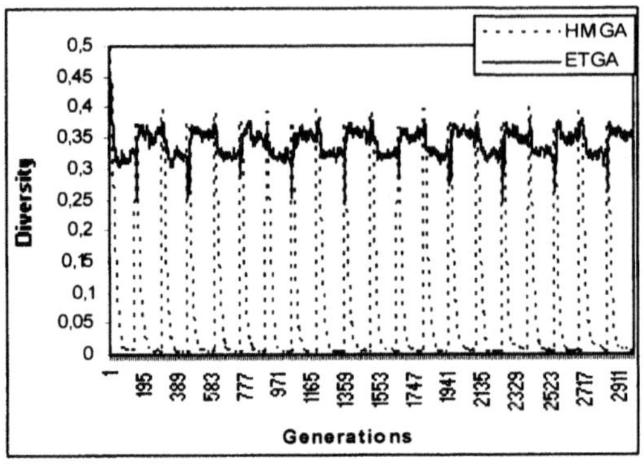

Fig. 8. Population's Diversity using HMGA and ETGA with Cycle Length = 300

4.3 Non-periodic changes between three values

The third type of change that we tested was non-periodic changes between three different values for the knapsack.

Figure 9 shows the graphical behavior of the four algorithms.

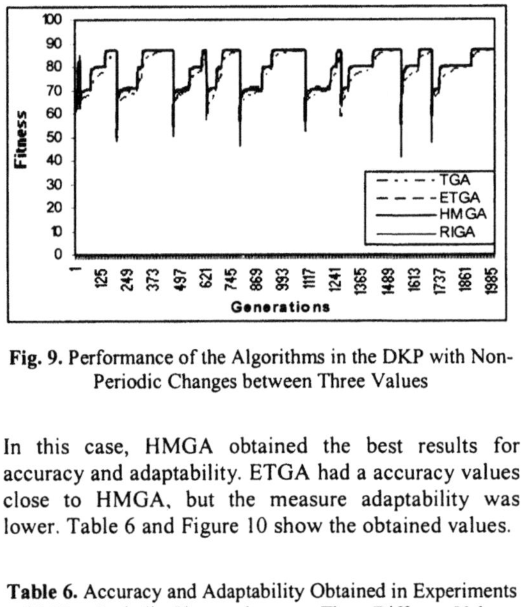

Fig. 9. Performance of the Algorithms in the DKP with Non-Periodic Changes between Three Values

In this case, HMGA obtained the best results for accuracy and adaptability. ETGA had a accuracy values close to HMGA, but the measure adaptability was lower. Table 6 and Figure 10 show the obtained values.

Table 6. Accuracy and Adaptability Obtained in Experiments with Non-Periodic Changes between Three Different Values

	TGA	ETGA	HMGA	RIGA
Accuracy	2.76	0.77	**0.54**	0.93
Adaptability	4.88	2.02	**1.13**	2.22

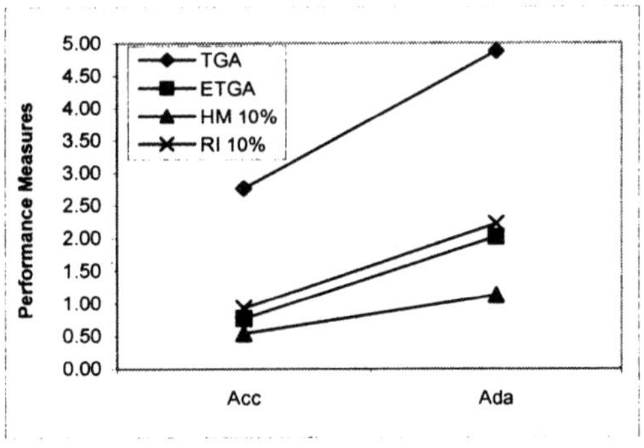

Fig. 10. Accuracy and Adaptability solving the Dynamic 0/1 KP with Non-periodic changes between Three Values

5 Conclusions

In this paper we compared the performance of three approaches based on promoting diversity in the population solving a classical non-stationary problem. One of the approaches was proposed by us and consists in the use of a new biologically inspired genetic operator called transformation, instead of crossover. A first form of transformation had already been used in dynamic and stationary problems (TGA). In this paper we also used an enhanced version of transformation resulting from the assessment of the used parameter setting (ETGA).

The study compared the performance of the GA using the two forms of transformation with two other wee known techniques based on the preservation of diversity: the triggered Hypermutation GA (HMGA) and the Random Immigrants GA (RIGA).

We used two measures to analyze the efficiency of the algorithms: accuracy and adaptability.

The results showed that making periodic changes with larger cycles, ETGA performed better than the other algorithms. Using smaller cycles, HMGA was the best choice. Observing the population's diversity we could see that with cycle lengths of 300, the HMGA at the end of each cycle achieves very low values of diversity. This may be the reason for the measured values.

Using the DKP with non-periodic changes, the HMGA was the best approach, but ETGA achieved very close results. RIGA had weak performance, but TGA was the worst of all.

We can also conclude that the new approach based on transformation is a good candidate to be used in situations where the environment is dynamic, particularly in cases where the cycle length is greater than 30.

Acknowledgements

This work was partially financed by the Portuguese Ministry of Science and Technology under the Program POSI.

References

[1] J. Branke (1999). Evolutionary Algorithms for Dynamic Optimization Problems - A Survey. Bericht 387, Februar 1999, AIFB, Universität Karlsruhe.

[2] J. Branke (1999). Memory EnhancedEvolutionary Algorithm for Changing Optimization Problems. In Proceedings of the 1999 Congress on Evolutionary Computation, pp. 1875-1881, IEEE, 1999.

[3] H. Cobb (1990). An Investigation into the Use of Hypermutation as an Adaptive Operator in Genetic Algorithms Having Continuous, Time-Dependent Nonstationary Environments. Technical Report AIC-90-001, 1990.

[4] H. Cobb, J. J. Grefenstette (1993). Genetic Algorithms for Tracking Changing Environments. In Proceedings of the Fifth International Conference on Genetic Algorithms, pp. 523-530. Morgan Kaufmann, 1993.

[5] K. A. De Jong (1975). Analysis of the Behavior of a Class of Genetic Adaptive Systems. Ph.D. Dissertation, Department of Computer and Communication Science, University of Michigan, 1975.

[6] D. E. Goldberg and R. E. Smith (1987). Nonstationary Function Optimization using Genetic Algorithms with Dominance and Diploidy. In J. J. Grefenstette (ed.), Proceedings of the Second International Conference on Genetic Algorithms, pp. 59-68. Laurence Erlbaum Associates, 1987.

[7] D. E. Goldberg (1989). Genetic Algorithms in Search, Optimization and Machine Learning. Addison-Wesley Publishing Company, Inc.

[8] J. J. Grefenstette (1992). Genetic Algorithms for Changing Environments. In R. Maenner, B. Manderick (eds.), Parallel. Problem Solving from Nature 2, pp. 137-144. North Holland, 1992.

[9] J. J. Grefenstette, C. L. Ramsey (1992) An Approach to Anytime Learning. In D. Sleeman and P. Edwards (eds.), Proceedings of the Ninth International Conference on Machine Learning, pp. 189-195. Morgan Kaufmann, 1992.

[10] B. Hadad, C. Eick (1997). Supporting Poliploidy in Genetic Algorithms using Dominance Vectors. In P. Angeline, R. G. Reynolds, J. R. McDonnell and R. Eberhart (eds.) Proceedings of the Sixth International Conference on Evolutionary programming, vol. 1213 of LNCS. Springer, 1997.

[11] K. P. Ng and K. C. Wong (1995). A New Diploid Scheme and Dominance Change Mechanism for Non-stationary Function Optimization. In Proceedings of the Sixth International Conference on Genetic Algorithms, pp. 159-166. Morgan Kaufmann, 1995.

[12] C. L. Ramsey and J. J. Grefenstette (1993). Case-based Initialization of Genetic Algorithms. In S. Forrest (ed.), Proceedings of the Fifth International Conference on Genetic Algorithms, pp. 84-91. Morgan Kaufmann, 1993.

[13] P. J. Russell (1998). Genetics. 5th edition, Addison-Wesley.

[14] A. Simões and E. Costa (2001). Using Biological Inspiration to Deal with Dynamic Environments. Proceedings of the Seventh International Conference on Soft-Computing (MENDEL'2001), Brno, Czech Republic, June 2001.

[15] A. Simões and E. Costa (2001). On Biologically Inspired Genetic Operators: Transformation in the Standard Genetic Algorithm. Proceedings of the Genetic and Evolutionary Computation Conference (GECCO'2001), San Francisco, USA, July 2001.

[16] A. Simões and E. Costa (2003). Improving the Genetic Algorithm's Performance when using Transformation. In the Proceedings of ICANNGA'2003.

Towards building computational agent schemes

Gerd Beuster, Pavel Krušina, Roman Neruda[1], Pavel Rydvan[2]

[1] Institute of Computer Science, Academy of Sciences of the Czech Republic, Pod vodárenskou věží 2, 18207 Prague 8, Czech Republic, email: bang@cs.cas.cz. [2]Faculty of Mathematics and Physics, Charles University, Malostranské náměstí 25, 11000 Prague 1, Czech Republic

Abstract

A general concept of representation of connected groups of agents (schemes) within a multi-agent system is introduced and utilized for automatic building of schemes to solve a given computational task. We propose a combination of an evolutionary algorithm and a formal logic resolution system which is able to propose and verify new schemes. The approach is illustrated on simple examples.

1 Introduction

Hybrid models, including combinations of artificial intelligence methods such as neural networks, genetic algorithms and fuzzy logic controllers, seem to be a promising and extensively studied research area [1]. We have designed a distributed multi-agent system [6] called Bang 3 that provides a support for an easy creation of hybrid AI models by means of autonomous software agents [3].

Besides serving as an experimental tool and a distributed computational environment [4], this system should also allow to create new agent classes consisting of several cooperating agents. The *scheme* is a concept for describing the relations within such a set of agents. The basic motivation for schemes is to describe various computational methods. It should be easy to 'connect' a particular computational method (implemented as an agent) into hybrid methods, using schemes description. The scheme description should be strong enough to describe all the necessary relations within a set of agents that need to communicate one with another in a general manner.

Example: The genetic algorithm itself, from this point of view, consists of several parts: the *Genetics* agent, which performs the basic genetic algorithm logic and glues all parts together, the *Fitness* agent, that evaluates the fitness function for each individual, the *Operators* agent, that provides genetic operators, metrics operators, and creation operators, the *Selection* agent, that provides the selection of individuals. There are also two optional agent types that can further optimize overall performance: the *Shaper* agent, that provides global processing of population individuals fitness function values — such as sigma scaling — and the *Tuner* agent, that by exploiting information about the genetic algorithm performance (like best individual fitness, average fitness, first and second derivatives of these etc) tunes genetic operators probabilities (cf. Fig 1). It is supposed that there will exist more rival agents implementing a particular function (such as fitness evaluating) and it will be possible to choose among them.

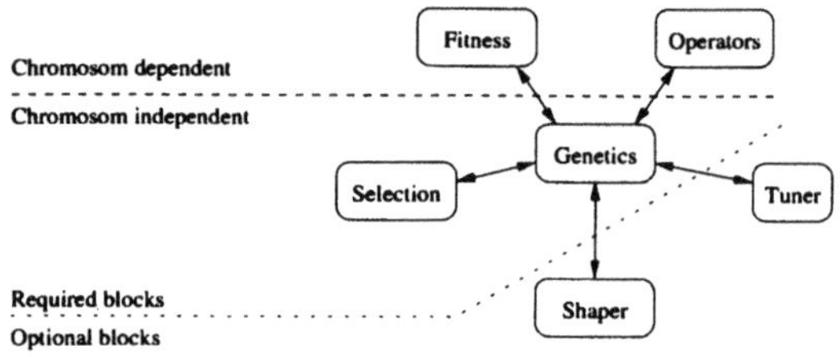

Fig. 1. Genetic algorithm as a multi-agent system.

This paper focuses on ways how to search the space of schemes representing a multi-agent system. The system consists of agents encapsulating individual computational methods, or their combinations. The behavior of the system is tested in the course of searching process by means of a given training dataset. In the case that the space is finite or 'small', variations of searching algorithms can be used (see e.g. [5]). We focused on employing an evolutionary algorithm together with the logics resolution system.

In the following section we present details on the scheme and evolutionary algorithm design. Next section treats a work on schemes as a logic constraint satisfaction problem. This approach can be used in two ways: either to generate new feasible solutions, or to verify solutions proposed by an evolutionary algorithm for their feasibility before they undergo the evolution. Such a hybrid approach neatly augments the evolutionary algorithm so it operates only on solutions that 'make sense'. Simple experiments and future work ideas conclude the paper.

2 Schemes

The scheme is a set of agents with a given topology of communication channels. The following mechanism for scheme specification has been designed.

The agents that comprise the scheme are called *building blocks*. Building blocks communicate through *incoming* and *outgoing gates*. Each agent can have any number of both incoming and outgoing gates.

One purpose of the schemes is that the hybrid computational methods designed in a form of scheme can be easily stored and used. Second, perhaps more interesting, challenge of the schemes concept is the *automatic scheme generation*. The scheme definition is a data structure consisting of the list of the building blocks and the interconnection among them. Actually, the scheme definition is a directed acyclic graph. This offers the possibility of automatic searching the space of schemes in order to find a suitable solution.

The proposed evolutionary algorithm operates on schemes definitions in order to find a suitable scheme solving a specified problem. The genetic algorithm has three inputs: First, the number and the types of inputs and outputs of the scheme. Second, the *training set*, which is a set of prototypical inputs and the corresponding desired outputs, it is used to compute the fitness of a particular solution. And third, the list of types of building blocks available for being used in the scheme.

We supply three operators that would operate on graphs representing schemes: *random scheme creation*, *mutation* and *crossover*.

The aim of the first one is to create a random scheme. This operator is used when creating the first (random) generation. The diversity of the schemes that are generated is the most important feature the generated schemes should have. The 'quality' of the scheme (that means whether the scheme computes the desired function or not) is insignificant at that moment, it is a task of other parts of the genetic algorithm to assure this. The algorithm for random scheme creation works incrementally. In each step one building block is added to the scheme being created. In the beginning, the most emphasis is put on the randomness. Later the building blocks are selected more in fashion so it would create the scheme with the desired number and types of gates (so the process converges to the desired type of function).

The goal of the crossover operator is to create offsprings from two parents. The crossover operator proposed for scheme generation creates one offspring. The operator horizontally divides the mother and the father, takes the first part from father's scheme, and the second from mother's one. The crossover is illustrated in Fig. 2.

The mutation operator is very simple. It finds two links in the scheme (of the same type) and switches their destinations. The mutation operator is illustrated in Fig. 3.

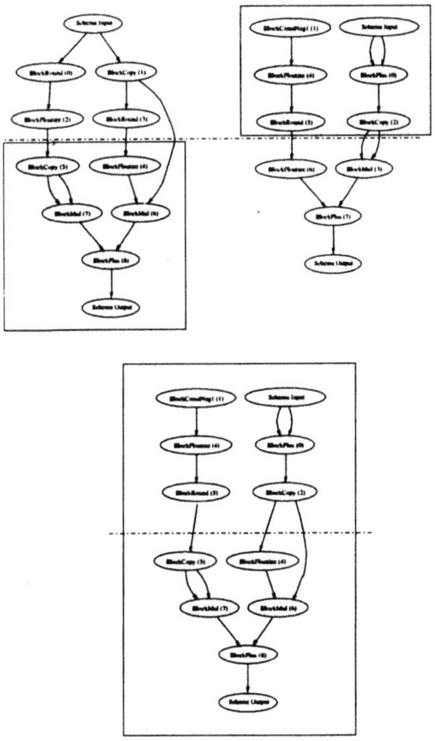

Fig. 2. Crossover of two schemes. The mother and father are horizontally divided and the offspring becomes a mixture of both.

3 Agent Constraints

Beside the genetic component, Bang also uses formal logics for the construction and evaluation of agent systems. Logics can be used both for the construction of new multi-agent systems, and for the verification of existing ones.

There are a number of applications for this:

- Sanity check of MAS

  When MAS configurations are generated automatically by a genetic algorithm, a lot of the system configuration will not work at all. Using the constraint satisfaction checking described before, it is possible to sort out these non-functioning systems without having to actual construct and test them.

- Fault Analysis

  When there are non-working part in user-constructed MAS, our constraint checking system can isolate the parts of the system that do not satisfy the constraints.

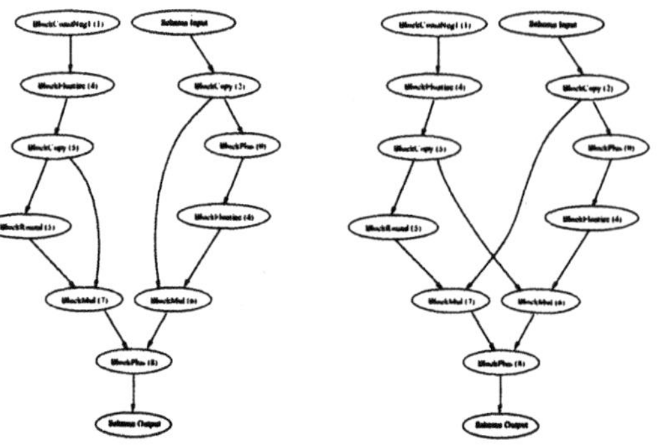

Fig. 3. Mutation on scheme. The destination of two links are switched.

- System Construction

  Given a — possible incomplete — description of a MAS and a set of constraints, our system can generate all MAS that satisfy the constraints. This can be used to automatically construct systems, or to assist the user: After the user has constructed a partial system, our system can aid the user in completing the system by showing possible extensions of the system

In order to apply formal logics to Bang, agent configuration is treated as a constraint satisfaction problem.[2] The logical description of a Bang 3 agent system consists of three parts: Descriptions of the agents, constraints on the individual agents, and constraints on the agent system as a whole.

Agent class definitions consist of a description of the agent's properties and of constraints on these properties, where agent descriptions are sets of terms and agent constraints are sets of horn-clauses over these terms:

Definition 1 (Agent Class Description) *An agent class description is a term* $\mathtt{agent\_class}(D, C)$, *where* D *is a description of the properties of the agent, and* C *is a set of constraints. The description of properties is a set of terms. Constraints are horn-clauses of the form*

$$\mathtt{constraint}(\mathtt{This}) \leftarrow C_1 \wedge C_2 \wedge \ldots \wedge C_n$$

with $C_1 \ldots C_n$ *terms, and* This *a free variable in* $C_1 \ldots C_n$.

Agents are derived from agent class descriptions by (partially) instantiating the terms of the agent class description.

A multi-agent system consists of a set of agents and constraints on the system as a whole:

Definition 2 (Multi-Agent System) *A multi-agent system description is a term* $\mathtt{mas}(A, C_s)$ *with* A *a set of agents,* C_s *a set of system wide constraints.* C_s *is a set of horn clauses of the form*

$$\mathtt{constraint}(\mathtt{This}) \leftarrow C_1 \wedge C_2 \wedge \ldots \wedge C_n$$

with $C_1 \ldots C_n$ *terms, and* This *a free variable in* $C_1 \ldots C_n$

When constraints are evaluated, variable This is unified with a description of the MAS. Reasoning is straight-forward: The terms can be transferred directly into a PROLOG-program. Valid configurations are generated by first attempting to satisfy the internal constrains of each agent and than satisfying the system-wide constrains. This is shown in algorithm 1.

Algorithm 1 Constraint Evaluation

Require: $\mathtt{mas}(A, C_s)$
  for all Agents $\mathtt{agent}(D, C) \in A$ **do**
    for all $\mathtt{constraint}(\mathtt{This}) \leftarrow C_1 \wedge C_2 \wedge \ldots \wedge C_n \in C$ **do**
      evaluate $\mathtt{constraint}(\mathtt{mas}(A, C_s))$
    end for
  end for
  for all $\mathtt{constraint}(\mathtt{This}) \leftarrow C_1 \wedge C_2 \wedge \ldots \wedge C_n \in C_s$ **do**
    evaluate $\mathtt{constraint}(\mathtt{mas}(A, C_s))$
  end for

So far, we have described a generic formalism for the definition of constraints onto multi-agent systems. In order to use this formalism with Bang 3, some standard terms and predicates have to be defined.

We start with a simple configuration problem, in which we are interested whether it is possible to connect agents to each other. For this, the following terms are used:

$\mathtt{name}(N)$ Where N is unique. This term is instantiated when an agent is created, and it serves as an identifier for the agent.

$\mathtt{gate}_{in}(I, T)$ Where I is a name, and T is a data type. This term means that an agent has an input gate (i.e. an interface for receiving data) called I which is of type T.

$\mathtt{gate}_{out}(I, T)$ Where I is a name, and T is a data type. This term means that an agent has an output gate (i.e. an interface for sending out data) called I which is of type T.

float, int, string These are basic data types.

$\mathtt{array}(T, A)$ A complex data type: Array of type T with arity A

A connection between two agents is valid if some output gate of the first agent matches an input gate of the second agent. This is expressed by the following formula:

$$\begin{aligned}&\mathtt{connects}(A, A_{out}, B, B_{in}, C) \leftarrow \\ &\quad \mathtt{in}(\mathtt{gate}_{out}(A_{out}, \mathtt{Type}), A) \wedge \\ &\quad \mathtt{in}(\mathtt{gate}_{in}(B_{in}, \mathtt{Type}), B) \wedge \\ &\quad C = \mathtt{conn}(A, A_{out}, B, B_{in})\end{aligned}$$

Here, A and B are agents, A_{out} is the name of an output gate of agent A, B_{in} is the name of an input gate of agent B, and in unifies the first argument with the appropriate term in the second argument.

A MAS is valid when all connections between input and output gates are valid. This can be checked recursively be the following predicate:

$$\begin{aligned}&\mathtt{check\_connections}(MAS) \leftarrow \\ &\quad MAS = \mathtt{mas}(A, C) \wedge \\ &\quad \mathtt{remove\_element}(F, R, C) \wedge \\ &\quad F = \mathtt{conn}([X, \mathtt{gate}_{out}(X_{out})], \\ &\qquad [Y, \mathtt{gate}_{in}(Y_{in})]) \wedge \\ &\quad \mathtt{connects}(X, X_{out}, Y, Y_{in}, F) \wedge \\ &\quad \mathtt{check\_connections}(\mathtt{mas}(A, R))\end{aligned}$$

Extending the system to more complex relationships is straightforward. As an example, we show how a notion of trust among agents can be described. The general idea is that an agent A trusts a agent B if it either knows directly that the agent is trustworthy, or if agent A trusts a third agent C, and C trusts B.

This is captured by the following definition:

Definition 3 (Trust) *Agent X trusts agent Y if it knows the agent is trustable, or if it knows an agent M which trusts agent Y:*

$$\begin{aligned}&\mathtt{trust}(X, Y) \leftarrow \\ &\quad \mathtt{agent}(X, \mathtt{trusts}(M)) \wedge \mathtt{trust}(M, Y)\end{aligned}$$

With this additional relationship, it is easy to formulate a constraint "all agent gates should be matched, and only agents should be connected who trust each other":[1]

$$\begin{aligned}&\mathtt{trusted\_MAS} = \\ &\quad (A, [\mathtt{check\_connections}, \\ &\qquad \mathtt{check\_trust}, \mathtt{Connections}])\end{aligned}$$

$$\begin{aligned}&\mathtt{check\_trust}(MAS) \leftarrow \\ &\quad MAS = \mathtt{mas}(A, C) \wedge \\ &\quad \mathtt{remove\_element}(F, R, C) \wedge \\ &\quad F = \mathtt{conn}([X, \mathtt{gate}_{out}(X_{out})], \\ &\qquad [Y, \mathtt{gate}_{in}(Y_{in})]) \wedge \\ &\quad \mathtt{trust}(X, Y) \wedge \\ &\quad \mathtt{check\_trust}(\mathtt{mas}(A, R))\end{aligned}$$

[1] check_connections and Connections appear both in the set of constraints, because Connections, the actual connections of agents within the MAS, is itself a constraint onto the system.

4 Experiments

This section describes the experiments we have performed with generating the schemes using the genetic algorithm described above.

The training sets used for experiments represented various polynomials. The genetic algorithm was generating the schemes containing the following agents representing arithmetical operations: *Plus* (performs the addition on floats), *Mul* (performs the multiplication on floats), *Copy* (copies the only input (float) to two float outputs), *Round* (rounds the incoming float to the integer) and finally *Floatize* (converts the int input to the float).

The selected set of operators has the following features: it allows to build any polynomial with integer coefficients. The presence of the *Round* allows also another functions to be assembled. These functions are the 'polynomials with steps' that are caused by using the *Round* during the computation.

The only constant value that is provided is -1. All other integers must be computed from it using the other blocks. This makes it more difficult to achieve the function with higher coefficients.

The aim of the experiments was to verify the possibilities of the scheme generation by genetic algorithms. The below mentioned examples were computed on 1.4GHz Pentium computers. The computation is relatively time demanding. The duration of the experiment depended on many parameters. Generally, one generation took from seconds to minutes to be computed.

The results of the experiments depended on the complexity of the desired functions. The functions, that the genetic algorithm learned well and quite quickly were functions like $x^3 - x$ or x^2y^2. The learning of these functions took from tens to hundred generations, and the result scheme precisely computed the desired function.

Also more complicated functions were successfully evolved. The progress of evolving function $x^3 - 2x^2 - 3$ can be seen in the Fig. 4 and 5. Having in mind, that the only constant that can be used in the scheme is -1, we can see, that the scheme is quite big (comparing to the previous example where there was only approximately 5–10 building blocks) — see Fig. 6. It took much more time/generations to achieve the maximal fitness, namely 3000 in this case.

On the other hand, learning of some functions remained in the local maxima, which was for example the case of the function $x^2 + y^2 + x$.

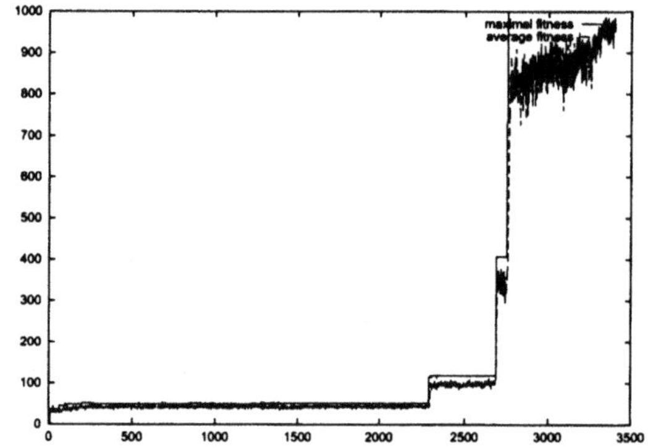

Fig. 4. Function $x^3 - 2x^2 - 3$. The history of the maximal and average fitness

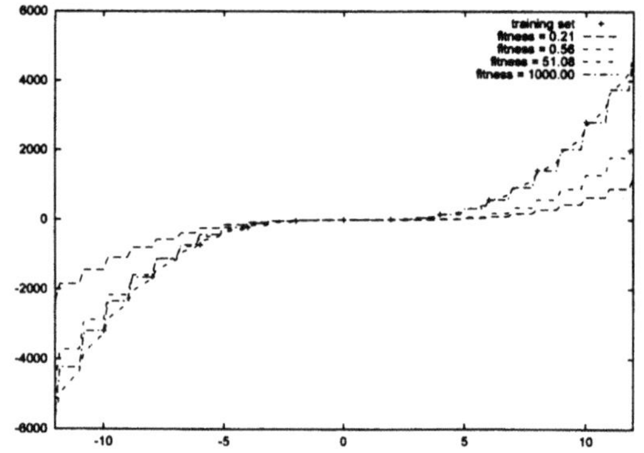

Fig. 5. Function $x^3 - 2x^2 - 3$. The best schemes from generation 0, 5, 200 and 3000

5 Conclusion

We have presented a hybrid system that uses a combination of evolutionary algorithm and a resolution system to automatically create and evaluate multi-agent schemes. So far, the implementation has focused on relatively simple agents computing parts of arithmetical expressions. Nevertheless, the sketched experiments demonstrate the soundness of the approach.

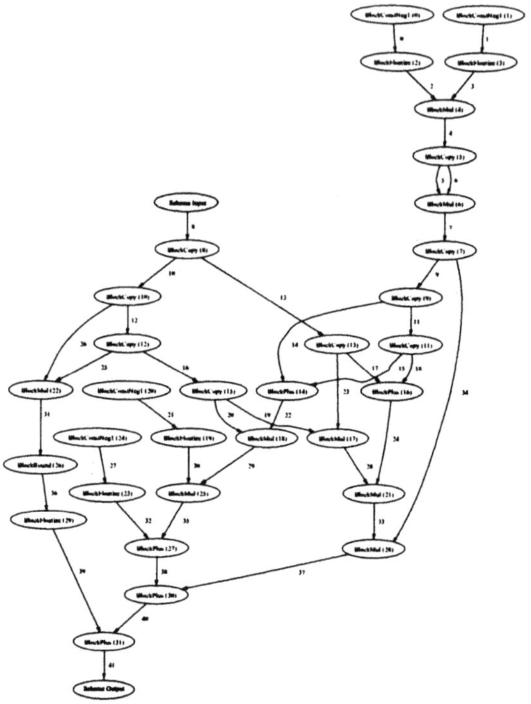

Fig. 6. Function $x^3 - 2x^2 - 3$. The scheme with fitness 1000 (out of 1000), taken from 3000th generation.

In our future work we plan to extend the system in order to incorporate more complex agents into the schemes. Our ultimate goal is to be able to propose and test schemes containing a wide range of computational methods from neural networks to fuzzy controllers, to evolutionary algorithms. While the core of the proposed algorithm will remain the same, we envisage some modifications in the genetic operators based on our current experience.

Namely, a finer consideration of parameter values, or configurations, of basic agents during the evolutionary process needs to be addressed. So far, the evolutionary algorithm rather builds the -3 constant by combining three agents representing the constant 1, than modifying the constant agent to represent the -3 directly. We hope to improve this behavior by introducing another kind of genetic operator. This mutation-like operator can be more complicated in the case of real computational agents such as neural networks, though. Nevertheless, this approach can reduce the evolutionary algorithm search space substantially.

We also plan to extend the capabilities of the resolution system towards more complex relationship types than the ones described in this paper. Our goal is to use ontologies for the description of agent capabilities, and have the CSP-solver reason about these ontologies.

Since the computations are very time consuming, our next implementation goal is to design a distributed version of our algorithm and run it on a cluster of workstations.

Acknowledgments

This work has been partially supported by Grant agency of the Czech Republic under grant number 201/02/0428. G. Beuster has been partially supported by a DAAD postgraduate grant in the framework of the common special academia program III of the federal states and the federal government of Germany.

References

[1] P. Bonissone. Soft computing: the convergence of emerging reasoning technologies. *Soft Computing*, 1:6–18, 1997.

[2] Eugene C. Freuder Daniel Sabin. Configuration as composite constraint satisfaction. In George F. Luger, editor, *Proceedings of the (1st) Artificial Intelligence and Manufacturing Research Planning Workshop*, pages 153–161. AAAI Press, 1996, 1996.

[3] S. Franklin and A. Graesser. "Is it an agent, or just a program?": A taxonomy for autonomous agents. In *Intelligent Agents III*, pages 21–35. Springer-Verlag, 1997.

[4] R. Neruda, P. Krušina, P. Kudová, and Z. Petrová. Multiagent environment for hybrid AI models. In *Artificial Neural Nets and Genetic Algorithms. Proceedings of the ICANNGA 2001 Conference*, Vienna, 2001. Springer-Verlag.

[5] J. S. Russel and P. Norvig. *Artificial Intelligence: A Modern Approach*. Prentice-Hall International, 1995.

[6] G. Weiss, editor. *Multiagent Systems*. The MIT Press, 1999.

The Design of Beta Basis Function Neural Network Using Hierarchical Genetic Algorithm

Chaouki AOUITI*, Adel M. ALIMI† and Aref MAALEJ‡

*University of 7 November, Carthage, Faculty of Sciences of Bizerta
BP W - 7021, Bizerta, Tunisia, E-mail: chaouki.aouiti@fsb.rnu.tn

†REGIM: Research Group on Intelligent Machines, University of Sfax, ENIS
Department of Electrical Engineering BP W - 3038, Sfax, Tunisia, E-mail: adel.alimi@ieee.org

‡LASEM: Laboratory of Electromechanical Systems University of Sfax, ENIS
Department of Mechanical Engineering BP W - 3038, Sfax, Tunisia, E-mail: aref.maalej@enis.rnu.tn

Abstract

We propose an evolutionary method for the design of Beta basis function neural networks (BBFNN). Classical training algorithms start with a predetermined network structure for neural networks. Generally speaking the neural network is either insufficient or over-complicated. This paper describes a hierarchical genetic learning model of the BBFNN. In order to examine the performance of the proposed algorithm, it is used for functional approximation problem. The results obtained have been encouraging.

Keyword: Genetic algorithms - Beta function - Neural Networks - Functions approximation - Learning - Beta bases function Neural Networks

1 Introduction

Classical training algorithms for artificial neural networks start with a predetermined network structure, and as such the response of NN depends strongly on its structure. Generally the network resulting from learning applied to predetermined architecture is either insufficient or over complicated. But in the last few years some researchers have developed learning algorithms, which incorporate structure selection mechanisms like constructive algorithms and pruning algorithms [7],[12],[13]. A constructive algorithm starts with a minimal network, an artificial neural network with a minimal number of hidden layer, hidden neurons and connections and adding new layers, neurons or connections whenever necessary, during the training phase. While, a pruning algorithm does the opposite, it starts with a maximal network structure and keeps deleting the unnecessary layers, nodes and connections during training.

Some researchers have developed learning GAs that incorporate structure selection mechanisms for neural networks, (Billings and Zheng, [5]; Samuel and Gisele, [11]; Yao and Liu [13]. Two approaches exist to evolve ANN architectures [13]. In the first approach, only the evolution of the architecture is considered. If a near optimal architecture is obtained, connection weights are trained. In the second approach, both the architectures and weights are optimized. For the first category, it is important to choose the information about an architecture that will be coded in the chromosome. An example is to encode all the information. This representation is the direct encoding. Another example is to encode only the most important parameters such that the number of hidden layers.

GA isn't used only for the design of artificial neural networks but they are also used in many other domains such as in fuzzy system design, as typical example is the work of Rahmouni et al. [9]. Our study is inspired from their work. It focuses on the evolutionary design of Beta Basis Function Neural Networks (BBFNNs) with a hierarchical GA. We propose a hierarchical genetic algorithm for the design of BBFNNs, which incorporates two key ideas. An outer GA to find the optimum number of neurons in the hidden layer of the BBFNN and an inner GA to find the parameter of the BBFNNs with the best number of neurons in the hidden layer.

The proposed methodology simultaneously derives optimal structure and optimal parameters for the BBFNN. The hierarchical genetic algorithm proposed was compared with a constructive algorithm. Simulation results provided here show the effectiveness of our proposed method.

The organization of this paper is as follows: section 2 is devoted to describe the Beta Basis Functions Neural Networks. In section 3, we present a hierarchical genetic algorithm for the design of BBFNNs. To illustrate the technique, numerical examples and discussions are presented in section 4 and concluding remarks are provided in section 5.

2 Beta Basis Function Neural Networks

The idea for using Beta functions for the design of Beta Basis Function Neural Network (BBFNN) that are generated versions of RBFNN was introduced by Alimi in 1997 [2]. The Beta function is used as a kernel function for many reasons, including but not restricted its great flexibility and its universal approximation characteristics. The Beta function is defined by:

$$\beta(x) = \beta_{x_0,x_1,p,q}(x) = \begin{cases} \left(\frac{x-x_0}{x_c-x_0}\right)^p \left(\frac{x_1-x}{x_1-x_c}\right)^q \\ \text{if } x \in [x_0, x_1] \\ 0 \\ \text{elsewhere} \end{cases} \tag{1}$$

Where $p > 0$, $q > 0$, x_0 and x_1 are real parameters, and:

$$x_c = \frac{px_1 + qx_0}{p+q} \tag{2}$$

Let

$$D = x_1 - x_0 \tag{3}$$

D is the width of the Beta function, So

$$\begin{cases} x_0 = x_c - \frac{Dp}{p+q} \\ x_1 = x_c + \frac{Dq}{p+q} \end{cases} \tag{4}$$

In the multi-dimensional case (dimension $= N$) and if $X = (x^1, ..., x^N)$, $X_0 = (x_0^1, ..., x_0^N)$, $X_1 = (x_1^1, ..., x_1^N)$, $P = (p^1, ..., p^N)$ and $Q = (q^1, ..., q^N)$ Then

$$\beta_{(X_0,X_1,P,Q)}(X) = \prod_{i=1}^{i=N} \beta_{(x_0^i,x_1^i,p^i,q^i)}(x^i) \tag{5}$$

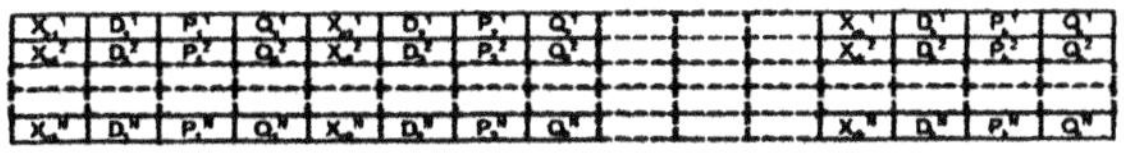

Figure 1: An N dimensional chromosome that represents a BBFNN with k hidden neurons

The output of BBFNN can be formulated as:

$$f(X) = \sum_{i=1}^{i=n_c} w_i \beta_i(X) \tag{6}$$

Where n_c is the number of neurons in the hidden layer.

Beta basis function neural networks were actively studied in the last few years ([1],[2],[3],[4]) and they have shown robust and interesting properties when compared to other neuro-fuzzy systems.

3 The hierarchical genetic model

Many articles have shown that genetic algorithms are able to find near-optimal solutions to complex problems [6]. In this paper we present a hierarchical genetic model for the design of both BBFNN. We use a discrete representation to code the center x_c, the width D, p and q of each Beta basis function.

Using genetic terminology in our data representation, the first step is to choose a representation for a possible solution. Each chromosome that represents a network in the BBFNN case is a matrix Fig(1). The number of lines in this matrix is equal to the number of variables in the function that we are approximating. For the number of columns, each chromosome is coded as a variable length because the hidden layer has a variable number of neurons in the BBFNN . As such a sequence of four genes codes a Beta. The first gene codes for the center, the second gene codes for the width and the other two gene code for p and q.

The second step is the choice of the objective function. In the BBFNN we choose:

$$f^*(x) = \sum_{j=1}^{j=n_c} w_j \beta_j(x) \tag{7}$$

Where n_c is the number of neurons in the hidden layer, β_j are the kernel functions of the jth neuron,

w_j is the jth weights and f^* is the output of the network. Given a set of data (x_i, y_i) $(i = 1, 2, ..., N)$. We can obtain the connection weights, the parameters of each neuron and the number of neurons by minimizing the following objective function:

$$f_1(Chrom) = \sum_{i=1}^{i=N} (y_i - f^*(x_i))(y_i - f^*(x_i))^T \quad (8)$$

If one uses this function as an objective function, the best structure that minimizes this objective function has N hidden nodes [5]. To provide a trade-off between network performance and network structure, the objective function can be amended to:

$$f_2(Chrom) = \frac{Log(N_{max} - n_c + 1) + Log(n_c - N_{min} + 1)}{1 + f_1} \quad (9)$$

Where N_{max} and N_{min} are the maximum and the minimum number of neurons in the hidden layer.

After choosing the objective function, we randomly choose an initial population. The number of columns in each chromosome is between $4N_{max}$ and $4N_{min}$. In the following, the genetic operators used in the outer and inner algorithms are described in detail.

In both outer and inner GA, we use the same selection operator, which is the truncation operator. In this selection operator with threshold T only the fraction T best individuals can be selected and they all have the same selection probability. This selection operator was introduced to the domain of genetic algorithm by Muhlenbein [8].

3.1 The outer genetic algorithm

In the outer genetic algorithm we have two crossover operator, the $\mathbb{R}^N$-Mating operator, the $\mathbb{R}^N$-Mutation operator, the $\mathbb{R}^N$- Addition operator and the $\mathbb{R}^N$-Elimination operator.

3.1.1 The Crossover operators

In the outer genetic algorithm we have used two-crossover operators. One of them changes the number of columns of each chromosome so it changes the number of neurons in the hidden layer for the BBFNN. The second operator does not change the number of columns of each chromosome.

(a) The $\mathbb{R}^N$-crossover1 operator: For the first operator, after the selection of the two chromosomes for which we will apply this crossover operator, we choose an arbitrary position a in the first chromosome and a position b in the second chromosome according to a. Then, we exchange the second part of the two chromosomes. If one child has more than N_{max} or less than N_{min} neurons in the hidden layer we choose another position b Fig(2).

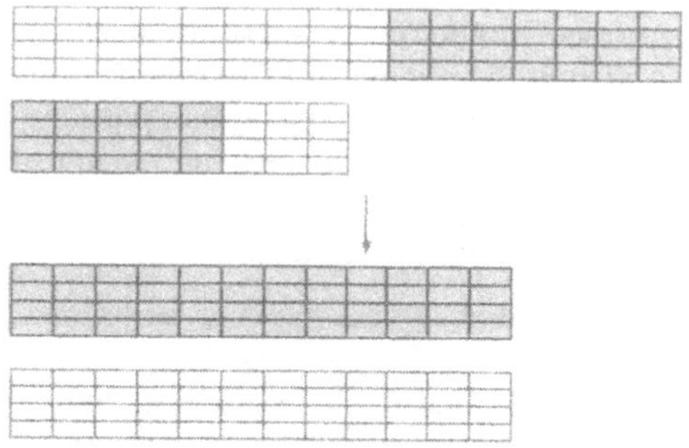

Figure 2: The $\mathbb{R}^N$-crossover1 operator

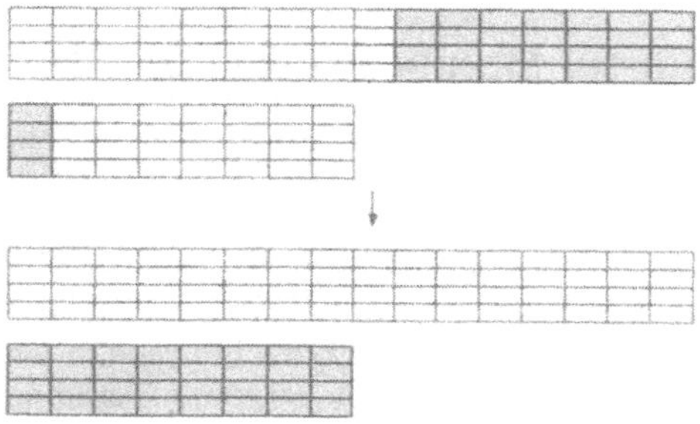

Figure 3: The $\mathbb{R}^N$-crossover2 operator

(b) The $\mathbb{R}^N$-crossover2 operator: For the second operator, after the selection of the two chromosomes for which we will apply this crossover operator, we choose an arbitrary position a in the first chromosome and a position b in the second chromosome according to a.

Let $Min\_point = Min(a, b)$. First we change the values of a and b to Min_point. Then, we exchange the second part of the two chromosomes. In this case, we have necessarily first children having the same length as the second chromosome and the second children having the same length as the first chromosome Fig(3).

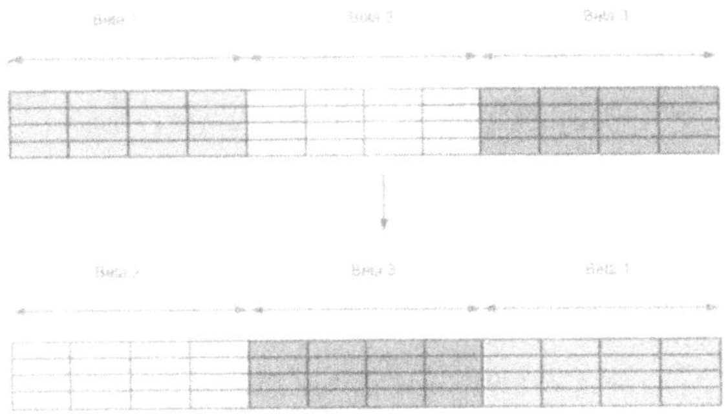

Figure 4: The $\mathbb{R}^N$-Mating operator

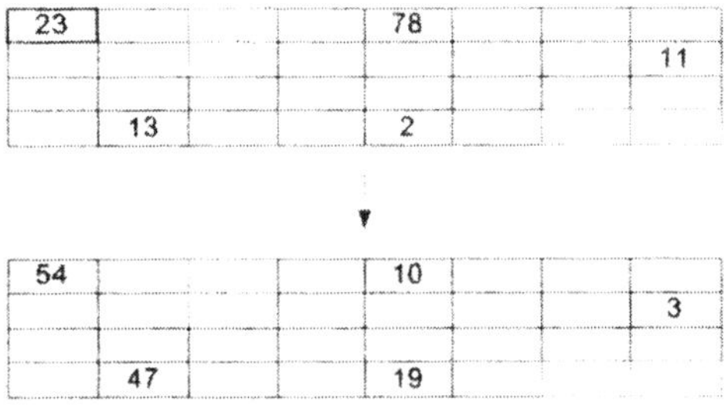

Figure 5: The $\mathbb{R}^N$-mutation operator

3.1.2 $\mathbb{R}^N$-Mating operator

In the BBFNN case, when we change the position of a neuron in the hidden layer the output of the network didn't change. For this reason before applying the crossover operator we apply the mating operator. So this operator changes the position of the neurons in the hidden layer in the BBFNN case Fig(4). With this operator the crossover operator will not be a one-point crossover but a modified multi-point crossover.

3.1.3 $\mathbb{R}^N$-Mutation operator

Generally, the initial population didn't have all the information that is essential to the solution. The goal of applying the mutation operator is to inject some new information into the population because. We begin by associating to each gene a real value between 0 and 1. If this value is less than the probability of mutation P_m than the operator mutation is applied in this position by changing the value of the gene Fig(5).

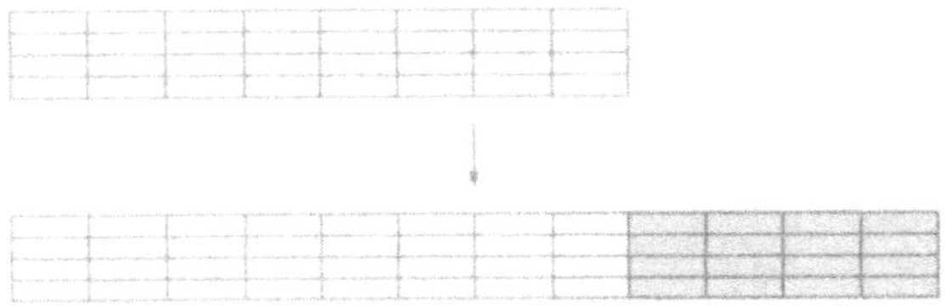

Figure 6: The $\mathbb{R}^N$-addition operator

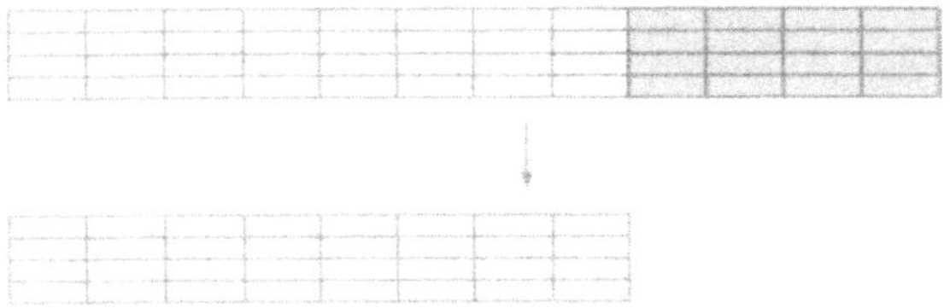

Figure 7: The $\mathbb{R}^N$-elimination operator

3.1.4 $\mathbb{R}^N$-Addition operator and $\mathbb{R}^N$-Elimination operator

Classic training algorithms for BBFNN start with a predetermined structure and the quality of the response of the BBFNN depends strongly on its structure. This problem will be solved when we use the Genetic Algorithms by the two operators R^N-Addition and R^N-Elimination. The first operator adds a neuron in the hidden layer Fig(6) while the second operator eliminates a neuron Fig(7).

3.2 The inner genetic algorithm

In the inner genetic algorithm, we have the $\mathbb{R}^N$-uniform crossover operator, the $\mathbb{R}^N$-Mutation operator and the $\mathbb{R}^N$-Addition/Elimination operators:

3.2.1 $\mathbb{R}^N$-Uniform Crossover operator

The R^N-Uniform crossover operator is radically different to the $\mathbb{R}^N$-crossover1 operator and $\mathbb{R}^N$-crossover2 operator. Copying the corresponding gene from one of the parents, chosen according to a randomly generated crossover mask, creates each gene in the child. Where there is a 1 in the crossover mask, the gene is copied from the first parent and where there is a 0 in the mask the gene is copied from the second parent Fig(8). The process is repeated with the parents exchanged to produce the second offspring. A new crossover mask is randomly generated for each pair of parents.

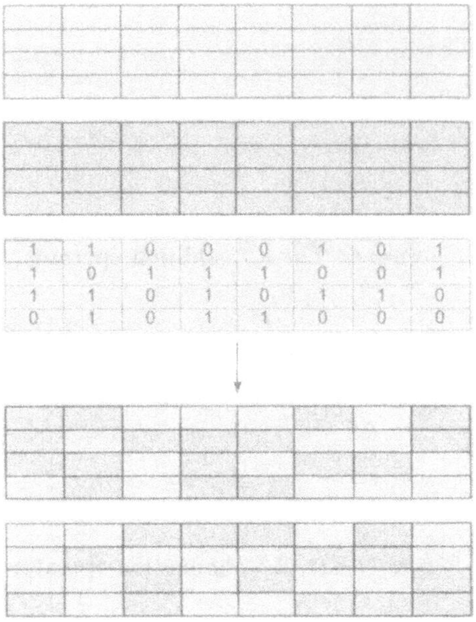

Figure 8: The $\mathbb{R}^N$-crossover operator

3.2.2 $\mathbb{R}^N$-Mutation operator

After the crossover operation strings are then subjected to mutation. Just as P_c controls the probability of a crossover, another parameter, P_m gives the probability that a bit will be flipped. The bits of a string are independently mutated, that is, the mutation of a bit does not affect the probability of mutation of other bits.

3.2.3 $\mathbb{R}^N$-Addition/Elimination operator

The last operator in the inner genetic algorithm, after the mutation operator, is The R^N-Addition/Elimination operator. Just as P_m controls the probability of mutation, another parameter, $Pad-el$ gives the probability that a chromosome will be flipped.

Let $X = (x_1, x_2, , x_N)$ be the N training points and $Z = (z_1, z_2, , z_N)$ be the desired output. If the chromosome represent a BBFNN with N_h neurons in the hidden layer, we find the vector of weights, $W = (w_1, w_2, , w_{N_h})$ between the first and the hidden layer by using the least squares method. Let $Y = (y_1, y_2, , y_N)$ the vector of the output of the BBFNN.

Let $Err = |Z - Y| = (|z_1 - y_1|, |z_2 - y_2|, , |z_N - y_N|)$.

Let $j_1 / \max_{1 \le i \le N} (Err_i) = Err_{j_1}$

and $j_2 \ / \min_{1 \le i \le N_h} (|w_i|) = |w_{j_2}|$

then the center of the j_2^{th} neuron in the hidden layer will be X_{j_1}.

3.3 The algorithm

1 Choose randomly the initial population.

2 Decode each chromosome in the population.

3 Compute the connection weights from the hidden layer to the output layer.

4 Find the fitness f_2 of each chromosome (find the best chromosome).

   if R iterations were completed in the external loop go to step(a) else go to step(5)

   a Choose randomly the initial population. Each chromosome defines a BBFNN that has the same number of neurons in the hidden layer us the best solution in the step 4 (this population contain the best chromosome of step(4)).

   b Decode each chromosome in the population.

   c Compute the connection weights from the hidden layer to the output layer.

   d Find the fitness of each chromosome that is the some square error between the desired output and the output of the network.

   e Apply the Truncation Selection (with threshold $T1$).

   f Apply Crossover, Mutation, and new/center operators.

   g If the number of generation is equal to an integer N_1 (N_1 is the maximum number of generation of the inner GA) then go to step(5), if (the sum square error of the best neural network is less then ε) then go out else return to step(b).

5 Apply the Truncation Selection (with threshold T).

6 Apply The $\mathbb{R}^N$-Crossover1 or the Crossover2, mutation, addition and elimination operators.

7 If the number of generation is equal to an integer N_2 (N_2 is the maximum number of generation of the outer GA) or ($f_2 > A$) then stop else return to step 2.

4 Simulation results and discussion

In order to examine the ability of the algorithm in producing networks and in generalizing well, we tested it in the approximation problem.

The HGADBBFNN will be compared with a constructive algorithm for the design of BBFNN. This algorithm is like the majority of neural construction method [14]. It starts from a Beta basis function neural networks with one neuron in the hidden layer and adds one node and connection as needed in order to learn the training set. The activation function used is with $p = q = 2$. Initially the BBFNN has no neurons in the hidden layer. We add a new neuron in the hidden layer until the network's sum of squared errors falls below GOAL or the maximum number of neurons is reached. In the constructive algorithm we use the same method used in the HGADBBFNN to find the weights vector between the hidden layer and the output layer.

We trained the BBFNN with different set functions to approximate different functions. We tested our hierarchical genetic algorithm on many approximands in the 1-D case, in the 2-D case, in the 3-D case and in the 4-D case.

We sampled 100 points of the function in the 1-D case to give a training set. The 2-D case used 10x10=100 samples. The 3-D case used 7x7x7=343 samples. The 4-D case used 5x5x5x5=625 samples.

Below is an example of test function we used as approximand in the 3-D case:

$$\begin{cases} f_3(x,y,z) = |\sin(\frac{\Pi x}{2})\cos(\frac{\Pi y}{2})cos(\frac{\Pi z}{2})| \\ x \in [-1,1], y \in [-1,1] \text{and } z \in [-1,1] \end{cases} \qquad (10)$$

Fig(9) shows the desired output f_3 for the last 100 points. The sum square error (SSE) between the desired solution and the solution proposed by the HGADBBFNN that has 34 neurons in the hidden layer is equal to 0.0401. Fig(10) shows the error between the desired output and the obtained output by our HGADBBFNN. If we use the constructive algorithm, the SSE between the desired solution and the solution proposed is equal to 0.037603 with 106 neurons in the hidden layer and 0.0620393 with 105 neurons in the hidden layer. Fig(11) shows the evolution of the maximum, minimum and best number of neurons in the hidden layer for f_3.

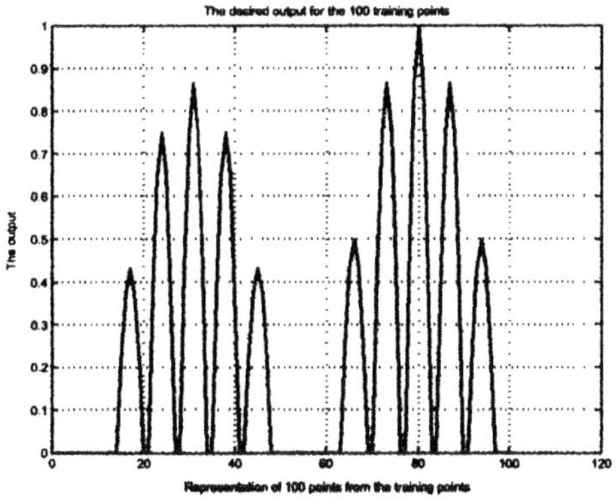

Figure 9: desired output

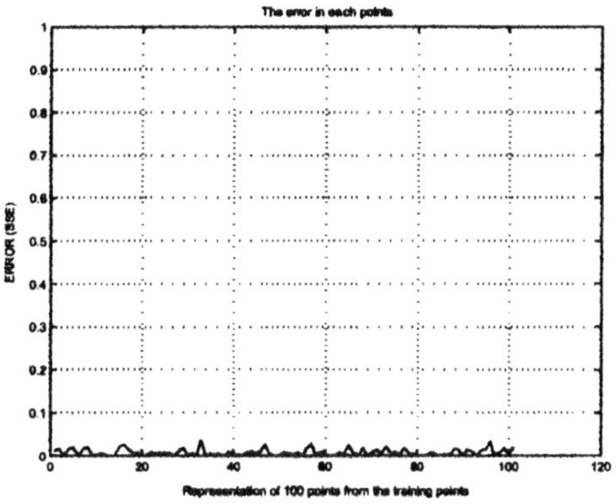

Figure 10: Error between the desired and the obtained output

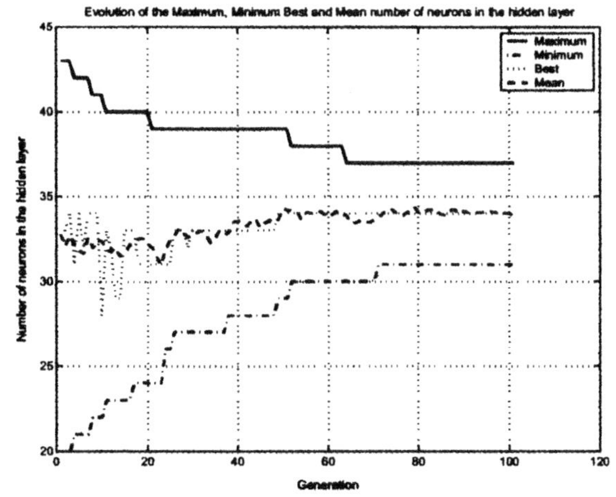

Figure 11: Evolution of the maximum, minimum and best number of neurons in the hidden layer

5 Conclusion

This paper presented a hierarchical GA for the design of BBFNN. The problem was to find the optimal network structure and the different parameters of the network.

Our genetic models can automatically determine appropriate structures and network parameters of beta basis function neural networks. In order to find the network's optimal structure, the process modifies the number of neurons in the hidden layer. The performance of the algorithm is achieved by evolving the initial population and by using operators that alter the sizes of the networks. This strategy attempts to avoid the Lamarckism [6].

Comparing to the constructive algorithms, the results show significant improvement in the resulting performance and topology. While we considered only BBFNN, the proposed hierarchical genetic algorithm can be easily extended to the design of other neural networks (i.e RBFNN, MLPNN, $\cdots$).

The originality of the genetic model resides on the fact that our hierarchical structure can automatically determine appropriate structures and network parameters for BBFNN.

References

[1] Alimi, A.M. Hassine, R. and Selmi, M *Beta Fuzzy Logic Systems: Approximation Properties in the SISO Case*, Int. J. Applied Mathematics and Computer Science, Special issue edited by D. Rutkowska and L.A. Zadeh, vol. 10, no. 4, pp. 101-119, (2000).

[2] Alimi, A.M. *The Beta Fuzzy System: Approximation of Standard Membership Functions* Proc. 17eme Journees Tunisiennes d'Electrotechnique et d'Automatique: JTEA'97, Nabeul, Tunisia, Nov., vol. 1, pp. 108-112, (1997).

[3] Aouiti, C. Alimi, A. M. and Maalej, A. *A Hierarchical Genetic Algorithm for the Design of Beta Basis Function Neural Network* ", Proceedings of the International Joint Conference on Neural Networks, IJCNN '02, Volume: 2, pp. 1246 -1251, (2002).

[4] Aouiti, C. Alimi, A. M. and Maalej, A. *A Genetic Designed Beta Basis Function Neural Network for Approximating Multi-Variables Functions*, in V. Kurkova et al. (eds.) Artificial Neural Nets and Genetic Algorithms, Springer-Verlag, Wien, pp. 383-386 (2001).

[5] Billings,S. A. and Zheng,G. L. *Radial Basis Function Network Configuration Using Genetic Algorithms*, Neural Networks, Vol. 8, No. 6, pp. 877-890, (1995).

[6] Goldberg, D.E. *Genetic Algorithms in Search, Optimization, and Machine Learning, Reading,* MA: Addison-Wesley (1989).

[7] Musavi, M.T., Ahmed, W., Chan, K.H, Faris, B.K. and Hummels, D.M. *On the training of radial basis function classifiers*, Neural Networks, Vol. 5, pp. 595-603, (1992) .

[8] Muhlenbein, H. and Schlierkamp, V. D. *Predictive models for the breeder genetic algorithm, Evolutionary Computation*, 1(1), (1993).

[9] Rahmouni, A. and Benmohamed, M. *Genetic Algorithm Based Methodology to Generate Automatically Optimal Fuzzy Systems*, IEE Proc-Control Theory Appl., Vol. 145, No. 6 pp. 583-586, (1998).

[10] Rosenblatt, F. *Representation of Functions by superpositions of a Step or Sigmoidal Function and their Applications to Neural Network Theory*, Project PARA, Cornell Aeronautical Lab. Rep., No. 85: 640-1, Buffalo, NY (1957).

[11] Samuel, P. and Gisele, L. *A Genetic Algorithm for Designing Distributed Computer Network Topologies*, IEEE Trans. Systems, Man, and Cybernetics-Part B: Cybernetics, Vol. 28, No. 2, pp. 249-258, (1998).

[12] Yao, X. *Evolving Artificial Neural Networks* Proceeding of the IEEE, 87, 9, September, 1423-1447 (1999).

[13] Yao X. and Liu, Y. *A New Evolutionary System for Evolving Artificial Neural Networks* IEEE Transactions on Neural Networks, 8, 3, (1997).

[14] Anderson, T. Rimer M. and Martinez, T. *Optimal Artificial Neural Network Architecture Selection for Bagging* Proceedings of the International Joint Conference on Neural Networks, IJCNN '01, Volume: 2, pp. 790 795, (2001).

[15] Demuth, H. and Beale, M. *Neural Network Toolbox, For Use with MATLAB* User's Guide, version 3.0., by The MathWorks, Inc. (1998).

Population Learning Algorithm for Resource-Constrained Project Scheduling

Piotr Jedrzejowicz, Ewa Ratajczak [1][1]

[1]Chair of Computer Science, Gdynia Maritime Academy, Poland, e-mail: {pj,ewra}@am.gdynia.pl

Abstract

The paper proposes applying the population-learning algorithm to solving a single mode resource constrained project scheduling problem with makespan minimization as an objective function. The paper contains problem formulation and a description of the proposed implementation of the population learning algorithm (PLA). To validate the approach a computational experiment has been carried. It has involved 1440 instances from the available benchmark data set. Experiment results show that the proposed PLA implementation is an effective tool for solving single mode resource constrained project scheduling problems. In a single run the algorithm has produced solutions with mean relative error value well below 1% as compared with available upper bounds for benchmark problems.

1 Introduction

The paper proposes applying the population-learning algorithm (PLA) to solving the resource constrained project scheduling problem (RCPSP) with makespan minimization as the objective function. In the paper a single-mode RCPSP is considered. In a single-mode case a project consists of a set of activities, where each activity has to be processed in a single, prescribed way (mode). Each activity requires some resources, availability of which is constrained. The discussed problem is computationally difficult and belongs to the NP-hard class. Because of its practical importance RCPSP has attracted a lot of attention and many exact and heuristic methods have been proposed for solving it (see for example [1, 5, 6, 7]). Exact algorithms seem suitable for solving smaller instances of RCPSP. On the other hand, heuristic approaches, used for solving its larger instances, can only be evaluated experimentally using sets of benchmark instances with known optimal solutions or upper bounds. Criteria for such an evaluation include, usually, two factors - quality of solutions obtained and computational resources required.

The proposed approach is an implementation of the population-learning algorithm which belongs to the class of population-based methods and which proved quite successful in solving some other difficult scheduling problems [9]. The paper is organized as follows: Section 2 includes problem formulation. Section 3 presents an idea of the population-learning algorithm. Section 4 contains details of the proposed implementation of PLA algorithm. Section 5 presents validating experiment and its results. Section 6 includes conclusions and suggestions for future research.

[1]The work was supported by grant KBN nr 8 T11 F020 19 2000

2 Problem formulation

A project consists of a set of n activities, where each activity has to be processed without interruption to complete the project. The dummy activities 1 and n represent the beginning and the end of the project. The duration of an activity j is denoted by d_j where $d_1 = d_n = 0$. There are r renewable resource types. The availability of each resource type k in each time period is $\bar{r}_k$ units, $k = 1, \ldots, r$. Each activity j requires r_{jk} units of resource k during each period of its duration where $r_{1k} = r_{nk} = 0$, $k = 1, \ldots, r$. All parameters are non-negative integers. There are precedence relations of the finish-start type with a zero parameter value (i.e., $FS = 0$) defined between the activities. In other words activity i precedes activity j if j cannot start until i has been completed. The structure of a project can be represented by an activity-on-node network $G = (SV, SA)$, where SV is the set of activities and SA is the set of precedence relationships. $SS_j(SP_j)$ is the set of successors (predecessors) of activity j. It is further assumed that $1 \in SP_j$, $j = 2, \ldots, n$, and $n \in SS_j$, $j = 1, \ldots, n-1$. The objective is to find a schedule S of activities i.e., a set of starting times $[s_1, \ldots, s_n]$, where $s_1 = 0$ and resource constraints are satisfied, such that the schedule duration $T(S) = s_n$ is minimized.

3 Population Learning Algorithm

Population-learning algorithm introduced originally in [8] is a population-based method inspired by analogies to a phenomenon of social education processes in which a diminishing number of individuals enter more and more advanced learning stages. PLA take advantage of the following features common to organized education systems [2].

- A huge number of individuals enter the system.
- Individuals learn through organized tuition, interaction, self-study, trials and errors.
- Learning process is inherently parallel (different schools, curricula, teachers, etc.).
- Learning process is divided into stages.
- More advanced stages are entered by a diminishing number of individuals from the initial population.
- At higher stages more advanced learning and improvement techniques are used.
- A final stage is reached by only a fraction of the initial population.

In PLA an individual represents a coded solution or part of it of the considered problem. Initially, a number of individuals, known as the initial population P (see also 2), is randomly generated or constructed using some construction heuristics.

Once the initial population has been generated, individuals enter the first learning stage ($LEARN_1(P)$). It involves applying some, possibly basic and elementary, improvement schemes. These can be based, for example, on some local search procedures. The improved individuals are then evaluated and better ones pass to subsequent stages. A strategy of selecting better or more promising individuals at each stage must be defined and duly applied ($SELSCT_i(P)$). At following stages the whole cycle is repeated. Individuals are subject to improvement and learning ($LEARN_i(P)$), either individually or through information exchange, and the selected ones are again promoted to a higher stage with the remaining ones dropped-out from the process. At the final stage the remaining individuals are reviewed and the best one represents a solution to the problem at hand.

At different stages of the process, different improvement schemes and learning procedures are applied. These gradually become more and more sophisticated and time consuming as there are less and less individuals to be taught. General idea of the PLA approach is shown in fig. 1.

Several PLA applications were described in [2, 3, 4, 8, 10].

PLA shares some features with evolutionary programs as discussed, for example, in [11]. The idea of refining population of solutions during the subsequent computation stages is common to PLA and memetic algorithms [12]. The latter, however, assume a constant population size and a single local search procedure and relay to a greater extend on typical genetic/evolutionary algorithms operators. There are also some similarities between PLA and cultural algorithms where an inheritance process operates at the micro and the micro-evolutionary levels [13].

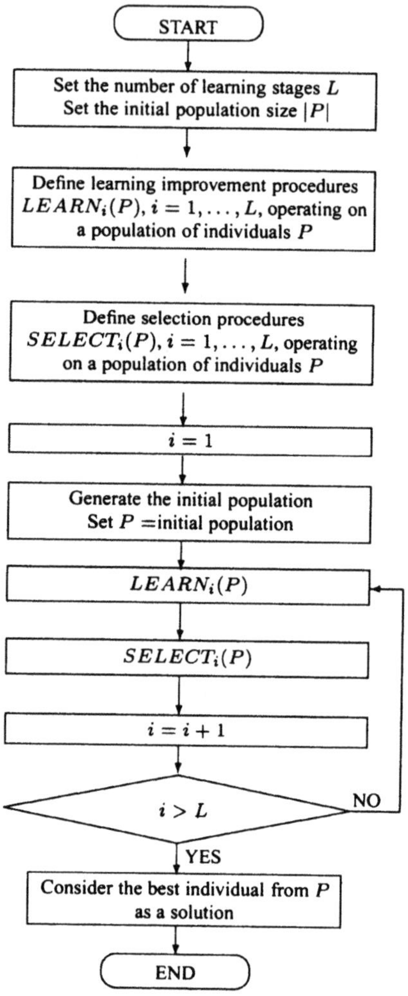

Fig. 1. General idea of the population learning algorithm.

4 PLA implementation

The PLA has been implemented with a view to solving the above formulated RCPSP problem. The implementation involves three learning and improvement stages. Value of the goal function is directly used as a measure of quality of individuals and hence as a selection criterion. An individual in the algorithm is represented as a vector of activities (schedule) $S = [a_1, \ldots, a_n]$, each activity a_j is an object consist of: starting time - s_j, duration - d_j, set of required units of resources, set of predecessors - SP_j and set of successors - SS_j.

The algorithm requires that values of the following parameters are set.

- p - multiplier used to calculate the size of an initial population
- $xi1$ - coefficient used to calculate the number of iterations at the first learning and improvement stage,
- $xm1$ - coefficient used to calculate the frequency of calls to a selection procedure within the first learning and improvement stage,
- $xi2$ - coefficient used to calculate the number of iterations at the second learning and improvement stage.

Values of the above control parameters are set at the algorithm fine-tuning phase. All random moves within the algorithm are drawn from the uniform distribution. All parameter values used in the following pseudo-code have been chosen by trials and errors during the fine-tuning phase.

The pseudo-code of the proposed algorithm is shown as Procedure PLA. In the pseudo-code P denotes population, and $|P|$ size of the population P.

```
Procedure PLA
begin
Set size of P, |P| ≐ p·n;
Set xi1, xm1, xi2;
Create initial population P;
{-- the first learning stage}
for it = 1 to xi1·n do
   for i = 1 to 40%·|P| do
      Crossover(Random(P), Random(P));
   for i = 1 to 5%·|P| do
      Mutation(Random(P));
   for i = 1 to 5%·|P| do
      LSA(Random(P),2,6,10);
   if it mod (xm1·n) = 0 then
      Selection(medium_makespan);
endfor
Selection(medium_makespan);
{-- the second learning stage}
for it = 1 to xi2·n do
   for i = 1 to 40%·|P| do
      Crossover(Random(P), Random(P));
   for 2 best solutions S ∈ P do
      EPTA(S,6,4);
   for i = 1 to 2%·|P| do
      LSA(Random(P),2,6,10);
endfor
Selection(medium_makespan);
{-- the third learning stage}
for every solution S ∈ P do
   EPTA(S,6,2);
   LSA(S,10,2,10);
endfor
end;
```

The algorithm creates an initial population by producing four individuals using simple construction heuristics and generating randomly the remaining ones. Heuristics are based on the following rules:

- shortest duration first,
- shortest duration last,
- longest duration first,
- longest duration last.

The first learning stage uses evolutionary operators and a simple local search algorithm (LSA). Three procedures - Crossover, Mutation and LSA are repeated $xi1 \cdot n$ times.

The Random(P) function chooses randomly an individual from the population P.

```
Procedure Crossover(S1,S2)
begin
Do simple (one point) crossover on two
randomly chosen individuals S1, S2 ∈ P;
end;
```

```
Procedure Mutation(S)
begin
Move an activity from randomly chosen
position in the schedule S to another
randomly chosen position in this
schedule, in such a way that no
successor of this activity can be
found before it.
end;
```

```
Procedure LSA(S, itNumber, iStep, fStep)
begin
for it = 1 to itNumber do
   for s = iStep to fStep do
      for j = 1 to scheduleLength − s do
         Exchange activity from position
         j with the activity from
         position j + s;
         if not the new solution is better
            then recall the exchange;
      endfor;
end;
```

The LSA procedure require four variables. The first, S denotes an individual, second ($itNumber$) is the number of iteration for the procedure. The last two indicate the initial and final distance between activities under exchange.

The second learning stage uses crossover and two heuristics - EPTA (exact precedence tree algorithm) and LSA. EPTA is based on the precedence tree approach proposed in [14]. It finds an optimum solution by enumeration for a partition of the schedule consisting of some activities, which number is denoted as *partExtent*. In the following pseudocode variable S denotes an individual, and $|S|$ its current size (number of activities). The *step* variable denotes the distance between starting points of the considered partitions.

Procedure EPTA(S, $partExtent$, $step$)
begin
$i = 1$;
while $i \cdot step + partExtent < |S|$ **do**
  Find an optimal solution for a part of the schedule beginning from activity on position $i \cdot step$ and ending in activity on position $i \cdot step + partExtent$.
end;

Finally, the third learning stage uses two heuristics EPTA and LSA, both with different value of parameters as compared with settings in the earlier stages, which results in more iterations and higher granularity of the neighbourhood explored.

5 Validating experiment

To validate the proposed approach computational experiment has been carried using 1440 benchmark instances of single-mode RCPSP. The benchmark data set used in the reported experiments includes 480 instances for each out of the three problem sizes (30, 60, 90 and 120 activities, respectively). The benchmark data set together with known upper bounds can be found at http://www.wior.uni-karlsruhe.de/rcpsp.

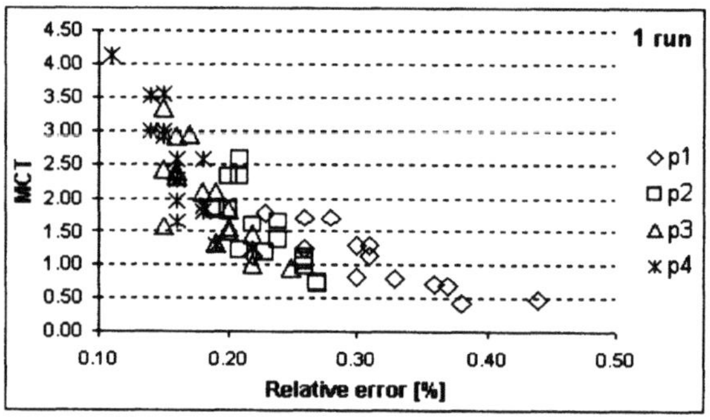

Fig. 2. Setting value of the multiplier p (single run)

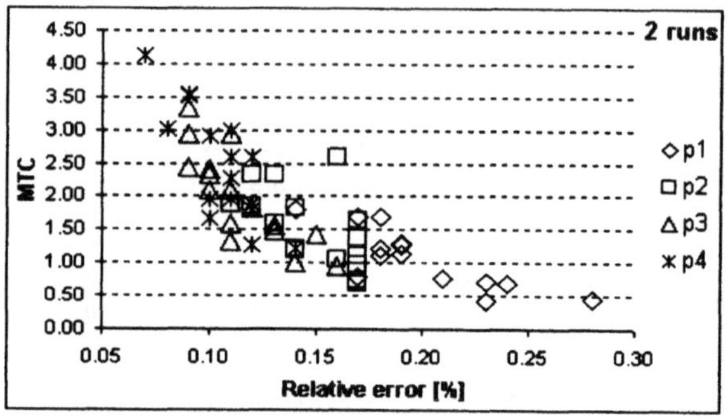

Fig. 3. Setting value of the multiplier p (two runs)

The fine-tuning phase of the experiment has been devoted to finding by trials and errors values of the PLA parameters assuring acceptable compromise between computation time and quality of solutions. This search has been carried using the subset of available benchmark instances consisting of all RCPSP with 30 activities. Figures 2 and 3 show example results with respect to setting value of the multiplier p. The final settings shown in table I include six selected combinations of parameter values.

set	p	$xi1$	$xm1$	$xi2$
a	3	$\frac{1}{2}$	$\frac{1}{3}$	$\frac{1}{3}$
b	3	$\frac{1}{2}$	$\frac{1}{3}$	$\frac{1}{2}$
c	3	$\frac{1}{2}$	$\frac{1}{2}$	$\frac{1}{2}$
d	4	$\frac{1}{2}$	$\frac{1}{3}$	$\frac{1}{3}$
e	4	$\frac{1}{2}$	$\frac{1}{3}$	$\frac{1}{2}$
f	4	$\frac{1}{2}$	$\frac{1}{2}$	$\frac{1}{2}$

Table I. Sets of the PLA parameter values used in the experiment.

The experiment involved solving all 1440 benchmark instances twice in two runs. The results are evaluated in terms of mean and maximum relative errors (mean RE, max RE), percent of solutions equal to respective upper bounds (equal UB) as well as mean computation time (MCT) for a single instance. Relative errors have been calculated with respect to available upper bounds. Experiment results are shown in tables II - V.

During experiment the PLA has contributed to finding an improved upper bound for one benchmark instance with 90 activities.

Experiment has been carried on the PC computer with the AMD XP 1600+ processor and 256 MB RAM for instances of problem size 30, 60 and 90. Computations for instances of problem size 120 has been carried on the PC computers with Pentium IV 1.7 (P1.7), 2.0 (P2.0) or 2.4 (P2.4).

combination	single PLA run			two PLA runs			MTC [s]
	mean RE	max RE	equal UB	mean RE	max RE	equal UB	
a	0.19%	5.17%	91.15%	0.11%	4.35%	94.17%	1.32
b	0.15%	4.31%	92.40%	0.11%	3.45%	93.96%	1.57
c	0.20%	5.18%	90.73%	0.13%	5.17%	93.54%	1.55
d	0.16%	4.61%	91.77%	0.10%	3.95%	94.38%	1.66
e	0.17%	4.84%	91.88%	0.10%	3.95%	94.38%	1.97
f	0.16%	5.06%	91.77%	0.11%	3.95%	93.96%	1.97

Table II. Experiment results (30 activities)

combination	single PLA run			two PLA runs			MTC [s]
	mean RE	max RE	equal UB	mean RE	max RE	equal UB	
a	0.63%	7.20%	74.79%	0.53%	6.06%	76.25%	14.75
b	0.59%	5.99%	73.85%	0.47%	4.55%	75.21%	18.25
c	0.58%	5.66%	75.42%	0.48%	5.05%	77.08%	18.28
d	0.59%	5.62%	75.10%	0.51%	5.05%	76.25%	18.71
e	0.53%	6.69%	75.94%	0.43%	5.05%	78.13%	23.07
f	0.55%	5.65%	75.52%	0.46%	5.05%	77.08%	23.14

Table III. Experiment results (60 activities)

combination	single PLA run			two PLA runs			MTC [s]
	mean RE	max RE	equal UB	mean RE	max RE	equal UB	
a	0.74%	6.88%	75.00%	0.64%	6.42%	75.83%	73.46
b	0.69%	6.85%	75.00%	0.58%	5.65%	76.04%	91.93
c	0.69%	6.90%	75.10%	0.58%	5.50%	76.04%	91.02
d	0.73%	6.42%	75.63%	0.63%	6.42%	76.04%	93.15
e	0.65%	5.85%	75.63%	0.54%	5.50%	76.46%	116.92
f	0.63%	6.29%	76.04%	0.52%	5.45%	76.88%	115.05

Table IV. Experiment results (90 activities)

combination	single PLA run			two PLA runs			MTC [s]
	mean RE	max RE	equal UB	mean RE	max RE	equal UB	
a	1.12%	8.09%	72.25%	1.05%	8.09%	74.92%	217(P 2.0)
b	1.45%	7.83%	72.35%	0.93%	6.85%	74.60%	242(P 2.0)
c	1.02%	8.55%	73.20%	0.92%	8.22%	75.42%	462(P 1.7)
d	1.07%	7.95%	74.00%	0.91%	6.86%	77.08%	384(P 2.0)
e	1.02%	8.22%	74.20%	0.87%	6.92%	78.04%	487(P 2.0)
f	0.99%	7.84%	76.58%	0.86%	6.69%	78.02%	437(P 2.4)

Table V. Experiment results (120 activities)

6 Conclusions

Experiment results show that the proposed PLA implementation is an effective tool for solving single mode resource constrained project scheduling problems. In a single run the algorithm has produced solutions with mean relative error value below 1% as compared with available upper bounds for benchmark problems. It should be noted that available upper bounds have not been obtained by applying one particular algorithm but rather a variety of different algorithms developed by various research teams over considerable period of time.

PLA results obtained in two runs are considerably better then single run ones. This proves that there is a room for further improvement of the PLA efficiency. Obvious solution would be a parallel PLA scheme. Future research will concentrate on developing a parallel PLA with possible information exchange between parallel processes of learning and improvement. Another worthwhile direction of research should be searching for effective procedures of setting values of the PLA parameters. The most critical one seems to be the initial population size. It should not, however, be considered in isolation to other important variables like selection criteria, number of iterations at learning and improvement stages and granularity of local search procedures employed. Identifying relations between these factors as well as their joint influence on the quality of results should be the subject of further studies.

General conclusion of the paper can be formulated as follows: Validating experiment results allow to consider PLA as a useful and promising framework to deal with solving computationally difficult combinatorial problems.

References

[1] N. Christofides, R. Alvarez-Valdes, J.M. Tamarit, Project scheduling with resource constraints: A branch and bound approach, European Journal of Operational Research, vol. 29, 1987, pp. 262-273.

[2] I. Czarnowski, P. Jedrzejowicz, E. Ratajczak, Population Learning Algorithm - example implementations and experiments, 4th Metaheuristics Internetional Conference, MIC 2001, Porto, Portugal 2001, pp. 607-612.

[3] I. Czarnowski, M. Forkiewicz, P. Jedrzejowicz, E. Ratajczak, A. Skakowski, I. Wierzbowska, Population-based scheduling on multiple processors, 4th Metaheuristics Internetional Conference, MIC 2001, Porto, Portugal 2001, pp. 613-618

[4] I. Czarnowski, W.J Gutjahr, P. Jędrzejowicz, E. Ratajczak, A. Skakowski, I. Wierzbowska, Scheduling multiprocessor tasks in presence of the correlated failures, The Proceedings of the Second International Workshop on Soft Computing Applied to Software Engineering, SCASE, Enschede, The Netherlands 2001.

[5] E.W. Davis, G.E. Heidorn, An algorithm for optimal project scheduling under multiple resource constraints, Management Science, vol. 17, 1971, pp. B803-B816.

[6] S. Hartmann, Project Scheduling with Multiple Modes: A Genetic Algorithm, Manuscripte aus den Instituten fr Betriebswirtschaftslehre, No. 435. University of Kiel, Germany.

[7] S. Hartmann, A. Drexl, Project Scheduling with Multiple Modes: A Comparison of Exact Algorithms, Manuscripte aus den Instituten fr Betriebswirtschaftslehre, No. 430. University of Kiel, Germany.

[8] P. Jedrzejowicz, Social Learning Algorithm as a Tool for Solving Some Difficult Scheduling Problems, Foundation of Computing and Decision Sciences, vol. 24, nr 2, 1999, pp. 51–66.

[9] J. Jedrzejowicz, P. Jedrzejowicz: Permutation Scheduling Using Population Learning Algorithm, in: Knowledge-Based Itelligent Information Engineering Systems and Allied Technologies, E.Damiani et al. (eds.), IOS Press, Amsterdam, 2002, 93 - 97.

[10] P. Jędrzejowicz, A. Skakowski, An island-based evolutionary algorithm for scheduling multiple-variant tasks, Proceedings of ICSC Symposium Engineering of Intelligent Systems – EEIS'2000, ICSC Academic Press, Paisley 2000, pp. 1–9.

[11] Z. Michalewicz, Genetic Algorithms + Data Structures = Evolution Programs, Springer, Berlin, 1992.

[12] P. Moscato, Memetic Algorithms: A short introduction. In: D.Corne, M.Dorigo, F.Glover (Eds.), New Ideas in Optimization, McGraw-Hill, New York, 1999, 219-234

[13] R.G. Reynolds: An Introduction to Cultural Algorithms. In: A.V. Sebald, L.J. Fogel (Eds.), Proceedings of the Third Annual Conference on Evolutionary Programming, World Scientific , River Edge, 1994, 131-139.

[14] A. Sprecher, A. Drexl, Solving multi-mode resource-constrained project scheduling problems by a simple, general and powerful sequencing algorithm. Part I: Theory, Part II: Computation. Manuscripte aus den Instituten fr Betriebswirtschaftslehre, No. 385, 386. University of Kiel, Germany.

Genetic Snakes: Application on Lipreading

Renaud Séguier, Nicolas Cladel [1]

Abstract

Contrary to Genetics Snakes, the current methods of mouth modeling are very sensitive to initialization (position of a snake or a deformable contour before convergence) and fall easily into local minima. We propose in this article to make converge two snakes in parallel via a genetic algorithm. The coding of the chromosome takes into account at the same time gradients and region type information contained in the image. In addition we introduce the use of STM (Sparse Template Matching) into the field of leapreading. Thanks to a temporal filter, word signatures (stored in Sparse Templates) make it possible to recognize various words pronounced several times at one week interval.

1 Introduction

Genetic Algorithms have been used for a many years in computer vision for their optimisation quality [1] and their aptitude to avoid local minima [2]. In particular, Genetics Snakes [3] [4] [5] make it possible to overcome the problem of snakes initialization.

This problem of initialization is crucial in lipreading. The potential applications (speech recognition, lip-synchronization in audio-video flows, synthesis of talking heads) require a robust algorithm of mouth modeling. This robustness is not achieved in current modeling techniques (shape model [6], dynamic contours [7], deformable models [8], cumulative histograms analysis [9]). For that reason certain teams turned to a low level mouth modeling (DCT Discrete Cosinus Transform) which requires very big learning bases [10]. These last techniques are restrictive: they do not analyze explicitly the shape of the mouth, and cannot thus be used in talking head type applications. We propose in the section 2 an implementation of Genetic Snakes (GS) to model the mouth in a robust way.

In the field of embedded system we need to implement simple and light memory classifiers. The section 3 introduce the use of such a classifier in lipreading: the STM (Sparse Template Matching) [11]. We will see that with the help of a temporal low-pass filter, the sparse matrices become robust with the possible temporal variation which appear when a person pronounces a word more or less quickly.

[1] Supélec - Team ETSN, Avenue de la Boulaie - BP28 35511 Césson Sévigné, France. E-mail: Renaud.Seguier@supelec.fr, Nicolas.Cladel@supelec.fr

2 Genetic Snakes

First of all we will present the preprocessing which enable us to generate images of differently filtered mouths. We will introduce afterwards the coding which bind the GS to the images, and then the energy that we will maximize using the genetic algorithm.

2.1 Image preprocessing

A face detector [12] locates the face in the image. We use then the same process as [13] to locate the mouth. In the V values (from YUV color coordinate system), the lips has a strong level of intensity while the teeth and the dark interior of the mouth are confused and rather dark (Fig. 1b). It is enough then to binarize the image to obtain a zone characterizing the interior of the mouth after a rapid morphological treatment (erosion and area growing starting from the center of the image, see Fig. 1c)). The same morphological treatment (erosion and area growing starting from points belonging to the upper and lower lips) is applied on light pixels to obtain the image d). We carry out finally the fusion of the images c) and d) to obtain the area of external lips contours (Fig. 1e)). The gravity center of the interior of the lips C_G is then evaluated starting from the image c). We apply then a Sobel filter to extract edges from these images and store the information of the direction of contour (see Fig. 2).

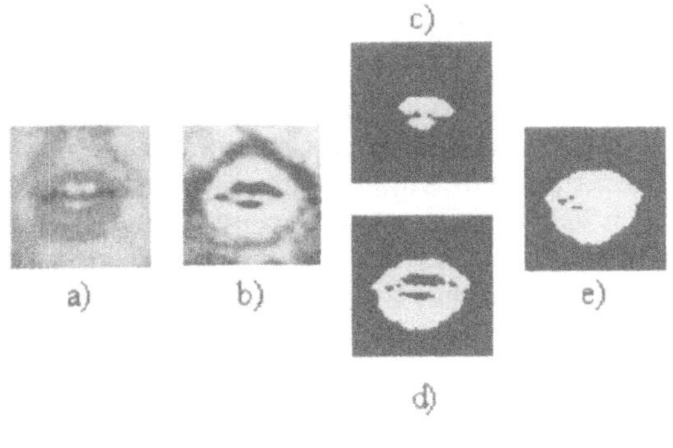

Fig. 1. Mouth preprocessing

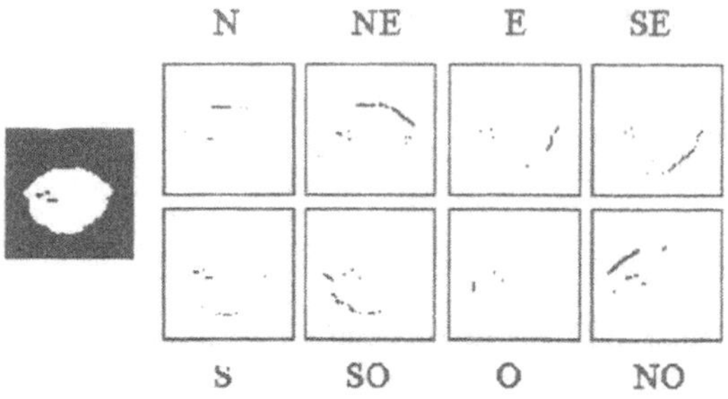

Fig. 2. Edges extraction

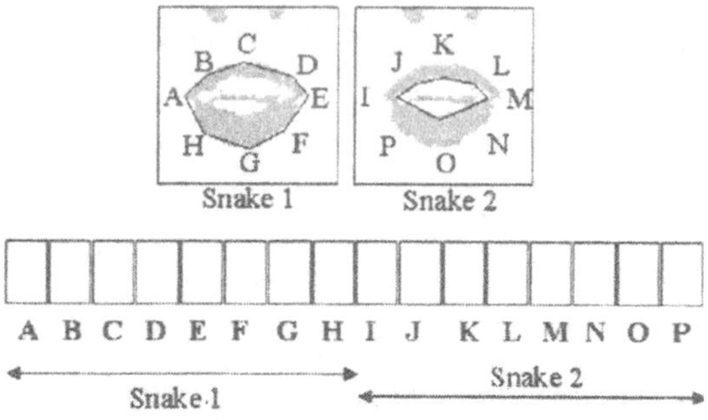

Fig. 4. Chromosome coding

2.2 Chromosome coding

Our aim is to find a first snake on the external lips contour and a second one on the interior contour. Each one of these snake is defined on eight nodes. To accelerate optimization, we make evolve the snakes only on contours points. Thus the node C (Fig. 4) will be defined only on North, North-East and North-West directed points in the area RC (see Fig. 3 and 3). This area is defined starting from the gravity centre C_G of the mouth previously given.

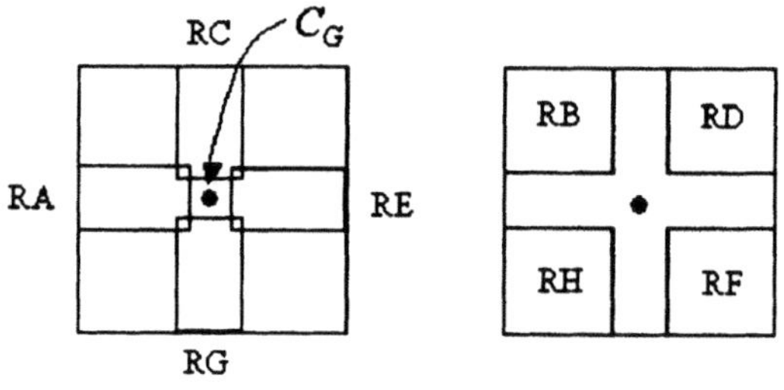

Fig. 3. Mouth regions

All the other nodes of the first snake are defined in the same manner by taking into account the area in which they must evolve and the contours orientation which characterize them. With regard to interior contour, insofar as the mouth is sometimes closed, it is difficult to define in a robust way the areas in which the nodes of the second snake must evolve. This is why we take into account only the contours orientation of the image d) of the figure 1 knowing that the nodes of the second snake can belong to any area of the image.
The position of each node of both snakes is coded on a chromosome gene as the figure 4 indicates it. Thus, the tenth gene codes the value $x + yL$ if L is the width of the image and (x, y) the coordinates of the node J belonging to the second snake.

2.3 Energie evaluation

The energy we wish to minimize is constituted by three terms controlled by three coefficients α_1, α_2 et α_3.

$$E = \alpha_1.E_{Bend} + \alpha_2.E_{Snake2} + \alpha_3.E_{InterSnake} \quad (1)$$

The first term makes it possible to control the rigidity of the curve. This constraint is evaluated on the whole set of the nodes of both snakes except for those corresponding to mouth corners $(A, E, I, and M, see Fig\ 4)$.

$$E_{Bend} = \sum_{i=2, i\neq 1+N/2}^{N} \| V1_{i-1} - 2V1_i + V1_{i+1} \|^2$$

$$+ \sum_{i=2, i\neq 1+N/2}^{N} \| V2_{i-1} - 2V2_i + V2_{i+1} \|^2 \quad (2)$$

where i is the number of the node on the curve knowing that a snake contains N nodes. The first node is on the left corner of the mouth (A or I in the figure 4), the node $1 + N/2$ is on the right corner of the mouth (E or M). Vj_i is i-th node of the snake j.

The second term E_{Snake2} takes into account the quantity of pixels $NbPix_{S2}$ belonging to the interior mouth (white pixels of the figure 1c) belonging to the area R1 of figure 5).

$$E_{Snake2} = \frac{1}{NbPix_{S2}} \quad (3)$$

The last term $E_{InterSnake}$ makes it possible both snakes to surround the lips. It takes into account at the same time the sum of the pixels levels belonging to the lips (V values, Fig 1b) in the area R2 of the figure 5):

$$Pix_{Lips} = \sum_{R2} V_{value} \quad (4)$$

and its complement evaluated as:

$$Pix_{nonLips} = NbPix_{Inter} - Pix_{Lips} \quad (5)$$

Where $NbPix_{Inter}$ is the amount of pixels between the two snakes (area R2 of the figure 5). Considering at the same time these two terms makes it possible to take into account the most clear pixels in the area R2 while minimizing the dark number of pixels in this same area.

We have thus:

$$E_{InterSnake} = Pix_{nonLips} + \frac{1}{Pix_{Lips}} \qquad (6)$$

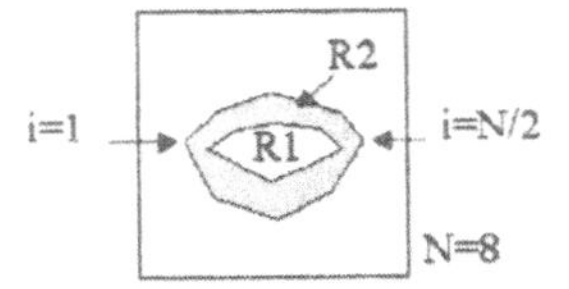

Fig. 5. Areas defined by both snakes

After minimization of energy E, the snake described by the chromosome of minimal energy evaluates contours of the lips. Because of the great variability of the lips between two images (acquired at 25Hz in our system), we do not implement a shape tracking like [14].

3 Lipreading System

The lipreading system includes a preprocessing and a classification module.

3.1 Preprocessing

Our team considers halfway parameters in term of level of analysis between DCT and hight-level models.

As figure 6 indicates it, six parameters only are sufficient to realize the lipreading: the height and the width of the mouth (H, L), their temporal derivative (δH, δL) and the percentage of light $\%W$ and dark pixels $\%B$ contained in the mouth [13]. The height and the width of the mouth are evaluated starting from the exterior snake:

$$H = G_y - C_y \qquad (7)$$

Where P_y represent the Y-coordinate of the point P in the figure 4. With regard to the width, the first node of the interior snake always does not correspond to node I since we did not restrict the area in which the nodes of the second snake evolve contrary to those of the first one (see section 2.2). Thus we evaluate the X-coordinate of I and M by detecting the X-coordinates min and max values of the points I to P:

$$L = maxV2_{ix} - minV2_{ix}, \text{with } i = 1..N \qquad (8)$$

where $V2_{ix}$ is the X-coordinate of the second snake point i.

Fig. 6. Preprocessing

In a small window of 10 pixels height centered on the mouth, we calculate an eight-bit grey level histogram of the pixels. The ratio between the number of pixels whose values are lower than 50 and the total pixel number gives us the percentage of dark pixels, that between the number of pixels whose values are superior to 150 and the total pixel number gives us the percentage of light pixels. We add to these four parameters (height, width, dark percentage, light percentage), the temporal derivate of the height and the width.

Those parameters are normalised taking into account their observed values all along the sequence.

3.2 Classification

The STM (Sparse Template Matching) [11] have a very weak complexity and require not much memory, reasons why they are well adapted to embedded applications. To be able to exploit them, we use the same generic preprocessing as in [13] which makes it possible to convert continuous signals to a representation in the form of a sparse matrix. This generic preprocessing is simple. During the phase of training a K-means identifies N prototypes or vector codes on parameters vectors $(H, L, \Delta H, \Delta L, \%W, \%B)$ taken at every instant. As shows it figure 7 the pronunciation of a word can then be characterized by a trajectory in the space of these vectors, trajectory itself characterized by a sequence of prototypes (learning sequence of figure 7: $P_1 \rightarrow P_1 \rightarrow P_3 \rightarrow P_4$).

This sequence which characterizes the pronounced word is identified using a Nearest Neighbour Classifier (NNC) algorithm (Fig. 8). At every instant, one compares the parameters vector with each prototype identified during the training phase, and one emits a unit value on the output associated with the nearest prototype (testing sequence of figure 7: $P_1 \rightarrow P_1 \rightarrow P_4 \rightarrow P_4$). A word is thus characterized by a sequence of prototypes, itself store in a binary sparse matrix.

Nevertheless a person always does not pronounce the same words at the same speed. To make a more robust detection we thus apply a temporal low-pass filter on this

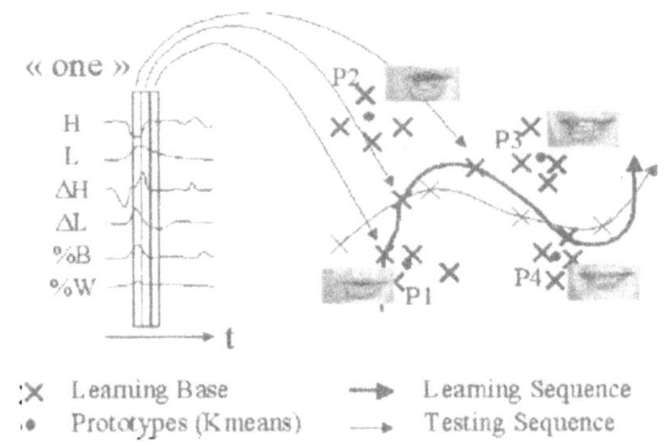

Fig. 7. Learning phase

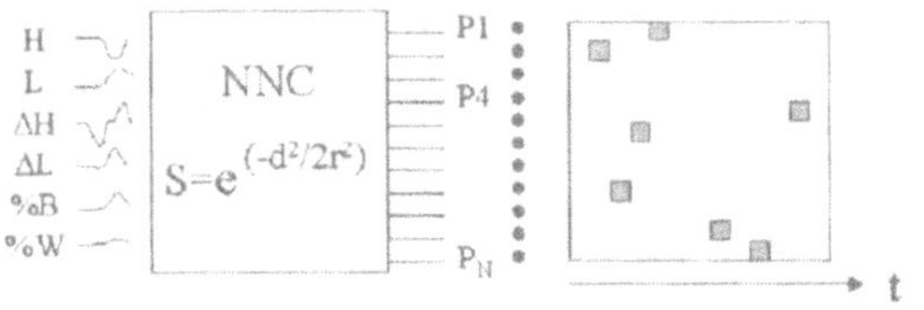

Fig. 8. The Nearest Neighbour Classifier wich produce a sparse matrix

matrix. Then, a sparse template is not represented any more by a sparse binary matrix but by a sparse multivaluated one as indicates it the figure 9.

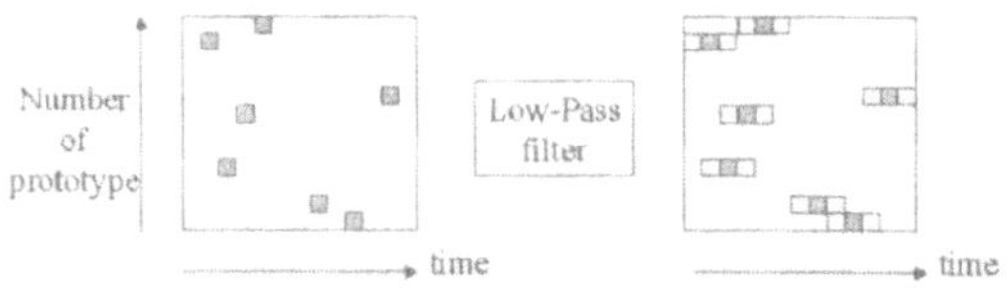

Fig. 9. Low-pass filtering along the time

Classification is extremely simple and takes as a starting point the techniques of tracking in image processing [15]. In the case of a one pass learning, the sparse matrix generated during the training represents the signature of the word which will have to be recognized. There is as many sparse template than words to be recognized. Classification is done simply by evaluating an Euclidean distance between the sparse matrix resulting from the new word presented and each sparse template. The smallest distance indicates the recognized word. If the training is done on several sequences, the sparse template characterizing a word is simply the average of the sparse matrices evaluated on each training sequence of the word to recognize. The figure 10 represents the whole classification treatment.

4 Results

Within the framework of separated word recognition, we tested our system on the first person of the European Data Base M2VTS [16] (Multi Modal Checking for Teleservices and Security applications). This base is dedicated to Audio-visual recognition and identification. The person pronounces four times (at one week interval) the digits from 0 to 9. We chose this base because it characterizes well the conditions of use in which the real time implementation of our system will have to function. The images were acquired at 25Hz with a weak resolution (288x360pixels in 4:2:2).

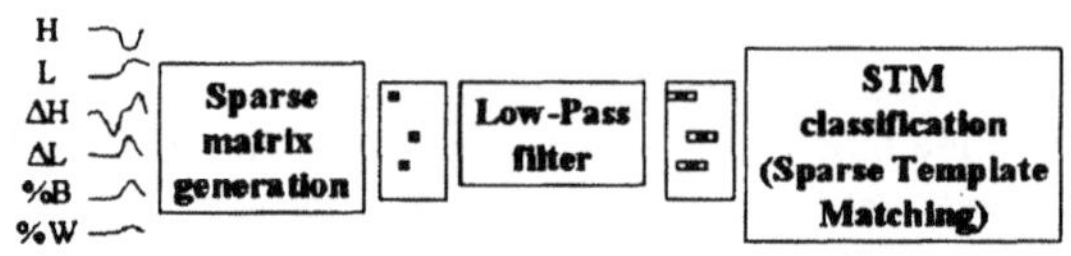

Fig. 10. Sparse Template Classification

The figure 11 makes it possible to visualize the difference between the real height of the mouth (dotted line) and that detected starting from the first snake (solid line). The variation between these two values is not significant. The important thing is that the detected height evolution follows well the real height one throughout sequence. Thus at instant 10, it seems that the snake is badly positioned contrary to the result provided at instant 114, whereas both exterior snakes are well positioned. On the other hand, at instant 71 it is clear that the first snake fell into a local minimum of the energy which induces a bad evaluation of the mouth height. On the four sequences, the mouth height is indeed badly estimated (as at instant 71, figure 11) in 1.2% of the cases only (in 8 images out of 669).

In the context of single-speaker lipreading, we tested two different configurations knowing that four sequences of ten digits were available: a one pass and a three sequences training. Ten prototypes were identified in training phase ($N = 10$, Fig. 8).

One pass learning The results are presented as an average performed on the four tests carried out starting from four different sequences of training; according to the number of the sequence used for the training, tests are made one the three other sequences. In this configuration, the system performed a correct classification in 68% of the cases. Let us note the temporal filter advantage (Fig. 9). If it is not implemented, the results are only of 58%,

Three sequences training The average result evaluated on the four various sequences of tests where obviously better: the classification was correct in 88% of the cases. Without the temporal filter, the results were only of 83%.

5 Conclusion

We proposed in this article an implementation of Genetics Snakes in lipreading. Two snakes converge in parallel towards lips contours. A very simple classifier (Sparse Template Matching) gives good results within the framework of this application and allows us to consider a real time implementation on an embeded system.

However the convergence of the GS is still too long (54s on P3-900Mz) even if the implementation was made in Matlab. Moreover, the coefficients α_1, α_2 and α_3 which control various energies to be optimized, were given at hand. It remains us to propose an automatic method who will allow us to identify the value of these coefficients on a person test and to check their robustness on the other people introduced in M2VTS.

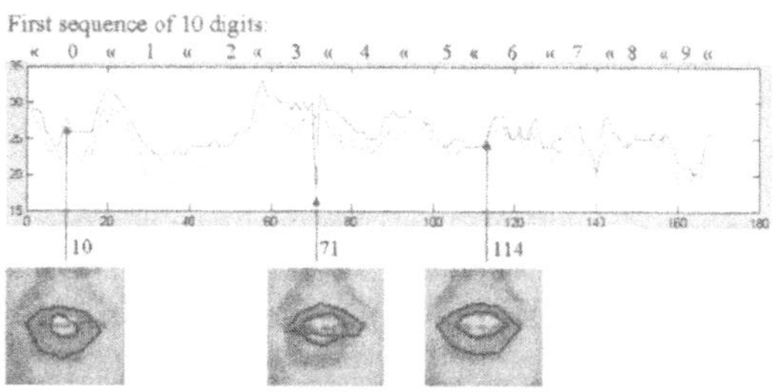

Fig. 11. Dotted line : real mouth height along the first sequence, solid line : first genetic snake height

6 Acknowledgment

A part of this research was supported by Brittany Region ("Région Bretagne") in France.

References

[1] C. Bounsaythip and J.T. Alander. Genetic algorithms applied to image processing - a review. In *Proc. of the 3rd Nordic Workshop on Genetic Algorithms (3NWGA)*, 1997.

[2] K. Sakaue, A. Amano, and N Yokoya. Optimization approaches in computer vision and image processing. *IEICE Trans. Inf. and Syst.*, 1999.

[3] A. Cagnoni, A. Dobrzeniecki, R. Poli, and J. Yanch. Genetic algorithm-based interactive segmentation of 3d medical images. *Image and Vision Computing, 17(12):881-895.*

[4] L. Ballerini. Genetic snakes for color images segmentation. *Lecture Notes in computer sciences 2037*, 2001.

[5] N. Covavisaruch and T. Tanatipanond. Deformable contour for brain mr images by genetic algorithm: From rigid totraining approaches. In *Proceedings, Image and Vision Computing New Zealand (IVCNZ '99)*, 1999.

[6] Michael T. Chan, You Zhang, and Thomas S. Huang. Real-time lip tracking and bimodal continuous speech recognition. In *Workshop on Multimedia Signal Processing*, 1998.

[7] P. Delmas, P.Y. Coulon, and V. Fristot. Automatic snakes for robust lip boudaries extraction. In *International Conference on Acoustics Speech and Signal Processing (ICASSP)*, 1999.

[8] Y. Tian, T. Kanade, and J. F. Cohn. Recognizing action units for facial expression analysis. *IEEE Trans. on Pattern Analysis and Machine Intelligence*, 2001.

[9] R. Séguier, N. Cladel, C. Foucher, and D. Mercier. Lipreading with spiking neurons: One pass learning. In *International Conference in Central Europe on Computer Graphics, Visualization and Computer Vision*, Feb 2002.

[10] G. Potamianos, J. Luettin, and C Neti. Hierarchical discriminant features for audio-visual lvcsr. In *International Conference on Acoustics Speech and Signal Processing (ICASSP)*, May 2001.

[11] G. Sullivan, K. Baker, A. Worrall, C. Attwood, and P. Remagnino. Model-based vehicle detection and classification using orthographic approximations. In *Proc of 7th British Machine Vision Conference*, 1996.

[12] R. Séguier, A. Le Glaunec, and B. Loriferne. Human faces detection and tracking in video sequence. In *Proc. 7th Portuguese Conf. on Pattern Recognition*, 1995.

[13] R. Séguier and D. Mercier. Audio-visual speech recognition: one pass learning with spiking neurons. In *International Conference on Artificial Neural Networks (ICANN)*, 2002.

[14] Michael T. Chan. Hmm-based audio-visual speech recognition integrating geometric and appearance-based visual features. In *Workshop on Multimedia Signal Processing*, 2001.

[15] Per-Erik Forssén. Window matching using sparse templates. In *http://www.isy.liu.se/cvl/ScOut/TechRep/*, 2001.

[16] S. Pigeon. M2vts. In *www.tele.ucl.ac.be/ PROJECTS/M2VTS/m2fdb.html*, 1996.

Applying Genetic Algorithms to Container Transhipment

Colin R.Reeves*, Daniel Scott*,
Andy Harrison†

*School of Mathematical and Information Sciences, Coventry University, UK, †Vector-TRI, Malvern, UK.

Abstract

The container transhipment problem involves scheduling a fleet of lorries to pick up or drop off containers of various sizes, while minimising the total distance travelled by the lorries. The problem originates in the need for logistics companies to solve this problem on a regular basis as part of their daily operations. In this paper we investigate the use of a genetic algorithm for the solution of the problem. Results are discussed and a general analysis and comparison of the validity and robustness of two approaches are given.

1 Introduction

The container transhipment problem (CTP) is a combinatorial optimization problem of direct relevance to the daily distribution scheduling of real-world logistics companies. In the UK, for the decade 1989 to 1999, the level of **empty running**—where lorries travel unladen—remained broadly steady at 30% [1]. Such a level of empty running is significant. When combined with **light running**, where lorries are not loaded to their full capacity, the result is increased costs, pollution and energy consumption along with unnecessary pressure on road space.

Through the use of optimization to pair outbound and inbound loads, substantial reductions can be made in empty running with corresponding cost savings. For example, the supermarket giant Tesco was able to reduce its empty running by 3 million miles per year, saving £720,000 in fuel costs [2]. While it is more difficult to determine the level of light running, it is believed to be substantial [2]. Significant reductions in light running should also be achievable through the use of optimization—for example, by the pairing of 20 foot loads.

Thus, practical near-optimal solutions to this problem could provide significant savings on the empty running costs associated with lorries scheduled for distribution tours (where a tour is the entire route that each lorry travels to complete the delivery and pick-up of all of the assigned containers).

1.1 The CTP

The CTP can be outlined as follows:

> *Given a number of containers to be delivered and picked up from various locations by a fleet of lorries, minimize the total distance travelled such that all orders are fulfilled subject to constraints including (but not limited to) a maximum capacity limit on a lorry. It is also assumed that the order in which the deliveries and pick-ups can be made is significant, and that the lorries always start and complete a tour at the main depot (of which there is only one).*

Clearly, variations on this theme are possible, but this formulation captures the essence of the problem. The CTP has links with many other well known optimization problems, such as bin-packing, transportation and assignment problems, all of which it extends in at least one way.

The specific assumptions made for this research were as follows:

- There are three types of container, designated by their lengths: 20ft, 30ft and 40ft.
- There is no limitation on the number of lorries that are available to transport containers (based on the fact that companies can subcontract work out).
- All lorries are of a standard size, enabling them to carry either one 40ft, one 30ft, or two 20ft containers.
- 30ft and 40ft containers can be considered to be identical in the sense that if a lorry is transporting either a 30 or 40ft container it is unable to carry any other containers.

- All lorries originate from and return to a single main depot.
- All distances are 'straight line' in that they do not accurately follow roads.
- There are no time windows for when deliveries or pick-ups have to be made.
- All containers are equally profitable, i.e. delivering a 20ft container is as valuable as delivering a 40ft container.
- Inbound and outbound journeys are separate and distinct, i.e. it is not possible to pick-up one container and drop it off before picking up and dropping off another *ad infinitum*.
- The Earth is assumed to be flat for the purpose of calculating distances (data are generated using towns and cities within the UK). Coordinates are projected onto a 2D-plane.

2 GA Representation

The basic principles of genetic algorithms (GAs) are well known, and need not be rehearsed at length here. It is sufficient to note that a GA is a population-based heuristic, which manipulates encoded representations of a solution ('chromosomes') by means of selection, recombination ('crossover') and mutation in order to generate new and potentially improved solutions. As usual there are different ways in which to implement them in different contexts. For a review of applications to combinatorial optimization problems, see [3].

GAs have been used for similar problems to the CTP before [4] but, as in these other cases, there is a fundamental question as to how best to represent a candidate GA solution. A simple approach would be to number the containers and represent the solution by a permutation that needs decoding by some 'online' algorithm. This approach was used to good effect in [5]. However, Falkenauer [4] disliked the permutation approach, which admittedly introduces some redundancy into the coding, arguing that for bin-packing and similar 'grouping problems' the representation should focus on the bins (in this context, the lorries), although it is not clear that this method is really superior to the other. Both approaches were therefore tried for the CTP.

To reduce redundancy in both representations the containers were partitioned into inbounds and outbounds. Each of these was then further partitioned into sectors. Hence, each solution comprises $2k$ chromosomes, for k sectors, $k \geq 1$. The genetic operators are applied to each chromosome independently. So, for illustrative purposes, it is only necessary to consider a single chromosome (sector), as the chromosomes are only brought into relation when the solution is decoded to calculate the fitness of the population.

$$\underbrace{2\,4\,3\,5\,6\,1\,3\,6\,7\,8\,9\cdots}_{\text{Container Part}} \quad | \quad \underbrace{1\,2\,3\,4\,5\,6\,7\,8\,9\cdots}_{\text{Lorry Part}}$$

Fig. 1. The 'lorry GA' representation. For example, containers 3 and 7 are both assigned to the 3rd lorry, which in this case has index 3.

Such a partitioning does not completely exclude invalid solutions, such as the assignment of a container to multiple lorries, from either representation. The genetic operators, however, have been designed to avoid generating invalid solutions.

The **permutation GA** represents assignments of containers within a sector by permutations of the container numbers and an appropriate decoder. This representation is designed to exploit the importance of order in the representation.

In contrast, the **lorry GA** focuses on the lorries into which the containers are packed. Each chromosome in this representation comprises two parts: a container part and a lorry part. In the container part, the ith locus represents the ith container and the value of the gene at this locus (the *allele*) represents the lorry to which the container is assigned. The lorry part is merely a list of all the lorries. The order of the lorries has no effect on the functioning of the genetic operators, although the significance of the lorry part will become clear when we describe the genetic operators. Figure 1 illustrates the lorry representation.

The implementation details of a GA is of course not limited to the choice of representation: the operators that the GA uses must 'fit' the coding, otherwise the approach may well be useless. This need is particularly acute for recombination.

2.1 Crossover

For the permutation GA, the C1 crossover [6] was applied to each chromosome independently. (Sectoring according to geographical location has already taken place.)

The genetic operators, crossover and mutation, for the lorry GA are more complicated and are designed to work only on the lorry part of each chromosome. A 'first fit' algorithm is used to assign displaced containers to lorries within the chromosome. This approach to the problem leans heavily on modelling the problem as bin packing. As such the crossover operator is based on the approach used by Falkenauer and Delchambre [7] and further developed in [4].

```
lorryCrossover(P1, P2)
{
  Save a copy of the P1 chromosome;
  Select one crossover point in the lorry part
  of P1;
  Select two crossover points in the lorry part
  of P2;
  Copy the lorries between the crossover
  points in P2;
  Inject them at the crossover point in P1;
  Find the containers assigned to the injected
  lorries from P2;
  FOR each injected container DO
    Find the lorry containing the duplicate
    container in P1;
    Record which other containers are
    assigned to it;
    Delete the lorry;
  ENDFOR
  Reinsert the displaced containers into P1;
  Renumber the lorries;
  Return the new P1;
}
```

Fig. 2. The lorry GA crossover operator.

Recombination in the lorry GA is achieved by applying the crossover operator described in figure 2 to each chromosome. Further details and an illustrative example can be found in [8].

2.2 Initial population

The two approaches also generate initial populations in differing ways. For the permutation GA the initial population was produced by creating a random permutation of all the containers in the data set, each having been assigned an index number. The permutation of all containers was then sorted into sectors making sure that the same order of containers was kept in relation to the overall chromosome. The lorry GA produces the population by assigning containers to lorries in order with a stochastic decision about assigning two 20ft containers to the same lorry. This method produces a less diverse initial population than the truly random method used for the permutation GA.

2.3 Sectoring

Methods of sectoring and their influence on the computational efficiency of the GA are discussed at length in [8]. Here we merely note that three methods were tried: *spatial homogeneity*, where the physical area of each sector was the same; *container homogeneity*, where contiguous physical areas had the same number of containers; and *fuzzy spatial sectoring*, which was like the first, except that sector 'membership', particularly for those locations near the sector edges, was defined not as either/or, but as a value between 0 and 1 in accordance with standard practice in the field of fuzzy logic.

2.4 Mutation

The mutation operator for the permutation GA performs a simple swap operation within sectors. It is only effective if it changes the order of the containers within container size: e.g., if the mutation operator swaps a 40 ft and a 20ft container that are next to each other then the mutation has no effect because of the mechanics of the first fit algorithm that is used to decode the chromosome.

The mutation operator for the lorry GA representation is more complicated and involves randomly deleting one lorry in a sector and reassigning the containers to other lorries if possible or creating a new lorry if not. Details can be found in [8].

The mutation rate for the lorry GA needs to be lower than for the permutation GA because the operators work on the lorries for each sector rather than the containers. For each sector the number of lorries will always be less than the number of containers, owing to the pairing of 20ft containers, unless each container is assigned to one lorry. The mutation rate should thus be proportional to the length of the strings that it is working on.

2.5 Fitness

The objective function is defined simply as the total distance that the lorry travels so that all the containers are delivered:

$$F = \sum_{i=1}^{n} d_i \tag{1}$$

where n = total number of lorries, d_i = total distance travelled by lorry i. This function is, of course, to be minimized. The permutation GA just used this objective function as its measure of fitness.

For the lorry GA the objective function in equation 1 could also be used. However, in keeping with the analogous problem of bin-packing, an alternative was tried, as in Equation 2 below.

$$F = \sum_{i=1}^{n} \left(d_i + k \left(1 - \frac{C_i}{C_{max}} \right) \right) \qquad (2)$$

where n and d_i are as before, k = weighting factor, C_{max} = the maximum capacity of a lorry (i.e., 80ft, counting both outbound and inbound deliveries), C_i = cumulative length of containers assigned to lorry i. The parameter k is chosen to be the median or mean of the individual lorry tour distances in the data set.

This function follows [7] closely, in rewarding lorries that are more closely filled to capacity with better fitness values. It is argued that a near optimal solution is likely to exhibit the following two characteristics: ideally, there will be a minimal number of lorries, and if not, lorries will be filled as much as possible.

2.6 Replacement

For both GAs, two children were created for each generation and inserted in the population in place of the two least fit members of the population.

2.7 Termination

The termination criterion for both GAs was set quite simply: stop on reaching 100 generations. This obviates the need for complicated tests for convergence.

3 Experiments

In order to test the GAs thoroughly it is necessary to have many different datasets. These were created on the basis of the geography of the UK, but the locations and their spatial distributions were randomized so that the results were not tuned to any particular pattern of locations. Details of the generating mechanisms are in [8]. The core sites used were the 50 largest towns in the UK, and the distribution of locations among these sites was either Poisson, binomial or negative binomial—these correspond to 'purely random', 'randomly regular', and 'randomly clustered' patterns respectively. Two sets were generated for each case with either 200 or 1000 locations for the containers. The smaller sets were used mainly in initial trials, but the larger datasets are reasonably representative of what happens in a typical day at a port such as Folkestone, where there could be up to 3000 container deliveries.

No. sectors	Perm GA		Lorry GA	
	Min	Mean	Min	Mean
Poisson data				
1	154676	155050	168390	173543
3	145535	145837	166061	169129
5	138922	139018	163888	166712
7	138519	138607	161276	164928
9	137019	137136	162809	167648
binomial data				
1	153800	153891	162440	164524
3	145797	146261	168466	170771
5	139333	139452	159664	162536
7	139333	139452	159815	161810
9	138252	138460	160047	162382
negative binomial data				
1	160825	161502	187034	189820
3	154741	154848	174185	178304
5	147151	147300	171362	174148
7	147334	147443	169032	171722
9	145385	145469	169569	172325

Table 1. Comparison of the two GAs on the 1000 container datasets.

For brevity, the results tabulated here (Table 1) relate only to the 1000 container datasets, and to the use of homogeneous spatial sectoring. These results are nonetheless representative of the overall picture. In each case we have tabulated the best (minimum) fitness and the average found from 3 independent runs of each GA. The influence of the number of sectors can also be seen.

There are two clear conclusions from these results: firstly, pre-processing the data by sectoring is clearly worthwhile, although there is a suggestion that the improvement is levelling off as the number of sectors increases. The second conclusion relates to the two GAs: it is clear that the permutation GA is uniformly better than the lorry GA.

The lorry GA was based on the 'grouping' principle identified as important by Falkenauer [4] for bin-packing and related problems. However, for the CTP it does not seem to perform as well as the permutation GA, whose basis Falkenauer questioned. In its defence, it should be pointed out that the lorry GA runs somewhat faster than the permutation GA—typically about 65-70% of the computational effort needed by the permutation GA. However, running it for longer does not reduce the whole surplus mileage, which may be as much as 20% more than that produced by the other approach. So runtime is not the whole story: probably the grouping principle is of less importance here where the maximum number of 'items' per 'bin' is only 2.

Another possibility is that the crossover operator used within the lorry GA is too disruptive—which, unfortunately, is the exact opposite of the role it is designed to play! Some credence is given to this idea by the fact that, while the variance of the fitness reduces steadily for the permutation GA, such behaviour was not observed for the lorry GA. Instead, the variance decreased very little, which suggests that the effect of selection is being negated by the effect of recombination as well as mutation.

Finally, before leaping to the conclusion that the lorry GA is inherently bad, we should realize that it forces several differences in the way genetic operators are used, and in which fitness is calculated. Thus, it is conceivable that the grouping principle is still valid, but that details of its implementation could be improved. From a practical standpoint, however, the fact that the simpler permutation GA appears to work satisfactorily meant we did not pursue this question further.

4 Conclusion

This paper has reported work on an application of genetic algorithms to the container transhipment problem. The usefulness of a pre-processing algorithm for sectoring was demonstrated, and two types of GA were implemented. The clear 'winner' of these GAs was not that predicted from previous research on related problems. This emphasizes the fact that GAs need to be tuned to the problem class on which we wish to operate, and that extrapolation from experience on other problems—even those which are apparently closely related—can be dangerous.

Some suggestions have been made as to why the 'grouping GA' approach performed poorly, but further work needs to be done to explore alternative explanations. As for the problem itself, it would be useful to investigate further the scalability of the algorithms implemented here. While the largest datasets considered approach the size of real problems in the UK, the Hong Kong port, for example, can have between 10,000 and 20,000 lorries transporting containers every day—an order of magnitude bigger.

References

[1] *Transport of Goods by Road in Britain 1999 (Annual report of the Continuing Survey of Road Goods Transport)* (2000), Dept. of the Environment, Transport and the Regions. ISBN 1-85112-389-X.

[2] *A New Deal for Transport: Better for Everyone (Government White Paper on the Future of Transport)* (1998), Dept. of the Environment, Transport and the Regions. ISBN 0-10139-502-7.

[3] Reeves, C.R. (1997) Genetic algorithms for the Operations Researcher. *INFORMS J on Computing*, **9**, 231-250.

[4] Falkenauer, E. (1998) *Genetic Algorithms and Grouping Problems*, John Wiley & Sons, Chichester.

[5] Reeves, C.R. (1996) Hybrid genetic algorithms for bin-packing and related problems. *Annals of Operations Research*, **63**, 371-396.

[6] Reeves, C.R. and Höhn, C. (1995) Integrating local search into genetic algorithms. In Rayward-Smith, V.J., Osman, I.H., Reeves, C.R. and Smith, G.D. (eds.) (1996) *Modern Heuristic Search Methods*, John Wiley & Sons, New York, 99-115.

[7] Falkenauer, E. & Delchambre, A. (1992). A genetic algorithm for bin packing and line balancing. In *Proc. of the IEEE 1992 International Conference on Robotics and Automation*, Nice, France, 1186-1192.

[8] Scott, D. (2002) *Genetic Algorithms for the Efficient Transhipment of Container Loads*, MSc Dissertation, School of Mathematical and Information Sciences, Coventry University, UK.

Finding routeways in airframe design using genetic algorithms

D. A. Nathan, R. M. Newman, C. R. Reeves
Coventry University, Priory St, Coventry, CV1 5FB, UK
email: r.m.newman@coventry.ac.uk

Abstract

Routing of wires and piping in airframe design is an arduous process. Current processes attempt to wire bundles of wires individually, resulting in a lengthy iterative process as attempts are made to satisfy the many constraints for each type of wire or bundle. This paper describes a new approach to this problem, in which pre defined routeways are established at an early stage of the design, so as to satisfy the constraints of the wiring and piping to be placed within. The work described here provides for automatic optimization of these routeways using genetic algorithms. Solutions produced this way are compared with those produced by a maze solver, and found to offer better use of the available space. Variations of the detail design of the genetic algorithm used have been prototyped and evaluated.

1 Introduction

In the past, wiring and piping for aircraft was done after the body structure and other on-board systems were completely defined. This was due to the fact that wires and pipes were considered to be flexible and easy to route in the available space.

This is however, no longer the case. As the electrical systems of aircraft grow both in size and complexity, electrical equipment and wire bundles begin to demand special space reservation as early as the conceptual phase of the aircraft design activity.The failure to 'respect' wiring and piping requirements in early stages of the design process can result in many expensive downstream changes and cause a significant decrease in the global design quality of the project. These concerns are serious enough that it is worthwhile researching ways of reserving and optimising space for routeways at a very early stage in the design process. This paper reports work that has been done to this end.

2 The Current Process

This description of the current process is compiled from a number of conversations with and documents furnished by Dr Elena Valfre of Alenia Aerospace, Milan, Italy. The currently defined design process consists of four main phases. The conceptual phase in which conceptual investigations are carried out on the structure, fluid systems, pneumatic systems, electrical systems, and others in order to establish feasibility of the design being produced. The definition phase comprises the basic definition and detail definition phases. Nowadays, it entails the production of a Digital Mock-Up (DMU) representing structure ,on-board systems and the approximate paths of wires and pipes that interconnect them. The output of this phase represents the starting point for the detail engineering phase, which is the final stage of the product design. As part of this stage, a prototype is produced, and on-board systems are refined and completed. The on-board systems are completely detailed, taking into account their requirements, functionalities, and their components which could include relays, switches, and connectors. A further sustaining phase is allowed for within the process, in order to cater for changes required by manufacturing, and these are managed within the digital mock-up.

In order to distribute and route wires effectively, it is very important to know about equipment locations and available space in the aircraft mock-up. It is however not feasible to locate wires in the structural model when bundles have not yet been composed. There is therefore a strong need for frequent information exchange between design and installation processes, contributing to the iterative nature of the work involved. As part of the design process, the designer needs to take into account the following factors:

There is a need to minimise the length of wires due to considerations involving resistance, heat, and propagation delays.

Holes are defined in the aircraft structure as a means to reduce the overall weight of the frame. When performing routing, it is recommended that the existing holes are utilised for the routing. Where these holes are unsuitable, the first step to consider is the possible enlargement of these holes to accommodate the routing. If this too is not satisfactory, the final resort is to bore a new hole.

Bend radius is a consideration for wiring and certain guidelines are adopted with respect to the bend radii for wiring. When routing wires, there is a need to consider accessibility in order to allow for the maintenance and servicing of the wiring. For instance, external bays are preferred for routing as opposed to internal bays since these allow a higher degree of accessibility.

Certain paths are preferred for routing. As an example, routing along longerons is preferred since they run the length of the aircraft, and provide effective anchorage or clipping points to secure the wire bundles.

It is important when routing electrical wires to ensure that they are never routed under any fuel lines, or flammable liquid lines, as any leakages which might arise could lead to fires and short circuits.

Other considerations include avoidance of crossover between wires and the minimisation of distance between equipment and routes, and the location of other installed routes.

Once this stage is complete, a preliminary 3D path is defined, firstly for routes passing through bulkheads, and then for other routes, "respecting" as much of the set requirements and other installation, assembly and maintainability-specific requirements as possible.

It was suggested by designers at Alenia that by establishing 'routeways' for wiring and piping at an early stage in the design process in such a way that the routeways themselves satisfied the constraints for the wiring or piping within them, it would be possible to obviate the need for late design changes necessitated by the detailed routing. One concern was the difficulty of finding an optimum layout for these routeways, leading to a belief that this solution would be a much less efficient use of aircraft space than individual, hand optimised routes. The work described here prototyped a simplified representation of the problem to establish whether genetic algorithms could be used to provide an efficient solution, given their successful use in some related problem areas [1].

3 Problem Representation

Based on the idea of computational grids in Computational Fluid Dynamics, which map a complex physical space into a computational space using simple shapes which nonetheless represent the essence of the problem [2] simple cuboids were used to represent the routing space. In this representation, a cube represents a routing opportunity, and not a physical space. This scheme allows for relationships to be established between adjacent cubes thus forming the wire and pipe routeways, and allows for greater flexibility in the manipulation of the model since the connectivity between cubes makes it possible to model routing opportunities which may exist. These relationships are important as it allows the optimisation algorithm to build and evaluate routing opportunities for particular subsystems.

Figure 1 shows this technique applied to the centre fuselage of the Typhoon fighter aircraft.

The space is divided into cubes of 250mm, which is the minimum distance between routes in this simple example.A cube is represented by a be integer chromosome as shown below.

Problems with the adjacency property can be addressed by defining operators (such as crossover and mutation) which are 'aware' of the adjacency configuration of the cubes. As an example, a mutation

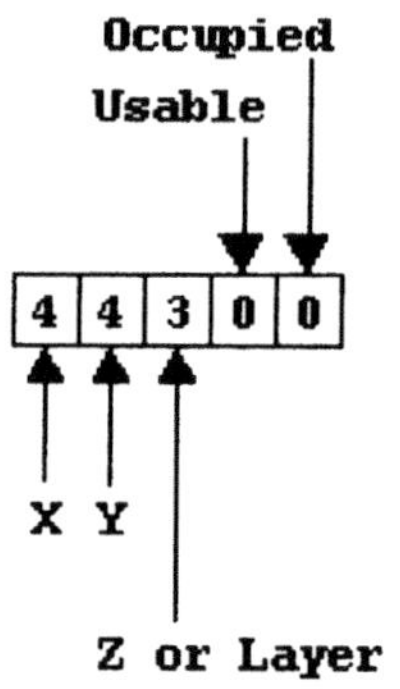

Figure 2: The chromosome

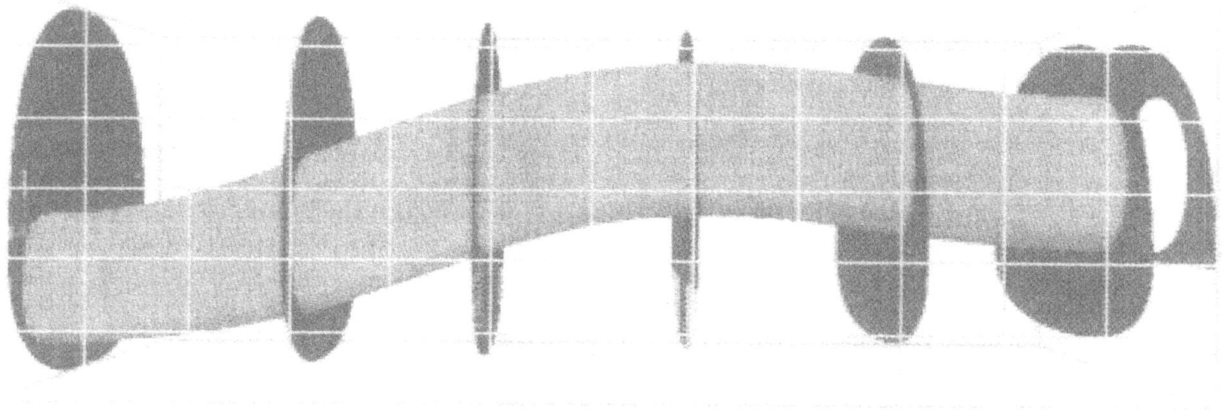

Figure 1: Grid representation of Typhoon centre fuselage

operator that attempts to mutate a particular cube in a routeway could call a function which would return the valid adjacent members for a cube.

Having addressed the minimum distance requirement of a route as a representation issue in the cube divisions, there are only two optimisation criteria involved , the length of the route and the cost of space used. The length of a routeway is given by to the sum of the number of cubes used. The cost is given by the sum of the associated costs of each of the cubes that make up a routeway allowing preference for location of the route. It is also possible to use a penalty scheme in which non-routable areas are not invalidated but rather avoided due to their higher cost association. In this simplified example, these are only two components of the cost or fitness function used to evaluate the routways found. The function is simply $f=\alpha l+(1-\alpha s)$ where l is the length of the route and s is the cost. α is a parameter determining the relative weighting between them.

4 Performance of a GA solution

The results presented here were based on trials a simple maze solver as a reference, and several variants of genetic algorithm, including mutation-only.

The GA type selected is of a class referred to as steady-state genetic algorithms [3]. The basic components which are required for a functioning GA for routing include a route generator, a basic population of individuals, crossover and mutation operators, a fitness function, and a selection and replacement strategy

The preliminary route generator is based on a simple recursive maze solver with an element of random choice built into it which produces a variety of routing solutions to a given routing problem with each call to it. This feature provides a level of diversity in the initial population. 1-point crossover has been used. Firstly, 2 parent chromosomes are chosen for crossover. The parents in this case are of differing lengths, since variable-length chromosomes are used. A random crossover point is selected between the second and the last-but-one gene since it makes no sense to perform crossover at the start or end points of a route. The two parents are then split at the crossover point thus yielding two head and tail segments. If the opposing head and tail segments line-up with each other, then these are joined to form new child chromosomes. Where the head and tail segments do not line-up with each other, a call is made to the preliminary route generator to generate the missing segment. The mutation operator selects a random mutation point and makes a call to the preliminary route generator to build a bypass route segment between them.

The performance (in terms of final fitness function) of the maze solver is shown in table 1 below. This forms a reference for the performance of different variants of the GA.

	Maze				Random				Strongest & Random			
	Bundle				Bundle				Bundle			
α	1	2	3	Total	1	2	3	Total	1	2	3	Total
0.25	86	92	107	285	60.4	58.4	59	177.8	59.2	58.6	60.8	178.6
0.5	65	68	77	210	48.8	48	43.8	140.6	44.2	48.2	48.8	141.2
0.75	43	43	47	133	33.8	34.6	30.4	98.8	34.2	33.6	30.8	98.6

Table 1: Performance of maze and GA solutions

α	Maze Solver
0.25	86
0.5	65
0.75	43

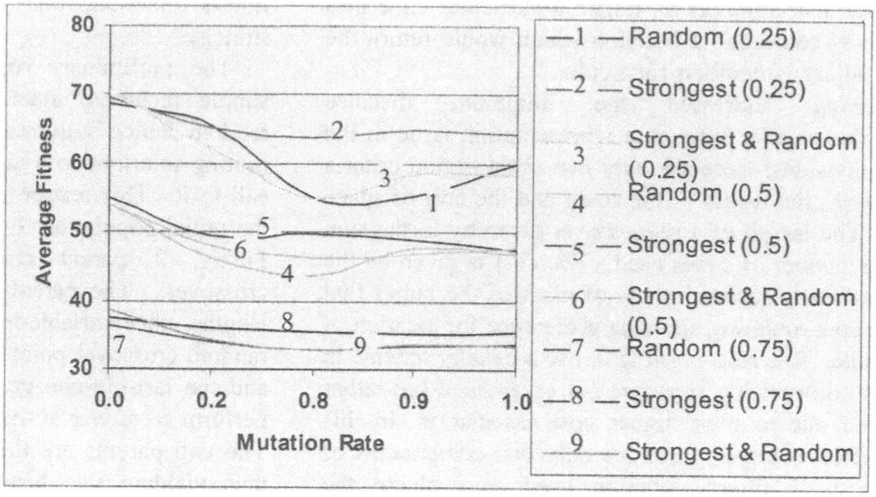

Figure 2: Performance of maze solver and GA solvers for single bundle problems

A summary of the various versions of the GA based routeway finder for the single route unconstrained case in shown in Figure 2. This shows the effect of three selection schemes (random only, strongest only and a combination of the two, based on a range of mutation rates and the same three values for α, the results given being the average of 1000 runs, as the GA solver is non-deterministic, unlike the maze solver. All variants of the GA solver give better results than the maze solver, with the best being the combined strongest and random selection scheme and a mutation rate of around 0.8.

Similar trials were conducted for a multiple multiple bundle problem. A strongest and random selection scheme and a mutation rate of 0.8 was used. The table below shows the results of the maze solver and the GA solver. Both versions of the GA solver strongly outperform the maze solver in this more complex situation, but there is little to choose between them.

5 Conclusions

The work presented here suggests that an approach based on the use of genetic algorithm optimisations to provide pre-designed routeways, which are designed to satisfy the constraints of the wires and pipes within them, could be a useful step in the design process for aircraft. GA solvers have been found to work better than simple maze solvers, particularly as complexity and constraints in the routing problem increases.

References

[1] Ali, M., Rammamurthy, B., Deogun, J. A genetic algorithm for routing in WDM optical networs with power considerations: The Unicast Case. Proc Globecomm '99, Symposium on high speed networks, IEEE, Rio de Janero, Brazi, 1999.

[2] Nagasu, H. Research in Computational Fluid Dynamics at National Aerospace Laboratory, Japan in Murphy T.K.S, Fielding J.P. (eds) Computer Applications in Aircraft Design and Operation, pp 195-208, Springer Verlag.

[3] Syswerda, G. Uniform Crossover in Genetic Algorithms, Proc 3rd International Conference on Genetic Algorithms, pp 2-9, Los Alamos, CA, Morgan Kaufman, 1989.

A Universal Knowledge Module and its Applications

R. F. Albrecht[1], G. Németh[2]

[1] Faculty of Natural Science, University of Innsbruck, 25, Technikerstr., 6020 Innsbruck, Austria, e-mail: Rudolf.Albrecht@uibk.ac.at, [2] Department of Telecommunications, Budapest University of Technology and Economics, 2, Magyar tudósok krt., 1117 Budapest, Hungary, e-mail: nemeth@hit.bme.hu

Abstract

We introduce a universal relational scheme of parameterized sets of objects of arbitrary types, taking also variables and topological structures into account, to represent precise and imprecise ("vague") knowledge. By operations applied to the scheme further knowledge can be deduced and by functional "queries" facts can be interrogated. As illustrative examples applications to databases, pattern recognition and logical deduction are considered.

1 Introduction

Knowledge is represented by sets of primitive or composite (structured) formal objects, symbolizing physical or mental objects, their relevant properties and their mutual relationships, which may be given explicitly or implicitly. Objects are parameterized by indices (identifiers, names, locations in space, instants of object occurrences in a model time, etc.) which may also be structured. The objects are collected in a general relational structure ("knowledge module") to which functional operations ("deduction rules") can be applied, involving objects and indices, their properties and structures, to produce further knowledge from current explicit knowledge. In addition, "queries" can be applied to the knowledge module, which for given query arguments produce a query result using the deduction rules and the data in the current module. Finally, we need composition and decomposition operations ("structors") for relational structures. The applications of the operation mentioned are subject to grammars. For a convenient representation, the module and the operations can be described by means of variables with fixed domains of variability. In this case, a query addresses all instances of the variables and thus the result can be expressed by variables, depending on the argument variables. Introducing valuated topological neighborhoods of objects and of indices (generalization of "fuzzy sets"), then a query yields a valuated approximate result.

Using concepts and reasoning given in earlier work of the authors, for example see ref. 1, 2, 3, 4, 5, the ideas are outlined in the following. Most of the presentation holds for general object and index sets. In practice, the sets are usually finite.

2 Relational Structures and Operations

2.1 Hierarchical relations

For any non-empty sets I, S and any function $\iota: I \to S$ we use the following notations: ι is an indexing, I is an index set, S is an object set, $s_{[i]} =_{\mathrm{def}} \iota(i)$ (which is an element of S), $s_i =_{\mathrm{def}} (i, s_{[i]})$ (which is an element of $I \times S$). $(s_i)_{i \in I}$ denotes a "family" (or parameterized set), $\iota(I) = \{s_{[i]} \mid i \in I\} \subseteq S$ is the set of this family.

Let there be given a family $(S_i)_{i \in I}$ of non-empty sets $S_{[i]}$, named also "types". Then the "product set" of this family is defined by $\prod_{i \in I} S_i =_{\mathrm{def}} \{(s_i)_{i \in I} \mid \bigwedge i \in I\, (s_{[i]} \in S_{[i]})\}$. If all $S_{[i]}$ are identical with S, we write S^I (exponentiation of S by I). This product is unordered. However, for linearly ordered finite I, e.g. $I = \{1, 2, \dots n\}$, the product is isomorphic to the Cartesian $\times_{i=1,2,\dots,n} S_i$. Classically, any $R \subseteq \prod_{i \in I} S_i$ is a "relation". More general, we define $R \subseteq \bigcup_{U \in \mathrm{pow}\, I} \prod_{i \in U} S_i$ as relation. An element of R is then of the form $(s_i)_{i \in U}$, where $U = \varnothing$ defines the empty family ε. R is a non parameterized set. It can be parameterized by an indexing $\iota: J \xrightarrow{\text{onto}} R$. The result is a family of families, one hierarchical level higher than that of the component families, denoted by $((s_{ji})_{i \in I[j]})_{j \in J}$ with well determined $I[j] \subseteq I$. We define $\bigwedge i \in I\, (J[i] =_{\mathrm{def}} \{j \mid j \in J \wedge i \in I[j]\})$. Then $R^c = ((s_{ji})_{j \in J[i]})_{i \in I}$ defines the conjugate relation to R. Denoting $I_j =_{\mathrm{def}} \{j\} \times I[j]$ and $J_i^c =_{\mathrm{def}} J[i] \times \{i\}$, we have $M =_{\mathrm{def}} \bigcup_{j \in J} I_j = \bigcup_{i \in I} J_i^c \subseteq J \times I$ with $J = \bigcup_{i \in I} J_{[i]}$ and $I = \bigcup_{j \in J} I_{[j]}$. Both, R and R^c can be concatenated to $(s_{ji})_{ji \in M}$. Similar, families of families of any hierarchical level and with multiple indices can be defined:

For $n = 1,2,...$ let there be given an index set $I(n) \neq \emptyset$, a set $\{R_{[i]}^{(n-1)} \mid i \in I(n-1)\}$ of non parameterized, non-empty relations of level n-1. $R_{[i]}^{(0)}$ is a set of "primitives" (families of level 0). We define $R^{(n)}$ on level n over the primitives by

$\emptyset \neq R^{(n)} \subseteq \prod_{i \in I(n)} R_i^{(n-1)}$, and the general relation

$\mathsf{R}[n] =_{def} \bigcup_{\nu=1,2,...n} R^{(\nu)}$.

2.2 Functions on relations

Let there be given a universe U of general, non parameterized relations R with families of form $(s_i)_{i \in I}$, $s_{[i]} \in \mathsf{S}$, S a fixed set, $I \subseteq \mathsf{I}$, I a fixed set. For any set $R \in \mathsf{U}, R \neq \emptyset$, a function $f: R \mapsto S \in \mathsf{U}$ with S being a singleton (one element set) is named a "concatenation", denoted by $\boldsymbol{K}$ (prefix-) or κ (infix-notation). If $S = \{s^{(n+1)} = (s_i^{(n)})_{i \in I},\}$ and if $s^{(n+1)}$ is a family of families $s_{[i]}^{(n)}$ of level n, then any function $\delta: S \mapsto \{s_i^{(n)} \mid i \in I\}$ is level lowering . Applied to above $\mathsf{R}[n]$, we define:

$\tilde{R}^{(n)} =_{def} R^{(n)}$,

for $\nu = n$-1, ..., 1 and for concatenations $\boldsymbol{K}$:

$\tilde{R}^{(\nu)} =_{def} R^{(\nu)} \cup \delta \boldsymbol{K} \tilde{R}^{(\nu+1)}$,

for $\nu = n, ..., 1$: $\tilde{\mathsf{R}}[\nu] =_{def} \tilde{\mathsf{R}}[\nu\text{-}1] \cup \tilde{R}^{(\nu)}$,

$\tilde{\mathsf{R}}[0] =_{def} \emptyset$.

We consider some examples:

(1) For a given $(s_i)_{i \in I}$ and an index set $J \subseteq \mathsf{I}$, the "projection" of $(s_i)_{i \in I}$ on J, $pr[J]\,(s_i)_{i \in I}$, is defined by $pr[J]\,(s_i)_{i \in I} = (s_i)_{i \in I \cap J}$, the "co-projection" with respect to J is defined by $cpr[J]\,(s_i)_{i \in I} = (s_i)_{i \in I \setminus J}$. For $pr[\{j\}]$ we write also $pr[j]$.

(2) $R = \{(s_{ji})_{i \in I[j]}) \mid j \in J\} \mapsto F = \{(s_{ji})_{ji \in M}\}$ with $M =_{def} \bigcup_{j \in J} I_j$ (i.e. $M \subseteq J \times I$) is a concatenation. In turn, a "cut" of F along j yields $F[j] = \{(s_{ji})_{i \in I[j]}\} = pr[V]\,R$ with $V = \{j\} \times I[j] \subseteq M$. Disregarding j, $pr[V]\,R$ and $\{(s_{[j]i})_{i \in I[j]}\}$ are isomorphic.

(3) If $\bigwedge j,j' \in J\,(j \neq j' \Rightarrow I[j] \cap I[j'] = \emptyset)$ and if we set $I = \bigcup_{j \in J} I[j]$, $s_{[j]i} = s_i$ for $i \in I[j]$, then $\{(s_{ji})_{i \in I[j]} \mid j \in J\} \mapsto \{(s_i)_{i \in I}\}$ is a concatenation.

(4) We assume: $R \subseteq \bigcup_{U \subseteq I} \prod_{i \in U} S_i$, $R = \{(s_{[j]i})_{i \in I[j]} \mid j \in J\}$, J finite, for each $S_{[i]}$ and each $U, \emptyset \neq U \subseteq J$, a (commutative) function $\#_i: S_{[i]}^{U} \to S_{[i]}$ is given with $\#_i((s)) = (s)$ for U being a singleton. Then R can be concatenated to $\boldsymbol{K}R = \{\#_i((s_{[j]i})_{j \in J[i]}) \mid i \in I\}$. If J is linearly ordered, the product is Cartesian and $\#_i$ is in general not commutative. For example, take a non-empty set D and $S =_{def} \text{pow } D$ (the power set of D) with the set operations $\cap, \cup$. We consider the relation $R = \{(s_{ji})_{i \in I[j]} \mid j = 1, 2,..n\}$, with all $s_{[ji]} \in S$. The conjugate is $\{(s_{ji})_{j \in J[i]} \mid i \in I\}$ with $J[i] \subseteq \{1, 2,..n\}$, $I = \bigcup_{j=1,2,..n} I[j]$. R can be concatenated to $\{\#((s_{[ji]})_{j \in J[i]}) \mid i \in I\}$, where # represents $\cap$ or $\cup$.

We consider the parameterized relation $R = ((s_{ji})_{i \in I[j]})_{j \in J}$ and a partitioning of all $I[j]$ into two parts, i.e.: $I[j] = I[j]' \cup I[j]''$ with $I[j]' \cap I[j]'' = \emptyset$, with $I[j]' \neq \emptyset$. We set $x_j =_{def} (s_{ji})_{i \in I[j]'}$, $y_j =_{def} (s_{ji})_{i \in I[j]''}$, $X =_{def} \{x_{[j]} \mid j \in J\}$, $Y =_{def} \{y_{[j]} \mid j \in J\}$ and assume $\bigwedge j,j' \in J\,(y_{[j]} \neq y_{[j']} \Rightarrow x_{[j]} \neq x_{[j']})$. In this case we say the partition is functional and yields a function $f: X \xrightarrow{onto} Y$ with $x_{[j]} \underset{R}{\mapsto} y_{[j]}$ according R. In the following, $\underset{R}{\mapsto}$ is replaced by $\mapsto$ if the meaning is clear from context. We use the same notation f for the functional relation of all pairs $(x_{[j]}, y_{[j]})$, $f \subseteq X \times Y$.

From $x_{[j]} \underset{R}{\mapsto} y_{[j]}$ and $y_{[j]} = u_{[j]} \,\kappa\, v_{[j]}$, $u_{[j]}$ and $v_{[j]}$ being disjoint and not ε, follows $x_{[j]} \underset{\tilde{R}}{\mapsto} u_{[j]}$, $x_{[j]} \underset{\hat{R}}{\mapsto} v_{[j]}$ and $x_{[j]} \,\kappa\, u_{[j]} \underset{R}{\mapsto} v_{[j]}$. $\tilde{R}$ and $\hat{R}$ are restrictions of R. A partition $(x_{[j]}, y_{[j]})_{j \in J}$ is "coarser" than a partition $(x'_{[j]}, y'_{[j]})_{j \in J}$ if $x_{[j]} \subseteq x'_{[j]}$ for all $j \in J$. By this a partial ordering of the partitions of R is defined.

Given $x_{[j]} \mapsto y_{[j]}$ with respect to f, then the reciprocal image of $y_{[j]}$ is $\overset{-1}{f}(y_{[j]}) =_{def} \{x_{[j']} \mid j' \in J \wedge f(x_{[j']}) = y_{[j]}\}$, written $y_{[j]} \underset{f}{\overset{-1}{\mapsto}} \overset{-1}{f}(y_{[j]})$.

Let a set measure μ be given on $s_{[j]} = (s_{[j]i})_{i \in I[j]} = x_{[j]} \,\kappa\, y_{[j]}$. Then $\mu(x_{[j]}) \,/\, \mu(s_{[j]})$ can serve as a determinacy measure of $s_{[j]}$ being determined by $x_{[j]}$.

Example 1: $I = \{1, 2, 3, 4, 5, 6, 7\}$, $J = \{1, 2, 3, 4\}$, F is shown in Table 1. $M = \{(1, 1), (2, 2), (2, 3),....(3, 7)\}$. $R = \{s_1, s_2, s_3, s_4\}$, $R^c = \{s_1^c, ...s_7^c\}$. We assume $s_{[11]} = s_{[41]} =$ a, $s_{[22]} = s_{[32]} = s_{[42]} =$ b, $s_{[23]} = s_{[33]} =$ c, $s_{[34]} =$ d, $s_{[44]} =$ d′, $s_{[35]} =$ e, $s_{[26]} =$ f, $s_{[46]} =$ f′, $s_{[37]} =$ g. $S_{[1]} = \{$a$\}$, $S_{[2]} = \{$b$\}$, $S_{[3]} = \{$c$\}$, $S_{[4]} = \{$d, d′$\}$, $S_{[5]} = \{$e$\}$, $S_{[6]} = \{$f, f′$\}$, $S_{[7]} = \{$g$\}$. For two partitions marked by "]" and "}", $x_{[j]}$ is right, $y_{[j]}$ is left of] or }, respectively, for instance $x_{[1]} = (a_{[1]1}) \mapsto y_{[1]} = ε$, $x_{[3]} = (d_{[3]4}, e_{[3]5}, g_{[3]7}) \mapsto y_{[3]} = (b_{[3]2}, c_{[3]3})$ for partition "}". A determinacy measure of $s_{[3]}$ determined by $x_{[3]}$ is card $x_{[3]}$ / card $s_{[3]}$ = 3/5. Partition "}" is coarser than partition "]". $(s_{[2]2}, s_{[2]3}) \underset{f}{\overset{-1}{\mapsto}} \{(s_{[2]6}), (s_{[3]4}, s_{[3]5}, s_{[3]7})\}$. According R^c, we have for example $(s_{1[1]}) \underset{R^c}{\mapsto} (s_{4[1]})$, $(s_{2[2]}) \underset{R^c}{\mapsto} (s_{3[2]}, s_{4[2]}), \ldots (s_{3[7]}) \underset{R^c}{\mapsto} \varepsilon$.

Table 1

J	s_1^c	s_2^c	s_3^c	s_4^c	s_5^c	s_6^c	s_7^c	
4	s_{41}	s_{42}		s_{44}	]}	s_{46}		s_4
3		s_{32}	s_{33}	]}s_{34}	s_{35}		s_{37}	s_3
2		s_{22}	]s_{23}			}s_{26}		s_2
1	]}s_1 1							s_1
	1	2	3	4	5	6	7	I
	a_1	b_2	c_3	d'_4	e_5	f_6	g_7	$\boldsymbol{K_\#}R$

Let there be given $\#_1$ with (a, a) $\mapsto$ (a), $\#_2$ with (b, b) $\mapsto$ (b), $\#_3$ with (c, c) $\mapsto$ (c), $\#_4$ with (d, d′) $\mapsto$ (d′), $\#_5$, $\#_6$ with (f, f′) $\mapsto$ (f),· $\#_7$. Then $\boldsymbol{K_\#}R = \{(a_1, b_2, c_3, d'_4, e_5, f_6, g_7)\}$.

2.3 Valuation of relations

In many applications the families $s^{(n+1)} = (s_i^{(n)})_{i\in I}$ of a hierarchical relation are "valuated", i.e. on level n a value $v_{[i]}^{(n)}$ taken of a domain $V^{(n)}$ is associated with $s_i^{(n)}$, written $(s_i^{(n)}, v_i^{(n)})$, and analogously $(s^{(n+1)}, v^{(n+1)})$. The valuations may be level dependent, but are i. g. independent of each other and usually assumed not to depend on the objects valuated. A particular case is the existence of a functional dependence $v^{(n+1)} = \varphi((v_i^{(n)})_{i\in I})$.

We explain this in detail: For levels $n = 0,1,2,..$ let there be given an index set $I^{(n+1)} \neq \varnothing$, for all $i \in I^{(n+1)}$ a set of relations $S_{[i]}^{(n)} \neq \varnothing$, and a set of valuations $V_{[i]}^{(n)}$. Any chosen $S^{(n+1)}$, $\varnothing \neq S^{(n+1)} \subseteq \prod_{i\in I^{(n+1)}} (S_i^{(n)} \times V_i^{(n)})$ is a relation on level n+1. We consider $S^{(n+1)}$ and assume, this relation is bijectively parameterized by the elements of a set J. Any pair (j,i) determines (s_{ji}) and v_{ji} uniquely, thus for given parameterizations, $(s_{ji}, v_{ji})_{i\in I[j]}$ can be represented by $(v_{ji})_{i\in I[j]}$.

Reducing the generality, we assume for all n and i that on level n all valuation types $V_{(i)}^{(n)}$ are equal to a type $V^{(n)}$. In many applications, the $V^{(n)}$ are complete lattices $(V^{(n)}, \leq_{(n)})$ with additional structures which we will study in the following:

For a complete lattice $(V, \leq)$ and an index set I, we consider $\tilde{V} =_{\text{def}} \bigcup_{\varnothing\subset I\subset \mathrm{I}} V^I$ with elements $v = (v_i)_{i\in I}$, $v_{[i]} \in V$. We want a reflexive and transitive extension "$\prec$" of "$\leq$" on V to $\tilde{V}$, i.e. $\prec\, = \,\leq$ on V. We define for $v = (v_i)_{i\in I}$, $v' = (v'_j)_{j\in J} \in \tilde{V} : v \prec v' \Leftrightarrow \bigwedge i \in I\, (\bigvee j \in J\, (v_{[i]} \leq v'_{[j]}))$. $\prec$ is in general no order relation, from $v \prec v'$ and $v' \prec v$ need not follow $v = v'$. A particular case is for example: if a bijection $\beta: I \rightarrow J$ exists and $\bigwedge i \in I\, (v_{[i]} \leq v_{[\beta(i)]})$, then $v \prec v'$. Given another complete lattice $(W, \leq)$, $W \cap V \neq \varnothing$ is admitted, then $\prec$ - homomorphisms $\lambda: \tilde{V} \rightarrow W$ (named "logic functions" in ref. 1) are considered. We have: $v \prec v' \Rightarrow \lambda v \leq \lambda v'$, $\bigsqcup v \leq \bigsqcup v' \Rightarrow \lambda\bigsqcup v \leq \lambda\bigsqcup v'$, $v \prec v'$ and $v' \prec v \Rightarrow \lambda v = \lambda v'$, $\bigsqcap v \prec v \prec v' \prec \bigsqcup v' \Rightarrow \lambda\bigsqcap v \leq \lambda v \leq \lambda v' \leq \lambda\bigsqcup v'$ and for $v = v'$: $\lambda\bigsqcap v \leq \lambda v \leq \lambda\bigsqcup v$. In case all components of v are equal to the element e, then $\lambda e = \lambda v$. Defining $v \prec' v' \Leftrightarrow \bigwedge i \in I\, (\bigvee j \in J\, (v_{[i]} \geq v'_{[j]})$, it follows $\bigsqcap v' \leq \bigsqcap v$. So far, for $w = \lambda((v_i)_{i\in I})$, λ depends on I. If only finite sets are admitted, λ may depend on card I only, i.e. $\lambda((v_i)_{i\in I}) = \lambda((v_{\beta(i)})_{i\in I})$ for a bijection ß: $I \rightarrow J$ with $v_{[i]} = v_{\beta([i])}$.

Example 2: $V = W = \overline{\mathbf{R}}_+$ (the non-negative real numbers completed with ➚), $\mathrm{I} = \mathbf{N}_0$ (the natural numbers with 0), I finite, card $I = n$. Using suitable relations $\prec$, examples for λ are: $\lambda = \max$, $(\Sigma(\ldots))/n$, $(*(\ldots))^{1/n}$. Because of $((\min v)_i)_{i\in I} \prec v \prec ((\max v)_i)_{i\in I}$ it follows: $n \min(v) \leq \Sigma(v) \leq n \max(v)$, $(\min(v))^n \leq *(v) \leq (\max(v))^n$. Using $\prec'$, an example is $\lambda = \min$.

Applied to the above valuated hierarchical relations $S^{(n)}$, the $V^{(n)}$ can be complete lattices with given logic functions $\lambda^{(n)}: (V^{(n)})^{I^{(n+1)}} \rightarrow V^{(n+1)}$.

Example 3 (propositional calculus): Let be $\mathrm{I} = \mathbf{N}_0$, for all levels, $V^{(n)} = (\{f, t\}, f < t)$ (type Boolean), $S^{(0)} = \{a, b, c,..\}$ a set of propositions. λ-functions are $\wedge$, $\vee$ (homomorphisms), $\neg\wedge$, $\neg\vee$ (antimorphisms). For example

level 0: $\{a, b, c, \ldots\}$,

level 1: $\{(a, t), (b, f), (c, t)\}$, card $I^{(1)} = 3$,

level 2: $\{(((a, t), (b, f)), \wedge(t, f) = f), (((a, t), (c, t)), \vee(t, t) = t)\}$, card $I^{(2)} = 2$. Indices are suppressed.

3 Variables and their control

We consider a non-empty family $(S_p)_{p\in P}$ of non-empty structures (general relations) S_p. For all p let be $S_{[p]} = \boldsymbol{K}_{[p]}(C, V_{[p]})$, $\boldsymbol{K}_{[p]}$ a concatenation of a p-independent component C and a p-dependent component $V_{[p]}$. C may be empty. To facilitate the representation of $\mathsf{S} = \{S_{[p]} \mid p\in P\}$ we define a variable var V on variability domain $\mathsf{V} = \{V_{[p]} \mid p \in P\}$ with respect to S, written var V : (V, S), and a variable S(var V) = $\boldsymbol{K}$(C, var V) on S. $\boldsymbol{K}$ depends on var V. We make the variables "controllable" by associating to S(var V) a function val with val: $P \times$ {var V} $\times$ {S(var V)} $\to \mathsf{V} \times \mathsf{S}$, with val: ($p$, var V, S(var V)) $\mapsto (V_{[p]}, S_{[p]})$, and to var V a function val: $P \times$ {var V} $\to \mathsf{V}$ with val: (p, var V) $\mapsto V_{[p]}$. The val-functions are named "control-" or "assignment" functions, P is the set of control-/assignment parameters. For assignment according parameter p we write also val(S(var V)): $p \mapsto S_{[p]}$ and val(var V): $p \mapsto V_{[p]}$, or S(var V) :=(p) $S(V_{[p]})$, var V :=(p) $V_{[p]}$. $S(V_{[p]})$ and $V_{[p]}$ are "instantiations" of S(var V) and var V, respectively. Consecutive assignments by parameters p, q result in a substitution $(V_{[q]}, S_{[q]})$ for $(V_{[p]}, S_{[p]})$.

A structure can contain variables (var $w_j)_{j\in J}$ with var w_j : $W_j = \{w_{j[q]} \mid q \in Q_j\}$. The composite variable var w = (var $w_j)_{j\in J}$ is in general defined on $W \subseteq \prod_{j\in J} W_j$. For $W \subset \prod_{j\in J} W_j$ assignments to the var w_j are in general dependent on each other, in this case admissible assignment parameters $q = (q_j)_{j\in J}$ are from a set $Q \subset \prod_{j\in J} Q_j$. This has to be taken into account in case of partial assignments to var w, resulting in the partial variable $(u_j)_{j\in J}$ with $u_j = w_{j[q(j)]}$ for $j \in J_a$, u_j = var w_j for $j \in J \setminus J_a$.

The parameters $(q_j)_{j\in J}$ can themselves be the result of a computation $\varphi_{[r]} : (p_i)_{i\in I} \to (q_j)_{j\in J}$, and $\varphi_{[r]}$ can be an instantiation of a variable var φ with control function val(var φ). Continuing this way we have a hierarchy of control functions.

More than one variable can be defined on the same domain V and elements of V can be subject to certain relations $\mathsf{S}_{[q]}$, $q \in Q$. (V, $(\mathsf{S}_{[q]})_{q\in Q}$) is then also named "type" of the variables. For example, $\mathsf{S}_{[q]}$ can be a functional relation.

For any hierarchical level n = 1,2,... we admit variables on domains with variables of level n-1. On level 0 are constants only. Assignments are applied from higher to lower level and reduce the level.

Example 4: Revisiting *Example 1*. The relation can be described by (var x, var f, var y := var f(var x). var f : $\{f_{[p]} \mid p \in P = \{"]", ")"\}$ corresponding to the two partitions}, var x : $\{x_{[pq]} \mid p \in P \wedge q \in Q_p = Q = J\}$.
var $x_{["]"]}$: $\{x_{["]", j} \mid j \in Q\}$, var $y_{["]"]}$: $\{y_{["]", j} \mid j \in Q\}$,
$x_{["]", 1]} = (s_{11})$, $y_{["]", 1]} = \varepsilon$,
$x_{["]", 2]} = (s_{23}, s_{26})$, $y_{["]", 2]} = (s_{22})$,
$x_{["]", 3]} = (s_{34}, s_{35}, s_{37})$, $y_{["]", 3]} = (s_{32}, s_{33})$,
$x_{["]", 3]} = (s_{46})$, $y_{["]", 3]} = (s_{41}, s_{42}, s_{44})$,
similar for the $x_{[")", j]}$.

Introducing variables for the s_{ji} and for the parameters, we have the composite variable var $x_{[\text{var } p, \text{var } j]}$ = (var $s_{[\text{var } p, \text{var } j]i})_{i\in \text{var } I}$ with (var p, var j) : $P \times J$, var $s_{[\text{var } p, \text{var } j]i}$: S_i independent of (var p, var j), var I : $\{I[j] \mid j \in J\} \subset$ pow I, depending on (var p, var j).

4 Topologized relations

To express neighborhood / similarity relations on the families of a relation $R \subseteq \bigcup_{U\in \text{pow } I} \prod_{i\in U} S_i$, we introduce topological structures by means of filter bases: Given a set S, then (pow S, $\subseteq$, $\cup$, $\cap$) is a complete Boolean lattice. Let $B_{[0]}$, $\emptyset \neq B_{[0]} \subseteq S$, be a set of points which by assumption are all equivalent with respect to the admitted accuracy. Further, let $\mathbf{B} = \{B_{[k]} \mid B_{[k]} \subseteq S$ and $k \in K\}$ be a set of upper sets of $B_{[0]}$ with the properties: from $B_{[k']} \in \mathbf{B}$ and $B_{[k'']} \in \mathbf{B}$ follows, it exists a $B_{[k]} \in \mathbf{B}$ with $B_{[k]} \subseteq B_{[k']}$ and $B_{[k]} \subseteq B_{[k'']}$. Then $\mathbf{B}$ is a "filter base". We assume: $B_{[0]} = \bigcap \mathbf{B}$ (denoted lim $\mathbf{B}$), $S \supseteq \bigcup \mathbf{B}$ (support of $\mathbf{B}$). We further assume, $\mathbf{B}$ is a complete lattice. Then for $s \in S$, $s^* \in$ lim $\mathbf{B}$, "generalized distances" $d_\cap(s^*, s) =_{\text{def}} \bigcap\{B \mid B\in\mathbf{B} \text{ and } s\in B\} \in \mathbf{B}$ and $d_\cup(s^*, s) =_{\text{def}} \bigcup\{B \mid B\in\mathbf{B} \text{ and } s\notin B\} \in \mathbf{B}$ can be defined. Let there be given a complete lattice $(V, \leq, \wedge, \vee)$. and a $\leq$-homomorphism $\phi: \mathbf{B} \to V$, defining a valuation of $\mathbf{B}$. We use $v(s^*, s) =_{\text{def}} \phi(d_\cap(s^*, s))$ as a distance measure of s from s^*. "Fuzzy sets" in the engineering literature are examples of filter bases.

Let $\{\mathbf{B}_{[l]} = \{B_{[lk]} \mid k \in K\} \mid l \in L\}$ be a set of isomorphic filter bases, and for all k, let $B_{[lk]} \in \mathbf{B}_{[l]}$ be uniformly with respect to l valuated by a $\subseteq$-homomorphism ϕ into V, i.e. $\phi(B_{[lk]}) = \phi(B_{[l'k]})$. We further assume, for $l \neq l'$ holds lim $\mathbf{B}_{[l]} \cap$ lim $\mathbf{B}_{[l']} = \emptyset$. For more details we refer to ref. 1, 2, 3, 4.

We apply these topological concepts to the relation R: Let $R^* \subseteq R$ be a non-empty subset. We set $S = R$, $L = R^*$. To each $r^* \in R^*$ let be adjoined a neighborhood system in form of a filter base $\mathbf{B}_{[r^*]}$ as defined above, with $r^* \in$ lim $\mathbf{B}_{[r^*]}$, and with all sets lim $\mathbf{B}_{[r^*]}$ being disjoint. A trivial case is, $R = R^*$, lim $\mathbf{B}_{[r^*]} = \{r^*\}$.

For fixed r^*, $v(r^*, r) =_{\text{def}} \phi(d_\cap(r^*, r))$ defines an abstract distance of any $r \in \bigcup \mathbf{B}_{[r^*]}$ from any $r^* \in \lim \mathbf{B}_{[r^*]}$. Hence, to any $r^* \in \lim \mathbf{B}_{[r^*]}$ exists a neighborhood $\mathbf{N}(r^*)$ with elements $u \in R$ "nearer" to r^* than to any other comparable $\hat{r} \in R^* \setminus \lim \mathbf{B}_{[r^*]}$. If the relation R is hierarchical, then the topological reasoning can be applied to each hierarchical level.

Let us consider a family $(r_i, v_i)_{i\in N}$, N being a non-empty segment of integers, with $r_{[i]} \in R$, $v_{[i]}$ the abstract distance of $r_{[i]}$ to a given $r_{[i]}^* \in R^*$. Then, for a logic function λ, $\lambda((v_i)_{i\in N})$ can be defined as distance of $\mathbf{K}((r_i)_{i\in N})$ from $\mathbf{K}((r_i^*)_{i\in N})$, $\mathbf{K}$ denoting a concatenation. Examples are normed vector spaces with λ being a norm: Let $\mathbf{R}$ be the field of real numbers and let $\mathbf{R}^n$ be an n-dimensional normed vector space. For $r_{[i]}^*, r_{[i]} \in \mathbf{R}$ a distance is $\lambda(r_i^*, r_i) = |r_{[i]}^* - r_{[i]}|$, for $\mathbf{K}(((x_i))_{i=1,\dots n}) = (x_i)_{i=1,\dots n}$, $x_i \in \{r_i^*, r_i\}$, a distance from $r^* = (r^*_i)_{i=1,\dots n}$ to $r = (r_i)_{i=1,\dots n}$ is $\lambda(r^*, r) = \|r^* - r\|$. Well known examples for norms are $\max\{|r_{[i]}^* - r_{[i]}| \mid i = 1,\dots n\}$, $(\sum_{i=1}^{n}(r_{[i]}^* - r_{[i]})^2)^{1/2}$.

5 Applications

On a universe $\mathbf{R}$ of parameterized relations (structures) let there be given:

(1) A set of *structural* operations, e.g. definition/deletion of a particular object structure or index structure, composition and decomposition of relations, especially concatenations, parsings, cuts, projections, definition of variables, their domains and assignments, definition of topological structures.

(2) A set of *functional* operations on objects of relations, e.g. equality, being part of, search for and quantified selection by properties, comparison of objects, enumeration, sorting, deduction rules to derive implicitely given structures.

(3) A "*query language*" based on the structural and functional operations, to provide "answers" to formalized "questions".

Structural, functional operations and queries are subject to rules and constraints (grammars), which we are not going to formalize here. We name a relation $R \in \mathbf{R}$ a "knowledge module".

Example 5: We revisit *Example 1* with $s_{j.} =_{\text{def}} (s_{ji})_{i\in I[j]}$, $s_{.i} =_{\text{def}} (s_{ji})_{j\in J[i]}$, and with all $s_{[ji]}$ equal to the above defined constants. Examples of queries are given in Table 2:

Table 2

given:	find (one, all):	answer:
$U = \{(3, 1), (2, 3), (4, 6)\}$	$pr[U]R$	$\{c_{23}, f'_{46}\}$
$x = (a_1)$	j with $x \subseteq s_{j.}$	$j = 1, 4$
$x = (a_1)$	$s_{j.}$ with $x = s_{j.}$	s_I
$x = (a_1)$	$s_{j.}$ with $x \subseteq s_j$	s_I, s_4
b, d	j with b∨c or b∧c ∈ set of s_j	$j = 2, 3, 4$ $j = 2, 3$
$I^* = \{2, 4\}$	j with $I[j] = I^*$	empty
$i = 4$	cut of R along i	$s_i^c = (d_{34}, d'_{44})$
R	fctl. decompos. and dependencies	see *Example 4*
any x, y	$x \kappa y \in R$?	by trials of concatenations
any z	$x, y \in R$ with $z = x \kappa y$?	by trials of parsings

Examples of queries for knowledge modules with topologies are given in Table 3:

Table 3

given:	find (one, all):	answer:
$r^* \in R^* \subseteq R$, $\mathbf{B}_{[r^*]}$ $x \in R$	$d_\cap(r^*, x)$, $d_\cup(r^*, x)$	see *Example 6*
$x \in R$, all $\mathbf{B}_{[r^*]}$	nearest $r^* \in R^*$ to x	comparison of the $v(r^*, x)$ see *Example 6*
var $\mathbf{B}$ on R $S \subseteq R$, all elements of S valuated	assignment to var $\mathbf{B}$, fitting the valuations on S	interpolation of the valuations on S, see *Example 7*

Example 6 (use of topologies): Considering the real numbers $\mathbf{R}$, let be $r^* \in \mathbf{R}$, $\mathbf{B}_{[r^*]} =_{\text{def}} \{[r^* - 0.5 * v, r^* + 0.5 * v] \mid v \in V\}$ with $V = \mathbf{N}_0$ or $\mathbf{N}$. $\mathbf{B}(r^*)$ is a monotone filter base on pow $\mathbf{R}$ with $\lim \mathbf{B}_{[r^*]} = \{r^*\}$ or $\{[r^* - 0.5, r^* + 0.5]\}$, respectively. ϕ: $[r^* - 0.5 * v, r^* + 0.5 * v] \mapsto v$ is a valuation. E.g. for $V = \mathbf{N}$, $r^* = 3$, $r = 4$, $v(r^*, r) = \phi(d_\cap(r^*, r)) = 2$, $\phi(d_\cup(r^*, r)) = 1$. For $R^* =_{\text{def}} \{r^* = (3_1, 4_2, 5_3), r^{**} = (1_1, 2_2, 5_3)\} \subset \mathbf{R}^3$, $\lambda = \max$, $V = \mathbf{N}$, $x = (3.2, 3.9, 4.5) \in \mathbf{R}^3$, $\lambda(r^*, x) = 1$, $\lambda(r^{**}, x) = 4$.

Example 7 (adaptation and fitting): Let var $\mathbf{B}$ on $R \subseteq \mathbf{R}^3$ to be adapted to a sample $(s_n = (s_{nk})_{k=1,2,3})_{n=1,2,\dots N}$ of size N, the components $s_{[nk]}$ of s_n, $s_{[nk]} \in S$, $S \subseteq R$, assumed to be independently valuated by the relative frequencies $v_{[nk]}$ of their occurrences in the sample.

We choose an (approximate) interpolation of the frequency function on a closed domain C, $S \subseteq C \subseteq R$, which is then linearly scaled to a function φ with $\varphi(C) \subseteq [0, 1]$ and with $1 \in \varphi(C)$, i.e. φ can be used as generating ("membership"-) function to define a filter base (neighborhood system) $\mathbf{B} = \{B_\alpha = \{ \overset{-1}{\varphi}([\alpha, 1]) \mid 0 \le \alpha \le 1\}\}$ with $\lim \mathbf{B} = \overset{-1}{\varphi}(\{1\})$. The set function $\overset{-1}{\varphi}$ is the reciprocal to the point function φ. $\varphi(C)$ is the set of valuations $v(r)$ of elements r of C.

In the preceding example we used a statistical property for grouping elements to neighborhoods. As well other properties or explicitly described grouping could be used.

An interpretation of our general approach is: I is a set of "properties", J is a set of "objects", object j has all properties of $I[j]$, property $i \in I[j]$ is valuated by $v_{[ji]} \in V_{[j]}$, often all $V_{[j]}$ are identical and equal V. In particular, $V = \{1\}$, $V = \{0, 1\}$.

Our approach is a generalization of the so-called "relational data base" concept, obtained for $R \subseteq \prod_{i \in I} S_i$, finite I representing the set of "attributes", $S_{[i]}$ the set of attribute values to attribute i. Neighborhood sytems are not considered.

Frequently, for a given $\mathbf{B}$, the elements of $\bigcup \mathbf{B}$ are arguments to a function $f\colon \bigcup \mathbf{B} \to Z$. If $r \in \bigcup \mathbf{B}$, $f(r) = z$, and if r is valuated by $v(r)$, then $f(r)$ can be valuated by $v(r)$. If more than one r is mapped onto the same z, then, if existing, lub $v(\overset{-1}{f}(z))$ (lub means least upper bound) can serve as a valuation of z.

The generation and adaptation of neighborhood systems (filter bases) fitting certain input values in a "learning"/"training" phase is a basic feature of all adaptive classifying (pattern recognizing) devices.

To compare two families v^*, v of valuations with finite and distinct index sets I^*, I, $J =_{\text{def}} I^* \cap I \neq \emptyset$, and a distance $\mathrm{d} = \mathrm{d}(pr(J)v^*, pr(J)v) \in \overline{\mathbf{R}}_{\mathbf{1}}$, i.e. $1 \le \mathrm{d} \le \infty$, we can for example use a measure $\delta =_{\text{def}} \text{card}(I^* \cup I) / \text{card}(I^* \cap I)$, $1 \le \delta < \infty$, to define $\mathrm{d}\delta$ as distance of (v^*, v).

To incorporate the knowledge about neighborhood systems, in the definition $R \subseteq \bigcup_{U \in \text{pow}\, I} \prod_{i \in U} S_i$ the set of factors sets $S_{[i]}$ can be extended by a factor set $S^* = \{\mathbf{B}_{[r^*]} \mid r^* \in R^*\}$, the $\mathbf{B}_{[r^*]}$ represented explicitly or implicitly by generating functions.

Introducing a linearly or partially ordered model time $(T, \le)$, processes of relations R can be considered. Examples are $(s_{\iota ji})_{(\iota ji) \in M \subseteq T \times J \times I}$ or $(((s_{\iota ji})_{i \in I[\iota j]})_{j \in J[\iota]})_{\iota \in T}$. This way the definition of continuously adapting knowledge modules is possible.

References

[1] Albrecht R. F. (1998) "On mathematical systems theory". In: R. Albrecht (ed.): Systems: Theory and Practice, Springer, Vienna-New York, pp. 33-86

[2] Albrecht R. F (1999) "Topological Approach to Fuzzy Sets and Fuzzy Logic". In: A. Dobnikar, N. C. Steele, D. W. Pearson, R. F. Albrecht (eds.): Artificial Neural Nets and Genetic Algorithms, Springer, Vienna-New York, pp. 1-7

[3] Albrecht R. F. (1999) "Topological Theory of Fuzziness". In: B. Reusch (ed.): Computational Intelligence, Theory and Applications, LNCS vol. 1625, Springer, Heidelberg-New York, pp. 1-11

[4] Albrecht R. F. (2001) "Topological Concepts for Hierarchies of Variables, Types and Controls". In: G. Alefeld, J. Rohn, S. Rump, F. Yamamoto (eds.): Symbolic Algebraic Methods and Verification Methods, Springer Vienna-New York, pp. 3-10

[5] Albrecht R. F., Németh G. (1998) "A Generic Model for Knowledge Bases", Periodica Polytechnica, Electrical Engineering, TU Budapest, vol. 42, pp. 147-154

Handling categorical data in rule induction

Martin Burgess,[1] Gareth J. Janacek,[2] Vic J. Rayward-Smith[3]

[1 2 3]School of Information Systems, University of East Anglia, Norwich, UK

Abstract

In this paper we address problems arising from the use of categorical valued data in rule induction. By naively using categorical values in rule induction, we risk reducing the chances of finding a good rule in terms both of confidence (accuracy) and of support or coverage. In this paper we introduce a technique called *arcsin transformation* where categorical valued data is replaced with numeric values. Our results show that on relatively large databases, containing many unordered categorical attributes, larger databases incorporating both unordered and numeric data, and especially those databases that are small containing rare cases, this technique is highly effective when dealing with categorical valued data.

1 Introduction

We have a database, D, of records which for simplicity we assume is in the form of a flat file. Each record comprises m attribute values, one of which is a designated output, or target attribute. Attributes can either be numerical or categorical. Whereas numerical data has an ordering, categorical data is often unordered.

Rule induction seeks to find a rule of the form $\alpha \Rightarrow \beta$, where the postcondition, β, is a constraint on the output attribute and the precondition, α, is an expression involving the other attributes. For example,

$$\text{Salary} \leq 30,000 \wedge \text{Mortgage} > 100,000 \Rightarrow \text{Status} = \text{debtor} \quad (1)$$

is a rule where the precondition is a conjunction of constraints on two numeric attributes (Salary and Mortgage) and the postcondition is a constraint on a categorical attribute (Status). Rule induction algorithms aim to find rules which appear to hold in the database. The quality of the rule is usually defined in terms of confidence (i.e. the proportion of records satisfying the precondition for which the postcondition holds) and support (i.e. the proportion of the database for which both the precondition and the post condition hold). The term 'coverage' is sometimes used as a synonym for support; however will use it here to describe the proportion of the records for which both the precondition and postcondition hold. More formally, if

$$A = \{r|\alpha(r)\} \quad (2)$$

denotes the set of records in the database for which the precondition holds and similarly

$$B = \{r|\beta(r)\}, \qquad C = \{r|\alpha(r) \wedge \beta(r)\} \quad (3)$$

then

$$\text{Confidence (or accuracy)} = \frac{|C|}{|A|}, \quad (4)$$

$$\text{Support} = \frac{|C|}{|D|} \quad (5)$$

$$\text{and Coverage} = \frac{|C|}{|B|}. \quad (6)$$

where D is the database. Note that we assume $|A|$, $|D|$, $|B| \neq 0$.

In this paper, we consider the case where the precondition is a conjunction of simple tests on various non-output attributes and the postcondition is a predefined test on the output attribute. The software 'Data Lamp' [1] was utilised to generate the output discussed in this paper. It uses simulated annealing to find an optimal rule of this form with a specified β value which maximises the quality measure

$$\lambda|C| - |A|. \quad (7)$$

By adjusting the real value, λ, the rule can be more accurate (and have less coverage) or be less accurate (and have greater coverage). Note that

$$\lambda|C| - |A| \geq 1 \text{ iff confidence} = \frac{|C|}{|A|} \geq \frac{1}{\lambda}$$

assuming $\lambda > 0$.

Thus, if we find a rule with a fitness level $\geq$1 using λ=1.11 (say), this means we will have found a rule with a confidence of at least 90%.

If an attribute, A, is numerical then a simple test is of the form $A > x$ or $A \leq x$, where x is a value in the associated domain. Such tests can also be used on ordered categorical data. However, if the attribute is categorical and unordered then the only tests we can have are $A = x$

or $A \neq x$. If, moreover, such an attribute, A, has several values of which one is x, then $A \neq x$ is unlikely to result in an appropriately accurate rule. Conversely, $A = x$, is a very limiting constraint and any rule containing such a test is likely to have low levels of support. Thus, in either case, inclusion of constraints involving such categorical values may generate poor quality rules. Data Lamp does not allow conjunction of multiple intervals of the same attribute. Thus, if a set of categorical values is ordered $x_1, x_2, ..., x_k$, then it cannot achieve the test $A = x_i$, by $A > x_{i-1}$ and $A \leq x_i$.

2 Ordering of categorical data

The problem with categorical data is that, unlike numerical data, categorical values in most cases, do not have an inherent order. On some occasions a sensible ordering can be inferred. For example, consider the Adult database [2], the field *Education*, though categorical valued, has an inherent order to it. For example, a child will start at pre-school then progress through his/her educational learning, following an order to his/her learning in terms of schooling. Such an order would be termed *natural*. On the other hand, if we talk about colours-red, yellow, blue etc., there is no such ordering.

When looking at large data sets, we wish to extract rules which can then be used for predictive purposes. These rules are conjuncts of tests on the attributes which help to predict the target class. When looking for rules in a large data set, unless we impose some ordering on unordered categorical data, we are very much limiting our choice of finding the best rule in terms of confidence and support.

Replacing unordered categorical data by ordered data is generally not effective unless the number of distinct values is relatively large. In fact, if the number is less than four, it is never necessary. Say an attribute, A, has values $\{x_1, x_2, x_3\}$, then

$A > x_1$ is equivalent to $A \neq x_1$

$A \leq x_1$ is equivalent to $A = x_1$

$A > x_2$ is equivalent to $A = x_3$

$A \leq x_2$ is equivalent to $A \neq x_3$

$A > x_3$ is invalid

$A \leq x_3$ is no constraint

Thus in this case the ordering has been detrimental since it has removed the option of tests of the form $A = x_2$ and $A \neq x_2$ unless we allow multiple tests on the attribute A.

2.1 Ordering categorical values

In this paper we consider two ways in which categorical field values may be ordered. We may wish to order according to $\hat{p}$ values or with u values (both are described later).

We claim that, ordering of unordered categorical data is better than no ordering at all. We illustrate this claim by experiments performed on three databases from the UCI repository [2], that is, *Mushroom*, *Auto Imports*, and *Adult*. With reference to section 3.1, with the *Mushroom* experiment, it is shown that with unordered categorical fields containing four or more values, ordering is advisable as shown by the results achieved. Where the database is small (see section 3.3–*Auto Imports* database) incorporating rare cases, not inferring some kind of ordering is unadvisable and can seriously hamper one's chances of discovering quality rules.

2.2 Proportion or p values

The technique of using $\hat{p}$ values is based on the proportion of records within the target class for a given categorical value. This technique generates $\hat{p}$ values without making any distributional assumptions about the populations. The justification of such an approach relies on asymptotic theory that is valid only if the sample sizes are reasonably large and well balanced across the populations. For small, sparse, skewed or heavily tied data, the asymptotic theory may not be valid. For some empirical results, and a more theoretical discussion see ref. [3], [4], [5]. The alternative approach based on u values addresses some of these concerns.

Let A be an unordered categorical attribute with k possible values, $case_1$, $case_2$, ..., $case_k$. Assume the output condition, β, is set. We first consider p_i=$P(\beta|case_i)$, where P is the probability. We define $\hat{p}_i = y_i/n_i$ where y_i is the number of cases where both A=$case_i$ and β hold, and n_i is the number of cases where A=$case_i$. We regard $\hat{p}_i$ as an estimate of the true value of p_i, where y_i is the outcome of a Binomial random variable with mean $n_i p_i$, and variance $n_i p_i(1-p_i)$. The difficulty is that, as we only see $\hat{p}_i$, we have some uncertainty about the actual value and we need to take this into account. In order to stabilise that variance we propose that we consider the transformed estimate

$$Z_i = \arcsin\sqrt{\hat{p}_i}. \tag{8}$$

It is then possible to show that Z_i is asymptotically Normal with mean (approximately) $\arcsin\sqrt{p_i}$ and variance $1/4n_i$. Thus, for large n, $\arcsin\sqrt{\hat{p}_i}$ has an approximate Normal distribution of mean $\arcsin\sqrt{p_i}$ and variance of $1/4n$, that is, a $\mathcal{N}(\arcsin\sqrt{p_i}, 1/4n)$ distribution if arcsin is measured in radians.

Using a result from ref. [6], we define

$$u_i = \arcsin(\sqrt{\hat{p}_i})\sqrt{4n_i} \qquad (9)$$

which is standard normal. Dividing through by the standard deviation is to allow for scaling onto a standard scale. Whereas, with the $\hat{p}_i$ values, we have some uncertainty concerning the actual values, the u values have standardised this uncertainty. In [7], it is suggested that the following angular modification is used in order to stabilise the variance for values of $n_i \leq 50$. We use this an alternative to u_i when n_i is small.

$$u_i' = \frac{\arcsin\left[\sqrt{\frac{n}{n+1}\hat{p}_i} + \sqrt{\frac{n}{n+1}\hat{p}_i + \frac{1}{n+1}}\right]}{2\sqrt{(1/n + 1/2)}} \qquad (10)$$

As before, we divide by standard deviation to allow for scaling onto a standard scale.

3 Multiple λ value experiment

3.1 Mushroom database

Our initial experiment used the Mushroom database [2]. The experiment was performed to illustrate the benefits of ordering categorical data and to investigate the differences between orderings achieved by $\hat{p}$ and u values.

Of the 8124 instances available, records containing unknown values were removed leaving 5644 records. The database contains 22 categorical fields with no continuous fields. The data was then randomly split into two parts, a training and a test set, with both sets containing 2822 records. The training set contains 1718 records of class *edible*, and 1104 records of class *poisonous*. The test set contains 1770 records of class edible, and 1052 of class poisonous.

Each field in our training data set is proportioned according to the target value.

Using the target class of *edible*, both $\hat{p}$ value and u value ordering produced the same fitness results producing the confidence and coverage levels of Table 3, which wre better than no ordering at all. For the target class of *poisonous*, both $\hat{p}$ value and u value ordering were again better than no ordering at all (refer to Table 4). In the following discussion we consider the more interesting case where the target class is poisonous.

If we refer to Table 1, we can see the breakdown of $\hat{p}$ and u values for the categorical values belonging to *cap_colour*. From the training data, from a total of 434 records there were for *white*, 160 records belonged to our target field of poisonous, and so the proportion of records is 160/434=0.369 (to three decimal places). For each categorical value in the field, *cap_colour*, the $\hat{p}$ value was calculated in a similar way. The u values were calculated using the formula from (9).

Table 1. Cap_colour from mushroom database

	$\hat{p}$		u
buff	1.000	gray	44.185
pink	0.904	yellow	42.306
yellow	0.627	white	27.187
gray	0.484	buff	25.523
white	0.369	brown	18.207
cinnamon	0.357	pink	18.108
brown	0.136	cinnamon	6.778
red	0.023	red	5.312

In Table 1, the values $\hat{p}$ and u are ordered according to their relevant values. The values are then replaced with integer values from 1=highest, ..., 8=lowest. The same technique was applied to the other 21 categorical valued fields. Similarly, the test set was set up using exactly the same encoding as the training set. The alternative u_i' technique (10) recommended when n_i is small results in a slightly different ordering as those achieved by the u values. The u_i' ordering results in *pink* and *brown* (Table 1) changing positions. These changes do not effect the overall results as shown in Table 2.

Table 2. Results from mushroom database

λ	$\hat{p}$	u	r
1.11	112(118)	110(115.3)	88(88)
1.25	252(265)	249(260)	198(199)
1.43	432(455)	426(445)	339(340)
1.67	671(707)	663(692)	527(529)
2	1007(1061)	994(1038)	790(794)
2.5	1511(1592)	1491(1557)	1185(1333)

Table 3. Confidence and coverage levels from mushroom test database, edible

	$\hat{p}$		u		r	
λ	Con.	Cov.	Con.	Cov.	Con.	Cov.
1.11	100	100	100	100	100	93.28
1.25	100	100	100	100	100	92.09
1.43	100	100	100	100	100	93.28
1.67	100	100	100	100	100	92.09
2	100	100	100	100	95.57	99.83
2.5	100	100	100	100	92.43	100

Table 4. Confidence and coverage levels from mushroom test database, poisonous

λ	$\hat{p}$		u		r	
	Con.	Cov.	Con.	Cov.	Con.	Cov.
1.11	100	95.72	100	94.5	100	75.10
1.25	100	95.72	100	94.5	100	75.10
1.43	100	95.72	100	94.5	100	75.10
1.67	100	95.72	100	94.5	100	75.10
2	100	95.72	100	94.5	100	75.10
2.5	100	95.72	100	94.5	100	75.10

For each λ value (see Table 2), there are a set of numbers which are called *fitness* values. These values describe how well each possible solution (rule) meets the criterion specified, in other words, we wish to maximise the quality measure given in (7). So for example, a λ value of 1.67 corresponds to a minimum confidence of 60%. The same applies to Tables 6 and 11 with the same λ values.

We first encoded all categorical data using integer values determined by the $\hat{p}$ values. We then ran the SA algorithm within Data lamp six times to find the best possible rule. For each value of λ, the fitness of this rule is reported in Table 2 on the column headed $\hat{p}$. The experiment was repeated but using the ordering defined by the u values, the corresponding fitness appearing in the column headed u. We compared our results with the results obtained using the raw data, that is the original unordered categorical data and the corresponding results are listed in the column headed r. The figures in brackets in Table 2, refer to results produced using the training database. The corresponding confidence and coverage levels for each value of λ (as shown in Table 2) are given in Table 4. In all cases, ordering the data proved effective.

3.2 Adult database

The second of our experiments used the Adult database [2]. This experiment compares the efficacy of ordering on $\hat{p}$ and u values with that of using a natural ordering. In the Adult database, the *educational-numbering* field is presented which provides a natural ordering, as an alternative to the field *Education* that was unordered. The Adult database is a database with both categorical and numerical attributes. However, some categorical fields have many values, for example, *occupation* has 14 values, and *education* has 16 values, and so, u values may still prove useful. The education field can be ordered in terms of academic progression although finding a linear ordering is not easy because we have both vocational and academic qualifications together; *education-numbering* assigns such a natural ordering. Using $\hat{p}$ values we get a slightly different ordering and u values give yet another quite different ordering that appears quite poor in places (for example, a doctorate is significantly lower than bachelor qualification). These orderings are compared in Table 5. In Table 6, experimental results are given where the field *education-numbering* is not used. Except for the smaller values of λ, the $\hat{p}$ ordering gives a much better result than the raw data. Since the database is large it is perhaps not surprising that $\hat{p}$ values prove better than u values.

Returning to Table 6, we repeat the experiment but use the *education-numbering* field instead of the *education* field. Comparing the raw values in Table 6, again we see that raw data but with the natural ordering of *education-numbering* gives much better results as the value of λ increases. However, in these cases the ordering of fields by $\hat{p}$ values is still the better option. The u ordering only gives good results as λ increases, that is, as the search moves towards widely applicable and less accurate rules.

Table 5. Ordering for educational field

Educational numbering	$\hat{p}$ ordering	u ordering
Doctorate	Prof-school	Bachelors
Prof-school	Doctorate	HS-grad
Masters	Masters	Some-college
Bachelors	Bachelors	Masters
Assoc-acdm	Assoc-voc	Prof-school
Assoc-voc	Assoc-acdm	Doctorate
Some-college	Some-college	Assoc-voc
HS-grad	HS-grad	Assoc-acdm
12th	12th	10th
11th	10th	11th
10th	7th-8th	7th-8th
9th	11th	12th
7th-8th	9th	9th
5th-6th	5th-6th	5th-6th
1st-4th	1st-4th	1st-4th
Preschool	Preschool	Preschool

The corresponding confidence and coverage levels for each value of λ (as shown in Table 6) are given in Tables 7 and 8.

Table 6. Results from Adult Database

No educational numbering				Educational numbering		
λ	$\hat{p}$	u	r	$\hat{p}$	u	r
1.11	62	63	63	62	63	63
1.25	149	149	150	149	149	149
1.43	260	260	260	260	260	260
1.67	492	408	413	502	407	422
2	1058	941	615	1012	919	1000
2.5	2152	1887	1404	2081	2149	2118

Table 7. Results from adult database with no educational numbering field

λ	$\hat{p}$		u		r	
	Con.	Cov.	Con.	Cov.	Con.	Cov.
1.11	99.18	16.41	99.34	16.22	99.19	16.59
1.25	98.89	16.81	99.04	16.73	99.04	16.81
1.43	98.89	16.81	98.89	16.81	98.89	16.81
1.67	75.39	39.08	98.89	16.81	94.34	18.46
2	70.66	48.81	68.63	46.89	98.73	16.84
2.5	65.07	60.41	55.98	71.38	56.43	52.16

Table 8. Results from adult database with educational numbering field included

λ	$\hat{p}$		u		r	
	Con.	Cov.	Con.	Cov.	Con.	Cov.
1.11	99.03	16.62	99.19	16.59	99.19	16.46
1.25	98.89	16.81	99.19	16.59	98.89	16.81
1.43	98.89	16.81	98.89	16.81	98.89	16.81
1.67	74.18	42.41	98.73	16.84	72.42	39.89
2	69.32	49.03	73.47	38.92	72.89	43.03
2.5	62.60	62.32	65.06	60.30	63.68	61.57

3.3 Auto imports database

Our final experiment used the Auto Imports database [2]. The experiment was set up to illustrate that, in small databases, where there are rare cases, the technique of u value ordering is very much preferred over $\hat{p}$ value ordering or no ordering at all.

With the Auto Imports experiments, we discretised the output field (*price*) into two clusters, *price* < 20970, and *price* ≥ 20970, with the aim of predicting membership of the first cluster. Because the database is so small, $\hat{p}$ values are often identical for particular attribute values. For the field *make*, all cars in the training database of *bmw, dodge, honda, mazda, mitsubishi, nissan, peugot, plymouth, saab, subaru, toyota, volkswagen* cost < 20970, and hence have $\hat{p}$=1. By chance, the cheaper *bmw* models have been selected in the training database. The total number of *bmw's* in the database is small and the $\hat{p}$ value has not been computed using a representative sample. In fact the sample is so small that even using u values is statistically unjustified. Whenever $n_i \leq 50$, the alternative u' values were used. These values introduced slight variations in ordering compared to that obtained using u values. In Table 10, position 5 is taken up with *volvo* with *plymouth* and *volkswagen* taking position 6. The u value ordering would have *plymouth* and *volkswagen* in position 5, and *volvo* in position 6. So the ordering in Table 10 is the ordering used in our experiments. Of the other categorical fields in the database–*body style, engine type,* and *fuel system*–the same ordering–*engine type*, and *fuel system* were ordered identically by u values and u' values. When both techniques were applied to the field *body style*, a different ordering is observed. The u' values generates the ordering *hatchback, sedan, wagon,* and *convertible*, whereas, the u values generates the ordering *sedan, wagon, hatchback,* and *convertible.*

Looking at Table 9, the $\hat{p}$ values only impose a partial ordering and cluster 12 makes of cars together. We encoded all these cars using the value 1. The u' values are more discriminatory and deliver the partial ordering of Table 10. Using u' or the u value technique (as appropriate), we always produce rules which outperform

Table 9. $\hat{p}$ value ordering from auto imports database

Position	Make
1	bmw,dodge,honda,mazda,mitsubishi,nissan, peugot,plymouth,saab,subaru,toyota, volkswagen
2	volvo

Table 10. Freeman value ordering from auto imports database

Make	Ordering
1	toyota
2	honda,subaru
3	mazda,nissan,peugot,saab
4	mitsubishi
5	volvo
6	plymouth,volkswagen
7	dodge
8	bmw

both those obtained using $\hat{p}$ ordering and those obtained from raw data. If we refer to Table 11, when λ=1.67 the u' and u values produced a fitness value of 46, which is impossible to obtain for both $\hat{p}$ and raw data values. The fitness value of 46 was achieved from the following rule:

- (make = toyota, honda, subaru, mazda, nissan, peugot, saab, mitsubishi, volvo, plymouth, volkswagen, dodge)
  AND (engine-location = front)
  AND (width $\leq$ 68.7657) $\Rightarrow$ output = 0

On the test database this has a coverage of 97.18%, and an accuracy of 100%. The attribute, *make*, in this rule is constrained to be the first twelve values from Table 10. If we were to use the ordering imposed by the $\hat{p}$ values, we would have to include, *bmw*, if it were to include these same twelve attribute values. This reduces the fitness of the rule. As for the raw data values, as stated previously in section 2, they could clearly not achieve the first part of the rule. Ordering with u' or with u values (as appropriate) produce the same result at each λ level as shown in Table 11.

Table 11. Results from auto imports database

λ	$\hat{p}$	u' or u	r
1.11	7.1	7.56	7.1
1.25	16	17.5	16.25
1.43	28.3	29.57	28.57
1.67	43.67	46	43.67
2	67	69	67
2.5	102.5	103.5	102.5

Table 12. Results from auto imports test database

λ	$\hat{p}$		u		r	
	Con.	Cov.	Con.	Cov.	Con.	Cov.
1.11	100	90.14	100	95.77	100	90.14
1.25	100	90.14	100	97.18	98.57	97.18
1.43	100	92.96	100	97.18	98.57	97.18
1.67	98.53	94.37	100	97.18	98.53	94.37
2	94.67	100	100	97.18	95.89	98.59
2.5	94.67	100	100	97.18	94.67	100

4 Conclusions

Unfortunately Data Lamp does not allow for multiple tests using the same attribute, but instead will search for the best rule containing only *one* attribute at the setting up stage. As used in our experiments, we allow for the number of attributes in out experiments to not exceed four. The *best* rule searched for may contain up to and including four attributes, of which some of the attributes may not in anyway contribute to the overall effect.

For the mushroom database, we have illustrated the benefits of ordering categorical data, and the differences that $\hat{p}$ and u value ordering make. With $\hat{p}$ ordering producing the best overall results, u value ordering is only slightly less better, and both are significantly better than those obtained from raw data.

As the database gets larger, a more obvious choice would be $\hat{p}$ values. In certain conditions, an ordering based on u values would be preferred, whereas in others $\hat{p}$ ordering or using raw data values is preferred.

Almost certainly in rare cases, using the u value technique is much better than both $\hat{p}$ value ordering and no ordering at all. The results produced only heighten the fact that with u value ordering, this technique is more adept at dealing with cases where there are very few values. An alternative approach would be to use the u'_i values when databases are particularly small ($n_i \leq 50$).

We have considered three techniques for ordering categorical data. We have argued that such ordering is advisable for rule induction techniques such as used by Data Lamp. We have shown that the technique based on u values is sometimes preferable to the more obvious ordering based on $\hat{p}$ values. We would expect this to be the case mostly when databases are small or when categorical fields have a large number of values, or where there are rare cases.

We believe $\hat{p}$ values, u values, and u'_i values may have much wider applications in knowledge discovery and are currently exploring their use in tree induction [8], all rule induction [9], and clustering [10].

References

[1] Lanner (2001) DataLamp and the templar framework. http://www.lanner.com/corporate. (2001)

[2] UCI repository of machine learning databases. http://www.ics.uci.edu/ mlearn/MLRepository.html.

[3] Agresti A. and Yang M. (1987) An empirical investigation of some effects of sparseness in contingency tables. Comm. Stat., 5:9-21.

[4] Reid R. C. and Cressie N. A. (1988) Goodness-of-fit statistics for discrete multivariate data. Springer-Verlag, New York.

[5] Haberman S. J. (1988) A Warning on the use of chi-squared statistics with frequency tables with small ex-

pected cell counts. volume 83, Issue 402. Journal of the american statistical association, pp. 555-560.

[6] Bishop Y. M. M., Fienberg S. E. and Holland P. W. (1975) Discrete multivariate analysis, MIT Press, Cambridge, Massachusetts, pp. 491-492.

[7] Freeman M. F. and Tukey J. W. (1950) Transformations related to the angular and the square root, volume 21, issue 4, Annals of mathematical statistics, pp. 607-611.

[8] Angoss knowledge engineering (1987) http://www.angoss.com.

[9] Richards G. and Rayward-Smith V. J. (2001) Discovery of association rules in tabular data, IEEE international conference on data mining, pp. 465-472.

[10] Kaufman L. and Rousseeuw P. (1990) Finding groups in data: An introduction to cluster analysis, John Wiley and Sons Inc.

Social Agents in Dynamic Equilibrium

Mark McCartney[1] and David Pearson[2]
[1]School of Computing and Mathematical Sciences, University of Ulster, Newtownabbey, County Antrim, Northern Ireland (m.mccartney@ulster.ac.uk) , [2]EURISE (Roanne Research Group), Jean Monnet University of Saint-Etienne, I.U.T. de Roanne, France (david.pearson@univ-st-etienne.fr)

Abstract

A simple model for social group interactions is introduced. The model is investigated for the minimal group size for the model N=3. Examples of model of model behaviour in terms of dynamic and static equilibrium are presented. Directions for further study of the model are considered.

1 Introduction

Sociodynamics is a relatively new field of interdisciplinary study which aims to provide quantitative models for how social groups evolve. It combines expertise form areas such as physics [1,2], mathematics [3], computer science [4] and sociology [5,6,7,8,9]. In this paper we present a model for sociodynamics based on fuzzy logic which builds upon earlier work [10,11,12].

Consider a social group of N agents who all interact with each other. Clearly, if the agents are people, the interactions between two agents j and k can occur in many different ways. At a psychological level we have, amongst others, verbal & visual interactions. A verbal interaction between j and k occurs via a conversation or written communication. A visual interaction between j and k may involve what the other person looks like. For example does j find k attractive (and vice versa)? Physical interaction could take on many forms. For example j might slap k in the face. Or in a panic situation individuals may find themselves, by their movement co-operating or competing to escape a disaster (see, for example, [2]).

Interactions within social groups are clearly dependant on a wide range of factors and are subtle. We however shall study a vastly simplified model of social interaction based on the *attitude* agents have to each other. Let $a_{jk}(t)$ be the time dependent function describing whether agent k *likes* agent j. The *like* function satisfies $a_{jk}(t) \in [0,1]$ where the larger the value of $a_{jk}(t)$ the more k likes j. Thus a value of 0 signifies that agent k definitely does not like agent j (j is an enemy) and a value of 1 signifies k definitely does like j (j is a close friend).

Obviously it is not necessary to have $a_{jk}(t) = a_{kj}(t)$ nor is it necessary for $a_{kj}(t)$ to have a value assigned to it even if $a_{jk}(t)$ does (you, for example, may have a view on Bob Dylan, but it is doubtful that Bob Dylan has a view on you). Thus if j does not know k we do not assign a value to $a_{kj}(t)$, but this does not mean that k does not know j and has a view on him/her. The term $a_{jj}(t)$ can be taken to signify j's self image, but we do not discuss that possibility here.

2 Evolving Attitudes

Initially let us assume that in the social group every agent has information about every other agent and has an opinion about how much they like them. A technique for generating this initial array of attitudes $[a_{ij}(0)]$ has been considered elsewhere [12].

If the group does not interact then we shall assume the attitudes do not change. (An alternative to this, which we do not consider in the current paper, would be factor into the model the effect that all the attitudes approach a (neutral) value of 0.5 with a characteristic 'half-life' or 'memory' in the absence of interaction.) Once the group does start to interact, attitudes begin to evolve. As noted above, in the real world our attitude to an individual will be influenced by a whole range of factors and interactions. In our simple model however we shall update how much k likes j by allowing k to ask a third party, m, their view of j. Then k updates their attitude to j to make it more, or less, like that of m's attitude to j, depending on k's attitude to m.

To formulate the model we first define the following step and ramp functions $H(x)$ and $R(x)$.

$$H(x) = \begin{cases} 1 & x > 0 \\ 0 & x = 0 \\ -1 & x < -1 \end{cases} \quad (1)$$

$$R(x) = \begin{cases} 1 & x > 1 \\ x & 0 \le x \le 1 \\ 0 & x < 0 \end{cases} \quad (2)$$

We then define the change function

$$G(x, y, \Delta, \sigma) = \begin{cases} x + \Delta & y \ge \frac{1+\sigma}{2} \\ x & \frac{1-\sigma}{2} \le y < \frac{1+\sigma}{2} \\ x - \Delta & y < \frac{1-\sigma}{2} \end{cases} \quad (3)$$

where $\Delta, \sigma \in [0,1]$.

We can then construct an attitude updating rule of the form,

$$a_{jk}(t+1) = R[G(a_{jk}(t), a_{mk}(t), \Delta_k H(a_{jm}(t) - a_{jk}(t)), \sigma_k)] \quad (4)$$

The effect of $G(a_{jk}(t), a_{mk}(t), \Delta_k H(a_{jm}(t) - a_{jk}(t)), \sigma_k)$ is to increase the difference between a_{jk} and a_{jm} by an amount Δ_k if $a_{mk} < \frac{1}{2}(1-\sigma_k)$, decrease the difference between a_{jk} and a_{jm} by an amount Δ_k if $a_{mk} \ge \frac{1}{2}(1+\sigma_k)$, and to leave the value of a_{jk} unchanged otherwise.

We can say that $\Delta_k \in [0,1]$ is a measure of k's *volatility*. Thus if Δ_k is small then when k changes his attitude to another individual he does so in a 'cautious' way, making only small changes. Whereas if Δ_k is close to one when his k changes his attitude to another individual he does so in a volatile way, changing dramatically from extreme dislike to extreme fondness (or vice versa) in a single step.

Further, we define $\sigma_k \in [0,1]$ as a measure of k's stubbornness, or resistance to change, where the larger the value of σ_k the more stubborn k is.

Equation (4) thus gives a simple model for group social dynamics, with each agent, k, being defined in terms of two characteristics, volatility, Δ_k , and stubbornness σ_k . Given an initial array of attitudes $[a_{ij}(0)]$ and characteristic column vectors $[\Delta_k]$ and $[\sigma_k]$ equation (4) generates the evolution of attitudes $[a_{ij}(t)]$.

3 Results

In this paper we consider the minimal group of three people. This simplification means that when an individual k has decided to update his/her attitude towards another person j the third party, m, of whom they seek advise is uniquely defined. Examples of sociological meaningful methods for choosing m when the group is larger than three people are given elsewhere, as are methods for choosing the number of a_{ij} which are updated in a given time step [12].

Obviously the model, even though it is simple, is parameter rich and in this paper we consider only one set of initial attitude attitudes:

$$a_{21}=0.9,\ a_{31}=0.8,\ a_{12}=0.6,\ a_{32}=0.5,\ a_{13}=0.2,\ a_{23}=0.1 \quad (5)$$

Further we assume each member of the group has identical characteristics of stubbornness and volatility i.e.

$$\sigma_j = \sigma\ \forall j \qquad \Delta_j = \Delta\ \forall j \quad (6)$$

Given these conditions the system will either evolve to a state of static or dynamic equilibrium, as illustrated in figure 1. Note that for values of stubbornness , $\sigma > 0.8$ the system always reaches static equilibrium. This is because if all members of the group are stubborn agents are unlikely to make any change to their attitudes i.e. $a_{jk}(t) = a_{jk}(0)\ \forall i, j$ and $t>0$. Apart from this trend, however, the results presented in figure 1 are not intuitively obvious. Thus we consider two illustrative examples of particular types of dynamic equilibrium.

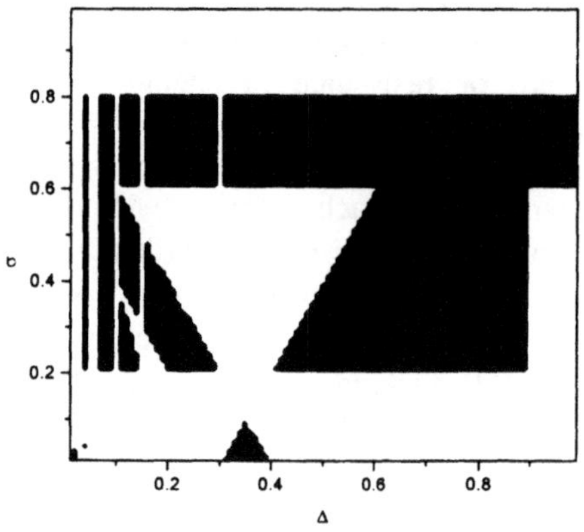

Fig. 1. In the shaded areas of (Δ,σ) space the group reaches dynamic equilibrium. In unshaded areas static equilibrium is reached. The initial conditions for the group are given by (5) and (6), (Δ,σ) points were chosen at intervals of 0.01 units.

One example is where the group evolves to a point where j wants to approach k's (stable) attitude towards m, but $|a_{mk} - a_{mj}|$ is not divisible by Δ and so a_{mj} oscillates between values above and below a_{mk}, overshooting on each occasion. This is illustrated in figure 2. A second type of dynamic equilibrium is where j and k both want to approach the other's attitude towards m, but because the group does not update until a full round of interactions is complete, they do not reach consensus, as is illustrated in figure 3.

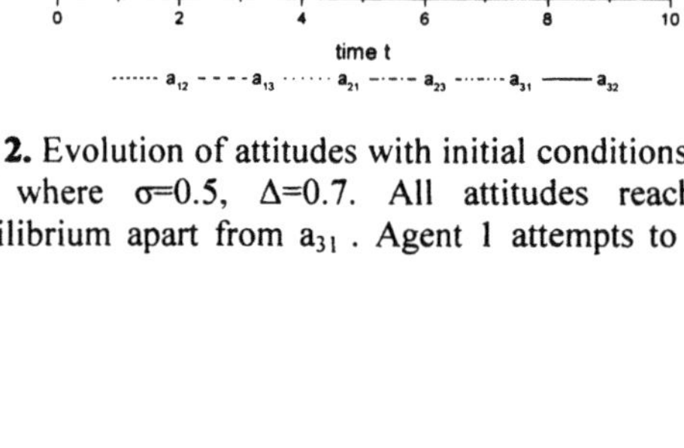

Fig 2. Evolution of attitudes with initial conditions (5) and (6) where σ=0.5, Δ=0.7. All attitudes reach static equilibrium apart from a_{31}. Agent 1 attempts to achieve consensus with agent 2 on their attitude to 3, but are unable to do so, because at each time step 1 over shoots 2's attitude to 3.

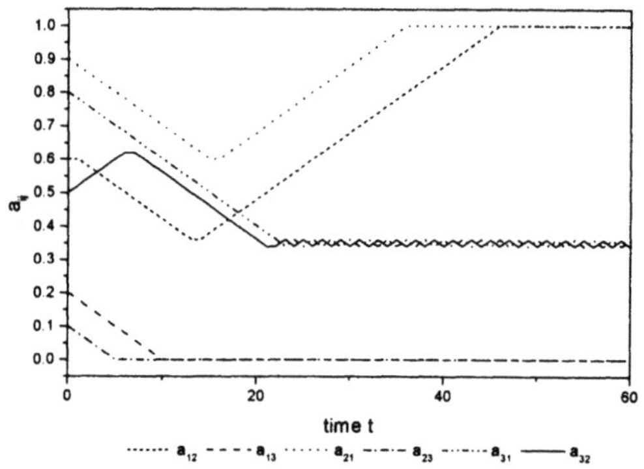

Fig.3 Evolution of attitudes with initial conditions (5) and (6) where σ=0.02, Δ=0.02. All attitudes reach static equilibrium apart from a_{31} and a_{32}. Agents 1 and 2 attempt to achieve consensus on their attitude to 3, but are unable to do so, because at each time step each updates to the others previous attitude. Agents 2 and 1 achieve a maximal attitude of 1 towards each other, and agent 3 concludes that s/he does not like the others.

This scenario could be avoided in some circumstances by building memory into our group. However, agents without memory are a real possibility. An example would be where agents represent large groups of individuals, such as political parties, which can sometimes exhibit little or no memory of past behaviour. A further way to avoid the behaviour illustrated in figure 3 is to allow only one agent to update their attitudes at each time step.

Perhaps surprisingly the simple model presented here is rich in behaviour. The broad categorisation of static and dynamic equilibria hides within it many possible scenarios, some of which are physically meaningful and others are artefacts of the model. In future work we intend to investigate and refine this model further and compare it with other models for social interaction.

References

1. Helbing D *Quantitative Sociodynamics,* Kluwer Academic Publishers 1995
2. Helbing D, Farkas IJ, and Vicsek T *Crowd Disasters and Simulation of Panic*

Situations in The Science of Disasters ed Bunde A, Kropp J and Schellnhuber HJ

3. Moss S *Game Theory: Limitations and an Alternative,* Journal of Artificial Societies and Social Simulation **4** 2001 http://www.soc.surrey.ac.uk/JASSS/4/2/2.html
4. Davidsson P *Agent Based Simulation: A computer Science View,* , Journal of Artificial Societies and Social Simulation **5** 2002 http://www.soc.surrey.ac.uk/JASSS/5/1/7.html
5. Conte R, Hegselmann R and Terna P (eds.) *Simulating Social Phenomena*, Springer Verlag, 1997
6. Gilbert N and Doran j (eds.) *Simulating Societies: The Computer Simulation of Social Phenomena*, UCL Press, London 1994
7. Hegselmann R, Mueller U and Troitzsch KG (eds.) *Modelling and Simulation in the Social Sciences from a Philosophy of Science Point of View*, Dordrecht, Boston (Kluwer) 1996
8. Liebrand WBG, Nowak A and Hegselmann R (eds) *Computer Modelling of Social Processes* Sage 1998
9. Weidlich W and Haag G *Concepts and Models of a Quantitative Sociology* Springer Verlag 1983
10. Pearson DW and Dray G *A Fuzzy Approach to Sociodynamical Interactions,* Proceedings of the International Conference on Artificial Neural Networks and Genetic Algorithms, Prague, Czech Republic, April 2001
11. McCartney M and Pearson DW *Fuzzy Logic and Sociodynamics, submitted to* Teaching Mathematics and its Applications 2001
12. McCartney M and Pearson DW *Sociodynamics Based on Commonality submitted to* IEEE Journal of Systems Man & Cybernetics 2002

Energy Management for Hybrid Electric Vehicles based on Fuzzy Clustering and Genetic Algorithms

Lucio Ippolito(*), Vincenzo Loia(**), Pierluigi Siano(*)
(*) Department of Electronic & Electrical Engineering
University of Salerno - Via Ponte Don Melillo, 1 - Fisciano (SA), 84084, ITALY
(**) Dipartimento di Matematica e Informatica
University of Salerno - Via S. Allende - Baronissi (SA), 84081, ITALY

Abstract - Throughout the world, there is a trend towards the growth of vehicle traffic and an increase in road transport. The number of cars is growing twice to three times faster than the rate of growth of the population and car use is increasing even more rapidly.
Growing congestion has led to ever greater environmental nuisance and pollution is causing lethal consequences for humans.
Reducing emissions from motor vehicles must involve a global strategy, bringing into play both technical innovation and strict regulations.
In this context Hybrid electric vehicles power train can offer a sensible improvement of the overall vehicle environmental impact, achieving at the same time a rational energy employment. The environmental challenges of this century require a new generation vehicles. Making the most of the modern control systems and design techniques, they will be more electrified and designed by total systems approaches, involving new materials, alternative fuels, and, above all, new powertrains.
Advanced energy flow management unit to split the instantaneous vehicle power demand between the internal combustion engine and the electric motor, will ensure that the power sources will be operated at high efficiency operating points and the related vehicle emissions will be minimized.
This led study to develop innovative methodologies for the real-time identification of the components operating points.
Starting from these considerations in the paper a knowledge-based control system for splitting the vehicle's power demand between the engine and motor is presented. The proposed approach by using fuzzy clustering and genetic algorithm permits to achieve better performance, mainly in terms of a reduced computational effort and an improved efficiency of the control system over various driving cycles.
To validate this approach, simulation tests and comparisons with other energy management strategies are discussed.

1. Introduction

Over the past several years, concerns have been raised over urban air contamination caused by exaust gas emmissions from public and private vehicles.
The public pressure has forced the state authorities to impose traffic restriction to avoid the overcoming of the pollution danger levels.
In this scenario, hybrid electric vehicles (HEVs) powertrain, combining electric motor with an auxiliary power unit, can contribute significantly to the improvement of vehicle performance and fuel economy, reducing at the same time the effects of the use of private cars on the air quality of the cities.
As well known, in a parallel-type hybrid electric vehicle (PHEV), the torque assisting and battery recharging control using the electric machine is the key point for efficient driving.
This consideration has led study to develop advanced control strategies for the energy flows management on board the vehicles [1,2,3].
Recently, the authors of the paper have proposed two different advanced control techniques based on a genetic-based control strategy, the former, and on a multi-objective optimization approach, the latter [4,5]. These two methodologies starting from the desired vehicle on road performance (design constrains), identify the instantaneous power flow distribution minimizing, by a real-time optimization procedure, a fixed scalar objective function descriptive of the control goals. Because the design objectives are conflicting with each other, the resolution of such a problem hasn't got a unique solution and a suitable trade off between the objectives must be identified. For the resolution of the problem the goal attainment method has been proposed since it is not subject to any kind of convexity limitations and it has the ability to solve a great number of complex engineering problems characterized by non linear, multimode, vectorial objective functions.
Even though the suggested methodology has proved its effectiveness and robustness, it must be evidenced its reduced applicability for real-time applications, today. This has led to develop a true real-time knowledge-based control strategy for splitting efficiently the vehicle's power demand between the engine and motor.
Making the most of this strategy the architecture of the energy management unit is designed. The energy management unit is based on a database in which the initial dataset is obtained solving the multi-objective problem. The dataset is constituted by the vehicle's state vector, the driver's power demand, and the optimal hybridization degree, which corresponds directly with the vehicle's energy flows.
The aim of the proposed knowledge-based control strategy is that of approximating with a high degree of accuracy the optimal power flows distribution identified with the multi-objective-based methodology, exhibiting quite the same performance in terms of both emissions and fuel consumption and at the same time assuring an extremely reduced computational time, paving the way for real-time applications.
Since the knowledge base of this control system consists of a database whose structure and size affects directly the computational effort, in order to reduce the

database size, without cutting the vehicle's performances, fuzzy-based clustering and genetic algorithms are applied for data analysis. Fuzzy clustering is used to discover the structure in the data set producing a new data set organized in distinct classes where the elements in each class, or cluster, are as similar as possible to each other and dissimilar to those in other classes. Genetic algorithms are used to identify the optimal number of classes of the data set.

2. Fuzzy clustering basic elements

Before the problem formulation, it is useful to introduce the basic concepts of fuzzy clustering methods for data analysis.

Fuzzy clustering plays an important role in solving problems in many applications, such as pattern recognition, classification, data mining, and image segmentation.

The fuzzy clustering method provides a better and more useful method to classify patterns, especially when the boundary between clusters are not clearly defined and, consequently, some pattern may belong to more than one cluster.

A variety of fuzzy clustering methods have been proposed [6,7] and most of them are based upon distance criteria. The most known method of fuzzy clustering is the Fuzzy c-Means method (FCM), initially proposed by Bezdek [8], and other authors [9,10].

2.1 Fuzzy C-Means algorithm

Given a set of unlabeled patterns $\mathbf{X}=(\mathbf{x}_1,\mathbf{x}_2,\dots,\mathbf{x}_N)$, $\mathbf{x}_i \in R^S$ where N is the number of patterns and S is the dimension of pattern vectors, the aim of the FCM algorithm is to find an optimal fuzzy c-partition and corresponding prototypes minimizing the objective function (total weighted mean-square error):

$$F_m(\mathbf{U},\mathbf{W}) = \sum_{j=1}^{C}\sum_{i=1}^{N}(\mu_{ij})^m d_{ij}^2 \qquad (1)$$

subject to the following constraints on U:

$$\mu_{ij} \in [0,1] \quad i=1,\dots..N \quad j=1,\dots.C \qquad (2)$$

$$\sum_{j=1}^{C}\mu_{ij} = 1 \quad i=1,\dots.N \qquad (3)$$

$$0 < \sum_{i=1}^{N}\mu_{ij} < N \quad j=1,\dots.C \qquad (4)$$

where:

N: the number of patterns in X

C: the number of clusters

$\mathbf{U}$: the membership function matrix

μ_{ij}: the value of the membership function of the ith pattern belonging to the jth cluster

d_{ij}: the distance, from $\mathbf{x}_i$ to $\mathbf{w}_j$, $d_{ij} = \left\|\mathbf{x}_i - \mathbf{w}_j^{(t)}\right\|$; where $\mathbf{w}_j^{(t)}$ denotes the cluster centre of the jth cluster for the tth iteration

$\mathbf{W}$: the cluster centre matrix

m: the exponent on μ_{ij}; to control fuzziness or amount of clusters overlapping

Objective function measures the quality of partitioning that divides a dataset into C clusters by comparing the distance from pattern $\mathbf{x}_i$ to the current candidate cluster centre $\mathbf{w}_j$ with the distance from pattern x_i to other candidate cluster centres [6,7].

Equation (1) describes a constrained optimization problem, which can be converted to an unconstrained optimization problem by using the Lagrangian multiplier technique.

To minimize the objective function under fuzzy constraints, given a fixed number C, m and ε, a small positive constant, the FCM algorithm, starts with a set of initial cluster centres (or arbitrary membership values). Then generate randomly a fuzzy c-partition and set iteration number t=0. A two step iterative process works as follows. Given the membership values $\mu_{ij}^{(t)}$, the cluster centre vector $\mathbf{W}$ is calculated by:

$$\mathbf{w}_j^{(t)} = \frac{\sum_{i=1}^{N}(\mu_{ij}^{(t-1)})^m \mathbf{x}_i}{\sum_{i=1}^{N}(\mu_{ij}^{(t-1)})^m} \qquad j=1,\dots.C \qquad (5)$$

Given the new cluster centres $\mathbf{W}^{(t)}$, the membership values $\mu_{ij}^{(t)}$ are updated by:

$$\mu_{ij} = \frac{1}{\sum_{l=1}^{C}\left(\frac{d_{ij}}{d_{il}}\right)^{\frac{2}{(m-1)}}} \qquad \begin{array}{l} i=1,\dots.N \\ j=1,\dots.C \end{array} \qquad (6)$$

where if d_{ij}=0 then μ_{ij} =1 and μ_{ij} =0 for $l \neq j$.

The process stops when $\left\|\mathbf{U}^{(t)} - \mathbf{U}^{(t-1)}\right\| \leq \varepsilon$, or a predefined number of iterations is reached.

3. Energy Flow Management based on Fuzzy Clustering and Genetic Algorithms

The multi-objective optimization of the energy flows management using the goal-attainment method, already

presented in [5], could imply a tremendous effort and therefore it could result inapplicable for real-time applications.
To overcome this intrinsic limitation, the research efforts were addressed to find a knowledge-based control strategy for the energy flows management problem.
The aim of such a kind of control is to approximate with an high degree of accuracy the optimal power flows distribution identified, over a specific drive cycle, by the multi-objective-based methodology, assuring at the same time an extremely reduced computational time.
In this control for real-time applications, the energy flows management makes use of a database, representing the knowledge of the system. The database consists of elements each of which is constituted by the state-vector of the vehicle and the power demand, which form the so-called "input vector", and by the optimum hybridization degree, which is the output variable.
In the present study, the state-vector of the vehicle consists of the state variables of the vehicle such as: the vehicle's speed, the state of charge of the battery (SOC), the catalyst temperature, the battery temperature, the internal combustion engine temperature, and so on.
The generation of data set stored is made solving the multi-objective power flows optimisation problem over a standard drive cycle and storing, at each time step, into a database the optimal hybridization degree together with the corresponding vehicle's state vector and power demand, as schematically depicted in fig.1.

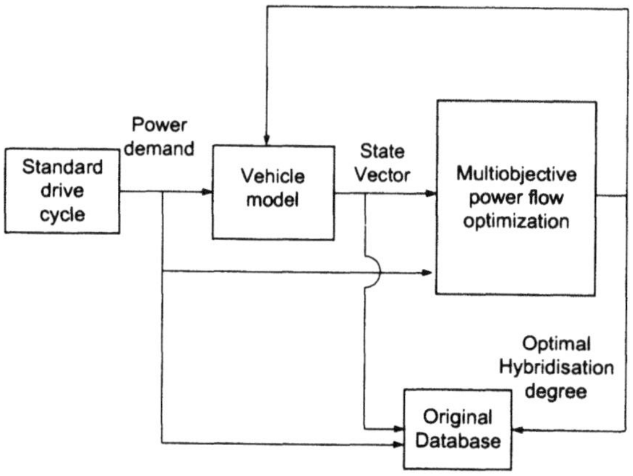

Fig. 1. Data set generation

The proposed control strategy is based on a similarity criterion between the stored input vector (state vector plus power demand), already existing in the database, and current input vector occurring at each time step of a given drive cycle.
As similarity criterion the Euclidean distance is used between two different input vectors. The smaller is the Euclidean distance between them, the higher is the degree of similarity of them.
At each time step, the Euclidean distances between the current input vector and all the stored input vectors into the database are calculated and a particular vector, characterized by the minimum distance, is selected. The output corresponding to the selected input vector is chosen as output control variable.
Obviously, such an approach, to guarantee the required performance, needs to find into the database at least one element - input vector - with an Euclidean distance from the current input vector lower than a prefixed one. This circumstance involves to use large and complex database, where the information relative to all driving cycles are merged.
Considering that, depending on the selected simulation time step, the number of element to be stored into the database could become extremely large, it is evident that the performance of the knowledge-based control strategy trained over all standard driving cycles could become worse, in terms of an increase of computational time.
For instance, developing Matlab® simulations for a given vehicle and assuming a simulation time step of one second, which represents a valid trade-off between simulation times and results accuracy, the number of elements which will form the database is reported in table 1.

Table 1: Number of elements and size of the database for standard drive cycles

Driving Cycle	Number of elements	Size [kbyte]
NEDC	1180	75,52
FTP	2478	158,59
US06	601	38,46
HWFET	766	49,02
For all cycles	5025	321,60

3.1 Fuzzy clustering and reduced database construction

In order to overcome the previous mentioned difficulty, clustering was applied to the "original" data set to group the element of database in c subgroups (clusters) of similar elements. With this algorithm, the system obtained from the data is transformed into a new system, organized in subsystems. This data elaboration permits to discover the structure of data obtaining a well-organized data set. Moreover, in order to reduce the database size, the basic idea in this work is to define a "radius of influence", r, of the centres of clusters.

In particular, given a set of C centres, $\mathbf{C} = (\mathbf{c}_1, \mathbf{c}_2, \ldots, \mathbf{c}_C)$ obtained with the FCM algorithm, where $\mathbf{c}_i = (x_{i1}, x_{i2}, \ldots, x_{iN}) \in R^N = R^{N-1} \times R$ and N is the dimension of pattern vectors, each centre c_i is associated to its projection in R^{N-1}, $\mathbf{c}_i^1 = (x_{i1}, x_{i2}, \ldots, x_{iN-1})$ and to its radius of influence given by $r(i) = \beta(i)\left\|\mathbf{c}^1{}_i - \mathbf{B}\right\|$, where $\left\|\mathbf{c}^1{}_i - \mathbf{B}\right\|$ is the Euclidean distance between $c^1{}_i$ and the barycentre $\mathbf{B} = (b_1, b_2, \ldots, b_{N-1})$ of the projections of the set of centres in R^{N-1} and $\beta(i)$ is an attenuation factor diverse for each centre, given by:

$$\beta(i) = \left(\frac{\sum_{\substack{j=1 \\ i \neq j}}^{C} \left\|\mathbf{c}^1{}_i - \mathbf{c}^1{}_j\right\|}{\sum_{k=1}^{C} \sum_{\substack{j=1 \\ k \neq j}}^{C} \left\|\mathbf{c}^1{}_k - \mathbf{c}^1{}_j\right\|} \right)^p \quad (7)$$

In the previous formula, p is an exponent varying in the interval [0.3, 0.6], which gives a degree of freedom that can be used to control the amplitude of $\beta(i)$.

The formulation of the radius of influence means that the bigger is the Euclidean distance between a centre projection $c^1{}_i$ and the barycentre, the bigger is its radius of influence. As a consequence, isolated centers will have bigger radius.

In particular, the database reduction is obtained deleting the elements to which correspond a distance between them and a specific centre lower than the radius of influence of the centre considered.

This permits to reduce drastically the number of elements stored into the database without worsen the vehicle's performance.

The exponent p used in the previous formula (7) determines the complexity of the knowledge base and the accuracy of the approximation. For bigger values of p, more centres will be created and more computational effort will be involved.

For the specific application, the FCM was selected amongst the clustering methods to analyze the data into the original database and a value of the exponent p was set to 0.3.

To apply the FCM algorithm the definition by the user of several parameters is required: the matrix norm, the fuzziness parameter m, the stopping criterion ε, the maximum number of iterations and the optimal number of clusters used. In order to clarify the construction of the reduced database, the following example is presented.

Example.

Assumptions:

- Dimensions of patterns vectors N: 3;
- Initial number of centres: 5.

as follows: $\mathbf{c}_i = (x_{1i}, x_{2i}, x_{3i}), i = 1,2,3,4,5$.

Case A.

Assume that an element of the initial database $\mathbf{z} = (z_1, z_2)$, with a distance from the projection $(\mathbf{c}_4^1)$ of the nearest centre less than the corresponding radius of influence, $\left\|\mathbf{z} - \mathbf{c}^1{}_4\right\| < r_4$ exists.

In this situation, because the element is inside the area of influence of $(\mathbf{c}_4^1)$, it is deleted and will not be stored into the database.

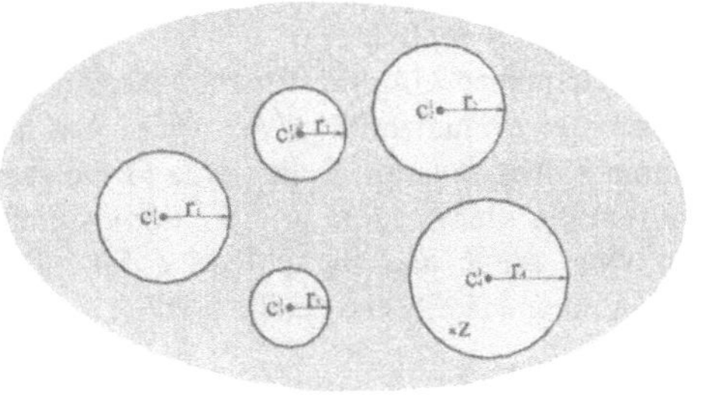

Fig. 2. Example of reduced database construction (case A)

Case B.

Assume that an element of the initial database $\mathbf{z} = (z_1, z_2)$, with a distance from the projection of all the centres bigger than the corresponding radius of influence: $\left\|\mathbf{z} - \mathbf{c}^1{}_i\right\| > r_i$, for $i=1,2,3,4,5$ exists.

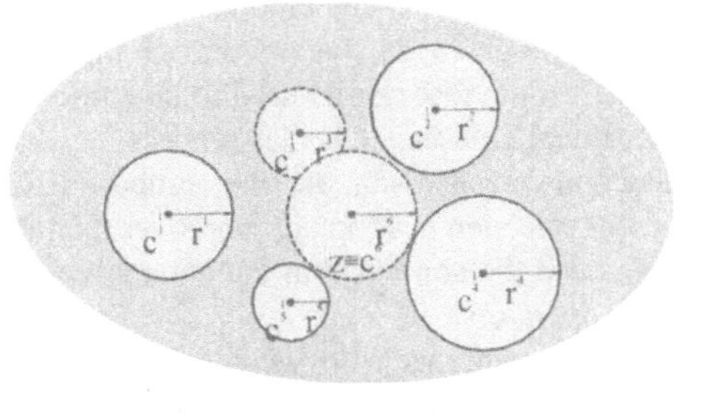

Fig. 3. Example of reduced database construction (case B)

In this situation the element is outside the area of

influence of all the centres and consequently, it will stored into the reduced database and will become a new centre as depicted in fig.3 in which the projections of all the centres with the corresponding radii of influence are shown.

3.2 GA for the selection of the optimal number of clusters

As before mentioned the selection of the optimal number of clusters, fitting a data set is particularly critical, affecting strongly the vehicle performance.

To date many quantitative methods [11,12] have been proposed for evaluation of clustering results, but, for the specific problem, these tools are not able to guarantee the respect of the overall vehicle's performance: a good clustering result could correspond to bad performance of the control system. Consequently, the number of clusters must be related to the vehicle performance and to the overall efficiency index.

In the present study, to identify the optimal number of clusters to be used in fuzzy C-Mean algorithm, a genetic approach is applied.

For the implemented GA, the fitness function of each chromosome is evaluated as $k/\tilde{F}$, where k is a very large number used to amplify the value of the function $(1/\tilde{F})$ so that the fitness value of the chromosomes will be in a wide range and the objective function that describes mathematically the design goal is:

$$\tilde{F}(N) = w_1\left|\int \frac{CO(t) - CO_{opt}}{CO_{opt}}dt\right| + w_2\left|\int \frac{HC(t) - HC_{opt}}{HC_{opt}}dt\right| + w_3\left|\int \frac{NO_x(t) - NO_{x,opt}}{NO_{x,opt}}dt\right| + w_4\left|\int \frac{FC(t) - FC_{opt}}{FC_{opt}}dt\right| + w_5\left|SOC_{final} - SOC_{opt}\right| \quad (8)$$

In the fitness function each term is the difference between the obtained and the desired performances normalized to the desired performances, over a given drive cycle. Moreover, each component of the objective function is weighted by factors w_i, giving a set of degree of freedom that can be used to take into account either technical aspects and social benefits.

As selection mechanism, in the proposed GA, a tournament selection is adopted; moreover the heuristic crossovers and the non uniform mutation are employed as genetic operators.

Adopting this approach to decide the number of clusters to be used, the FCM algorithm is applied to the original dataset, obtaining a reduced database that will be used to implement a real time control of power flows on board the vehicle.

4. Case Study

In order to evaluate the effectiveness of the proposed knowledge-based energy flows management strategy various comparative simulations have been developed. In particular, the results furnished by the proposed strategy and the multi-objective-based one are compared.

The characteristics of the vehicle used in the simulations are derived applying the optimal sizing design procedure proposed by the authors in [13,14]. The power train characteristics are summarized in table 2.

Table 2 : PHEV power train characteristics

Parameter	Value
Vehicle data	
Coefficient of aerodynamic drag	0.335
Frontal area, [m^2]	2.0
Coefficient of rolling resistance	0.009
Glider mass, [kg]	592
Cargo mass, [kg]	136
Motor data	
Type	IM
Maximum power, [kW]	22
Maximum torque, [N-m]	80
Maximum speed, [rpm]	10000
Specific power, [W/kg]	824
Peak efficiency	92%
Energy storage system data	
Type	VRLA
Modules number	14
Voltage, [V]	12
Module's weight [kg]	11
Peak power, [kW]	3.5
Energy capacity, [Ah]	25
Engine data	
Type	ICE
Maximum power, [kW]	32
Specific power, [W/kg]	313
Peak efficiency	34%

Using Matlab, the test vehicle has been simulated over various driving cycles and in particular:

- the New European Driving Cycle (NEDC)
- the Federal Test Procedure (FTP)
- the US EPA's Supplemental Federal Test procedure cycle (US06)
- the Highway Fuel Economy Test (HWFET).

The selection of the optimal number of clusters, fitting each data set for the four different drive cycles mentioned above was obtained applying the GA described in the previous paragraph.

For tuning the GA, the population size was set to 10 and the number of generations to 50, these values as shown from simulation results can guarantee the convergence of the algorithm to a satisfactory solution. For each individual of a population the fitness function is evaluated running the following steps:

1. the fuzzy clustering algorithm is applied to the native dataset and the reduced database is derived;
2. the performances of the vehicle equipped with the knowledge-based control system, working with the reduced database, is evaluated over a given driving cycle.

In the present case study FCM method was selected for data clustering. The Euclidean norm was used as matrix norm, the fuzziness parameter *m*, was set to 2.0, the value of ε was set to 10^{-5}, the maximum number of iterations was set to 100, the exponent *p* used for the definition of the radius of influence *r* was set to 0.3. By the FCM method a sensible reduction of the database size was obtained, as shown in table 3.

Table 3: Number of elements of the database

Driving Cycle	Number of elements	
	Original database	Reduced database
NEDC	1180	184
FTP	2478	465
US06	601	131
HWFET	766	156
For all cycles	5025	936

The vehicle, equipped with the control strategy considered before, is then simulated on various drive cycles in order to evaluate the relative performances in terms of pollutant emissions, fuel consumption and CPU time consumption during the cycle.
The obtained results are summarized in table 4.
The analysis of the simulated data reveals as the knowledge-based control system approximates with a good accuracy the optimal power flows distribution identified with the multi-objective-based methodology, over all the driving cycles considered.
The effectiveness of the obtained results is emphasized, furtherly, examining the computational time required by the knowledge-based control.
Measuring the CPU time consumption, using a Matlab embedded function, it can be evidenced that with the proposed strategy the CPU time is reduced of about 85% respect with that one required by multi-objective-based methodology, in the worst case considered (FTP cycle).

It is worth to note that, applying the proposed methodology, satisfying performance, in terms of emissions, could be obtained quite over all the drive cycles.

Table 4: Simulation results

Driving Cycle	Monitored Parameters	Multi-objective strategy	Knowledge-based strategy	Deviation %
NEDC	*HC [g/km]*	0.247	0.2485	+0.63%
	CO [g/km]	1.469	1.4642	-0.36%
	NO_x [g/km]	0.134	0.1337	-0.59 %
	Fuel Consumption *[liters/km*100]*	3.593	3.5992	+0.17 %
	Final *SOC*	0.557	0.5584	-0.16 %
	CPU Time Consumption during the cycle [s] (N=184)	965.86	70.09	-92.74%
FTP	*HC [g/km]*	0.173	0.1713	-0.75%
	CO [g/km]	0.996	0.8444	-15.23%
	NO_x [g/km]	0.151	0.1501	-0.6%
	Fuel Consumption *[liters/km*100]*	3.963	3.6926	-6.82%
	Final *SOC*	0.547	0.5219	-4.65%
	CPU Time Consumption during the cycle [s] (N=465)	2018.3	287.45	-85.75%
US06	*HC [g/km]*	0.216	0.217	0.49%
	CO [g/km]	2.048	2.054	0.31%
	NO_x [g/km]	0.251	0.247	-1.69%
	Fuel Consumption *[liters/km*100]*	5.127	5.165	0.006%
	Final *SOC*	0.508	0.5077	0.75%
	CPU Time Consumption during the cycle [s] (N=131)	747.75	35.5	-95.25%
HWFET	*HC [g/km]*	0.163	0.1603	-1.4%
	CO [g/km]	0.913	0.8853	-3.0%
	NO_x [g/km]	0.143	0.1381	-3.7%
	Fuel Consumption *[liters/km*100]*	3.446	3.142	-0.99%
	Final *SOC*	0.557	0.5523	-0.86%
	CPU Time Consumption during the cycle [s] (N=156)	833.39	52.24	-93.73%

For CPU time consumption it makes reference to a PC with a Pentium III processor, 733 MHz, and 256 Mb RAM

5. Final Remarks

Starting from the consideration that the energy flows management on board a PHEV is an activity involving many general aspects and variables, the paper has dealt with the proposal of a knowledge-based control strategy, to be used in the design of the energy management unit.

The knowledge base of the control system is created off-line using a multi-objective power flow optimization strategy. The elements of the database are obtained simulating the vehicle over one, or more, test cycles. The multi-objective strategy has permitted to identify the optimal hybridization degree corresponding to each vehicle's state during the cycle.

In order to make the control system faster, without deteriorating the vehicle's performance, data set is analyzed with the FCM. In particular, introducing a new definition of the "radius of influence" for each centre of the clusters, a reduction of database size and complexity without worsen the vehicle's performance is obtained.

Simulations have confirmed the expectations, revealing good performance of the proposed knowledge-based control strategy. It has proved its suitability for the specific application, exhibiting quite the same performances in terms of both emissions and fuel consumption on a given cycle, and at the same time assuring an extremely reduced computational time. It is obvious, that the control system performance depends on the definition of the radii of influence of the centres, This means it can be improved the vehicle's performance adopting a self-learning approach.

6 References

[1] W.C. Morchin, (1998) Energy management in hybrid electric vehicles, in Proceedings of 17th Digital Avionics Systems Conference, vol. 2, pp. I41/1-I41/6.

[2] S.D. Farrall and R.P. Jones, (1993) Energy management in an automotive electric/heat engine hybrid powertrain using fuzzy decision making, in Proceedings of the International Symposium on intelligent control, pp. 463-468.

[3] M.R. Cuddy and K.B. Wipke, (1997) Analysis of the Fuel Economy Benefit of Drivetrain Hybridization, in Proceedings of SAE International Congress and Exposition, Detroit, Michigan.

[4] V. Galdi, L. Ippolito, A. Piccolo, and A. Vaccaro, (2001) Optimisation of Energy Flow Management in Hybrid Electric Vehicles via Genetic Algorithms, in Proceedings of 2001 IEEE/ASME International Conference on Advanced Intelligent Mechatronics, Como, Italy, pp. 434-439.

[5] V. Galdi, L. Ippolito, A. Piccolo and A. Vaccaro, (2001). Multiobjective optimization for fuel economy and emissions of HEV using the goal-attainment method, in Proceedings of 18th International Electric Vehicle Symposium, Berlin.

[6] S. Nascimento, B. Mirkin and F. Moura-Pires (2000) A Fuzzy Clustering Model of Data and Fuzzy c-Means Fuzzy Systems, 2000. FUZZ IEEE 2000. The Ninth IEEE International Conference, Volume: 1Page(s): 302 -307 vol.1.

[7] Ming-Chuan Hung and Don-Lin Yang (2001) An Efficient Fuzzy C-Means Clustering Algorithm Data Mining, 2001. ICDM 2001, Proceedings IEEE International Conference, Page(s): 225 -232

[8] J. Bezdek and R. Hathaway, (1988) Recent convergence results for fuzzy c-means clustering algorithms, Journal of Classification, 5(2), pp. 237-247.

[9] G. Gustefson and W. Kessel, (1979) Fuzzy clustering with a fuzzy covariance matrix, in Proc. IEEE CDC, San Diego, pp 761-766.

[10] F. Klawonn and Keller, (1999) Fuzzy clustering based on modified measures, J. Kok, D. Hand and M. Berthold (eds.), Advances in Intelligent Data Analysis, Third International Symposium,(IDA'99), Lecture Notes in Computer Science, 1642, Springer-Verlag, pp. 291-301.

[11] Rajesh N. Dave. (1996) Validation fuzzy partitions obtained through c-shells clustering, Pattern Recognition Letters, Vol. 17, pp 613-623.

[12] Gath I., Geva A.B. (1989) Unsupervised optimal fuzzy clustering, IEEE Transaction on Patterns Analysis and Machine Intelligence, Vol. 11(7).

[13] V. Galdi, L. Ippolito, A. Piccolo and A. Vaccaro, (2000) Evaluation of Emissions Influence on Hybrid Electric Vehicles Sizing, in Proceedings of International Conference on Power Electronics, Electrical Drives, Advanced Machines and Power Quality, Ischia, Italy.

[14] V. Galdi, L. Ippolito, A. Piccolo and A. Vaccaro, (2001). A genetic based methodology for Hybrid Electric Vehicle Sizing Soft Computing. vol 5, issue 6, pp 451-457.

SpringerComputerScience

V. Kurkova, N. C. Steele, R. Neruda, M. Karny (eds.)

Artificial Neural Nets and Genetic Algorithms

Proceedings of the International Conference in Prague, Czech Republic, 2001

2001. XIII, 504 pages. 375 figures.
Softcover EUR 115,90
(Recommended retail price)
Net-price subject to local VAT.
ISBN 3-211-83651-9

The papers in this volume present theoretical insights and reports on successful applications of articifical neural networks and genetic algorithms. A dual affinity with biology is shown as several papers deal with cognition, neurocontrol, and biologically inspired brain models and others describe successful applications of computational methods to biology and environmental science. Theoretical contributions cover a variety of topics including nonlinear approximation by feedforward networks, representation of spiking perceptrons by classical ones, recursive networks and associative memories, learning and generalization, population attractors, and proposal and analysis of new genetic operators or measures. These theoretical studies are augmented by a wide selection of application-oriented papers on topics ranging from signal processing, control, pattern recognition and times series prediction to routing tasks.

SpringerWienNewYork

A-1201 Wien, Sachsenplatz 4–6, P.O. Box 89, Fax +43.1.330 24 26, e-mail: books@springer.at, Internet: **www.springer.at**
D-69126 Heidelberg, Haberstraße 7, Fax +49.6221.345-229, e-mail: orders@springer.de
USA, Secaucus, NJ 07096-2485, P.O. Box 2485, Fax +1.201.348-4505, e-mail: orders@springer-ny.com
Eastern Book Service, Japan, Tokyo 113, 3–13, Hongo 3-chome, Bunkyo-ku, Fax +81.3.38 18 08 64, e-mail: orders@svt-ebs.co.jp

SpringerComputerScience

Zenon Waszczyszyn (ed.)

Neural Networks in the Analysis and Design of Structures

2000. VIII, 307 pages. 128 figures.
Softcover **EUR 70,–**
Recommended retail price
Net-price subject to local VAT.
ISBN 3-211-83322-6
CISM International Centre for Mechanical Sciences.
Courses and Lectures, No. 404

Neural Networks are a new, interdisciplinary tool for information processing. Neurocomputing being successfully introduced to structural problems which are difficult or even impossible to be analysed by standard computers (hard computing). The book is devoted to foundations and applications of NNs in the structural mechanics and design of structures.

Contents:

- Fundamentals of Artificial Neural Networks
- Genetic Algorithms and Neural Networks
- Applications of Neural Networks in Modeling and Design of Structural Systems
- The Neural Network Approach in Plasticity and Fracture Mechanics
- Neural Networks in Advanced Computational Problems
- Neural Networks and Fuzzy Logic in Active Control of Mechanical Systems

SpringerWienNewYork

A-1201 Wien, Sachsenplatz 4–6, P.O. Box 89, Fax +43.1.330 24 26, e-mail: books@springer.at, Internet: **www.springer.at**
D-69126 Heidelberg, Haberstraße 7, Fax +49.6221.345-229, e-mail: orders@springer.de
USA, Secaucus, NJ 07096-2485, P.O. Box 2485, Fax +1.201.348-4505, e-mail: orders@springer-ny.com
Eastern Book Service, Japan, Tokyo 113, 3–13, Hongo 3-chome, Bunkyo-ku, Fax +81.3.38 18 08 64, e-mail: orders@svt-ebs.co.jp

GPSR Compliance

The European Union's (EU) General Product Safety Regulation (GPSR) is a set of rules that requires consumer products to be safe and our obligations to ensure this.

If you have any concerns about our products, you can contact us on ProductSafety@springernature.com

In case Publisher is established outside the EU, the EU authorized representative is:

Springer Nature Customer Service Center GmbH
Europaplatz 3
69115 Heidelberg, Germany

Batch number: 10732720

Printed by Printforce, the Netherlands